ASTRONOMY, COSMOLOGY AND FUNDAMENTAL PHYSICS

ASTROPHYSICS AND SPACE SCIENCE LIBRARY

A SERIES OF BOOKS ON THE RECENT DEVELOPMENTS
OF SPACE SCIENCE AND OF GENERAL GEOPHYSICS AND ASTROPHYSICS
PUBLISHED IN CONNECTION WITH THE JOURNAL
SPACE SCIENCE REVIEWS

VOLUME 155
PROCEEDINGS

ASTRONOMY, COSMOLOGY AND FUNDAMENTAL PHYSICS

PROCEEDINGS OF THE THIRD ESO-CERN SYMPOSIUM,
HELD IN BOLOGNA, PALAZZO RE ENZO, MAY 16-20, 1988

Edited by

MICHELE CAFFO
Istituto Nazionale di Fisica Nucleare, Bologna, Italy

ROBERTO FANTI
Dipartimento di Astronomia, Università di Bologna, Italy

GIORGIO GIACOMELLI
Dipartimento di Fisica, Università di Bologna, Italy

and

ALVIO RENZINI
Dipartimento di Astronomia, Università di Bologna, Italy

KLUWER ACADEMIC PUBLISHERS
DORDRECHT / BOSTON / LONDON

Library of Congress Cataloging in Publication Data

ESO-CERN Symposium (3rd : 1988 : Bologna, Italy)
    Astronomy, cosmology, and fundamental physics : proceedings of the
  Third ESO-CERN Symposium, Bologna, Palazzo Re Enzo, May 16-20, 1988
  / edited by Michele Caffo ... [et al.].
        p.   cm. -- (Astrophysics and space science library)
    Includes index.
      ISBN-13:978-94-010-6923-6
    1. Astronomy--Congresses. 2. Cosmology--Congresses. 3. Nuclear
  physics--Congresses. 4. Particle (Nuclear physics)--Congresses.
  I. Caffo, Michele. II. Title. III. Series.
  QB1.E78   1988
  520--dc20   89-32245

ISBN-13:978-94-010-6923-6   e-ISBN-13:978-94-009-0965-6
DOI: 10.1007/978-94-009-0965-6

Published by Kluwer Academic Publishers,
P.O. Box 17, 3300 AA Dordrecht, The Netherlands.

Kluwer Academic Publishers incorporates
the publishing programmes of
D. Reidel, Martinus Nijhoff, Dr W. Junk and MTP Press.

Sold and distributed in the U.S.A. and Canada
by Kluwer Academic Publishers,
101 Philip Drive, Norwell, MA 02061, U.S.A.

In all other countries, sold and distributed
by Kluwer Academic Publishers Group,
P.O. Box 322, 3300 AH Dordrecht, The Netherlands.

TABLE OF CONTENTS

POSTERS

FOREWORD

In the development of Fundamental Physics on one side, and of Astronomy/Cosmology on the other side, periods of parallell, relatively independent progress seem to alternate with others of intense interaction and mutual influence. To this latter case belong the very beginnings of Modern Physics, with Galileo and Newton. There is now a widespread feeling that another of such flourishing periods may have started some ten years ago, with the advent of Unified Theories and the introduction of Inflationary Cosmologies. The interaction between the two disciplines has become tighter ever since, spurring studies of e.g. astronomical and particle *Dark Matter* candidates, Superstrings and Cosmic Strings, phase transitions in the Early Universe, etc. etc. Then the recent birth of Neutrino Astronomy has added further flavor to this splendid *conjunction*.

It was indeed with the clear perception of this trend that six years ago CERN and ESO decided to jointly organize a series of symposia focusing on the interactions between Astronomy, Cosmology, and Fundamental Physics, to be held about every two years. The aim of these meetings is to bring together astronomers, cosmologists, and particle physicists to exchange information, to discuss scientific issues of common interest, and to take note of the latest devolopments in each discipline that are relevant to the other.

The First ESO-CERN Symposium was held at CERN (Geneva) on November 21-25, 1983. Then for its Second edition the ESO-CERN Symposium moved to Garching bei München, where ESO headquarters are located, and took place on March 17-21, 1986. The Third Symposium was then held in Bologna in the frame of a very special occasion: the Ninth Centenary of the University of Bologna, that according to the tradition started in 1088 AD. The University of Bologna is believed to be the oldest in the western world, and perhaps the oldest in the world if by *University* we mean an institution based on the dual principle of teaching, and link of such teaching to independent research. The Symposium was therefore intended to be the contribution of the international community of the astronomers and the physicists to the 1988 celebrations. As organizers of the Symposium and members of the University of Bologna we would like to express our appreciation to the Directors of CERN and ESO for having enthousiastically accepted our invitation to hold the 1988 Symposium in Bologna, where it then took place from May 16 to 20, in Palazzo Re Enzo - Salone del Podestà, kindly made available by the Council of the City of Bologna.

As for the first two Symposia, the programme was structured around a limited number of invited lectures followed by discussion periods, during which pertinent shorter contributions were also presented. A poster session was also organized, thus allowing the presentation of several other interesting contributions. These Proceedings contain most of the main lectures and of the short contributions in the order of their presentation, and several of the poster papers.

At the end of the Symposium the *Rector Magnificus* of the University of Bologna conferred to Professors Léon Van Hove and Lodewijk Woltjer the *Laurea Honoris Causa*, respectively in Astronomy and in Physics, in the recognition of their outstanding achievements in scientific research, and of the impulse that they have provided to the development

of the two European organizations CERN and ESO, of which they have been General Di
rectors for many, crucial years. A brief summary of their acceptance address is reported
in the following pages.

In conjunction with the Symposium a scientific exhibition on the current status and
the main future projects of the research in Astronomy and Elementary Particle Physics
was also organized. The display *Da Infinito a Infinito: dal Cosmo al Quark e Ritorno* was
based on materials kindly provided by ESO, CERN, INFN, CNR and the Astronomical
Observatory of Bologna. The very successful exhibition was open to the public during
three weeks, and over thirty thousand visitors had a chance of admiring e.g. models of the
LEP accelerator now being completed at CERN, of the ESO Very Large Telescope, and of
the Gran Sasso Underground Laboratory, as well as some Bolognese historical specimens
of over thirty years ago, such as one of the first Bubble Chambers and the first Zenithal
Multimirror Telescope. It is again a pleasure to thank CERN, ESO, INFN and the local
institutions for their essential support in the organization of the exhibition.

As organizers of the Symposium we would like to express our gratitude to the many
people who contributed to its success, and in particular to the members of the Scientific Organizing Committee for their assistance in the definition of the programme, to the
invited lecturers for their excellent presentations, and to the chairpersons for their vigorous handling of the various sessions. Then we would like to thank all the technical and
administrative personnel of the Physics and Astronomy Departments, of the Institute of
Radioastronomy of CNR, of the Astronomical Observatory of Bologna, and of the local
section of INFN, for having assisted in a variety of ways our organizational work before,
during, and after the meeting. A special thank goes to the many students in Physics and
Astronomy who greatly helped with the practical arrangements in the Conference Hall and
assisted the public at the scientific exhibit.

Finally, we would like to acknowledge the generous support of the Council of the City
of Bologna, of the National Institute of Nuclear Physics (INFN), of the National Council
of Research(CNR), and of the University of Bologna.

Michele Caffo, Roberto Fanti, Giorgio Giacomelli, and Alvio Renzini

Bologna, December 10, 1988

The index which covers the invited reviews, has been prepared by Roeland van der Marel at the
Sterrewacht Leiden.

SCIENTIFIC ORGANIZING COMMITTEE

ELLIS J., CERN, Geneva, Switzerland
GIACOMELLI G., Physics Department, University of Bologna , Italy
KOSHIBA M.T., CERN, Geneva, Switzerland, and Tokai University, Japan
NANOPOULOS D., Physics Department, University of Wisconsin, USA
REES M.J., Institute of Astronomy, University of Cambridge, UK
RENZINI A., Astronomy Department, University of Bologna, Italy
SALAM A., ICTP, Trieste, Italy, and Imperial College, London, UK
SCIAMA D.W., SISSA, Trieste, Italy, University of Oxford, UK
SETTI G., ESO, Garching, Fed. Rep. Germany
SHAVER P., ESO, Garching, Fed. Rep. Germany
VAN HOVE L., CERN, Geneva, Switzerland
ZEL'DOVICH Y.B., Inst. for Physical Problems, USSR Acad. Sci., Moscow, USSR

SESSION CHAIRMEN

Monday, May 16.
B. PONTECORVO, JINR, Dubna, USSR
H. VAN DER LAAN, ESO, Garching, Federal Republic of Germany

Tuesday, May 17.
N. STRAUMANN, Theoretical Physics Institute, University of Zurich, Switzerland
T. KIBBLE, Blakett Laboratory, Imperial College, London, UK

Wednesday, May 18.
G. GIACOMELLI, Physics Department, University of Bologna , Italy

Thursday, May 19.
D. SCIAMA, University of Oxford, UK, and SISSA, Trieste, Italy
V. TRIMBLE, Physics Department, University of California, Irvine, USA

Friday, May 20.
G. SETTI, ESO, Garching, Federal Republic of Germany

LOCAL ORGANIZING COMMITTEE

GIACOMELLI G., Physics Department, University of Bologna (Co-chairman)
RENZINI A., Astronomy Department, University of Bologna (Co-chairman)
BERGIA S., Physics Department, University of Bologna
CAFFO M., INFN, Bologna
FANTI R., Astronomy Department, University of Bologna
MESSINA A., Astronomy Department, University of Bologna

The *Rector Magnificus* Prof. Fabio Roversi-Monaco reads the motivation of the *Laurea Honoris Causa* in Astronomy to be conferred to Prof. Léon Van Hove.

Acceptance Address, Laurea Honoris Causa, Bologna, 20 May, 1988

Rector Magnificus, Members of the Faculty, Ladies and Gentlemen,

To be declared *Doctor Honoris Causa* of your University as it celebrates its 900th anniversary is such eminent privilege that I do not quite know how to express my gratitude. That this degree is awarded me at the close of ESO/CERN Symposium which was so generously hosted by Bologna is an additional gesture which I greatly appreciate. Over the last three decades I had many contacts with Bologna scientists, and it is a pleasure for me to thank all concerned for the honour just bestowed upon me.

Among the many celebrities who studied or taught here, name struck me, Erasmus of Rotterdam, the great humanist who lived and worked all over our continent, also near Brussels where there is still a "Maison d'Erasme" close to where I was born. I always regarded him as the European scholar "par excellence", and it is surely no accident that he put the University of Bologna on his intellectual itinerary.

Twenty eight years ago I left University teaching to join CERN, the European Laboratory for High Energy Physics. Somewhat surprisingly, it is at CERN that I fully realized the importance of the Universities for the advancement of science, because as I discovered it still is the University system which best provides the lasting intellectual symbiosis between research and learning, between eager students and experienced teachers, a symbiosis which is the lifeblood of scientific progress. I feel that this is my strongest reason to thank you, Mr. Rector and Members the Faculty, for the degree awarded to me.

Léon Van Hove

The *Rector Magnificus* Prof. Fabio Roversi-Monaco congratulates Prof. Lodewijk Woltjer while conferring to him the *Laurea Honoris Causa* in Physics.

Acceptance Address, Laurea Honoris Causa, Bologna, 20 May, 1988

Rector Magnificus, Members of the Faculty, Ladies and Gentlemen,

It is a great honor for me to accept today the *Laurea* which you have awarded to me. The honor is the greater if I look at the venerable names which have been associated with your University in the past – names like Copernicus and Galvani in the Sciences, like Petrarca and Dante in the Humanities. However, your University is not only famous for its past, but it has succeded in renewing itself and in remaining one of the important universities in Europe. The occasion today is also a particular pleasure to me because I see here several friends with whom have had a chance to collaborate in scientific investigations like Giampietro Puppi and Giancarlo Setti.

For many years now I have been engaged in furthering the growth of European science through cooperative ventures. The enthusiasm for such ventures in Italy has been a very rewarding experience. But coming here to Bologna one has to wonder how progress has really been made. At a Coulcil of ministers assembled a few years ago in Paris to discuss mobility European researchers, one of the main results was the decision set up "doctoral networks" which would allow European students from one university to get credit for studies done at another. Obviously this was an important step. But looking at the of nationalities here in the medieval period – the Flemish, Spanish and other colleges – one wonders if after all recent efforts one may not be still behind the situation six centuries ago. When ESO was founded complex negotiations took place to give it exemptions from customs duties, taxes etc.. Even though problems still occur in our "common" market Europe. Again it is interesting to see that in 1158 Friedrich I made a decree protecting from interference all scholars who had to travel to Bologna. Apparently, many aspects of European integration are nothing new – but rather the imperfect recovery of what has existed long ago.

Your University has had a glorious past. I wish to you that Bologna will be also in the future a center of excellence and one of the key places in the intellectual development of Europe.

Lodewijk Woltjer

LIST OF PARTICIPANTS

ABRAMOWICZ M., SISSA, TRIESTE, ITALY
AICHELBURG P.C., DEPARTMENT OF THEORETICAL PHYSICS, UNIV. WIEN, AUSTRIA
ALBANO F., DIPARTIMENTO DI ASTRONOMIA, UNIVERSITÀ DI PADOVA, ITALY
ALLES W., ITALIAN EMBASSY, BONN, FRG
ALLKOFER O.C., INSTITUT FÜR KERNPHYSIK, KIEL, FRG
AMICO P., DIPARTIMENTO DI ASTRONOMIA, UNIVERSITÀ DI PADOVA, ITALY
ANDREANI P.M., DIPARTIMENTO DI FISICA, UNIVERSITÀ "LA SAPIENZA", ROMA, ITALY
BANDIERA R., OSSERVATORIO ASTROFISICO DI ARCETRI, FIRENZE, ITALY
BARBARINO G., INFN, NAPOLI, ITALY
BARBIERI C., DIPARTIMENTO DI ASTRONOMIA, UNIVERSITÀ DI PADOVA, ITALY
BARBOUR J.B., COLLEGE FARM, OXON, UK
BARLETTA A., INFN, BOLOGNA, ITALY
BARONE E.M., DIPARTIMENTO DI FISICA, UNIVERSITÀ DI BOLOGNA, ITALY
BARROW J.D., ASTRONOMY CENTRE, UNIVERSITY OF SUSSEX, BRIGHTON, UK
BARTOLI A., DIPARTIMENTO DI FISICA, UNIVERSITÀ DI BOLOGNA, ITALY
BARTOLINI C., DIPARTIMENTO DI ASTRONOMIA, UNIVERSITÀ DI BOLOGNA, ITALY
BASILE M., DIPARTIMENTO DI FISICA, UNIVERSITÀ DI BOLOGNA, ITALY
BASSANI L., ISTITUTO TESRE - CNR, BOLOGNA, ITALY
BATTISTINI P., DIPARTIMENTO DI ASTRONOMIA, UNIVERSITÀ DI BOLOGNA, ITALY
BEL N., OBSERVATOIRE DE MEUDON, FRANCE
BELLOTTI E., LABORATORI DEL GRAN SASSO - INFN, ASSERGI, ITALY
BEMPORAD C., DIPARTIMENTO DI FISICA, UNIVERSITÀ DI PISA, ITALY
BENDINELLI O., DIPARTIMENTO DI ASTRONOMIA, UNIVERSITÀ DI BOLOGNA, ITALY
BENVENUTI P., ESO, GARCHING, FRG
BERGAMINI R., ISTITUTO DI RADIOASTRONOMIA - CNR, BOLOGNA, ITALY
BERGIA S., DIPARTIMENTO DI FISICA, UNIVERSITÀ DI BOLOGNA, ITALY
BERNARDI G., LPNHE, UNIVERSITÉ DE PARIS, FRANCE
BERSANI F., DIPARTIMENTO DI FISICA, UNIVERSITÀ DI BOLOGNA, ITALY
BERSHADY M., ASTRONOMY AND ASTROPHYSICS CENTRE, CHICAGO, USA
BERTIN G., SCUOLA NORMALE SUPERIORE, PISA, ITALY
BERTOLA F., DIPARTIMENTO DI ASTRONOMIA, UNIVERSITÀ DI PADOVA, ITALY
BIERMANN P., MAX-PLANCK-INSTITUT FÜR RADIOASTRONOMIE, BONN, FRG
BIGNAMI G., ISTITUTO DI FISICA COSMICA - CNR, MILANO, ITALY
BOBISUT F., DIPARTIMENTO DI FISICA, UNIVERSITÀ DI PADOVA, ITALY
BONETTI A., INFN, FIRENZE, ITALY
BONIFAZI A., OSSERVATORIO ASTRONOMICO, UNIVERSITÀ DI BOLOGNA, ITALY
BONOLI F., DIPARTIMENTO DI ASTRONOMIA, UNIVERSITÀ DI BOLOGNA, ITALY
BONOMETTO S., DIPARTIMENTO DI FISICA, UNIVERSITÀ DI PERUGIA, ITALY
BORG A., CEN SACLAY, GIF-SUR-YVETTE, FRANCE
BÖRNER G., MAX-PLANCK-INSTITUT FÜR PHYSIK UND ASTROPHYSIK, GARCHING, FRG
BRACCESI A., DIPARTIMENTO DI ASTRONOMIA, UNIVERSITÀ DI BOLOGNA, ITALY
BRAGAGLIA A., DIPARTIMENTO DI ASTRONOMIA, UNIVERSITÀ DI BOLOGNA, ITALY
BRANDENBERGER R., DEPARTMENT OF PHYSICS, BROWN UNIVERSITY, PROVIDENCE, USA
BRUNI G., DIPARTIMENTO DI FISICA, UNIVERSITÀ DI BOLOGNA, ITALY
BURCKHART H.J., CERN, GENÈVE, SWITZERLAND
BUZZONI A., OSSERVATORIO ASTRONOMICO DI BRERA, MERATE, ITALY
CAFFO M., INFN, BOLOGNA, ITALY
CALOI V., ISTITUTO DI ASTROFISICA SPAZIALE, FRASCATI, ITALY

CAMPANELLA M., INFN, MILANO, ITALY
CAMPANINI R., DIPARTIMENTO DI FISICA, UNIVERSITÀ DI BOLOGNA, ITALY
CAMPORESI T., CERN, GENÈVE, SWITZERLAND
CANNATA F., INFN, BOLOGNA, ITALY
CAPILUPPI P., DIPARTIMENTO DI FISICA, UNIVERSITÀ DI BOLOGNA, ITALY
CAPPI A., DIPARTIMENTO DI ASTRONOMIA, UNIVERSITÀ DI BOLOGNA, ITALY
CAROLI E., ISTITUTO TESRE - CNR, BOLOGNA, ITALY
CAVALIERE A.G., DIPARTIMENTO DI FISICA, UNIVERSITÀ "TOR VERGATA", ROMA, ITALY
CAVALLO F.R., DIPARTIMENTO DI FISICA, UNIVERSITÀ DI BOLOGNA, ITALY
CECCHETTI G., INFN, PAVIA, ITALY
CELECHINI E., DIPARTIMENTO DI FISICA, UNIVERSITÀ DI FIRENZE, ITALY
CELMINI S., ISTITUTO TESRE - CNR, BOLOGNA, ITALY
CELNIKIER L., OBSERVATOIRE DE MEUDON, FRANCE
CHEW CHONG K., DEPARTMENT OF PHISYCS, UNIVERSITY OF SINGAPORE, SINGAPORE
CHIOSI C., DIPARTIMENTO DI ASTRONOMIA, UNIVERSITÀ DI PADOVA, ITALY
CORTIGLIONI S., ISTITUTO TESRE - CNR, BOLOGNA, ITALY
COSTA G., DIPARTIMENTO DI FISICA, UNIVERSITÀ DI PADOVA, ITALY
CRANE P., ESO, GARCHING, FRG
CRISTIANI S., OSSERVATORIO ASTRONOMICO DI PADOVA, ITALY
CROZON M., LDPC, COLLEGE DE FRANCE, PARIS, FRANCE
CUFFIANI M., DIPARTIMENTO DI FISICA, UNIVERSITÀ DI BOLOGNA, ITALY
DA VIÀ C., CERN, GENÈVE, SWITZERLAND
DAL FIUME D., ISTITUTO TESRE - CNR, BOLOGNA, ITALY
DALLAPORTA N., DIPARTIMENTO DI ASTRONOMIA, UNIVERSITÀ DI PADOVA, ITALY
DANESE L., DIPARTIMENTO DI ASTRONOMIA, UNIVERSITÀ DI PADOVA, ITALY
DAPERGOLAS A., NATIONAL OBSERVATORY OF ATHENS, GREECE
DE MARZO C., DIPARTIMENTO DI FISICA, UNIVERSITÀ DI BARI, ITALY
DE RUITER H.R., ISTITUTO DI RADIOASTRONOMIA - CNR, BOLOGNA, ITALY
DE SABBATA V., DIPARTIMENTO DI FISICA, UNIVERSITÀ DI FERRARA, ITALY
DELPINO F., OSSERVATORIO ASTRONOMICO DI BOLOGNA, ITALY
DENINNO M.M., DIPARTIMENTO DI FISICA, UNIVERSITÀ DI BOLOGNA, ITALY
DERADO I., MAX-PLANCK-INSTITUT FÜR PHYSIK UND ASTROPHYSIK, MÜNCHEN, FRG
DI COCCO G., ISTITUTO TESRE - CNR, BOLOGNA, ITALY
DI IORIO A., OSSERVATORIO DI COLLURANIA, TERAMO, ITALY
DRAGONI G., DIPARTIMENTO DI FISICA, UNIVERSITÀ DI BOLOGNA, ITALY
DRESSLER A., MOUNT WILSON & LAS CAMPANAS OBSERVATORIES, PASADENA, USA
EINASTO J., TARTU ASTROPHYSICAL OBSERVATORY, TORAVERE, USSR
ELLIS G., SISSA, TRIESTE, ITALY
FABBRI F., DIPARTIMENTO DI FISICA, UNIVERSITÀ DI BOLOGNA, ITALY
FANTI C., DIPARTIMENTO DI ASTRONOMIA, UNIVERSITÀ DI BOLOGNA, ITALY
FANTI R., DIPARTIMENTO DI ASTRONOMIA, UNIVERSITÀ DI BOLOGNA, ITALY
FEDERICI L., OSSERVATORIO ASTRONOMICO DI BOLOGNA, ITALY
FERETTI L., DIPARTIMENTO DI ASTRONOMIA, UNIVERSITÀ DI BOLOGNA, ITALY
FERRARI A., ISTITUTO DI FISICA GENERALE, UNIVERSITÀ DI TORINO, ITALY
FERRARI E., DIPARTIMENTO DI FISICA, UNIVERSITÀ "LA SAPIENZA", ROMA, ITALY
FERRARO F., DIPARTIMENTO DI ASTRONOMIA, UNIVERSITÀ DI BOLOGNA, ITALY
FERRERO M.I., ISTITUTO DI FISICA, UNIVERSITÀ DI TORINO, ITALY
FLIN P., JAGIELLONIAN UNIVERSITY OBSERVATORY, KRAKOV, POLAND
FOCARDI P., DIPARTIMENTO DI ASTRONOMIA, UNIVERSITÀ DI BOLOGNA, ITALY
FORT B.P., ESO, GARCHING, FRG
FRANCESCHINI A., DIPARTIMENTO DI ASTRONOMIA, UNIVERSITÀ DI PADOVA, ITALY

FRONTERA F., ISTITUTO TESRE - CNR, BOLOGNA, ITALY
FUSI PECCI F., DIPARTIMENTO DI ASTRONOMIA, UNIVERSITÀ DI BOLOGNA, ITALY
GALLI D., OSSERVATORIO DI ARCETRI, FIRENZE, ITALY
GALLI M., ENEA, BOLOGNA, ITALY
GALLI M., DIPARTIMENTO DI FISICA, UNIVERSITÀ DI BOLOGNA, ITALY
GALLINO R., ISTITUTO DI FISICA, UNIVERSITÀ DI TORINO, ITALY
GAMBERINI G., INFN, BOLOGNA, ITALY
GAZZARUSO L., DIPARTIMENTO DI FISICA, UNIVERSITÀ DI BOLOGNA, ITALY
GELLER M., CENTER FOR ASTROPHYSICS, CAMBRIDGE, USA
GIACCONI P., DIPARTIMENTO DI FISICA, UNIVERSITÀ DI BOLOGNA, ITALY
GIACOMELLI G., DIPARTIMENTO DI FISICA, UNIVERSITÀ DI BOLOGNA, ITALY
GIACOMELLI P., THE ROCKFELLER UNIVERSITY, NEW YORK, USA
GIBBONS G.W., DAMTP, UNIVERSITY OF CAMBRIDGE, UK
GIOVANNINI A., ISTITUTO DI FISICA, UNIVERSITÀ DI TORINO, ITALY
GIOVANNINI G., DIPARTIMENTO DI ASTRONOMIA, BOLOGNA, ITALY
GIURICIN G., DIPARTIMENTO DI ASTRONOMIA, UNIVERSITÀ DI TRIESTE, ITALY
GOLDHABER G., LBL, BERKELEY, USA
GONZALES-MESTRES L., LAPP, ANNECY-LE-VIEUX, FRANCE
GOTZ G., MAX-PLANCK-INSTITUT FÜR PHYSIK UND ASTROPHYSIK, GARCHING, FRG
GREGGIO L., DIPARTIMENTO DI ASTRONOMIA, UNIVERSITÀ DI BOLOGNA, ITALY
GREGORINI L., DIPARTIMENTO DI ASTRONOMIA, UNIVERSITÀ DI BOLOGNA, ITALY
GRUEFF G., DIPARTIMENTO DI ASTRONOMIA, UNIVERSITÀ DI BOLOGNA, ITALY
GUARNIERI A., DIPARTIMENTO DI ASTRONOMIA, UNIVERSITÀ DI BOLOGNA, ITALY
HACK M., DIPARTIMENTO DI ASTRONOMIA, UNIVERSITÀ DI TRIESTE, ITALY
HARGROVE C., HIGH ENERGY PHYSICS SECTION, OTTAWA, CANADA
HELD E., OSSERVATORIO ASTRONOMICO DI BOLOGNA, ITALY
HILL A., DIPARTIMENTO DI FISICA, UNIVERSITÀ DI BOLOGNA, ITALY
HILLEBRANDT W., MAX-PLANCK-INST. FÜR PHYSIK UND ASTROPHYSIK, GARCHING, FRG
HOYLE F., BOURNEMOUTH, UK
HUBER M.C.E., ESA/ESTEC, NORWIJK, THE NETHERLANDS
HUZITA H., INFN, PADOVA, ITALY
JØRGENSEN C., CHIMIE MINERALE, UNIVERSITÉ DE GENÈVE, SWITZERLAND
JULVE J., INSTITUTO DE ESTRUCTURA DE LA MATERIA, MADRID, SPAIN
JUNOR B., ISTITUTO DI RADIOASTRONOMIA - CNR, BOLOGNA, ITALY
KIBBLE T., BLAKETT LABORATORY, IMPERIAL COLLEGE, LONDON, UK
KIBBLEWHITE E., NOAO, TUCSON, USA
KOLLER K., INSITUT FÜR THEORETISCHE PHYSIK, UNIVERSITÄT MÜNCHEN, FRG
KOSHIBA M.T., CERN, GENÈVE, SWITZERLAND
KRANJC A., DIPARTIMENTO DI FISICA, UNIVERSITÀ DI MILANO, ITALY
KREYSA E., MAX-PLANCK-INSTITUT FÜR RADIOASTRONOMIE, BONN, FRG
KRON R., DEPARTMENT OF ASTRONOMY, UNIVERSITY OF CHICAGO, USA
LARI C., ISTITUTO DI RADIOASTRONOMIA - CNR, BOLOGNA, ITALY
LAURENZI D., DIPARTIMENTO DI FISICA, UNIVERSITÀ DI BOLOGNA, ITALY
LAYTER J., DEPARTMENT OF PHYSICS, UNIVERSITY OF CALIFORNIA, RIVERSIDE, USA
LE COULTRE P., INSTITUTE OF HIGH ENERGY PHYSICS, ZÜRICH, SWITZERLAND
LEDERMAN L., FERMILAB, BATAVIA, USA
LIEBUNDGUT B., ASTRONOMISCHES INSTITUT, UNIVERSITÄT BASEL, SWITZERLAND
LYNDEN-BELL D., INSTITUTE OF ASTRONOMY, CAMBRIDGE, UK
LYTH D., DEPARTMENT OF PHYSICS, UNIVERSITY OF LANCASTER, UK
MADSEN M. S., ASTRONOMY CENTRE, UNIVERSITY OF SUSSEX, BRIGHTON, UK
MAINARDI F., DIPARTIMENTO DI FISICA, UNIVERSITÀ DI BOLOGNA, ITALY

MANDOLESI N., ISTITUTO TESRE - CNR, BOLOGNA, ITALY
MANNO V., ESA, PARIS, FRANCE
MANTOVANI F., ISTITUTO DI RADIOASTRONOMIA - CNR, BOLOGNA, ITALY
MARANO B., OSSERVATORIO ASTRONOMICO DI BOLOGNA, ITALY
MARCAIDE J., INSTITUTO DE ASTROFISICA DE ANDALUCIA, GRANADA, SPAIN
MARCELLINI S., DIPARTIMENTO DI FISICA, UNIVERSITÀ DI BOLOGNA, ITALY
MARGIOTTA A., DIPARTIMENTO DI FISICA, UNIVERSITÀ DI BOLOGNA, ITALY
MARZARI CHIESA A., DIPARTIMENTO DI FISICA, UNIVERSITÀ DI TORINO, ITALY
MASANI A., DIPARTIMENTO DI FISICA, UNIVERSITÀ DI TORINO, ITALY
MASIERO A., INFN, PADOVA, ITALY
MATTEUZZI P., INFN, BOLOGNA, ITALY
McCREA W., ASTRONOMY CENTRE, UNIVERSITY OF SUSSEX, BRIGHTON, UK
MERIGHI R., OSSERVATORIO ASTRONOMICO DI BOLOGNA, ITALY
MESSINA A., DIPARTIMENTO DI ASTRONOMIA, UNIVERSITÀ DI BOLOGNA, ITALY
MONACELLI P., DIPARTIMENTO DI FISICA, UNIVERSITÀ "LA SAPIENZA", ROMA, ITALY
MONETI G., DEPARTMENT OF PHYSICS, SYRACUSE UNIVERSITY, USA
MONTANARI A., DIPARTIMENTO DI FISICA, UNIVERSITÀ DI BOLOGNA, ITALY
MONZONI V., DIPARTIMENTO DI FISICA, UNIVERSITÀ DI FERRARA, ITALY
MOSCARDINI L., DIPARTIMENTO DI ASTRONOMIA, UNIVERSITÀ DI BOLOGNA, ITALY
MOSCOSO L., DPHPE, CEN-SACLAY, GIF-SUR-YVETTE, FRANCE
MOSKOWITZ B., DEPARTMENT OF PHYS. AND ASTR., UNIVERSITY OF ROCHESTER, USA
NANOPOULOS D., DEPARTMENT OF PHYSICS, UNIVERSITY OF WISCONSIN, MADISON, USA
NATALUCCI L., ISTITUTO TESRE - CNR, BOLOGNA, ITALY
NERI D., DIPARTIMENTO DI FISICA, UNIVERSITÀ DI BOLOGNA, ITALY
NIELSEN BORGE S., CERN, GENÈVE, SWITZERLAND
NOTZOLD D., MAX-PLANCK-INSTITUT FÜR PHYSIK, MÜNCHEN, FRG
OCCHIONERO F., OSSERVATORIO ASTRONOMICO DI ROMA, ITALY
OSTERBROCK D.E., LICK OBSERVATORY, UNIVERSITY OF CALIFORNIA, SANTA CRUZ, USA
PACINI F., OSSERVATORIO ASTROFISICO DI ARCETRI, FIRENZE, ITALY
PADRIELLI L., ISTITUTO DI RADIOASTRONOMIA - CNR, BOLOGNA, ITALY
PALLOTTI G., DIPARTIMENTO DI FISICA, UNIVERSITÀ DI BOLOGNA, ITALY
PALLOTTINO G.V., DIPARTIMENTO DI FISICA, UNIVERSITÀ "LA SAPIENZA", ROMA, ITALY
PALMONARI F., DIPARTIMENTO DI FISICA, UNIVERSITÀ DI BOLOGNA, ITALY
PALUMBO G., DIPARTIMENTO DI ASTRONOMIA, UNIVERSITÀ DI BOLOGNA, ITALY
PANEK M., COPERNICUS ASTRONOMICAL CENTER, WARSZAWA, POLAND
PANTANO O., SISSA, TRIESTE, ITALY
PARMA P., ISTITUTO DI RADIOASTRONOMIA - CNR, BOLOGNA, ITALY
PARMEGGIANI G., OSSERVATORIO ASTRONOMICO DI BOLOGNA, ITALY
PARTRIDGE R.B., DEPARTMENT OF ASTRONOMY, HAVERFORD COLLEGE, USA
PATRIZII L., INFN, BOLOGNA, ITALY
PECCEI R., DESY, HAMBURG, FRG
PEROLA G.C., ISTITUTO ASTRONOMICO, UNIVERSITÀ "LA SAPIENZA", ROMA, ITALY
PERSIC M., OSSERVATORIO ASTRONOMICO DI TRIESTE, ITALY
PICCIRILLO L., DIPARTIMENTO DI FISICA, UNIVERSITÀ "LA SAPIENZA", ROMA, ITALY
PIZZICHINI G., ISTITUTO TESRE - CNR, BOLOGNA, ITALY
POLI B., DIPARTIMENTO DI FISICA, UNIVERSITÀ DI BOLOGNA, ITALY
PONTECORVO B., JINR, DUBNA, USSR
POULSEN J.M., ISTITUTO TESRE - CNR, BOLOGNA, ITALY
PREDAZZI E., DIPARTIMENTO DI FISICA TEORICA, UNIVERSITÀ DI TORINO, ITALY
PREDIERI F., DIPARTIMENTO DI FISICA, UNIVERSITÀ DI BOLOGNA, ITALY
REALE A., DIPARTIMENTO DI FISICA, UNIVERSITÀ DELL'AQUILA, ITALY

REEVES H., SECTION D'ASTROPHYSIQUE, CEN-SACLAY, GIF-SUR-YVETTE, FRANCE
REMIDDI E., DIPARTIMENTO DI FISICA, UNIVERSITÀ DI BOLOGNA, ITALY
RENO M. H., FERMILAB, BATAVIA, USA
RENZINI A., DIPARTIMENTO DI ASTRONOMIA, UNIVERSITÀ DI BOLOGNA, ITALY
RESSEL M.T., ASTRONOMY & ASTROPHYSICS CENTER, CHICAGO, USA
RIFATTO A., DIPARTIMENTO DI ASTRONOMIA, UNIVERSITÀ DI PADOVA, ITALY
RIMONDI F., DIPARTIMENTO DI FISICA, UNIVERSITÀ DI BOLOGNA, ITALY
ROOD R. T., UNIVERSITY OF VIRGINIA, CHARLOTTESVILLE, USA
RYDBECK O.E.H., SPACE OBSERVATORY, UPPSALA, SWEDEN
SAGLIA R., SCUOLA NORMALE SUPERIORE, PISA, ITALY
SALAM A., ICTP, TRIESTE, ITALY
SALVATI M., OSSERVATORIO ASTROFISICO DI ARCETRI, FIRENZE, ITALY
SANCISI R., KAPTEYN LABORATORY, GRONINGEN, THE NETHERLANDS
SANZANI G., DIPARTIMENTO DI FISICA, UNIVERSITÀ DI BOLOGNA, ITALY
SARKAR S., RUTHEFORD APPLETON LABORATORY, OXON, UK
SATZ H., DEPARTMENT OF PHYSICS, UNIVERSITÄT BIELEFELD, FRG
SCARAMELLA R., OSSERVATORIO ASTRONOMICO DI ROMA, ITALY
SCHMIDT-KALER T., ASTRONOMISCHES INSTITUT, UNIVERSITÄT BOCHUM, FRG
SCHRODER U.E., INSTITUT FÜR TEORETISCHE PHYSIK, UNIVERSITÄT FRANKFURT, FRG
SCIAMA D. W., SISSA, TRIESTE, ITALY
SCIONI M., DIPARTIMENTO DI FISICA, UNIVERSITÀ DI BOLOGNA, ITALY
SCOTT S., SOMERVILLE COLLEGE, OXFORD, UK
SEITTER W., ASTRONOMISCHES INSTITUT, UNIVERSITÄT MÜNSTER, FRG
SEMERIA F., DIPARTIMENTO DI FISICA, UNIVERSITÀ DI BOLOGNA, ITALY
SERRA LUGARESI P., DIPARTIMENTO DI FISICA, UNIVERSITÀ DI BOLOGNA, ITALY
SERRA R., DIPARTIMENTO DI FISICA, UNIVERSITÀ DI BOLOGNA, ITALY
SETTI G., ESO, GARCHING, FRG
SHAPIRO M.M., UNIVERSITY OF MARYLAND, COLLEGE PARK, USA
SHAVER P., ESO, GARCHING, FRG
SHEN B., DEPARTMENT OF PHYSICS, UNIVERSITY OF CALIFORNIA, RIVERSIDE, USA
SINIGALLIA G., DIPARTIMENTO DI FISICA, UNIVERSITÀ DI BOLOGNA, ITALY
SIROLI G., CERN, GENÈVE, SWITZERLAND
SIRONI G., DIPARTIMENTO DI FISICA, UNIVERSITÀ DI MILANO, ITALY
SIVARAM C., INSTITUTE OF ASTROPHYSICS, BANGALORE, INDIA
SPILLANTINI P., DIPARTIMENTO DI FISICA, UNIVERSITÀ DI FIRENZE, ITALY
SPINETTI M., INFN, FRASCATI, ITALY
SPOONER N.J.C., RUTHERFORD APPLETON LABORATOY, OXON, UK
SPURIO M., DIPARTIMENTO DI FISICA, UNIVERSITÀ DI BOLOGNA, ITALY
STAROBINSKI A.A., LANDAU INSTITUTE FOR THEORETICAL PHYSICS, MOSCOW, USSR
STEPHEN J., ISTITUTO TESRE - CNR, BOLOGNA, ITALY
STIAVELLI M., SCUOLA NORMALE SUPERIORE, PISA, ITALY
STIRPE G., OSSERVATORIO ASTRONOMICO DI BOLOGNA, ITALY
STRAUMANN N., INST. FÜR THEORETISCHE PHYSIK, UNIVERSITÄT ZÜRICH, SWITZERLAND
SUZUKI S., INFN, PADOVA, ITALY
SVENSSON B.E.Y., DEPT. OF THEORETICAL PHYSICS, UNIVERSITY OF LUND, SWEDEN
TANAKA Y., ISAS, TOKYO, JAPAN
TENNER A., NIKHEF - H, AMSTERDAM, THE NETHERLANDS
TINAZZI M., DIPARTIMENTO DI FISICA, UNIVERSITÀ DI PADOVA, ITALY
TOSI M., OSSERVATORIO ASTRONOMICO DI BOLOGNA, ITALY
TRIMBLE V., DEPARTMENT OF PHYSICS, UNIVERSITY OF CALIFORNIA, IRVINE, USA
TRINCHIERI G., OSSERVATORIO ASTROFISICO DI ARCETRI, FIRENZE, ITALY

TRUMPER J., MAX-PLANCK-INST. FÜR EXTRATERRESTRISCHE PHYSIK, GARCHING, FRG
TURNER M.S., FERMILAB, BATAVIA, USA
TURRINI S., DIPARTIMENTO DI FISICA, UNIVERSITÀ DI BOLOGNA, ITALY
UBERTINI P., ISTITUTO DI ASTROFISICA SPAZIALE - CNR, FRASCATI, ITALY
ULRICH M.H., ESO, GARCHING, FRG
VAN DER LAAN H., ESO, GARCHING, FRG
VAN HOVE L., CERN, GENÈVE, SWITZERLAND
VAUCLAIR S.D., OBSERVATOIRE DE TOULOUSE, FRANCE
VELO G., DIPARTIMENTO DI FISICA, UNIVERSITÀ DI BOLOGNA, ITALY
VENTURI G., DIPARTIMENTO DI FISICA, UNIVERSITÀ DI BOLOGNA, ITALY
VENTURINO C., DIPARTIMENTO DI FISICA, UNIVERSITÀ "LA SAPIENZA", ROMA, ITALY
VETTOLANI G., ISTITUTO DI RADIOASTRONOMIA - CNR, BOLOGNA, ITALY
VIGOTTI M., ISTITUTO DI RADIOASTRONOMIA - CNR, BOLOGNA, ITALY
VITTORIO N., DIPARTIMENTO DI FISICA, UNIVERSITÀ DELL'AQUILA, ITALY
WALDNER F., ISTITUTO DI FISICA, UNIVERSITÀ DI UDINE, ITALY
WATSON P., PHYSICS DEPARTMENT, CARLETON UNIVERSITY, OTTAWA, CANADA
WEIDEMANN V., INSTITUT FÜR THEORETISCHE PHYSIK, UNIVERSITÄT KIEL, FRG
WIDROW L., ASTRONOMY AND ASTROPHYSICS INSTITUTE, CHICAGO, USA
WOLTJER L., ESO, GARCHING, FRG
YANNICK G.H., LDPC, COLLEGE DE FRANCE, PARIS, FRANCE
ZAMORANI G., OSSERVATORIO ASTRONOMICO DI TRIESTE, ITALY
ZAVATTI F., DIPARTIMENTO DI ASTRONOMIA, UNIVERSITÀ DI BOLOGNA, ITALY
ZEILINGER W.W., ASTRONOMY DEPARTMENT, UNIVERSITÄT WIEN, AUSTRIA
ZEMBOWICZ R., COPERNICUS ASTRONOMICAL CENTER, WARSZAWA, POLAND
ZIOUTAS K., PHYSICS SECTION, UNIVERSITY OF THESSALONIKI, GREECE
ZITELLI V., DIPARTIMENTO DI ASTRONOMIA, UNIVERSITÀ DI BOLOGNA, ITALY
ZUCCHELLI S., DIPARTIMENTO DI FISICA, UNIVERSITÀ DI BOLOGNA, ITALY

REGISTERED PRESS OBSERVERS

CHOWN M., NEW SCIENTIST, LONDON, UK
FRASER G., CERN COURIER, GENÈVE, SWITZERLAND
LINDLEY D., NATURE MAGAZINE, WASHINGTON DC, USA

REGISTERED STUDENTS

ARZARELLO F., DIPARTIMENTO DI FISICA, UNIVERSITÀ DI BOLOGNA, ITALY
BARDELLI S., DIPARTIMENTO DI ASTRONOMIA, UNIVERSITÀ DI BOLOGNA, ITALY
BEDESCHI M., DIPARTIMENTO DI ASTRONOMIA, UNIVERSITÀ DI BOLOGNA, ITALY
BELARDI C.M., DIPARTIMENTO DI ASTRONOMIA, UNIVERSITÀ DI BOLOGNA, ITALY
BELLAGAMBA L., DIPARTIMENTO DI FISICA, UNIVERSITÀ DI BOLOGNA, ITALY
BELLOTTI R., DIPARTIMENTO DI FISICA, UNIVERSITÀ DI BARI, ITALY
BENCIVENNI S., DIPARTIMENTO DI FISICA, UNIVERSITÀ DI BOLOGNA, ITALY
BERSANI F., DIPARTIMENTO DI FISICA, UNIVERSITÀ DI BOLOGNA, ITALY
BERTOLA E., DIPARTIMENTO DI ASTRONOMIA, UNIVERSITÀ DI BOLOGNA, ITALY
BETTELLI L., CERN, GENÈVE, SWITZERLAND
BOATTO S., DIPARTIMENTO DI FISICA, UNIVERSITÀ DI BOLOGNA, ITALY
BOFFI F.R., DIPARTIMENTO DI ASTRONOMIA, UNIVERSITÀ DI BOLOGNA, ITALY

BONDI M., DIPARTIMENTO DI ASTRONOMIA, UNIVERSITÀ DI BOLOGNA, ITALY
BORDONI M.A., DIPARTIMENTO DI ASTRONOMIA, UNIVERSITÀ DI BOLOGNA, ITALY
BRUNI G., DIPARTIMENTO DI ASTRONOMIA, UNIVERSITÀ DI BOLOGNA, ITALY
BRUZZI A., DIPARTIMENTO DI ASTRONOMIA, UNIVERSITÀ DI BOLOGNA, ITALY
CAFAGNA F., DIPARTIMENTO DI FISICA, UNIVERSITÀ DI BARI, ITALY
CALFAPIETRA A., DIPARTIMENTO DI FISICA, UNIVERSITÀ DI BOLOGNA, ITALY
CAPPI M., DIPARTIMENTO DI ASTRONOMIA, UNIVERSITÀ DI BOLOGNA, ITALY
CARRETTA E., DIPARTIMENTO DI ASTRONOMIA, UNIVERSITÀ DI BOLOGNA, ITALY
CESARINI R., INFN, FIRENZE, ITALY
CHIERICI F., DIPARTIMENTO DI FISICA, UNIVERSITÀ DI BOLOGNA, ITALY
CIMATTI A., DIPARTIMENTO DI ASTRONOMIA, UNIVERSITÀ DI BOLOGNA, ITALY
COCCHERI S., DIPARTIMENTO DI FISICA, UNIVERSITÀ DI BOLOGNA, ITALY
COCCHI A., DIPARTIMENTO DI FISICA, UNIVERSITÀ DI BOLOGNA, ITALY
COMASTRI A., DIPARTIMENTO DI ASTRONOMIA, UNIVERSITÀ DI BOLOGNA, ITALY
CREMONINI C., DIPARTIMENTO DI FISICA, UNIVERSITÀ DI BOLOGNA, ITALY
CRISTALLO M.R., DIPARTIMENTO DI ASTRONOMIA, UNIVERSITÀ DI BOLOGNA, ITALY
CURTI F., DIPARTIMENTO DI ASTRONOMIA, UNIVERSITÀ DI BOLOGNA, ITALY
D'AURIA S., DIPARTIMENTO DI FISICA, UNIVERSITÀ DI BOLOGNA, ITALY
DALLA CASA D., DIPARTIMENTO DI FISICA, UNIVERSITÀ DI BOLOGNA, ITALY
DALLA CECA R., DIPARTIMENTO DI ASTRONOMIA, UNIVERSITÀ DI BOLOGNA, ITALY
DE GRANDI S., DIPARTIMENTO DI ASTRONOMIA, UNIVERSITÀ DI BOLOGNA, ITALY
DE LUCA S., DIPARTIMENTO DI FISICA, UNIVERSITÀ DI BOLOGNA, ITALY
DI LORENZO M., DIPARTIMENTO DI FISICA, UNIVERSITÀ DI BOLOGNA, ITALY
FABBRI M.G., DIPARTIMENTO DI FISICA, UNIVERSITÀ DI BOLOGNA, ITALY
FLORIS F., DIPARTIMENTO DI ASTRONOMIA, UNIVERSITÀ DI BOLOGNA, ITALY
FRANCHINI E., DIPARTIMENTO DI ASTRONOMIA, UNIVERSITÀ DI BOLOGNA, ITALY
GALLI D., DIPARTIMENTO DI FISICA, UNIVERSITÀ DI BOLOGNA, ITALY
GHETTI R., DIPARTIMENTO DI FISICA, UNIVERSITÀ DI BOLOGNA, ITALY
GIUSTI G., DIPARTIMENTO DI FISICA, UNIVERSITÀ DI BOLOGNA, ITALY
GRECO C., DIPARTIMENTO DI ASTRONOMIA, UNIVERSITÀ DI BOLOGNA, ITALY
GROSSI D., DIPARTIMENTO DI ASTRONOMIA, UNIVERSITÀ DI BOLOGNA, ITALY
GUARNIERI M.D., DIPARTIMENTO DI ASTRONOMIA, UNIVERSITÀ DI BOLOGNA, ITALY
IMPICCIATORE N., DIPARTIMENTO DI FISICA, UNIVERSITÀ DI BOLOGNA, ITALY
KELM B., DIPARTIMENTO DI ASTRONOMIA, UNIVERSITÀ DI BOLOGNA, ITALY
LAZZARI L., DIPARTIMENTO DI ASTRONOMIA, UNIVERSITÀ DI BOLOGNA, ITALY
MALAGUTI G., ISTITUTO TESRE - CNR, BOLOGNA, ITALY
MALFERRARI L., DIPARTIMENTO DI FISICA, UNIVERSITÀ DI BOLOGNA, ITALY
MARROI R., DIPARTIMENTO DI ASTRONOMIA, UNIVERSITÀ DI BOLOGNA, ITALY
MILAZZO P.M., DIPARTIMENTO DI FISICA, UNIVERSITÀ DI BOLOGNA, ITALY
NICASTRO L., DIPARTIMENTO DI ASTRONOMIA, UNIVERSITÀ DI BOLOGNA, ITALY
ODORICI F., DIPARTIMENTO DI FISICA, UNIVERSITÀ DI BOLOGNA, ITALY
PALAZZI E., ISTITUTO TESRE - CNR, BOLOGNA, ITALY
PAPINI G.P., DIPARTIMENTO DI ASTRONOMIA, UNIVERSITÀ DI BOLOGNA, ITALY
PAVAN V., DIPARTIMENTO DI FISICA, UNIVERSITÀ DI BOLOGNA, ITALY
PERIA A., DIPARTIMENTO DI ASTRONOMIA, UNIVERSITÀ DI BOLOGNA, ITALY
PIAN E., DIPARTIMENTO DI FISICA, UNIVERSITÀ DI BOLOGNA, ITALY
PICCIONI D., DIPARTIMENTO DI FISICA, UNIVERSITÀ DI BOLOGNA, ITALY
PIETRINI P., ISTITUTO DI ASTRONOMIA, FIRENZE, ITALY
POLUZZI W., DIPARTIMENTO DI ASTRONOMIA, UNIVERSITÀ DI BOLOGNA, ITALY
POLZINE K., ASTRONOMISCHES RECHEN INSTITUT, HEIDELBERG, FRG
PRADERIO G., DIPARTIMENTO DI FISICA, UNIVERSITÀ DI BOLOGNA, ITALY

QUATTRINI E., DIPARTIMENTO DI FISICA, UNIVERSITÀ DI BOLOGNA, ITALY
REBECCHI S., DIPARTIMENTO DI ASTRONOMIA, UNIVERSITÀ DI BOLOGNA, ITALY
ROSSETTI E., DIPARTIMENTO DI FISICA, UNIVERSITÀ DI BOLOGNA, ITALY
SANTOSASI G., DIPARTIMENTO DI FISICA, UNIVERSITÀ DI BOLOGNA, ITALY
SBARRA C., DIPARTIMENTO DI FISICA, UNIVERSITÀ DI BOLOGNA, ITALY
SCARNATO A., DIPARTIMENTO DI ASTRONOMIA, UNIVERSITÀ DI BOLOGNA, ITALY
SILINGARDI R., DIPARTIMENTO DI ASTRONOMIA, UNIVERSITÀ DI BOLOGNA, ITALY
SILVOTTI R., DIPARTIMENTO DI ASTRONOMIA, UNIVERSITÀ DI BOLOGNA, ITALY
SPAZZOLI O., DIPARTIMENTO DI FISICA, UNIVERSITÀ DI BOLOGNA, ITALY
STANGHELLINI C., DIPARTIMENTO DI FISICA, UNIVERSITÀ DI BOLOGNA, ITALY
TAMPELLINI V., DIPARTIMENTO DI ASTRONOMIA, UNIVERSITÀ DI BOLOGNA, ITALY
TOMA G., DIPARTIMENTO DI FISICA, UNIVERSITÀ DI BOLOGNA, ITALY
TRASIBONDI M., DIPARTIMENTO DI ASTRONOMIA, UNIVERSITÀ DI BOLOGNA, ITALY
VIEZZER R., DIPARTIMENTO DI ASTRONOMIA, UNIVERSITÀ DI BOLOGNA, ITALY
VOLPONI S., DIPARTIMENTO DI FISICA, UNIVERSITÀ DI BOLOGNA, ITALY
ZAMAGNI M., DIPARTIMENTO DI ASTRONOMIA, UNIVERSITÀ DI BOLOGNA, ITALY
ZANICHELLI A., DIPARTIMENTO DI ASTRONOMIA, UNIVERSITÀ DI BOLOGNA, ITALY
ZELLERMAYER C., DIPARTIMENTO DI ASTRONOMIA, UNIVERSITÀ DI BOLOGNA, ITALY
ZINANI A., DIPARTIMENTO DI ASTRONOMIA, UNIVERSITÀ DI BOLOGNA, ITALY
ZUBER K., ASTRONOMISCHES RECHEN INSTITUTE, HEIDELBERG, FRG
ZUCCA E., DIPARTIMENTO DI ASTRONOMIA, UNIVERSITÀ DI BOLOGNA, ITALY

ASTROPARTICLE PHYSICS (1988)

Abdus Salam
International Centre
      for Theoretical Physics,
Trieste, Italy

and

Imperial College
London, U.K.

Abstract

A brief review of the present situation in particle physics, astrophysics and cosmology is presented in a unified manner.

1. Introduction

I am honoured to have been asked to give this introductory lecture on the subject of Particle Physics, Cosmology and Astrophysics.

The twentieth century has been called the century of Science. There have been four standard models which have been developed during the second half of this century. These are:

1) The Plate Tectonics model in geology;
2) The Double Helix model in Biology;
3) The Hot Big Bang model in Astrophysics and Cosmology; and
4) The Standard $SU_c(3) \times SU_L(2) \times U(1)$ model in Particle Physics.

The major development during the last fifteen years has been the realisation that models 3) and 4) have converged. I shall speak mainly about this aspect of the subject in Part I of this talk.

In Part II, I shall concentrate on non-baryonic dark matter and searches for it. There are problems here of the greatest moment, common to both Cosmology and Particle Physics.

PART I

2. Historical Gifts

How models 3) and 4) have historically influenced each other may be seen in the following table:

1

M. Caffo et al. (eds.), Astronomy, Cosmology and Fundamental Physics, 1–22.
© *1989 by Kluwer Academic Publishers.*

Gifts of Particle Physics to Astrophysics and Cosmology		Gifts of Astrophysics and Cosmology to Particle Physics
Nucleosynthesis	$\Rightarrow$	Cosmic abundance of $H, D, H^3, He^3, He^4, Li^7$ $\Leftarrow$ # of $\nu < $ MeV in mass
Spontaneous symmetry breaking (see Note 1) in:		$\Leftarrow$ Temperature dependance of phase transitions
1) the Electroweak unification	$\Rightarrow$	$T_c \approx 270$ GeV, $t_{cosmic} \sim 10^{-12}$sec
2) the Electronuclear unification (grand unification, G.U.T.)	$\Rightarrow$	1) $\begin{cases} T_c \approx 10^{14} \text{GeV}, t_{cosmic} \sim 10^{-35}\text{sec} \\ \text{proton decay and } \bar{p} \text{ slaughter} \end{cases}$ 2) Cosmic strings predicted by GUTS as seeds of galaxies. 3) "Paleogeny vs. Neogeny" relevant for the problem of the "Large scale structure of the universe", i.e. did the large scale of the universe get determined by initial fluctuations when the universe was 10^{-35} secs old (in the epoch of electro–nuclear (GUT) breaking)?
Inflation Superstrings T.O.E. $\Big\}$ (Theory of Everything)	$\Rightarrow$	Cosmological relics (like monopoles) diluted by inflation
Non–accelerator experiments ν–oscillations (MSW–Bethe mechanism) to resolve the missing solar ν mystery	$\Rightarrow$	$\Leftarrow$ 1) Relics: Dark matter (the missing light problem); Wimps (weakly interacting particles); Shadow matter and gravitational–wave detection. 2) γ–astronomy at very high energies e.g. $\approx 10^8$ GeV from CYGNUS-3X or similar extra–galactic sources. 3) ν–astronomy with SN87; clues to ν–masses, ν life–time, # of ν species as well as limits on Axion coupling

TABLE 1

3. The Three Eras

I shall divide the history of the subject of Astroparticle Physics into Three Eras. Tables 2–6 and particularly the "Remarks" column in the tables will give a description of the Physics situation and the open problems.

1) The SPECULATIVE ERA (including the Super–String epoch, Inflation, G.U.T., Supersymmetry–breaking, up to the cosmic time when electroweak transition took place)	$10^{-43} sec < t_{cosmic} < 10^{-12} sec$ Both Physics and Cosmology unknown
2) The ELECTROWEAK ERA (up to the end of the Big Bang when matter came to dominate over radiation)	$10^{-12} < t_{cosmic} < 10^{12} sec$ Both Physics and Astrophysics known and in accord with the standard models (3) and (4)
3) The LARGE SCALE MATTER ERA	$10^{12} sec < t_{cosmic} \leq 10^{18} sec$ Physics known but Astrophysics unknown

TABLE 2

4. The Speculative Era

EPOCH	TEMPERATURE	MODALITY	REMARKS	
A) Quantum Birth of the Universe as a Quantum Fluctuation M^2 (universe) = 0 B) Two-dimensional (d=2) superstrings, string size $\leq 10^{-33}$ cms	$\geq 10^{19}$ GeV (Planck mass)	Number of Bose matter fields = 26 (alternatively 10 Bose fields + Fermi fields are needed to cancel the conformal anomalies)	Riemann surfaces traced by closed strings (first rate mathematics of Riemann surfaces needed for Physics Research)	Important ingredient → Fermi-Bose Equivalence → $\Psi = e^{i\phi}$ in $d = 2$
C) Birth of space-time ("outer space") Epoch of one Force	$\geq 10^{19}$ GeV	d-dimensional space-time arises as zero modes of d spin-zero Bose fields	The theory describes $N = 1$ supergravity with nearly unique super-symmetric (SS) "inner space" $E_8 \times E_8'$ (or $SO(32)/Z_2$) for $d = 10$. The Yang-Mills fields corresponding to these groups arise miraculously as composite solitonic objects	One unified force (T.O.E.) String theory most likely to be finite (not just renormalizable)
D) Descent to four space-time (d=4)	10^{19} GeV	Kaluza-Klein-like (e.g. orbifold or Calabi-Yau) compactification takes one down from $d = 10$ to $d = 4$. Gauge symmetry broken by Higgs in the adjoint representation associated with Hosotani-Wilson loops	Tragedy ⇒	1) One loses the uniqueness of the theory when $d = 10$ descends down to $d = 4$ (there can be more than several million theories in four dimensions)
		Massive pyrgons appear, where m^2 (pyrgon) = $N(i^2 M_{planck}^2)$ where $N = 1, 2, 3, \ldots$ generically	Suggested experimental tests for this string theory (T.O.E.) are rather meagre, some of these are: A) extra one or two Z^0 B) fractionally charged dyons, even of zero (small) mass	2) No descent yet gives the standard particle model $SU_c(3) \times SU_L(2) \times U(1)$ uniquely
λ = the cosmological constant; the most outstanding unsolved problem in Astro-Particle Physics: explain why $\lambda_0 \approx 0$				$\lambda \approx M_{planck}^4$ (in general) radiatively; $\lambda = 0$ for exact (supersymmetry) $\lambda \propto M_s^4$ if SS is broken at M_{ss}. At present $\lambda_0 = 10^{-122} M_{planck}^4$ empirically

TABLE 3

5. Speculative Era (continued)
THE INFLATIONARY EPOCH (Non Accelerator Physics)

EPOCH	TEMPERATURE	MODALITY	REMARKS	OPEN PROBLEMS
E) Inflationary epoch (see Note 2) Use scalar field to motivate inflation (see Note 3)	10^{19}GeV?	Inflation solves problems of 1) Horizon 2) Flatness 3) Rotation 4) Fluctuations 5) Overabundance of Monopoles and Relic particles	Two forces now; Electro-nuclear (G.U.T.) + (gravity);	1) End of inflation? When? 2) The nature of phase transition which decouples gravity?
F) local electro-nuclear G.U.T. breaking $T_c \approx M_{GUT}$	10^{14}GeV An ideal GUT theory would have a gauge group which is not semi-simple (i.e. does not contain U(1) factors) in order to ensure quantization of all gauge charges.	1) Proton-decay and $\bar{p}$ slaughter, through X-decays where $\begin{cases} X \Rightarrow \text{Higgs or gauge particles} \\ \text{in } SU(5)GUT \\ qq \to \bar{X} \; \bar{q}\bar{\ell} \text{ i.e.} \\ \text{proton} \Rightarrow qqq \to \bar{\ell} \end{cases}$ (other decay channels for other GUTS) 2) However, no experimental evidence for $p \to \pi^0 e^+$ decay up to $\tau_p \approx 10^{32}$ years. If $\tau_p \approx 10^{33}$ years, experiments will need to be done on the moon.	Topological defects when GUT breaks; 1. Monopoles (π_2) 2. Cosmic strings (π_1) 3. Domain Walls (π_0)	⇒ Theory without inflation gives them as too abundant for comfort; mass of GUT monopole $\approx \frac{M_{GUT}}{\alpha}$ ⇒ Good for galaxy seeding ⇒ nuisance and unwanted at present

TABLE 4

TABLE 4 (continued)

EPOCH	TEMPERATURE	MODALITY	REMARKS	OPEN PROBLEMS
Examples of possible GUT breaking: $E_6 \Rightarrow SO(10) \Rightarrow$ $SU(5) \times U(1) \Rightarrow$ $SU(3) \times SU(2) \times U(1)$	Pati and Salam suggested that quark and lepton matter should not be treated separately and suggested placing (?) into one multiplet of four colours $SU_c(4)$, the fourth colour being $B - L$. This construction was generalised by Georgi and Glashow who suggested putting (?) and (?) into one multiplet of $SO(5)$, Georgi and Fritzsch later generalised $SO(5)$ to $SO(10)$.	ν_R predicted in $SO(10)$ GUT	4. Note that if density of monopoles is ≈ Parker limit, the GUT monopoles of mass $10^{16} GeV$ would close the universe by themselves (Barish)	Experiment with MACRO in Gran Sasso will give 4 monopole events/yr. at Parker limit
Other partial GUT schemes like $SU(4) \times SU_L(2) \times SU_R(2)$ are possible		If ν_R mass is m_R one obtains, $m_{\nu_L} = \dfrac{m^2 (\text{charged lepton})}{m_R}$ from a see-saw mechanism		
G) Global GUT breaking at $m_{global} \approx 10^{11} GeV$	Goldstone Particles arise when a global symmetry is broken spontaneously.	Examples: 1) Axions may accompany Peccei–Quinn global symmetry breaking. (This symmetry was introduced to solve the problem of CP violation in QCD (The Peccei–Quinn global symmetry breaking leads to a potential whose imaginary part is tiny and may give rise to a weak "fifth" force $\approx 10^{-5}$ times weaker than gravity, with a very long range.) 2) Majorans and 3) Familons may be the same particle in a properly ordered GUT theory. (Majorans would bring about spontaneous breaking of lepton # and lead to double β-decay; familons could be responsible for global breaking of the family symmetry).	New mass-scale for axions, majorans and familons?	
			Goldstone particles must have derivative coupling $\sim \dfrac{m_{global}}{g}$ with m_{global} large, perhaps $\simeq 10^{11}$ GeV	Goldstone particles must have zero mass due to spontaneous global symmetry breaking. They acquire small mass instantonically if global chiral anomaly is present.
	3) Supersymmetry breaking mass M_{ss}. $\approx 10^5 - 10^{11}$ GeV		SS breaks at all Temperatures, $T > 0$, no T_c for this symmetry.	Local SS = supergravity → super Higgs effect i.e. spin 3/2 gravitino acquires a mass $\approx m_{SS}^2 m_{planck}^{-1}$ (m_{SS} is spontaneous SS breaking mass)

6. The Electroweak Era

The Second Era extends from just before electroweak transition sets in (which happens at 10^{-12} secs), up to just after "Big Bang" ends and matter dominates (10^{12}secs) over radiation.

4 EPOCHS	TEMPERATURE	t_{cosmic} assuming the standard model of cosmology)	Remarks
Continuation of broken G.U.T. and broken supersymmetry epochs	1. $10^5 - 10^8$ GeV	10^{-24} to 10^{-18} secs	New accelerator ideas (like Plasma beatwave) needed; new technology must be developed if accelerator Particle Physics is to survive beyond the year 2010 (say)
	2. $10^2 - 10^5$ GeV	$10^{-18} - 10^{-12}$ secs (see Note 5).A minimal $SU(3) \times SU(2) \times U(1)$ supersymmetric standard model needs two Higgs doublets i.e. (three live neutral Higgs + $H^{\pm}$) (in addition to photinos, gluinos, winos, zinos, higgsinos, squarks and sleptons).	Present technology for pp (and possibly for e^+e^- linear colliders) will carry through. Discoveries beyond the Standard Model of Particle Physics expected below 10^3 GeV, c.o.m. energies. Particularly of Supersymmetric partners of known particles produced in pairs (due to conservation of R quantum #. In general $R = -1$, for the new particles)
Standard Epoch; Broken electroweak theory (see Note 4) (up to quark-lepton transition around 1/10 GeV).	3. $1/10 - 10^2$ GeV	$10^{-12} - 10^{-4}$ sec	Standard model of particle physics $SU_c(3) \times SU_L(2) \times U(1)$ ($T_c \approx G_F^{-1/2} = 270$ Gev) Electroweak symmetry breaking yields 1. W and Z masses ≈ 100 GeV; this and spin (one) of W and Z verified directly. 2. $SU_c(3)$ quark-hadronic transition (CERN) 3. Three families of quarks and leptons (except Top Quark all discovered) 4. Higgs not found yet

TABLE 5

4 EPOCHS

	TEMPERATURE	t_{cosmic} assuming the standard model of cosmology)	Remarks
Family mystery 20 parameters (see Note 6) → a mystery; Hope that GUT and eventually the string theory (T.O.E.) will determine these in terms of m_{planck} alone.		$sin^2\theta$—(the mixing between γ and Z^0) cannot be determined from standard model theory (need GUTS)	
$1/10$ GeV → 1 ev	4. $10^{-4} - 10^{12}$sec		Nucleosynthesis "Big Bang ends" with emission of Penzias–Wilson background radiation; matter begins to dominate over radiation

TABLE 5 (continued)

7. The Third Era

THE LARGE SCALE STRUCTURE OF THE UNIVERSE

	COSMIC TIME	TEMPERATURE	REMARKS
Physics and Astrophysics known	$\approx 10^{-4}s$	$\sim 10^2$MeV	$\pi\&\mu$ annihilation colour confinement
	$\approx 1s$	1 MeV	Neutrino decoupling
	$\approx 4s$	0.5 MeV	e^+ (slaughter)
	≈ 3 min	0.1 MeV	D bottleneck, He synthesis
	$\approx 3 \times 10^4$ years	2 eV	non–relativistic matter domination
	$\approx 4 \times 10^5$ years	0.3 eV	Atomic H formation "recombination"
Galaxies, Clusters, Super–clusters, form between 10^6 years and 10^{10} years. Physics known but astrophysics cloudy	≈ 15 G years	10^{-3} eV $\approx (3^0$k)	Present

TABLE 6

8. The Third Era (continued)

THE LARGE SCALE STRUCTURE IN THE UNIVERSE
1980s REVOLUTION

1. It was previously believed that galaxies, clusters and superclusters of galaxies are uniformly distributed in space.

However, the new picture evolving during the 1980s is that the 3– dimensional plots of redshifts are finding clusters of galaxies distributed on the surface of large "empty" bubbles. These have diameters 20 to 50 Mpcs, one to two orders of magnitude larger than the thickness of the surface of the bubbles.

2. It seems that billions of massive stars exploded becoming super novae, the blast waves from these explosions formed the empty bubbles. Galactic clusters formed where bubbles intersected.

3. One must be cautious however.

"Since many of the known highest–redshift objects were found by accident, and in any case their properties have not been uniquely predicted by any physical model, one must recognize that there is difficulty in making credible generalizations from these biased samples about events in the distant universe."

R. Kron

NOTES TO PART I

NOTE 1

I was at Cambridge when the theory of nucleo-synthesis was worked out by Fred Hoyle and others. At that time I thought that this was as it should be – Astrophysics should naturally receive inputs from Nuclear Particle Physics. When during the 1980s the converse happened with the numbers of ν species predicted by cosmologists (presumably) agreeing with the laboratory determination of this number, then the Particle Physicists sat up. Already during the 1970s, when it was realised that gauge symmetry restoration takes place at a critical temperature T_c, and one recalled that such high temperatures (of the order of 300 GeV for the electro–weak transition) had occurred in the early universe in the Hot Big Bang model, most Particle Physicists felt that they had to learn about the early universe and the phase transitions therein.

NOTE 2

TESTS OF INFLATION

1) $\Omega = 1 + 10^{-BIG\#}$.
2) Adiabatic density perturabations with the Harrison- Zeldovich spectrum.
3) Expected spectrum of gravitational waves with $\lambda \sim 1$ km up to 10^{28}cm. No spectrum of relic gravitational waves $< 1K$.

"In inflationary scenarios, primeval gravitons like any other pre–inflationary relic are exponentially diluted during inflation".

Starobinsky, Turner

4) The above remark about (zero–mass) gravitons may have relevance for any surviving zero–mass shadow matter which may be part of the second E_8' in the heterotic $E_8 \times E_8'$ string theories. Such shadow matter is supposed to interact with normal matter only through its shared gravitational interaction.

NOTE 3

INFLATIONARY CHAOTIC COSMOLOGY

According to A.D. Linde (Physics Today, September 1987 issue):

"The orthodox version of inflation assumed it to be a modest variation on the standard hot Big Bang theory. It was still assumed that there was an initial singularity at t = 0, that after the Planck time (about 10^{-43} seconds) the universe became hot, and that inflation was just a brief interlude in the evolution of the Universe".

This has changed. For example, in Linde's theory of chaotic inflation, consider a field φ which satisfies the Einstein-Friedman equations:

$$\ddot{\varphi} + 3H\dot{\varphi} = -m^2\varphi \qquad (1)$$

$$H^2 + \frac{k}{R^2} = \frac{4\pi}{3M_p}(\dot{\varphi}^2 + m^2\varphi^2) \qquad (2)$$

(Here $H = \frac{\dot{R}}{R}$.)

(1) It can be shown that if the initial value of the field $\varphi = \varphi_0 > 1/5\ M_p$, where $M_p = M_{Planck}$, the friction term in Eq.(1) makes the variation of the field φ very slow, so that one can neglect $\ddot{\varphi}$ in Eq.(1) and $\dot{\varphi}$ in (2). Making these approximations, one can solve

$$\varphi(t) = \varphi_0 - \frac{mM_p}{2\sqrt{3}\pi}t$$

$$R(t) = R_0\ exp\left(\frac{2\pi}{M_p^2}[\varphi_0^2 - \varphi^2(t)]\right)$$

where

$$R(t) = R_0\ exp(Ht)$$

and the Hubble "constant" H is given by

$$H(\varphi) \approx \sqrt{\frac{4\pi}{3}\frac{m\varphi}{M_p}}$$

(2) If the field φ is smaller than $1/5\ M_p$, the friction term in Eq.(1) becomes small, and φ oscillates rapidly near its equilibrium value of zero.
(3) For m of the order of $10^{-4}M_p$ this implies that in our simplest model the inflationary domains of the universe typically expand to 10^{10^8} times their original size!
(4) After expansion by a factor 10^{10^8} all initial inhomogeneities, monopoles and domain boundaries have been swept beyond the horizon.

The average amplitude of such perturbation generated during a time interval H^{-1} (in which the universe expands by a factor of e) is given by:

$$\frac{\delta\varphi}{\varphi} \approx \frac{H}{2\pi\varphi} = \frac{m}{3\pi M_p}$$

Perturbations of the field lead to perturbations of density that are just right for subsequent galaxy formation if m, the mass of the quantum of φ, is around $10^{-4}M_p$.

NOTE 4

NOTE ON THE STANDARD MODEL OF PARTICLE PHYSICS AND THE ROLE OF FERMI TRANSITION TEMPERATURE $T_c \approx 270\ GeV$

The standard model of today's particle physics describes three replicated families of quarks and leptons. The first family consists of the so-called up and down quarks (u_L, d_L) and (u_R, d_R) (L and R stand for left and right chirality of spin-1/2 particles). Each quark comes in three colours: red, yellow and blue. There are, in addition, three colourless leptons, (e_L, v_L) and e_R. Thus this family has 12 quarks and 3 leptons (altogether 15 two-component objects) with $30 \approx 2 \times 15$ degrees of freedom.

The second family has charm and strange quarks (c, s) (replacing the up and down (u, d) quarks) while the electron and its neutrino are replaced by the muon and its neutrino. Like the first family, there are 15 two-component objects. The third family likewise consists of top and bottom (t, b) quarks plus the tauon and its neutrino.

In addition to these $45(= 3 \times 15)$ spin-(1/2) two-component objects, there are the 12 Yang-Mills-Shaw gauge spin-1 messengers corresponding to the symmetry $SU_c(3) \times SU_L(2) \times U(1)$ – the photon $\gamma, W^\pm, Z^0$ and eight (confined) gluons. Nine of these (γ and eight gluons) are massless. In addition, there should at least be one physical spin- zero Higgs H_0 giving a total minimum degrees of freedom ($118 = 3 \times 15 \times 2 + 9 \times 2 + 3 \times 3 + 1$) for the particles in the standard model. All particles except the top quark and the Higgs in this list have been discovered and their masses and spins determined (even though the colour–carrying quarks and gluons are confined). (In this context, it is worth remarking that CERN data from $Sp\bar{p}S$ have confirmed the *theoretical (tree diagram) expectation* of $W_\pm, Z^0$ masses to within 1%. (Experiments give $81.8 \pm 1.5\ Gev$ for $W^\pm$ and $92.6 \pm 1.7\ GeV$ for Z^0 masses.) The model is unified: the γ and Z^0 mix, but the magnitude of the mixing is expressed as a parameter $(sin^2\theta)$ which remains to be fixed by experiment. The unification happens when the temperature is higher than the Fermi mass scale $G_F^{-1/2} \approx 270\ GeV$ which, according to the standard cosmological model occurred when the Universe was 10^{-12} sec. old. Before this phase transition occurred, there were three fundamental forces (electroweak, strong and gravitational). Afterwards, the electroweak force separated into electromagnetism and the weak nuclear force, with $W^\pm$ and Z^0 becoming massive.

NOTE 5

SUPERSYMMETRY (MATTER–FORCE–SYMMETRY)
Astounding Symmetry Discovered Theoretically in 1974

Astounding because: it connotes symmetry between fermions and bosons: i.e. symmetry between fermionic matter of spin– $\frac{1}{2}$ or $\frac{3}{2}$, (objects which are *individualists* obeying the Pauli Exclusion Principle) and bosonic force messengers of spins–0 or 1 or 2 which are *collectivists* and like to congregate.

MINIMAL (BROKEN) SUPERSYMMETRY MODEL has two Higgs multiplets plus higgsinos.

No evidence has been found yet for the existence of partners of quarks or leptons up to $\approx 50\ GeV$. The most crucial open problem in particle physics is to discover if these particles exist (expectedly below 1000 GeV centre of mass energy).

NOTE 6

ACCELERATORS NOW AND IN THE FORESEEABLE FUTURE

YEAR	ACCELERATORS	$\sqrt{s}(GeV)$ centre of mass energy	CONSTITUENT (peak-Max, GeV) $\sqrt{s}$	LUMINOSITY $(cm^{-2}sec^{-1})$	LOCALITY
1987	$Sp\bar{p}S$	900	$100 - 300\ q\bar{q},q\bar{q}$	10^{30}	CERN
1987	Tevatron	2,000	$200 - 600\ q\bar{q},q\bar{q}$	10^{30}	FERMILAB
1987	TRISTAN	$60(e^+e^-)$	60	8×10^{31}	Japan
1987	SLC	$100(e^+e^-)$	100	6×10^{30}	Stanford
1987	Bepc	$4(e^+e^-)$	4	5×10^{30}	Beijing
1989	LEP I	$100(e^+e^-)$	100	1.6×10^{31}	CERN
1995	LEP II	$200(e^+e^-)$	200	5×10^{31}	CERN
1991	UNK	3,000	$300\text{-}900\ qq,q\bar{q}$	10^{31}	Serpukhov
1991	HERA (ep)	320	100-170	5×10^{31}	Hamburg
?	LHC(pp)	20,000	2,000-3,000	10^{33}	CERN
?	SSC(pp)	40,000	4,000-5,000	10^{33}	USA
?	CLIC(e^+e^-)	4,000	4,000	$10^{33} - 10^{34}$	CERN
?	VLLP(e^+e^-)	4,000	4,000	$10^{33} - 10^{34}$	Serpukhov
?	ELOISATRON (pp)	100,000	10,000-12,000	$10^{33} - 10^{34}$	Sicily

The SSC is meeting opposition in the US because alleged of high costs - $5 billion over 5 years of construction. Compare this with the sums of monies already spent on a project like SDI which have amounted to $12.7 billion so far.

NOTE 6 (continued)

1) For the circular accelerators, the bending magnet may be improved by Superconductivity Technology, but the real limitation is due to synchrotron radiation $\propto (E^4)$. The cost and size of the accelerator increase as E^2. Here E is the c.m. energy.

2) For linear accelerators, the highest *Electric Field* gradients $\mathcal{E}$ achievable with today's technology, are at most around 1/10 GV per meter. Twenty years hence (when, for example, we may have mastered the technology of laser beat–wave plasma accelerators) the gradient may go up by a factor of 1000 – i.e. 1/10 TV per metre. This may mean that a 30 km long accelerator would produce center of mass energy $(\sqrt{s}) \simeq 10^4\ TeV$.

3) Chen and Noble have shown that if one can use longitudinal electron plasma waves in a metal, the electron density is of the order of $10^{22}cm^3$ (versus normal plasma densities of the order of $10^{14} - 10^{18}cm^3$) and we gain a factor of $\sqrt{n} \simeq 10^2 - 10^3$ (with the maximum energy limited to $10^5\ TeV$, on account of channeling radiation).

To be crazy, an accelerator around the moon may generate $10^6\ TeV$; an accelerator around the earth – as Fermi once conceived – may be capable of $\sqrt{s} \simeq 10^7\ TeV$, while an accelerator extending from earth to the sun would be capable of $\sqrt{s} \simeq 10^{11}\ TeV$ (with $\mathcal{E} \sim 1/10$ TV/metre). In the same crazy strain, for an accelerator to be capable of generating $\sqrt{s} \simeq 10^{16}\ TeV$ (the theoretically favoured, Planck Energy) one would need an accelerator 10 light years long.

NOTE 7

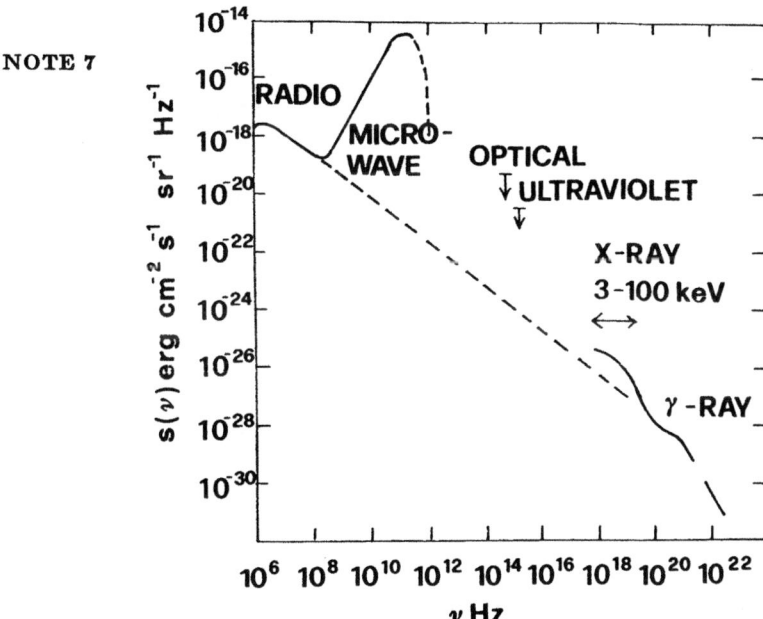

The isotropic sky flux, $s(\nu)$ versus ν

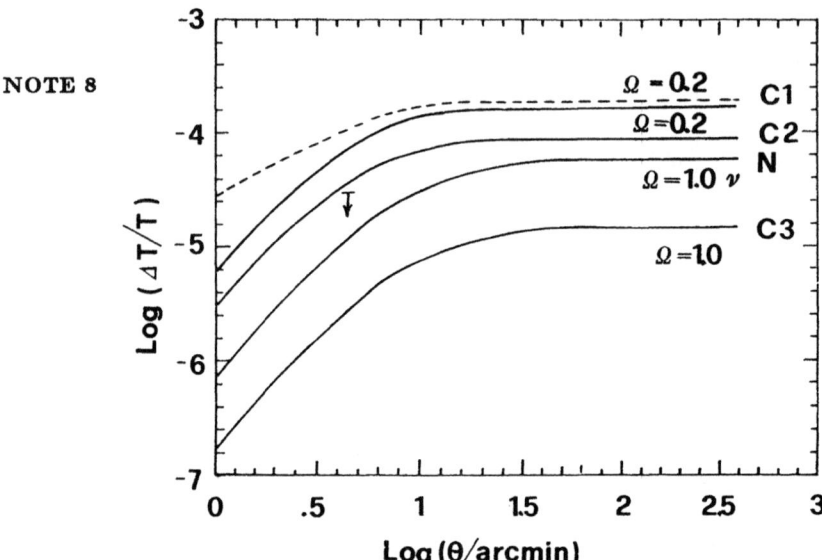

An example of the predictions of $\Delta T/T$ for different assumptions about the nature of the matter content of the universe, particularly the "dark matter" ("C" is for cold particles; "N" is for massive neutrinos.) Note that the single data point shown excludes some models and that measurements at comparable sensitivity, but $\theta \sim 30'$, would exclude most. Also, $H_0 = 75$ km/sec per Mpc, except for $Cl(H_0 = 50)$. (From Bond and Efstathiou, 1984).

PART II

"If all the matter of the universe were evenly scattered ..., and every particle had an innate gravity toward all the rest, ... matter could never convene into one mass ... but it would make an infinite number of great masses, scattered at great distances from one another ... and thus might sun and fixed stars be formed ..."

Newton

9. Dark Matter

The existence of dark matter was speculated upon 50 years ago by F. Zwicky. He showed that the visible mass of the galaxies in the coma cluster was inadequate to keep the cluster bound. Oort showed that the mass necessary to keep our own galaxy together was at least three times that concentrated into the observable stars. In recent years this has emerged as the major open problem of Cosmology and Particle Physics.

15

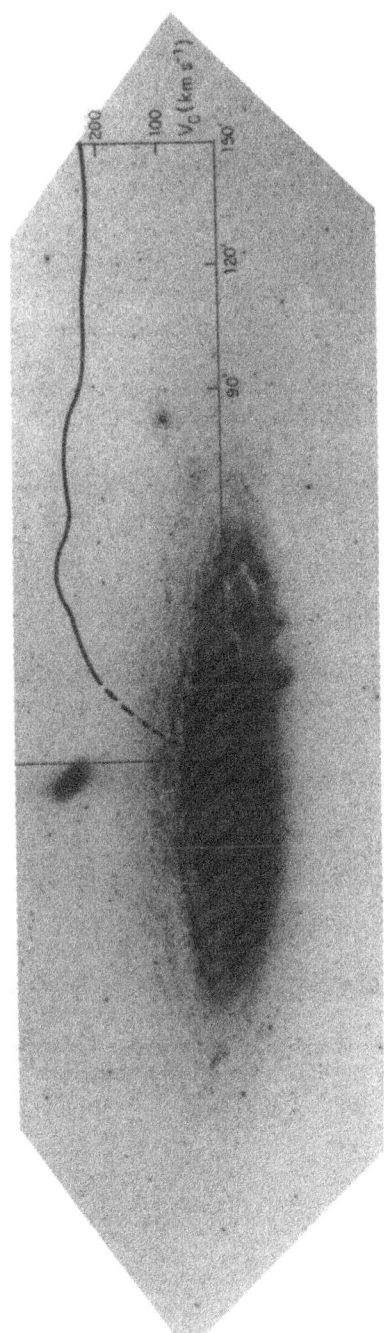

Figure 1

The Andromeda Galaxy M31 is shown with, superimposed on it, the rotation velocity of neutral hydrogen, inferred from 21 cm line radio studies. The rotation curve remains "flat" even beyond the optical outer limits of the galaxy, implying that the outlying gas is "feeling" the gravitational field of dark matter around the galaxy. (Courtesy of Morton Roberts.)

DARK MATTER (continued)

Define $\Omega = \rho/\rho_{\text{critical}}$ where $\rho_{\text{critical}}\ (= \rho_c) = \frac{3H^2}{8\pi G}$

Empirically $\Omega_{\text{photonic}} = 3 \times 10^{-5}\frac{T}{2.7K°}1/h_0^2$ where $H_0 = h_0\ 100$ km/sec/MPC $\Omega_{\text{baryon}} \approx .014$

First hypothesis :

$\Omega_{DM} = \Omega_{\text{baryonic}} = .014$ (At most this may be $\sim .15$ Limited by abundance of H^2, He^3, He^4, Li^7)

Baryonic Dark Matter could exist in the form of:

1. White dwarfs
2. Neutron Stars
3. Black holes
4. Jupiters

Second hypothesis :

$\Omega_{DM} > \Omega_{\text{baryonic}}$;

At best, empirically, $\Omega_{\text{spiral galaxies} < 30\ \text{MPC}} + \Omega_{\text{group of galaxies} < 30\ \text{MPC}} = .2 \pm .1$

Assume that

$\Omega_{\text{smooth}} = .7 + .1$ to make up $\Omega_{\text{total}} = 1 \rightarrow$ respecting the inflationary hypothesis

---

TABLE 7

10.   If Dark Matter Is Not Baryonic, What Is It?

"Not only is man not the Centre of the universe physically (Copernicus) or biologically (Darwin) but we and all we see are not even made of the predominant matter variety in the universe."

Martin Rees

(If dark matter is not baryonic).

There are three classes of Dark Matter candidates, Hot, Warm and Cold:

10.1 HOT DARK MATTER PARTICLES (LIKE NEUTRINOS) STILL IN THERMAL EQUILIBRIUM:

1) Cosmological number density comparable to microwave background
$\Rightarrow$ i.e. mass $\approx$ few tens of eV;

2) *Fluctuation Spectrum.* The spectrum of fluctuations at late times in a hot dark matter model is controlled mainly by free streaming.

3) *Free streaming* destroys any fluctuation smaller than supercluster size. This gives top-down scale structure if dark matter is hot, i.e. galaxies and clusters form after superclusters.

(This is *not* the case for warm or cold dark matter.)

4) If $H_0 = h_0\ 100\ km/sec\ Mpc$;

$\Sigma_\nu\ (1/2\ g_\nu)M_\nu \approx 100\ eV \frac{\rho}{\rho_c}\ h_0^{-2}$;

Thus, required neutrino mass,

$M_\nu \sim (25\ \text{to}\ 100)\ eV/\text{species}.$

Present experimental limits

$M_{\nu_\tau} < 35\ MeV, \quad M_{\nu_\mu} < 250\ KeV,$

$M_{\nu_e} < 18\ eV\,(\text{Zurich}),\quad 23\ eV\,(\text{Los Alamos}),\quad 30\ eV\,(\text{Tokyo}),\quad \approx 19\ eV\,(\text{ITEP})$

from SN 1987a
$$M_{\nu_e} < 15eV.$$

10.2 POTENTIAL PROBLEMS WITH *HOT* DARK MATTER

1) Galaxy formation presumably took place before $z = 3$. If QSOs are associated with galaxies, as suggested by galactic luminosity around nearby QSOs, abundance of QSO at $z > 2$ is inconsistent with "Top–Down" neutrino dominated scheme.

2) X-rays from the shock–heated pancakes are missing.

"These (serious) problems, however, may not be fatal for the hypothesis that neutrinos are the dark matter."

J. Primack

10.3 CANDIDATES FOR WARM DARK MATTER

1) Supersymmetric partners, like the light gravitino $M = M_{SUSY}^2 \ M_{pl}^{-1}$ (spontaneous SS breaking) so $M \approx 1 \ keV$ if $M_{SUSY} \approx 10^6 GeV$.

2) A hypothetical light right-handed neutrino ν_R (predicted, for example, by GUT SO(10)) could be a warm–dark matter candidate but Particle Physics provides no reason why ν_R should be light.

10.4 CANDIDATES FOR COLD DARK MATTER

1. *Quark Nuggets* (Witten)

   i.e. Ultra Dense Matter with $\#u \sim \#d \sim \#s$

2. *Massive Neutrinos*

   Few GeV.

3. *Axions*

   Light scalar Goldstone bosons needed to conserve CP in strong nuclear interactions.

4. *Supersymmetric Relics*

   Lightest one (perhaps photinos of a few GeV in mass) expected to be stable due to the conservation of R quantum number (which in general $= -1$ for the new supersymmetric partners).

10.5 CONSTRAINTS ON AXION MODELS (CLOSING WINDOW)

1. <u>Laboratory Experiments</u>

   (axion $\to 2\gamma$, and assuming the # of axions $\approx 10^3$ times time # photons)
   $f_a > 10^3$ GeV (where f_a is defined from the Lagrangian term $f_a^{-1} \ \phi_a \ F_{\mu\nu} \ \tilde{F}_{\mu\nu}$ and ϕ_a is the axion–field)

2. <u>Standard Solar Model</u>

   $f_a > 10^7$ GeV

3. <u>Solar Axio $-$ Electric Effect</u>

   search in underground Ge detector (ahlen et al.)
   $f_a > .5 \times 10^7$ GeV

4. <u>Red Giants, White Dwarfs, Neutron Systems</u>

5. Supernova 1987

   new limit $f_a > 10^{11}$ GeV

6. Cosmological Limit

   $f_a < 10^{12}$ GeV

11. Cosmic Strings In Relation to TOE Strings

1. Any string produced before inflation is exponentially diluted.
2. Cosmic strings may be *superconducting*, large currents $(\approx 10^{20} A)$ release energies $\approx 10^{60}$ ergs; trigger *explosive* galaxy formation.
3. *"It is argued that, in fundamental string theories, as one traces the universe back in time, a point is reached when the expansion rate is so fast, that the rate of string creation due to quantum effect balances the diluition of the string density due to the expansion. One is therefore led into a phase of constant string density and an exponentially expanding universe. Fundamental strings therefore seem to lead naturally to inflation."*

Turok

12. Laboratory Tests For Dark Matter

A.   1) Axion detectors $a \rightarrow 2\gamma$ in an inhomogeneous magnetic field find
     $m_a \leq 10^{-5}$ eV;

2) Bolometric detectors $\rightarrow$ with $\sigma_{scatt} \sim \sigma_{weak}$ deposit keV energies;

3) Monopole detectors (like the MACRO) in Gran Sasso (which is a truly impressive laboratory).

   "Laboratory schemes for detecting a halo population of exotic particles most worthwhile and exciting high–risk experiment in Physics or Astrophysics (as important as the discovery of microwave background in the 1960s)."

M. Rees

   Mean velocity of halo particles relative to the detector would have an *annual variation* (because of the earth's motion around the sun).
   The variation in amplitude $\sim$ few % and peaking in June would provide *evidence against spurious background.*
B. A variety of detection principles such as superheated superconducting granules (SSG), bolometers, ballistic phonons, rotons in superfluid helium, transition edge thermometers and superconducting tunnel junctions have recently been (theoretically) investigated for SSG devices. Since the involved energy quanta for these detectors are so much smaller ($\sim 1/1000$ eV for breaking a Cooper pair in a superconductor for example) than for conventional ionisation (~ 20 eV) or semiconductor (~ 1 eV) detectors, in principle very low energy thresholds and very good energy resolution can be expected ...
   "For solar neutrino detection, the coherent neutral current neutrino–nucleus scattering method is used. This method has the advantage that the cross- section is three orders of magnitude larger than the cross-section of other processes, like, for example, inverse beta–decay. Thus, an SSG detector with a weight of a few kilograms would measure the same event rate as a multi–ton detector based on other processes. The second advantage is that the SSG detector responds to all neutrino flavours equally."

K. Pretzl

Main Dark Matter Candidates in Particle Physics (Gonzalez-Mestres and Perret-Gallix, 1987)

PARTICLE	MASS	PRESENCE NEAR EARTH	ABUNDANCE	INTERACTION WITH MATTER	PROPOSED DETECTION TECHNIQUES
LIGHT NEUTRINO	$m < 30eV$ cosmology $m < 18 - 32eV$ (Experiment) (Supernova ?)	COSMIC GALACTIC	$\Omega \sim 1$ if $5eV < m_\nu < 30eV$	COHERENT SCATTERING IF DIRAC MASS	??
	$m > 10^{-5} eV$ (Cosmology)	GALACTIC	$\Omega \sim 1$ if $m_a \sim 10^{-5}$ eV?	$a \to \gamma$ conversion in a strong emf.	LOW TEMPERATURE ELECTROMAGNETIC CAVITIES
AXION	$m < 10^{-2} ev$ (Stars)	SOLAR	flux on earth: 10^5 to $10^{11} cm^{-2} s^{-1}$	$a \to \gamma$ conversion in atoms.	SILICON DIODES LOW TEMPERATURE DETECTORS
HEAVY NEUTRINOS	$m > 3$ GeV	GALACTIC	Eventually, $\Omega \sim 1$	WEAK (COHERENT IF DIRAC MASS)	IONIZATION DETECTORS FOR HEAVY PARTICLES
LSP (Lightest supersymmetric particle)	Model-dependent $(1 - 100 GeV?)$	GALACTIC	Eventually, $\Omega \sim 1$	SUPERSYMMETRIC (spin–dependent in most models)	LOW TEMPERATURE DETECTORS FOR ENERGY DEPOSITS BELOW 1 keV: SSG STJ Bolometers
COSMIONS	$4\,GeV < m < 10\,GeV$	SOLAR and GALACTIC	$\Omega \sim 1$ possible	$\sigma \sim 10^2 \sigma_{weak}$	
MONOPOLES	$m \sim 10^{16}$ GeV (GUTS)	GALACTIC TRAPPED AROUND SUN	PARKER BOUND BOUNDS FROM RUBAKOV EFFECT	ELECTROMAGNETIC	CONVENTIONAL SUPERCONDUCTING
QUARK NUGGETS	HEAVY (UNKNOWN)	GALACTIC	Eventually, $\Omega \sim 1$	ATOMIC COLLISIONS	ACCORDING TO MASS

TABLE 8

13. Envoi

I started this lecture by speaking of the four standard models elaborated during this century. In closing I would like to mention how the biological standard model is being influenced by recent advances in particle physics.

It is well known that one of the basic problems in biology is the left– handedness of naturally occurring amino–acids and the right–handedness of sugars. In the laboratory these are produced as racemic mixtures with left and right molecules equal in numbers. (Thalidomide was one such laboratory racemic mixture which led to tragic results for the new–born babies.) A happier case is that of penicillin (D–type) which splits open the D–type skins of bacteria.

In this respect it has recently been suggested by S. Mason at King's College in London and others (New Scientist, 19 January 1984) that the clue to a solution of this mystery may lie in electroweak unification and the appearance of the neutral (left– handed) weak interactions in the chemical Hamiltonian. This is shown to make for a small preponderance of left–handed amino–acids (and right–handed sugars) – 1 part in 10^{17}. This preponderance, plus the longevity of the biological epoch, *apparently* explains the occurrence in natural environment of the stated forms of chiral molecules.

14. Conclusions

14.1 OPEN QUESTIONS BEYOND THE STANDARD MODEL

PARTICLE PHYSICS

1. Are there supersymmetric particles?

2. The dark matter, does it exist? Its composition?

3. Are quarks & leptons composite? (not elementary at energies in excess of 1000 GeV)

4. Do gauge-bosons like W_R (mediating weak $V + A$) or $SU_A(3)$ (strong axial gluons) or string-inspired $Z^{0'}$ exist? likewise the existence of axions, familons, majorons at a new mass scale.

5. The near zeroness of the cosmological constant

14.2 OPEN PROBLEMS IN COSMOLOGY

1. The dark matter. Does it exist? Its composition?

2. Photon-to-baryon ratio

Grand unification theories suggest that this ratio can be explained in terms of baryon non-conservation processes, and GUT parameters. Is that so?

3. Fluctuation spectrum $(\Delta\rho/\rho)$

4. The near zeroness of the cosmological constant

I shall conclude by showing you Glashow's picture of the Universe redrawn by Prof. Giacomelli. This should show "generalized" gravity as it emerges from string theory as the Theory of Everything (T.O.E.) uniting all things "great and small". I am indebted to John Ellis, Martin Rees, Dennis Sciama and Donald Lynden–Bell for help with cosmology as well as the contributors to this Conference who responded to my request to give me a preview of what they were going to speak about.

DA INFINITO A INFINITO

DAL COSMO AL QUARK E RITORNO

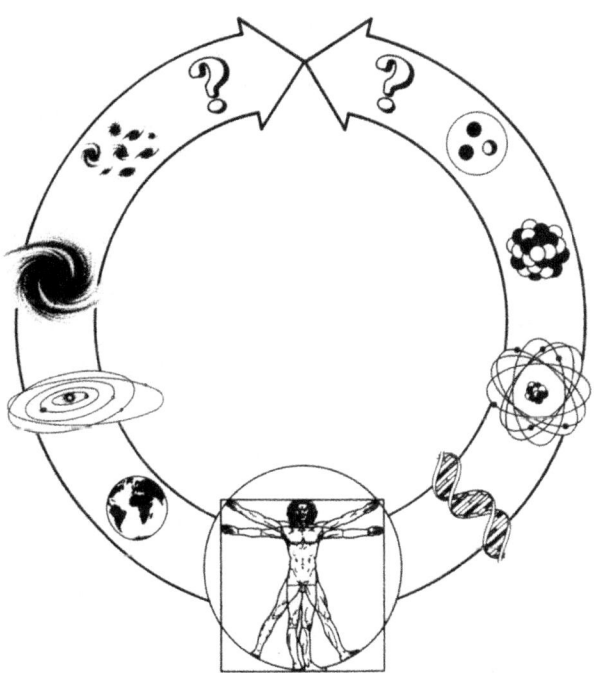

Bologna, Palazzo Re Enzo
dal 7 al 29 Maggio - Orario della mostra 10-19

The Powers of Ten

Measuring Instruments	Distance m.	Systems		Forces
Telescopes	10^{26}	Universe		Gravity
Telescopes	10^{22}	Cosmic Strings Voids / Supercluster of Galaxies / Cluster of Galaxies		Gravity
Telescopes	10^{19}	Galaxy		Gravity
Telescopes	10^{12}	Solar System		Gravity
Eye	1	Man		Electromagnetism
Microscopes	10^{-8}	Molecule		Electromagnetism
Microscopes	10^{-10}	Atom		Electromagnetism
Accelerators	10^{-14}	Nucleus		Strong
Accelerators	10^{-15}	Proton		Strong
Accelerators	$<10^{-17}$	Quarks Leptons	?	Strong

Fermilab 1987

COSMOLOGICAL PARAMETERS OF THE UNIVERSE

Alan Dressler
Mount Wilson and Las Campanas Observatories
Carnegie Institution of Washington
813 Santa Barbara Street
Pasadena, CA 91101
USA

ABSTRACT. A primary goal of observational cosmology is the measurement of the parameters H_0, q_0, Ω_0, Λ_0 and T_0 which characterize a Friedmann model of the universe. This article compares and reviews recent work in this field, with emphasis on the Hubble constant H_0, that describes the expansion rate at the present epoch. Present data favor $H_0 \approx 60 - 80$ km/sec/Mpc and $\Omega_0 \approx 0.1 - 0.2$. Within observational errors, these values are consistent with $T_0 \sim 15$ Gyr and $\Lambda_0 = 0$, but the low value of Ω_0 appears incompatible with the inflation paradigm of the early Big Bang.

1. INTRODUCTION

I thank the organizers for inviting me to participate in this historic occasion and to review the dynamical parameters of the universe. As an observer, I will interpret this task to be a brief summary of where we stand in measuring the parameters H_0, q_0, Ω_0, T_0 and Λ_0 that characterize a Friedmann model of the universe. Sandage and Tammann's (1984) review at the first ESO-CERN Conference on Cosmology and Fundamental Physics gave a balanced presentation of both the theory and observations of dynamical parameters of the universe. I will concentrate on the status of observationals, in order to include as many of the new results as possible.

I'd like to think of myself as an objective reviewer, one who does not really care what these values finally turn out to be. However, I do have to confess to publishing one paper on the value of the Hubble constant, so beware! I will do my best to ignore this and to summarize and evaluate the evidence as presented in the many papers I have read in preparation for this talk.

M. Caffo et al. (eds.), Astronomy, Cosmology and Fundamental Physics, 23–40.
© *1989 by Kluwer Academic Publishers.*

2. THE HUBBLE CONSTANT

The Hubble constant, or Hubble ratio as Gerard de Vaucouleurs likes to call it, is arguably the most important number in cosmology. Regrettably, it has been among the most uncertain numbers as well. Although it has been recognized for some fifty years that there is a linear relation between distance and recessional velocity that characterizes the expansion field for galaxies, determinations of the constant of proportionality have changed by an order-of-magnitude since Hubble's first measurements. For the last twenty years, though, measurements of the slope of this linear relationship have settled in somewhere between 50 and 100 km/sec/Mpc. Unfortunately, this factor of two uncertainty is intolerable for most matters of cosmological interest such as the age of the expansion or the scale size of clustering of galaxies (the correlation length), thus frustrating attempts to rule in and out various models of the early universe, the growth of structure, histories of stellar evolution in galaxies, for example.

The Astrophysical Journal has been the battlefield upon which two camps have clashed over the value of the Hubble constant. Gerard de Vaucouleurs of the University of Texas has long advocated what is called the "short distance scale" with a Hubble constant of ~100 km/sec/Mpc. Allan Sandage of MWLCO and Gustav Tammann of the University of Basel have championed the "long distance scale" with a Hubble constant of 50. Both sides have followed a similar procedure which involves three major steps: (1) obtain reliable distances by what are called primary means to at least a handful of local galaxies; (2) use secondary or even tertiary indicators of relative distance to "bootstrap" to some hundreds of more distant galaxies, preferably far enough out that local turbulence in the Hubble flow is not a problem; and (3) correct for statistical biases that inevitably creep in because galaxy samples are usually flux limited rather than volume limited. There are other troublesome corrections to be reckoned with, most notably, dust absorption in the sample galaxies and our own Galaxy.

In his book "The Cosmological Distance Ladder", Rowan-Robinson (1985) describes a fascinating feature of this long-standing disagreement between the Sandage-Tammann and de Vaucouleurs' camps - that the error estimates assigned by either side rigorously exclude even the possibility that the other could be correct. Rowan-Robinson concludes, I think correctly, that both sides have consistently underestimated the systematic and random errors associated with this difficult measurement. However, I think that this logjam is about to be broken, in fact, it may already have been broken. I believe that evidence is accumulating that the Hubble constant lies in between the two legendary values. Rowan-Robinson suggested this, but a good deal of the evidence that leads me to this conclusion is more recent than his forward-looking analysis. Much of the activity in the last 10 years can be traced to the entry into the game of new players and new techniques, for example, the Tully-Fisher method of obtaining relative distances of spiral galaxies, exploited by the late Marc Aaronson,

Jeremy Mould, John Huchra and their collaborators. Initially this new method seemed to confirm de Vaucouleurs' short distance scale, but, as I will discuss in a few minutes, improved measurements, coupled with a better understanding of the preferred analysis procedure, has brought the Hubble constant determined in this manner down to an intermediate value.

When reading the literature on this subject, one becomes unhappily aware of the large range of values adopted for important input parameters, for example: (1) the distance to the Hyades star cluster (from trigonometric means); (2) dust absorption within our Galaxy; (3) absorption by enshrouding dust of the light of Cepheid variable stars, crucial to the calibration of distances; (4) the amplitude of deviations from a linear Hubble flow that arise due to clumpiness of the mass distribution. The distances to primary calibrators obtained by Sandage and Tammann as compared to de Vaucouleurs' typically vary 30-40% with some cases differing by a factor of 2! John Huchra (1987) has provided a good review of these matters. Also, important papers are too often riddled with obscure corrections which have been made more-or-less subjectively - measurements made by eye, fits to data made by hand - adjustments as large as 30% that cannot, in general, be reproduced by other investigators. I have found, in preparation for this review, that when a set of standard values are adopted for these parameters, and objective criteria for measuring and fitting data replace subjective judgements, almost all techniques give H_0 in the range 60-80. This seems consistent with the errors of these methods which are not yet completely understood, even after so many years of effort. Though we may not know the final values or best procedures, it is encouraging to find that if we make the same assumptions, different techniques are giving similar answers, varying by much less than a factor of two.

It all begins with the distances to primary calibrators like the Andromeda galaxy M31, and other spirals like M33, M81, and NGC 2403. Distances to these galaxies depend on the use of Cepheid variable stars whose intrinsic luminosities are tightly correlated with the time interval over which these stars vary in brightness. There is a clear physical basis of this relation that is at least partially understood. As stressed by Rowan-Robinson, this fact lends to them extra credibility. (The peak luminosities of supernovae explosions share this excellent feature and have the additional advantage of observability over much greater distances due to their extraordinary luminosities. Unfortunately, we have, as yet, too few data to fully investigate and exploit this technique.) The period-luminosity relation for Cepheids is calibrated chiefly by stars in our nearest neighbor galaxy, the Large Magellanic Cloud, in comparison with Cepheids in Galactic star clusters. By studying LMC Cepheids, an ensemble at a common distance, one obtains not only a reliable calibration but also can assess the errors of the technique. I would like to add this as another criterion by which to ascribe a high weight to a technique of distance determination.

A chief area of concern in the use of Cepheids has been the absorption of light by the interstellar dust in which the Cepheids are embedded, but great progress is now being made by measuring luminosities in infrared light for which absorption is small. By this technique, Welch, McAlary, Freedman, Madore, and others are securing this important foothold of the distance scale (see, e.g., McAlary, et al. 1984, Madore et al. 1985, Freedman 1988). A good example of this progress concerns the distance to the galaxy M81. Until recently, there has been no Cepheid-based distance to M81, so its distance was linked through secondary means to a companion spiral galaxy NGC 2403 whose distance is based on Cepheids. However, Sandage has adopted a distance to M81 almost twice as far as NGC 2403, which has contributed substantially to several of his recent studies which derive a low value of H_0. Now Freedman and Madore (1988) have just reported the first Cepheid measurements for this galaxy which, when coupled with abundant new Cepheid data for NGC 2403, confirm the association of these two at the previously estimated distance. Only a handful of galaxies have distances known from Cepheids, but they are, I believe, becoming quite reliable. Here the Hubble Space Telescope will make its biggest contribution to the distance scale by allowing us to increase this number to as many as 20.

Even with our few well measured calibrating galaxies we should be able to determine the Hubble constant with a random error of less than ten percent. To do this, we need reliable secondary distance indicators, more luminous than Cepheids, which can vault us well into the Local Supercluster and beyond. There have been many candidates for so-called "standard candles", - the size of HII regions, the luminosity of the brightest red and blue stars or of entire clusters of stars - most have perished when subjected to careful scrutiny. Instead, today's research centers on comparing the luminosities of entire galaxies. Since large galaxies have a range in brightness of 100, one must restrict the range in some way, or first predict each galaxy's luminosity in a distance-independent fashion.

An example of the latter is a most promising technique of measuring relative distances to spirals based on a empirical correlation between a spiral galaxy's luminosity and its rotation speed, the Tully-Fisher (TF) relation. Rotation speed can be measured independent of a galaxy's distance, so such a relation provides the necessary tool. That there is such a correlation is not surprising, but that these two quantities are so well correlated implies that there is an intrinsicly small scatter in surface density, something that must relate back to galaxy formation. Researchers Pierce and Tully (1988) have recently refined and tested the method using galaxies in the Ursa Major and Virgo clusters of galaxies. As with the Cepheid variables, it is desirable to apply the TF method of predicting distances to a sample at a common distance, like a cluster of galaxies, since this improves the distance estimate by $\sqrt{N}$ and allows us to determine the intrinsic errors. (Cluster samples are relatively immune to the dread Malmquist bias which has plagued determinations of H_0. When dealing with a flux limited sample with a large intrinsic scatter in luminosity, one systematically underestimates the distance to the rare, more luminous and more distant galaxies that contaminate the closer sample.) Pierce and

Tully, in this new study, correct the major problem in previous studies of this type - they now have accurate, consistent measures of the observed brightnesses of the spirals in the sample. When combined with good measures of the rotation speeds, as determined from the HI radio line widths, a small scatter is found, equivalent to an uncertainty of 15-30% in distance per galaxy. Thus, a cluster distance can be measured to much better than 10%, although since Pierce and Tully rely on only three local calibrating galaxies, the final accuracy of H_0 is limited by these. The distances they find for these clusters are then divided by the measured recessional velocity of each cluster, yielding H_0 = 85 ±11 km/sec/Mpc. However, when they apply the necessary correction to the recessional velocity for the "infall" our Galaxy toward these clusters, Pierce and Tully adopt a value of 300 km/sec. I'll consistently adopt what I think is a better value of 250 km/sec which would lower their value to H_0 = 81. (A change of 100 km/sec in this infall velocity will change all the results I quote by 7%.) As I'll mention later, there is another potential systematic error which I believe could lower this value by another 10%.

Aaronson et al. (1986) and coworkers also have data for about 10 galaxies in each of 10 additional clusters. These measurements give similar results compared to the new work, but the uncertainties are greater because of luminosities that are not as well measured, and larger systematic errors associated with fitting procedures because of sample incompleteness. Again, however, a small scatter is found for the Tully-Fisher relation in these cluster samples.

I am encouraged by the small scatter in these data, but Kraan-Korteweg, Cameron, and Tammann (1988) have analyzed a much larger Virgo cluster sample and claim that the Tully-Fisher relation has a factor of 2 more scatter than found by these other groups. However, this appears to be the result of their enlarging the sample to include galaxies whose adherence to a tight Tully-Fisher relation has not been demonstrated, and in fact, would not be expected. The samples previously discussed are typically populated by Sc galaxies of intermediate-to-high luminosity, this largely determined by the requirement of an HI detection. On the other hand, fully 30% of the new Kraan-Korteweg et al. sample are low luminosity Magellanic-type irregular galaxies. From dynamical considerations, and due to complications of geometry and internal absorption, it is unlikely (and certainly undemonstrated) that these are suitable. It is not surprising, then, that Kraan-Korteweg et al. find a larger scatter for the TF relation, mostly reflected in a morphological type dependence which is particularly strong in the blue luminosity passband they use. Furthermore, their data lack the homogeneity of the studies by Pierce and Tully and Aaronson and collaborators; clearly it is always easier to destroy a correlation than improve it.

Since Kraan-Korteweg et al. claim that the true scatter of the TF relation in Virgo is much larger than originally claimed, they must explain why the TF relations for other clusters studied by Aaronson et al. (1986) also show a small scatter. They suggest that at each rotation speed the other group has detected only the most luminous spirals. Though they do not point to examples of the spirals that were left out of the Aaronson et al. study, I suspect they are again referring to dwarfish, irregular galaxies which they now include in the Virgo sample. I find nothing inconsistent in what Aaronson et al. have done in defining their sample and calibrators in a more restricted way so as to achieve a smaller scatter, so I believe the criticism by Kraan-Korteweg et al. has little merit. However, their provocative paper raises a number of interesting points that should be pursued further. For example, there is a 50% incompleteness in some of the distant cluster samples because the HI radio fluxes are too low to allow measurement of the rotation speed. It is important to push harder on some of these objects, and to model incomplete data from complete samples, to insure that the incompleteness does not correlate with IR luminosity. By the way, Kraan-Korteweg et al. find a low H_0, but replacing the aberrant distance for M81, their study still gives $H_0 \sim 60$ +/- 7, on the low end of the other studies but not at odds with them.

This is an appropriate time to discuss my own determination of H_0, based on a secondary distance indicator like the Tully-Fisher relation but applied to spheroidal components of galaxies. A group of us, called the Seven Samurai, discovered for elliptical galaxies a good relationship between an unusually defined diameter Dn and the characteristic stellar velocity, referred to as the velocity dispersion (Dressler et al. 1987). The scatter of this relation is such as to give rise to ~20% uncertainty in distance per galaxy measured, about as accurate as the TF relation. Again, we validated this relation by testing in clusters where many ellipticals at the same distance could be compared. This relative distance indicator had no absolute calibrators for H_0, however, because there are no elliptical galaxies with distances determined by primary means. I knew from previous work of my own that bulges of disk galaxies share many properties of elliptical galaxies so I tested the so-called Dn - σ relation on them and found that they follow the same relationship as the ellipticals - for 50 disk galaxies in the Virgo and Coma clusters the slope and zero-point are the same (Dressler 1987). A Hubble constant can then be measured since the distances to M31 and M81, two local galaxies with large bulges, are known by primary means. I get a value of $H_0 = 63 \pm 10$, again in my favored range of 60-80. I believe that this provides a useful complement to the Tully-Fisher method, and it will benefit from more studies of bulges of galaxies in other clusters, and the addition of more local calibrators through the HST Cepheid programs.

In such a brief review I cannot be complete but I will mention two more studies that arrive at intermediate values of the Hubble constant. Pritchet and Van den Bergh (1987) have compared novae in M31 to those in Virgo cluster galaxies. Adopting the same input parameters that I have for the other studies, their value would be $H_0 = 63 \pm 13$. Finally, in a series of important papers concentrating on Malmquist and related biases, Bottinelli et al. (1986,1987) have reanalyzed the Tully-Fisher data by Aaronson et al. (1982) and also re-analyzed de Vaucouleurs' "sosie" approach. They find H_0 ranging from 50-80 depending on the calibrator distances and other input parameters.

Sandage (1988b) has also reanalyzed the Aaronson et al. field sample
and published H_0 = 55 ±13, but this rises to H_0 = 64 when the proper
distance to M81 is used. This should probably be raised another 10%
because the correction for Malmquist bias Sandage applies is too large
due to the poorly supported assumption that the scatter in the TF is
much larger than quoted.

Thus, many studies with different techniques now find H_0 in the
range 60-80. We cannot be content with this - the systematic errors
are still poorly understood and the random errors due to too few
calibrators must be reduced. It is encouraging, however, to find some
convergence after all these years. I must, however, finish where I
began, with the two poles of Sandage and de Vaucouleurs, still so far
apart. Sandage (1988a) has recently published a paper in which he
finds the lowest value yet, $H_0 \approx 42$. Sandage has returned to
comparing the luminosities of Sc galaxies, selected by their optical
appearance to be the most luminous examples, van den Bergh luminosity
class I. Not surprisingly, galaxies selected this way have a large
scatter in intrinsic luminosity, and they suffer the problems of
subjective criteria discussed earlier. The method has not been
validated by objective criteria applied to cluster sample (and
probably cannot easily be), and can only be applied to field Sc
galaxies by making certain assumptions about their properties and the
linearity of the local Hubble flow. In this paper Sandage properly
stresses the importance of correcting for the Malmquist effect
whenever a flux limited sample is built around a class of objects that
have a large spread in intrinsic luminosity. The unusually low value
of H_0 =42 rests not on this, however, but on Sandage's excessive
distance for M81 and his adoption of extraordinary reddening in our
Galaxy in the direction of M31. Replacing these with conventional
values raises H_0 to 52 ±14, and it appears to me that by replacing
the eye fits with objective techniques the value will go higher still.
Moreover, the ScI's undoubtedly provide one of the poorer tools
available, one which is demonstrably inferior to the Tully-Fisher
method in terms of the size of random errors. Thus, with the
abandonment of earlier standard candles like of HII regions diameters
and brightest stars, I find little support for a value of $H_0 < 60$
km/sec/Mpc.

On the other end, de Vaucouleurs has adopted a unique approach
that consistently leads him to $H_0 \approx 100$ (see, e.g., de Vaucouleurs
and Peters 1986, de Vaucouleurs and Corwin 1986). Following Paturel
(1984), he looks for galaxies that are similar in morphological type
and rotation velocity to the local calibrators - these are called
"sosies". Then by various methods he compares luminosities to derive
Hubble ratios for each field galaxy. Unfortunately, there is again no
cluster sample with which to test this technique. The high value
de Vaucouleurs derives comes from two features of his analysis: (1)
smaller distances for the calibrators; (2) his belief that his method
does not suffer a bias from the Malmquist effect. He verifies this by
noting the lack of a smooth trend of increasing Hubble ratio with
distance, the signature of a Malmquist bias. Rather he observes a
sharp discontinuity which he believes is a true description of the

local Hubble flow, as distinct from the smooth infall models preferred by others.

We can all be grateful to Sandage and de Vaucouleurs for pioneering this crucial and difficult research. It is both appropriate and encouraging that others are forging ahead on the path they have opened. I doubt that Sandage and de Vaucouleurs will ever agree on a value for the Hubble constant, but I see others progressing toward a resolution of their disagreement. Rowan-Robinson derived a most likely value for the Hubble constant $H_0 = 67 \pm 15$ after adopting a consistent set of input parameters and carefully evaluating the errors of each tool. In my opinion, the most recent work I have discussed today supports his conclusion, and I too believe that the Hubble constant will be pinned down within the next 10 years to be in the range 60-80.

In this discussion I have not mentioned anything about the age of the universe inferred from the Hubble constant and its comparison with ages derived from radioactive dating or stellar evolution studies. I agree with Aaronson, Huchra, and Mould that the determination of the Hubble constant should be done as best we know how, without trying to match it to other, quite independent data. I'll address the question of how well these numbers do agree later.

Before leaving the Hubble constant, it's appropriate to say a few words about the departures from a smooth, linear Hubble flow that are attributed to a lumpy distribution of matter in the local universe. Aaronson et al. (1982) pioneered in this field by using the Tully-Fisher relation to map out the local deceleration field produced by the overdensity of the Virgo cluster and the Local Supercluster that contains it. They found clear evidence for such a deceleration pattern with an amplitude of $V \approx 250$ km/sec at the position of the Local Group. That is, the Local Group's recession has been slowed by this amount due to this overdensity. This is still small compared to the departure from a pure Hubble flow as implied by the Local Group's motion of 600 km/sec with respect to the microwave background radiation. Sandage and Tammann (1984) speculated that this might be largely due to the pull of the Hydra-Centaurus supercluster, the next nearest overdensity. The Seven Samurai, of which I am a member, took up the task of enlarging the sphere of measurements of deviations from pure Hubble flow to include this more distant supercluster. With our sample of ~400 elliptical galaxies we compared observed velocities with estimated distances to investigate the amplitude of departures from a smooth, isotropic Hubble flow. We found that the deceleration pattern responsible for the microwave dipole anisotropy extends beyond the Hydra-Centaurus supercluster, some 70% further to an enormous excess in the galaxy distribution in the Supergalactic plane. We have nicknamed this entire region (including Hydra-Centaurus) "the Great Attractor". Futher investigation of galaxy counts and a redshift survey of this region (Dressler 1988) have confirmed that this previously unidentified supercluster is at the proper distance, direction, and of the proper amplitude to provide the necessary deceleration.

The Great Attractor model is still controversial, but it seems clear that there are significant departures from pure Hubble flow over large scales. Sandage and Tammann have argued that because the scale is so large, the deviations do not appreciably affect the linearity of the Hubble flow on the small scale of the Local Supercluster, where galaxies used for the determination of the Hubble constant are located. This is clearly not the case, since the Virgo induced velocity is on the order of 200 km/sec and a tidal shear field of comparable magnitude is expected from the more distant deceleration that accounts for the microwave dipole anisotropy. These effects are too large to be ignored, and any study that constrains the local Hubble field to be linear will err because of it. Unfortunately, some circular proofs have started this way. Because the Hubble flow is assumed by some to be linear, large errors are derived for such techniques as the Tully-Fisher method applied to field spirals. One can then find that the Hubble flow is indeed linear, within these large errors. That the whole process is self-consistent should give little comfort.

I do not despair about the complication to the determination of H_0 added by deviations from a pure Hubble flow. This is because I think we are already able to model the flow pattern rather well. To show this, I will refer to a remarkable illustration made by Burstein and included in Burstein et al. (1988). He uses the velocity field predicted by the Great Attractor model to investigate the scatter in the Tully-Fisher relation for field spirals in the Aaronson et al. (1982) sample. Burstein shows the TF relation between intrinsic luminosity and rotation speed when luminosities are calculated from apparent brightnesses and distances based on four different models: (1) pure Hubble flow, (2) adding Virgo infall, (3) adding a bulk flow in the direction of the Great Attractor, (4) the full velocity field predicted by the Great Attractor model. I want to re-emphasize that the elliptical galaxy sample was used to make the model, and, as such, is completely independent of the data use the spiral galaxy sample used to test the model. What is obvious is that the assumption of a linear velocity field causes one to derive a noisy Tully-Fisher relation. Including the expected deceleration field of the Local Supercluster improves things somewhat, and adding bulk flow improves it further. But, most remarkably, using the full model reduces the scatter to less than 0.5 mag, the same scatter found for the TF cluster samples but this time for the field galaxies. This final improvement arises mainly because of the addition of the quadrupole field first noted by Lilje et al. (1986). It is easy to destroy a correlation, but difficult to improve one as is the case here. I believe this demonstrates that we have a good hold on the local deviations from pure Hubble flow and, therefore, can reliably turn local determinations of the Hubble constant into global values. This is excellent evidence, also, that the TF relation as applied by Aaronson et al. (1982, 1986), and our Dn - σ relation for ellipticals, do have a small scatter that make them very powerful relative distance indicators. Attempts to discredit these correlations are strongly contradicted by these data.

3. THE MEASUREMENT OF q_0

In Friedmann cosmologies, the deceleration paramater q_0 describes the geometry of the universe. Assuming that the cosmological constant Λ_0, the repulsion term invented by Einstein to create a model of a static universe, is vanishingly small, $0 < q_0 < 1/2$ covers the range from an empty, hyperbolic space to one of critical density that will eventually coast to a near standstill. If measuring the Hubble constant has been frustrating to observational cosmologists, determining q_0 has been downright exasperating. The most ambitious attempt was piloted by Allan Sandage (see Sandage et al. 1976, and Kristian et al. 1978 and references therein). Using the brightest galaxies in clusters as standard candles, Sandage and coworkers pushed the Hubble diagram to remote distances in order to see a departure in the linear relation due to a deceleration in the expansion rate over time. Both the redshifts and photometry of these distant galaxies required heroic efforts with instrumentation that was an order-of-magnitude less efficient than that available today, and for that reason, barely adequate for the task.

Sandage showed that the dispersion in the luminosity of brightest cluster galaxies was sufficiently small that the goal could be reached, and his last papers on the subject indicated a high value of q_0 favoring the model of a closed universe. However, doubts in the efficacy of the approach were raised by Beatrice Tinsley (1972, see also Tinsley and Gunn 1976), whose stellar population models implied that uncertainties in the luminosity evolution of galaxies seen so far in the remote past would frustrate attempts to obtain a secure value. These cluster galaxies were probably more luminous in the past, but by how much? Ostriker and Hausman (1977) pointed to a correction in the opposite sense that might also be needed, due to the increasing brightness with time of central cluster galaxies as they cannibalized their neighbors. Eventually, discouraged by the uncertainties in these evolutionary corrections, Sandage declared that "the road to q_0", by this technique at least, "is blocked."

Simultaneously J. B. Oke and Jim Gunn were also using the Hale telescope for an independent investigation, also based on galaxies in remote clusters. This program, which continues at the present time, is aimed at reducing the selection biases and understanding the evolutionary corrections that must be made. Gunn and Oke's (1975) published value of q_0 is low, so low as to imply an unphysical solution, but Gunn believes that formerly there were observational problems in measuring the brightnesses of these galaxies which were too faint to actually see, even with the Hale telescope. Gunn, Oke, and John Hoessel have new data not yet completely analyzed that represents the most ambitious attempt to find q_0 through this technique.

Other tests have been proposed and tried, like the angular size of galaxies as a function of redshift, or using luminosity indicators to turn quasars into standard candles. Radio galaxies at redshifts $z >$ 1 have been used to push the Hubble diagram much further than possible with "ordinary" galaxies (Spinrad and Djorgovski 1987). Such methods also appear to be limited by evolutionary corrections that are poorly understood. It is conceivable that observations of a set gravitational lenses that are sufficiently easy to model will someday provide a definitive test of the geometry of the universe over large scales, and give H_0 as well. For the present, however, no technique of measuring q_0 has proven very reliable.

It is against this backdrop of frustration that Ed Loh and Earl Spillar described a new study to measure the geometry of space up to redshifts approaching unity (Loh and Spillar 1986a,b). They chose a sample of faint field galaxies with the intention of measuring the comoving volume element as a function of redshift that encloses a fixed number of galaxies. For this first attempt they needed redshifts of nearly a thousand very faint galaxies, just beyond what can be done with state-of-the-art spectroscopy. Therefore, they chose to estimate redshifts by fitting six narrow band colors to the spectral energy distributions measured for low-redshift galaxies. In principle, these photometric redshifts are accurate enough for the task, but there is still a question about the use of low-z templates, and, as yet, an insufficient number of these photometric redshifts have been checked by actual spectroscopy. I will assume, however, that these data are reliable.

Loh and Spillar's technique is to fit each of three redshift intervals to a Schechter luminosity function, extrapolating to the total number of galaxies down to some cutoff and measuring the total luminosity. This technique is independent of luminosity evolution (if the luminosities of all galaxies evolve by the same amount!), an advantage, it would appear, over the test using the Hubble diagram. Like the Hubble diagram, however, this technique measures the evolution of the metric, so it is sensitive to any influence, including a uniform matter component or a non-zero cosmological constant (unlike the local tests of Ω_0, discussed below). Loh and Spillar claim that their data show a clear preference for a critical density corresponding to $q_0 = 1/2$. (They actually express their result in terms of the density parameter Ω_0, but Ω_0 is more clearly associated with a measurement of density, and their measurement is more a test of geometry).

A crucial problem with this work has been pointed out by Sandage (1988a), who notes that Loh and Spillar have not estimated the incompleteness of their sample as a function of redshift. Their selection procedure seems to ignore the almost certain selection effect that many high-z galaxies are missing because decreasing surface brightness, which falls as $(1+z)^4$, results in an increasing dropout rate. A good knowledge of the surface brightness distribution for present-epoch galaxies (ignoring the fact that this distribution is itself likely to evolve with cosmic time), combined with the signal-to-noise selection criterion in the fixed aperture, is needed to properly model the incompleteness. An shortfall of only 30% at z = 0.8 will change Loh and Spillar's result from q_0=0.5 to q_0=0.1. Sofi Bahcall and Scott Tremaine (1988) have pointed out, in addition, that the method is in peril because, according to what little we do know about galaxy evolution, the shape of the luminosity function should not remain constant with lookback time, as Loh and Spillar assume. Loh (1988) has commented on these and other issues in an enhanced analysis of the study, and it appears that there is progress in understanding the problems involved in this approach. However, from both the points of view of data and interpretation, it is clearly much too early to claim a reliable measurement of q_0 by this technique.

In summary, there are no measurements of q_0 to date that can be taken very seriously. If they have succeeded in anything, it has been in the study of the evolution of galaxies, a very important subject in its own right.

4. THE MEASUREMENT OF Ω_0

Faced with the seeming intractability of measuring q_0, many have turned to a more local measurement, Ω_0 - the mass density of the present universe normalized to the critical density required for closure. In the case of a zero-cosmological constant, $\Omega_0 = 2q_0$ so $\Omega_0 = 1$ corresponds to the critical density that is favored in models of the early universe which have an era of exponential growth referred to as inflation.

Several techniques have been used to estimate Ω_0, and with few exceptions, they all seem to imply $\Omega \sim 0.1-0.2$, far short of the closure density. These tests, which use galaxies as test particles to make kinematic measurements of matter density, have an advantage over the q_0 tests described above in that they do not require corrections for evolution. On the other hand, they have the disadvantage that they are sensitive only to the gravitational mass clustered on a scale smaller than that sampled by the galaxies.

Classical baryosynthesis models produce the correct amount of deuterium and helium-3 if the baryons supply about 3-15% of the closure density. This constraint has suggested that baryons might only supply ~10% but that the remaining 90% might be in non-baryonic form, for example, massive neutrinos or cold-dark-matter particles like photinos. At present, the only way to look for this exotic matter is by its gravitation field. The baryons in galaxies seem to fall far short of closure density, $\lesssim 1\%$, but when we add the probable amount of dark matter seen in galactic halos we again find something like $\Omega_0 = 0.1-0.2$ This result agrees well with what we get when clusters of galaxies are "weighed".

Perhaps these results for the mass density on scales appropriate to galaxies are not surprising. If there is extra non-baryonic matter that raises Ω_0 to unity, it is probably spread more uniformly than the baryonic component. This supposed tendency for the baryonic matter to clump more than the non-baryonic matter, called biasing, is an essential feature of models of the formation of structure with cold-dark-matter. So it makes sense to look $\Omega_0 = 1$ on a bigger scale, somewhat larger than the correlation length of galaxies, say, which is somewhere between 5-10 Mpc (Groth and Peebles 1977). The first such measurements at this scale were of the infall of the Local Group toward Virgo, the center of the Local Supercluster (Yahil, Sandage and

Tammann 1980, Davis and Huchra 1982). These groups estimated the excess luminosity in a spherical volume reaching to the Local Group to be $\Delta L/L \sim$ 2-3. If this is representative of the mass excess as well, the infall velocity of several hundred km/sec again implies $\Omega_0 \sim$ 0.1-0.2. Additional studies of the motions of galaxies in the Local Supercluster have given similarly low values, as reviewed by Tully (1987). Conceivably, however, biasing might be important. If the mass excess was less than that implied by the luminosity excess by a factor of two or three, high values of Ω_0 could still pertain. This may be a stronger biasing than one is comfortable with in, say, the cold dark matter model, because the scale is already somewhat larger than a correlation length, but it is perhaps a viable model.

Measurements of the random velocities of field galaxies can also be used to estimate the mass density. This test, called the cosmic virial theorem, was pioneered by Jim Peebles (1976). He found that the correlation length, coupled with random velocities of 200-300 km/sec also implies $\Omega_0 \sim$ 0.1, and like numbers have been derived in later attempts by others. Finally, Ford et al. (1981), and recently, Postman, Geller and Huchra (1988) have been "weighing" whole superclusters and again finding $\Omega_0 \sim$ 0.2.

In summary, all of these studies which have attempted to estimate the matter density on scales of several correlation lengths or smaller obtain similar values of about 10-20% of the closure density. If closure density is to be reached, the remaining matter must be distributed over scales of 20 Mpc or larger. The first attempts to measure such scales give conflicting results. In my study of the Great Attractor, I repeat the "infall to Virgo" test on a scale some three times larger. I find that a value of $\Omega_0 \sim$ 0.2 comes quite naturally out of my analysis, even on this enormous scale. However, I had to assume that the matter outside the sphere centered on the Great Attractor and reaching to the Local Group is uniformly distributed, a dubious proposition. The point is that a good measure of Ω_0 requires knowledge of the velocity field and galaxy distribution over a scale several times larger than the one that is being attempted. When we have mapped such a large region more completely, I am confident that we can eventually obtain quite a good measure of Ω_0 for the clustered component over a large scale.

In their study of the galaxy distribution as mapped by IRAS, Meiksin and Davis (1986) and Villumsen and Strauss (1987) have concluded that the observed anisotropy of the galaxy distribution on the $\sim$ 50 Mpc scale is quite small, and if this is responsible for the 600 km/sec motion of the Local Group with respect to the microwave background, the implied Ω_0 is significantly closer to unity than found for these other studies. This might be the first claim of a measurement of $\Omega_0 = 1$ which can be taken seriously, though Yahil (1988) points to several uncertainties in the method and cautions against early adoption of this result. If correct, however, this result indicates that we may have finally reached a scale over which mass traces light.

Yahil (1987) has stressed the remarkable agreement of many attempts to estimate Ω_0 on scales of $\lesssim$20 Mpc that find values of

0.1-0.2. I certainly agree that this consistency cannot be ignored. Whatever prejudice we may have to find Ω_0 = 1, there is very little evidence for this at present and a fair amount against. Biasing may be the answer, but it, too, is only a supposition. As an observer, I must conclude that the data support a value of Ω_0 of 10-20% of the closure density, regardless of what I or my physicist friends may wish to find. I was amused by a preprint from G.F.R. Ellis (1988) who discusses the circumstances under which inflation can produce a universe with Ω_0 = 0.1. I am afraid that his solution would be in conflict with the amazing isotropy of the microwave background, but this paper serves as a reminder to observers that the subject of inflationary models is far from exhausted.

5. THE COSMOLOGICAL CONSTANT AND THE AGE OF THE UNIVERSE

The three dynamical parameters H_0, q_0, and Λ_0 define T_0, the age of a Friedmann model. Thus, one can solve for Λ_0 with reliable values of H_0, q_0, and an independent way of estimating the age of the universe. The two widely discussed ways of age dating are (1) comparing stellar evolution models with observational data for the oldest stellar populations, and (2) observing the line strengths of radioactive elements with half-lives comparable to the age of the universe.

For many years ages of 15-16 Gyr were derived by comparing stellar evolution tracks to HR diagrams of metal-poor globular star clusters, thought to be the among the oldest stellar populations in the galaxy. Recently, Vandenbergh (1988) has produced new models that properly recognize the fact the oxygen abundance may not track the low iron abundance in such clusters but instead be significantly higher. This brings the age of the oldest clusters down to some 13-14 Gyr, a significant decrease. Sandage (1988a) notes that, allowing 1-2 Gyr from the Big Bang until the Galaxy forms these stars, the derived age of the universe is nearly 14-16 Gyr, consistent with Ω_0 = 1 and Λ_0 = 0 only if H_0 ~ 40 km/sec/Mpc. As I have said previously, I am unconvinced by the evidence for so low a value of H_0. I note, however, that in the range H_0 = 60-80 that I have been endorsing, there is little conflict, at least at the low end of the range, with the low Ω_0 that observations seem to imply. Nor should one consider the new globular cluster ages as the final value (Rood 1987). We may put too much faith in these models, perhaps so far as to prejudice our evaluation of what values of H_0 are acceptable. Such matters as mass loss on the main sequence and the puzzle of the missing solar neutrinos should give us some pause. In my opinion, it is still too early to rule out the possibility that the oldest globulars have ages of only ~10 Gyr.

Another method of determining the age of the Galaxy is to compare the luminosity function for white dwarf stars to predictions from theoretical cooling curves (Winget et al. 1987). The ages found are typically ≲10 Gyr. Leaving aside possible uncertainties in these models and their interpretation, I note that this procedure yields an age for the disk of the Galaxy. Because the disk may have formed well after the birth of the Galaxy, it is not very useful to compare this measurement, by itself, with the age of the cosmological expansion.

Time scales based on nuclear processes have been well reviewed by Meyer and Schram (1986) and Cowan et al. (1987). New data for this type of analysis has been provided by Butcher (1987) who has derived new ages for the oldest stars by comparing a spectral line from radioactive thorium with a half-life of 14 Gyr to that of stable neodymium. His data suggest that the Galaxy is at most 10 Gyr in age, considerably younger than the stellar evolution timescale. Willy Fowler has also been applying his expertise in matters nuclear to these problems, and he, too, is convinced that a young age is appropriate for the Galaxy (Fowler 1987). He likes a Hubble constant of 50, an $\Omega_0 = 1$ and $\Lambda_0 = 0$. However, the studies by Meyer and Schram and Cowan et al. indicate that the uncertainties in the theory and its application are still large. Furthermore, on the observational side, the spectral lines of these rare elements are extremely weak and therefore very difficult to measure. Even the identifications may be in doubt. There is also some question as to how the supernova birth rate, an important input parameter, varied over the early history of the Galaxy. Thus, it seems that this new approach offers great potential, but is far from mature.

In summary, I agree with others who have found no pressing reason to adopt a non-zero cosmological constant. The most secure techniques seem to suggest an observed age of the Galaxy which may be uncomfortably long for an intermediate value of $H_0 = 60-80$ km/sec/Mpc and a universe at critical density, but the latter is not mandated by the observations and there is still considerable uncertainty in the application of stellar evolution to age-dating. It is too soon to panic.

6. SUMMARY

The observed values $H_0 = 60-80$ km/sec/Mpc, $\Omega_0 = 0.1-0.2$, and an age of the Galaxy of 13-14 Gyr, which implies $T_0 = 14-16$ Gyr, seem to be mutually consistent with a Friedmann model with $\Lambda_0 = 0$. There may be theoretical considerations that prompt others to adopt different values to these, but I, as an observer, prefer to evaluate the evidence gathered by my colleagues. In astronomy, observation has driven theory at least as often as the other way around. I look forward to watching theory and observation dance their way, sometimes gracefully and sometimes awkwardly, toward an acceptable cosmological model.

38

7. REFERENCES

Aaronson, M., Huchra, J., Mould, J., Schechter, P. L., and Tully,
    R. B. 1982, Ap. J., 258, 64.
Aaronson, M., Bothun, G. J., Mould, J., Huchra, J., Schommer, R. A.,
    and Cornell, M. E. 1986, Ap. J., 302, 536.
Bahcall, S. R., and Tremaine, S. 1988, Ap. J. Lett., 326, L1.
Bottinelli, L., Gouguenheim, G., Paturel, G., Teerikorpi, P. 1986,
    Astr. Ap., 156, 157.
Bottinelli, L., Fouque, P., Gouguenheim, L., Paturel, G., Teerikorpi,
    P. 1987, Astr. Ap., 181, 1.
Burstein, D., Davies, R. L., Dressler, A., Faber, S. M., Lynden-Bell,
    D., Terlevich, R. J., and Wegner, G. 1988, in Large Scale Motions
    in the Universe, Proc. Pont. Acad. Sci. Study Week 27, in press.
Butcher, H. R. 1987, Nature, p. 127
Cowan, J. J., Thielemann, F.-K., and Truran, J. W. 1987, Ap. J.,
    323, 543.
Davis, M., Huchra, J. 1982. Ap. J., 254, 437.
de Vaucouleurs, G., and Peters, W. L. 1986, Ap. J., 303, 19.
de Vaucouleurs, G., and Corwin, H. G. 1986, Ap. J., 308, 487.
Dressler, A. 1987, Ap. J., 317, 1.
Dressler, A., Lynden-Bell, D., Burstein, D., Davies, R. L., Faber,
    S. M., Terlevich, R. J., and Wegner, G. 1987a, Ap. J., 313, 42.
Ellis, G. F. R. 1988, preprint.
Ford, H. C., Harms, R. J., Ciardulo, R., Bartko, F. 1981,
    Ap. J. Lett., 245, L53.
Fowler, W. A. 1987, Q. Jl. R. Astr. Soc., 28, 87.
Freedman, W. L. 1988, Ap. J., 326, 691.
Freedman, W. L, and Madore, B. F. 1988, preprint.
Groth, E. J., and Peebles, P. J. E. 1977, Ap. J., 217, 385
Gunn, J. E., and Oke, J. B. 1975, Ap. J., 195, 255.
Huchra, J. P. 1987, in Relativistic Astrophysics: 13th Texas
    Symposium, (Singapore: World Scientific Publishing), pp. 1-7.
Kraan-Korteweg, R. C., Cameron, L. M., and Tammann, G. A. 1988,
    Ap. J., in press.
Kristian, J., Sandage, A., and Westphal, J. A. 1978, Ap. J., 221,
    383.
Lilje, P. B., Yahil, A., and Jones, B. J. T. 1986, Ap. J., 307, 91.
Loh, E. D., and Spillar, E. J. 1986a, Ap. J., 303, 154.
Loh, E. D., and Spillar, E. J. 1986b, Ap. J. Lett., 307, L1.
Loh, E. D. 1988, Ap. J., 329, 24.
Madore, B. F., McAlary, C. W., McLaren, R. A., Welch, D. L.,
    Neugebauer, G., and Matthews, K. 1985, Ap. J., 294, 560.
McAlary, C. W., Madore, B. F., and Davies, L. E. 1984, Ap. J., 276,
    487.
Meiksin, A. R., and Davis, M. 1986, A. J., 91, 191.
Meyer, B. S., and Schram, D. N. 1986, Ap. J., 311, 406.
Ostriker, J. P., and Hausman, M. A. 1977, Ap. J. Lett., 217, L125.
Paturel, G. 1984, Ap. J., 282, 382.

Peebles, P. J. E. 1976, Ap. J. Lett., 205, L109.
Pierce, M. J., and Tully, R. B. 1988, Ap. J., 330, 579.
Postman, M., Geller, M. J., and Huchra, J. P. 1988, A. J., 95, 267.
Pritchet, C. J., and van den Bergh, S. 1987, Ap. J., 318, 507.
Rood, R. T. 1987, in 13th Texas Symposium on Relativistic
    Astrophysics, ed. M. P. Ulmer, (Singapore: World Scientific
    Publishing Co.), pp. 422-429.
Rowan-Robinson, M. 1985, The Cosmological Distance Ladder, (San
    Francisco: Freeman and Co.).
Sandage, A., 1988a, Ap. J., 331, 583.
Sandage, A., 1988b, Ap. J., 331, 605.
Sandage, A., Kristian, J., and Westphal, J. A. 1976, Ap. J., 205,
    688.
Sandage, A., and Tammann, G. A. 1982, Ap. J., 256, 339.
Sandage, A., and Tammann, G. A. 1984, in Large Scale Structure of the
    Universe: Cosmology and Fundamental Physics, eds. G. Setti and L.
    van Hove, (Geneva:ESO-CERN), pp. 127-149.
Spinrad, H., and Djorgovski, S. 1987, in Observational Cosmology, IAU
    Symp. 124, eds. A. Hewitt, G. Burbidge, and L. Z. Fang,
    (Dordrecht: Reidel), pp. 129-141.
Tinsley, B. M. 1972, Ap. J. Lett, 173, L93.
Tinsley, B. M. 1976, Ap. J., 203, 52.
Tully, R. B., 1987, in Observational Cosmology, IAU Symp. 124, eds.
    A. Hewitt, G. Burbidge, and L. Z. Fang, (Dordrecht, Reidel), pp.
    207-215.
Vandenberg, D. A. 1988, in Globular Clusters in Galaxies, IAU Symp.
    126, (Dordrecht, Kluwer), pp. 107-120.
Villumsen, J. V., and Strauss, M. A. 1987, Ap. J., 332, 37.
Winget, D. E., Hansen, C. J., Liebert, J., van Horn, H. M., Fontaine,
    G., Nather, R. E., Kepler, S. O., and Lamb, D. Q. 1987,
    Ap. J. Lett., 315, L77.
Yahil, A. 1987, in Nearly Normal Galaxies, From the Plank Time to the
    Present, ed. S. M. Faber, (New York:Springer-Verlag), pp.
    332-342.
Yahil, A. 1988, in Large Scale Motions in the Universe, Proc. Pont.
    Acad. Sci. Study Week 27, in press.
Yahil, A., Sandage, A., Tammann, G. A. 1980, Ap. J. 242, 448.

DISCUSSION.

Brandenberger: What would be the mass of the Great Attractor?
Dressler: A few times $10^{16} M_\odot$, comparable to the largest known superclusters.

Börner: Is there actual observational evidence that a homogeneous density distribution is reached, in view of the Large Scale Structure observed?
Dressler: No, not really. The good agreement of the normalizations for the CFA survey in the north and SSRS survey in the south suggests at least an approach to a homogeneous scale. However, Margaret Geller would say, based on the CFA "slices", that the largest surveys (R $\simeq$ 10000 Km/sec) show inhomogeneities that cover the entire volume.

Jørgensen: Supposing Butcher is perfectly correct in identifying thorium and neodymium absorption lines, is there any good reason to believe that all supernovae at all times produce a constant thorium/neodymium abundance ratio?
Dressler: This is outside my area of expertise. You may wish to look at the articles by Meyer and Schramm, and Cowan *et al.* I cited in the talk.

Reeves: The uncertainties on nucleocosmochronology are easily underestimated. The recent case of neodymium is typical. Do you have any comment on the recent clock based on white dwarfs statistics and cooling curves?
Weidemann: This age is very uncertain, depending on white dwarf cooling theory, between 6 and 14 Gyrs (see Winget and van Horn, 2nd Conference on Faint Blue Stars, 1987, A. G. Davis Phillip *et al.* eds., p. 363)

Turner: 1) I was surprised that you are not more optimistic about using the IR Hubble diagram to determinate q_o; it seems to offer great potential as preevolutionary effects are minimized. 2) I wish to emphasize the importance of the difference between $H_o = 60$ and 66 Km/sec Mpc as a critical test of the inflationary paradigm. If we accept that the age of the universe is at least 10 Gyr, than inflation cannot tolerate a Hubble constant higher than 66 Km/sec Mpc.
Dressler: At present, the evolutionary corrections appropriate for intermediate age populations like AGB stars are not well known, and have been generally excluded from the models. Thus, the potential of this method is not clear at this time. The smaller sensitivity of the IR Hubble diagram to evolutionary corrections could be illusory.

Electroweak Physics in 1988

R.D. Peccei

Deutsches Elektronen Synchrotron DESY

Hamburg, Fed. Rep. Germany

Abstract

I review some salient features of the standard electroweak model and describe the impact of some recent experimental results on family issues and flavor mixing. Tests of the standard model beyond tree level and prospects for detecting Higgs bosons are also discussed.

1 Introduction

In 1988, before the start up of SLC and LEP, the $SU(2) \times U(1)$ model of Glashow, Salam and Weinberg [1] continues to be the paradigm for describing the electroweak interactions. This does not mean that the model has remained static. Rather, in the last year, some issues have sharpened. Most notably, it now looks much more likely that there are only three families of quarks and leptons. At the same time, some other issues have become better defined, but in a sense more intriguing. For instance, radiative corrections appear to be on line with standard model expectations, provided that the top quark mass is not too large. Yet, the substantial flavor mixing seen in the B system argues for a sizable top mass. Finally, some issues continue to be as mysterious as ever, particularly those connected with the symmetry breaking sector of the theory. Given this state of affairs, in this talk I shall try to give a snapshot of the status of the standard electroweak theory, as it is today - a theory not fossilized, but one which is still in a settling process.

2 Three Families and No More?

In the standard electroweak model the fundamental fermions have repetitive $SU(2) \times U(1)$ properties: all left handed fields are in $SU(2)$ doublets, while their right handed counterparts are $SU(2)$ singlets. To date, with the exception of only one excitation (the top quark), we have evidence for three such "families" of quarks

41

M. Caffo et al. (eds.), Astronomy, Cosmology and Fundamental Physics, 41–65.

© *1989 by Kluwer Academic Publishers.*

and leptons. Although we have no <u>direct</u> experimental evidence for the top quark, we have excellent indirect confirmation that there exist an $SU(2)$ partner for the left handed bottom quark and hence, inferentially, for the top quark. The best indication for this [2] comes from the measurement of the axial charge of b quarks in e^+e^- experiments at PEP and PETRA. The differential cross section for the process $e^+e^- \to b\bar{b}$ receives, in lowest order, contributions from both photon and Z exchange, as shown in Fig. 1. This latter graph gives rise to an asymmetry in the number of b ($\bar{b}$) quarks produced in the hemisphere along the direction of the incident e^- (e^+) in the CM system. The forward-backward asymmetry - with forward being along the direction of the incoming e^- - for b quarks can be well approximated, in the PEP/PETRA energy range, by retaining only the interference term between the photon and Z contributions and one finds:

$$A^b_{FB}(s) = \frac{9G_F a^b}{16\pi\sqrt{2}\alpha} \left[\frac{sM_Z^2}{M_Z^2 - s} \right] \tag{1}$$

where s is the square of the CM energy and a^b is the axial charge of the b quark. This charge takes different values, depending on the $SU(2)$ assignments of the b quark. For instance:

$$a^b = \begin{cases} -1 & b_L \ doublet; \ b_R \ singlet \\ 0 & b_L \ singlet; \ b_R \ singlet \\ 1 & b_L \ singlet; \ b_R \ doublet \end{cases} \tag{2}$$

The forward-backward asymmetry for b-quarks has been measured by a number of experiments at PEP and PETRA, most notably by JADE [3], whose results are shown in Fig. 2. The b-quarks are identified through the large transverse momentum leptons they produce in their semileptonic decays. The values for the asymmetry obtained by the various PEP and PETRA experiments have been compiled by Greenshaw and Marshall [4] and reviewed by Wu [5]. The averaged value obtained [4]

$$a^b = -0.84 \pm 0.21 \tag{3}$$

needs to be corrected for $B - \bar{B}$ mixing, which dilutes somewhat the signal. This has been done by Wu [5], who arrives at a final value of

$$a^b = -1.08 \pm 0.29 \tag{4}$$

Although the errors are still large, this result is in perfect agreement with the supposition that the left handed b quark has a partner. So the top quark exists; the only question remaining is what is its mass!

There are two lower bounds on the top mass, one experimental and the other theoretical, which suggest that top is at least a factor of ten heavier than bottom[1].

---

[1]There is a safer, but smaller, experimental lower bound from TRISTAN [6], $m_t \gtrsim 26.4$ GeV which requires very little theory input.

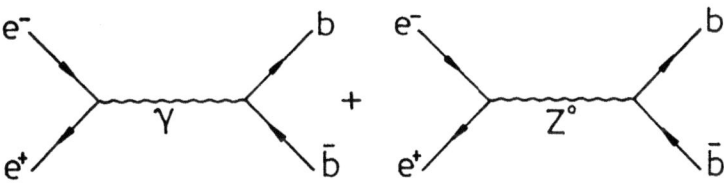

Figure 1: Diagrams contributing to $e^+e^- \rightarrow b\bar{b}$ in lowest order.

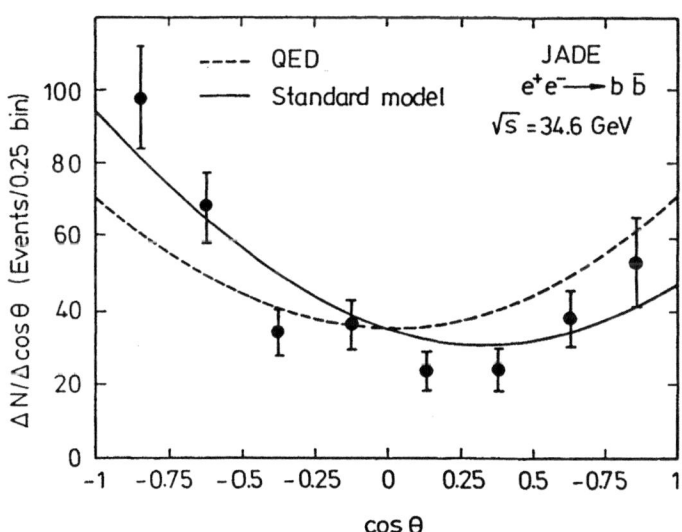

Figure 2: The measured angular distribution for the process $e^+e^- \rightarrow b\bar{b}$ measured by the JADE collaboration [3].

The experimental bound on top comes from the UA1 collaboration, working at the Sp$\bar{p}$S collider [7]. In p$\bar{p}$ collisions top can be produced either as a byproduct of W decay, if its mass is below the $W \rightarrow t\bar{b}$ threshold, or in association with a $\bar{t}$, by gluon-gluon fusion. The first process has a well determined rate, which can be inferred from the experimentally measured W production and the known $W \rightarrow t\bar{b}$ branching fraction in the standard model. The rate for the second process, however, is more uncertain, since it depends on the knowledge of the gluon structure functions and of higher order QCD corrections. The signal for both these sources of top is an isolated muon with large transverse momentum, from the semileptonic decay of top, plus jets resulting from the accompanying debris. This signal, although characteristic, is not without background and a careful Monte Carlo study is needed to ascertain the presence of top.

Fig. 3, taken from [7], shows the expected signal, as a function of m_t, for the cuts imposed by the UA1 collaboration. As can be seen, for $m_t \geq 50$ GeV the main source of the signal is the gluon-gluon fusion process. The signal in Fig. 3 lies above the calculated background for $m_t \geq 56$ GeV (at 90% confidence) and this allows the setting of a lower bound on m_t. Because the predicted rate is dominated by the (somewhat) theoretically uncertain QCD process, this bound is not the most conservative one can set. UA1 sets a more cautious lower bound of $m_t \geq 44$ GeV [7] by varying the QCD inputs for the theoretically predicted top cross section and selecting the lowest value among these. A similar result has also been obtained recently by Altarelli et al. [8], who incorporated the effects of higher order QCD corrections for heavy quark production [9] and obtained $m_t \geq 41$ GeV. In summary, from the nonobservation of a clear signal in high energy p$\bar{p}$ scattering one can deduce - subject to the above mentioned theoretical ambiguities - an experimental lower bound for the top mass, in the range

$$m_t \geq (41 - 56) \ GeV \tag{5}$$

The observation of large $B_d - \bar{B}_d$ mixing by the ARGUS collaboration [10] allows one to infer, theoretically, a lower bound of a comparable magnitude for m_t to that given in (5). As I shall discuss this topic below in some detail, I will not comment further on this bound here. It is also possible, theoretically, to establish an upper bound for m_t, from the consistency of the calculations of electroweak radiative corrections with high precision neutral current experiments and with the W/Z mass measurements. The upper bound one obtains - which will be more fully examined in the next Section - lies in the range

$$m_t \leq (180 - 200) \ GeV, \tag{6}$$

depending on the value of the Higgs mass. The range bracketed by Eqs. (5) and (6) will be, hopefully, further reduced (or top will be found!) by the 1988/89

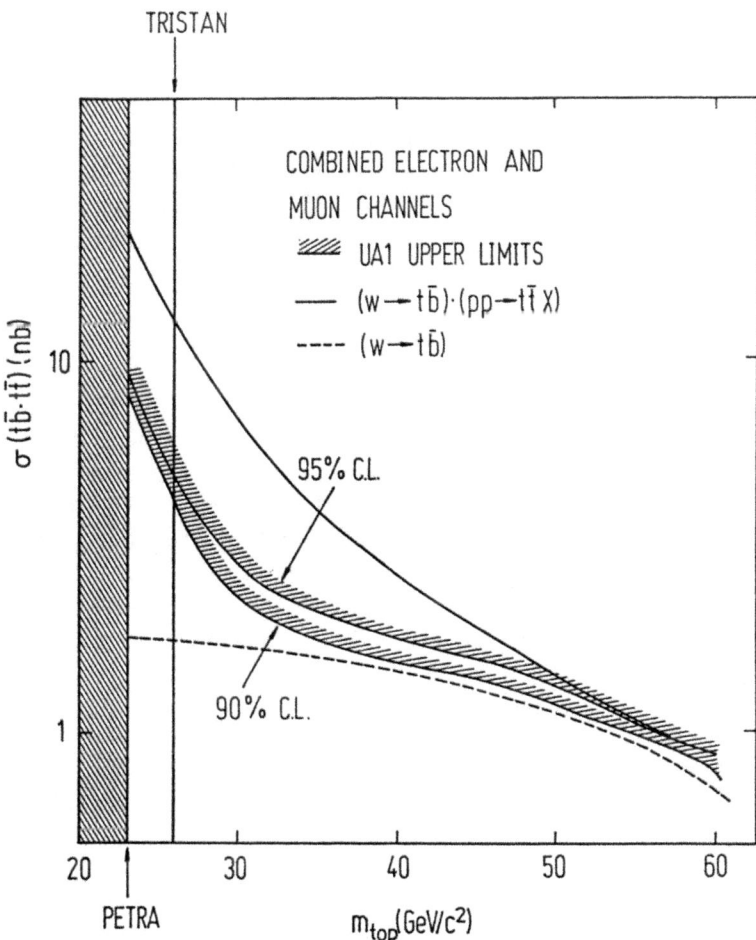

Figure 3: Expected top signal for the UA1 cuts [7], as a function of m_t . The dashed line is the signal from $W \to t\bar{b}$.

operation of the Tevatron and the upgraded CERN collider. These experiments should be sensitive to top masses up to around 80 GeV.

Given the reasonable certitude of the existence of top, it is natural to ask if there are more than three families of quarks and leptons. My personal prejudice is that the answer to this question is no. This is based on a, mostly unscientific, feeling that the masses of quarks and leptons should be bounded by roughly M_W and that, since top is already heavy, there is no room left. Experimentally, however, one is beginning to tackle this question more quantitatively. On the one hand, there are some lower bounds. from UA1 [7], on the masses of 4th generation leptons and charge -$\frac{1}{3}$ quarks.

$$m_{l'} \geq 41 \ GeV \qquad (90\% \ C.L.) \tag{7}$$

$$m_{b'} \geq 32 \ GeV \qquad (90\% \ C.L.) \tag{8}$$

On the other hand, rather sharp limits are beginning to emerge from e^+e^- experiments and from the CERN collider on the number of light neutrinos. These limits, which are becoming of comparable quality to the cosmological bound from nucleosynthesis [11] ($N_\nu \leq 4$), can be reasonably taken to be limits on the number of families. Although we do not really understand the origin of masses at all, it would seem to be a perverse world in which the fourth generation neutrinos would pick up a mass greater than $\frac{1}{2} M_Z$, while their lighter generation cousins are essentially massless!

PETRA and PEP have set limits on the number of neutrinos, N_ν, by looking for the process $e^+e^- \rightarrow \gamma \ Nothing$, where "$Nothing$" corresponds here to a neutrino-antineutrino pair. Since the Z couples universally to neutrinos of each generation, the rate for $e^+e^- \rightarrow \gamma \ Nothing$ is proportional to N_ν [2]. Even though in these experiments the CM energy is well below the Z mass, one obtains quite good limits on the number of neutrinos. The most sensitive results come from ASP at PEP and CELLO at PETRA, giving 90% C.L. limits of $N_\nu \leq 7.5$ [12] and $N_\nu \leq 8.7$ [13], respectively. An analysis of all e^+e^- data, done by the CELLO collaboration [13], gives a stronger combined limit of

$$N_\nu < 4.6 \quad (90\% \ C.L.) \tag{9}$$

Neutrino counting limits from the S$p\bar{p}$S collider make use of an upper bound on the ratio R for the probability of producing W's and observing them in their $e\nu_e$ mode, to the probability of producing Z 's and observing them in their e^+e^- mode:

$$R = \frac{\sigma_W B(W \rightarrow e\nu_e)}{\sigma_Z B(Z \rightarrow e^+e^-)} \tag{10}$$

---

[2]This is not precisely correct, since for the process $e^+e^- \rightarrow \gamma\nu_e\bar{\nu}_e$ there is also an additional W-exchange contribution.

Experimentally, the UA1 and UA2 collaborations obtain the values $R = 9.1 {+1.7 \atop -1.2}$
[14] and $R = 7.2 {+1.7 \atop -1.2}$ [15], respectively, for this ratio, which gives an average [16]

$$R = 8.4 {+1.2 \atop -0.9} \qquad (11)$$

and a 90% C.L. upper bound

$$R < 10.1 \qquad (90\% \ C.L.) \qquad (12)$$

It is this upper bound that can be used to bound N_ν.

By definition, the branching fraction ratio is given by

$$\frac{B(W \to e\nu_e)}{B(Z \to e^+e^-)} = \frac{\Gamma(W \to e\nu_e)}{\Gamma(Z \to e^+e^-)} \frac{\Gamma^Z_{tot}}{\Gamma^W_{tot}} \qquad (13)$$

Both ratios above are calculable in the standard model. However, the ratio of the total widths depends on both N_ν and the top quark mass. Clearly Γ^Z_{tot} will go up if there are more neutrinos [3]. With the present limits on m_t, it is unlikely that the $t\bar{t}$ mode can contribute to Γ^Z_{tot}. However, it is quite possible that the decay $W \to t\bar{b}$ is kinematically allowed and, obviously, the lighter top is, the larger the W total width will be. The behaviour of the branching fraction ratio (13) is plotted in Fig. 4, as a function of m_t and the number of neutrinos.

To bound N_ν from the bound on R of Eq. (12), one has to know the production ratio σ_W/σ_Z of W's to Z's at the collider. This ratio can be estimated theoretically, but there are some uncertainties related to the structure functions one uses and to higher order QCD corrections [4]. A recent analysis by Colas, Denegri and Stubenrauch [16], which considers both BCDMS and EMC structure functions, gives for this ratio

$$\frac{\sigma_W}{\sigma_Z} = 3.25 \pm 0.10 \qquad (14)$$

Using the lower 1σ value for the ratio σ_W/σ_Z above and the branching fractions plotted in Fig. 4, the 90% C.L. of Eq. (12) bounds $N_\nu \leq 5$, if $m_t \leq 70$ GeV, but for heavier m_t this bound is $N_\nu \leq 3$. These bounds are shown in Fig. 5. To keep the theoretical uncertainty in the above bounds in perspective, one should note that a 5% decrease in the 1σ lower limit for the ratio σ_W/σ_Z corresponds to an additional neutrino species. So, even though it is likely that there are only three families, it is not sensible to turn the above argument around and try to put a bound on m_t this way!

---

[3] Each extra neutrino species contributes approximately 170 MeV to the total Z width.

[4] These latter uncertainties, however, are not very large since one is dealing with a cross section ratio.

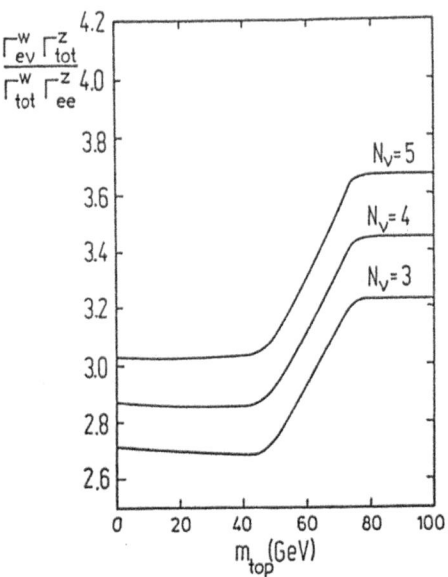

Figure 4: Branching fraction ratio as a function of m_t and N_ν.

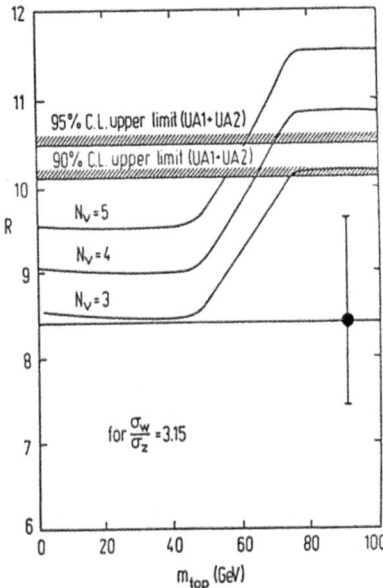

Figure 5: Bounds on N_ν as a function of m_t. The experimental point shown is the average UA1-UA2 result. From [16].

3   Probing Electroweak Radiative Effects

The measurements of electroweak parameters are now sufficiently precise that one can begin to test the presence of radiative corrections. The Weinberg angle Θ_W can be extracted from high precision ν_μ and $\bar{\nu}_\mu$ deep inelastic scattering data. In lowest order, if the scattering is done off an isoscalar target, this angle is directly related to the ratio R_ν of ν_μ neutral current scattering to ν_μ charged current scattering:

$$R_\nu = \{[\frac{1}{2} - sin^2\Theta_W^0] + \frac{5}{9}sin^4\Theta_W^0[1 + r]\} \tag{15}$$

where r is the ratio of ν_μ to ν_μ charged current scattering. A global analysis of all high precision experiments [CDHS, CHARM, CCFRR, FMM] done by Amaldi et al. [17], including all corrections, except electroweak radiative corrections, gives

$$sin^2\Theta_W^0 = 0.242 \pm 0.006 \tag{16}$$

Electroweak radiative corrections - adopting the convenient definition of $sin^2\Theta_W$ [18], $sin^2\Theta_W = 1 - \frac{M_W^2}{M_Z^2}$ - change the value (16) by an amount Δs^2. This shift depends very mildly on the (as yet unknown) Higgs mass M_H, as long as $M_H \leq$ TeV. Its dependence on m_t, however, is stronger and one has [19]

$$(\Delta s^2)_{top} \simeq 4.4 \times 10^{-4}[\frac{m_t}{M_W}]^2 \tag{17}$$

Using as canonical values $M_H = 100$ GeV and $m_t = 45$ GeV, one finds [17]

$$\Delta s^2 = -0.009 \pm 0.001 \tag{18}$$

and the result of the global analysis of Amaldi et al. [17] yields

$$(sin^2\Theta_W)_{DIS} = 0.233 \pm 0.003 \pm [0.005] \tag{19}$$

Here the last error in brackets is an estimate of the theoretical error of the whole analysis. Comparing (19) to (16), one sees that the radiative corrections have lowered the value of $sin^2\Theta_W$ by about 4%.

Radiative effects also change the predictions for M_W and M_Z . Using the same definition of the Weinberg angle as above [18], one has [20]:

$$M_W^2 = \frac{\pi\alpha}{\sqrt{2}G_F sin^2\Theta_W}[\frac{1}{1 - \Delta r}] = \frac{(37.281 GeV)^2}{sin^2\Theta_W[1 - \Delta r]} \tag{20}$$

where Δr is a theoretically calculated radiative correction. The measured values of the W and Z masses by the UA1 and UA2 collaborations, in principle, determine $sin^2\Theta_W$ directly, via its definition. However, the errors are such that this is not the most accurate value of $sin^2\Theta_W$ one can obtain from these measurements. Rather, it is better to extract $sin^2\Theta_W$ from Eq. (20) directly, imputing a value for the radiative correction Δr. This quantity, like Δs^2, depends very mildly on M_H , for $M_H \leq$ TeV, but more significantly on m_t [19], with the latter dependence going roughly as:

$$(\Delta r)_{top} \simeq -0.007(\frac{m_t}{M_W})^2 \tag{21}$$

For the canonical values of M_H and m_t previously adopted, one finds [17] [20]

$$(\Delta r) = 0.0713 \pm 0.0013, \tag{22}$$

which provides a significant 7% change upwards in the value of $sin^2\Theta_W$ extracted from the W (and Z) masses.

Using the most recent values of the W and Z masses obtained by the UA1 and UA2 collaborations [21], Eq. (20) yields for $sin^2\Theta_W$ the value [17]

$$(sin^2\Theta_W)_{M_W/M_Z} = 0.228 \pm 0.007 \pm [0.002], \tag{23}$$

with the bracketed error again being theoretical. Several remarks are in order:

1. The radiative corrections help to bring the value of $sin^2\Theta_W$ obtained in deep inelastic scattering and the one gotten through the measurement of the W and Z masses in better agreement with each other. This can be seen in the Table below

Table I: Radiative effects for Θ_W

	$sin^2\Theta_W^0$	$sin^2\Theta_W$
Deep inelastic	0.242 ± 0.006	$0.233 \pm 0.003 \pm [0.005]$
W/Z Masses	0.212 ± 0.007	$0.228 \pm 0.007 \pm [0.002]$

2. A global fit of all neutral current experiments, of which the most precise are the ones we have discussed, finds a consistent value for $sin^2\Theta_W$. For the canonical choices of M_H and m_t adopted one has [17]

$$sin^2\Theta_W = 0.230 \pm 0.0048 \tag{24}$$

3. One cannot, however, allow the value of m_t to become too large since the corrections $(\Delta r)_{top}$ and $(\Delta s^2)_{top}$ of Eqs. (17) and (21) act in opposite direction to each other. Thus, if m_t is too big, the excellent agreement shown in Table I above will be spoiled. This phenomenon has been analyzed in detail by Amaldi et al. [17] and it provides a theoretical upper bound for m_t. This bound is in the range indicated in Eq. (6), with the uncertainty in the exact 90% C.L. limit obtained depending on the value of the Higgs mass, as shown in Fig. 6

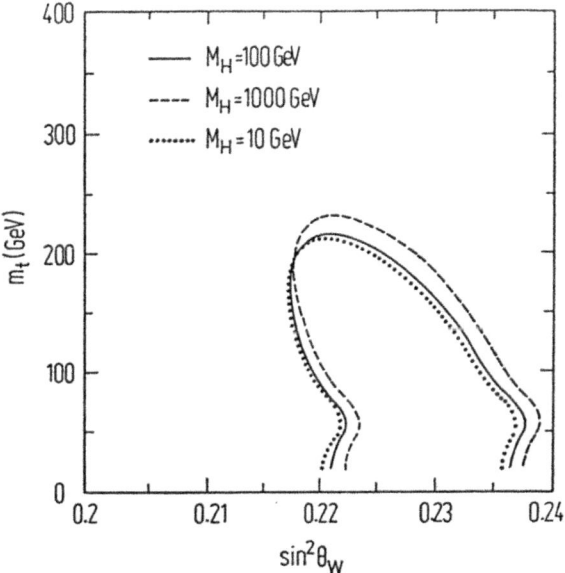

Figure 6: Allowed 90% C.L. regions in the $sin^2\Theta_W$-m_t plane, for different values of M_H. From [17].

4. One can extract a value of (Δr) by using the measurement of the W mass and the accurate value of $sin^2\Theta_W$ obtained in deep inelastic scattering. The value obtained

$$(\Delta r)_{exp} = 0.077 \pm 0.037 \qquad (25)$$

is in nice agreement with the theoretical expectation (22), but is only 2σ away from zero. One will have to wait for the turn on of SLC and LEP to obtain a more accurate value. I note that the accuracy expected there [22], $\delta(\Delta r) \leq 0.01$, will begin to be sensitive to the effects of very heavy Higgs boson, since [19]

$$(\Delta r)_{Higgs} \sim 0.0024\ln(\frac{M_H}{M_Z})^2 \qquad (26)$$

4 Flavor Physics Constraints

1987 brought three important new experimental inputs for the physics of flavor:

1. Large mixing was seen in the B_d system [10]

2. The ARGUS collaboration reported [23] the first evidence for decays involving a direct $b \rightarrow u$ quark transition

3. The NA31 collaboration at CERN presented the first indication for $\Delta S = 1$ CP violation [24]

In 1988 we are still digesting this information. Here I want to make some remarks on its theoretical impact [5]. I will assume in what follows, for the most part, that there are only three families of quarks and leptons since, although the data does not exclude more than 3 families, all data is consistent with the quark charged current weak interactions being described by a 3x3 unitary matrix V: the Cabibbo Kobayashi Maskawa matrix.

It is useful to adopt a parametrization of the CKM matrix first suggested by Maiani [27]. In this parametrization there is a natural hierarchy between the mixing angles and one can write, following Wolfenstein [28]:

$$sin\theta_1 = \lambda; \quad sin\theta_2 = A\lambda^2; \quad sin\theta_3 = A\rho\lambda^3, \tag{27}$$

where the new parameters A and ρ are of $0(1)$ and $\lambda \simeq 0.22$ is, essentially, the Cabibbo angle. To $O(\lambda^4)$, V takes the simple form:

$$V = \begin{vmatrix} 1 - \frac{1}{2}\lambda^2 & \lambda & A\rho\lambda^3 e^{-i\delta} \\ -\lambda & 1 - \frac{1}{2}\lambda^2 & A\lambda^2 \\ A\lambda^3(1 - \rho e^{i\delta}) & -A\lambda^2 & 1 \end{vmatrix} \tag{28}$$

The parameter A is reasonably well fixed from measurements of the B lifetime. The parameter ρ and the CP violating phase δ, on the other hand, are still quite uncertain. However, their ranges are more restricted as a result of the new 1987 data. I will briefly summarize our present knowledge of all these quantities below.

Because $|V_{ub}|^2 << |V_{cb}|^2$, the B-lifetime and the B semileptonic branching ratios can be used to fix A. A typical recent analysis, by Altarelli and Franzini [29], gives

$$|V_{bc}|^2 = A^2\lambda^4 = \frac{(2.9 \pm 0.6) \times 10^{-13}}{\tau_B(10^{-12}sec)} \tag{29}$$

which, using the world average for τ_B [30], yields

$$A = 1.05 \pm 0.17 \tag{30}$$

That is, one knows A at the 10-20 % level.

Semileptonic B decays show, at present, no evidence for the direct quark transition $b \rightarrow u$. Hence one can use existing bounds on charmless semileptonic B decays to infer a bound on ρ. What is usually quoted are bounds on

$$R = \frac{\Gamma(b \rightarrow u)}{\Gamma(b \rightarrow c)} \simeq 2\frac{|V_{ub}|^2}{|V_{cb}|^2} = 2(\lambda\rho)^2, \tag{31}$$

---

[5]Some new experimental information became available at the time of the writing of this report. The CLEO collaboration at CESR confirmed the large mixing seen by ARGUS [25]. On the other hand, CLEO did not observe the decays $B_d \rightarrow p\bar{p}\pi(\pi)$ [26] and their 90% C.L. on these modes lies about a factor of 2 below the ARGUS observation [23].

where the factor of 2 above is a phase space factor. Using the bound from CLEO [31], $R \leq 0.08$ [6], this implies

$$\rho \leq 0.9 \tag{32}$$

The observation by ARGUS of charmless B decays [23]

$$BR(B^+ \rightarrow p\bar{p}\pi^+) = (3.7 \pm 1.3 \pm 1.4) \times 10^{-4}$$
$$BR(B^0 \rightarrow p\bar{p}\pi^+\pi^-) = (6.0 \pm 2.0 \pm 2.2) \times 10^{-4} \tag{33}$$

provides the first evidence that $V_{ub} \neq 0$ and hence gives a lower bound for ρ. Unfortunately, translating an exclusive branching ratio into a value of V_{ub} is difficult. The ARGUS collaboration [23] gives a (very) conservative lower bound from (33) of $\frac{|V_{ub}|}{|V_{cb}|} \geq 0.07$, which implies

$$\rho > 0.3 \tag{34}$$

In fact, it is relatively easy to get a much bigger answer for ρ than (34) [7]. For present purposes, however, I will assume that ρ lies in the range indicated by Eqs. (32) and (34).

The parameter space allowed for the CP violating phase δ has been impacted by the ARGUS measurement of $B_d - \bar{B}_d$ mixing [10]. What was measured by ARGUS is the ratio of wrong sign dileptons in B_d decay:

$$r_d = \frac{\Gamma(B_d \rightarrow l^- X)}{\Gamma(B_d \rightarrow l^+ X)} = 0.21 \pm 0.08 \tag{35}$$

This ratio is related directly to the mixing parameter $x_d = \left(\frac{\Delta m}{\Gamma}\right)_{B_d}$ via [8]

$$r_d \simeq \frac{x_d^2}{2 + x_d^2} \tag{36}$$

The ARGUS result (35) implies $x_d = 0.78 \pm 0.18$. If one takes into account of the existing CLEO upper bound on r_d [35], the combined ARGUS and CLEO results yield a 90% C.L. range for x_d:

$$0.44 \leq x_d \leq 0.78 \tag{37}$$

Theoretically, the mixing parameter x_d in the Cabibbo Kobayashi Maskawa model arises from the box graph of Fig. 7, with the dominant contribution coming from the t quark loop. A straightforward evaluation of this graph yields for x_d a

---

[6]This bound is rather conservative, as emphasized by [32]. Similar bounds have been also obtained by ARGUS and the XTAL BALL [33].

[7]For instance, Shifman's [34] first analysis of the ARGUS data suggested $\frac{|V_{ub}|}{|V_{cb}|} \geq 0.3$, which implies $\rho > 1.3$ - which is above the upper bound (32)! Thus the non observation by CLEO [26] of charmless B decays, at the level claimed by ARGUS, may help alleviate a minicrisis.

[8]This formula assumes that $\Delta\Gamma \ll \Gamma$, for the B_d system. Thus in (36) I have altogether neglected the contributions of $\Delta\Gamma$ relative to Δm.

Figure 7: Box graph contributing to x_d

formula which is quadratically dependent on both m_t and V_{td} [9]

$$x_d \sim |V_{td}|^2 m_t^2 \qquad (38)$$

Thus, one of the byproducts of the observation of a rather large $B_d - \bar{B}_d$ mixing by ARGUS [10] is that m_t cannot be too small. Indeed, soon after the announcement of the ARGUS result a flood of theoretical papers all, more or less, agreed that the range (37) necessitated $m_t \geq 50$ GeV. However, Eq. (38) also makes it clear that the ARGUS measurement has implications on ρ and the phase δ, since

$$|V_{td}|^2 = A^2 \lambda^6 [1 + \rho^2 - 2\rho \cos\delta] \qquad (39)$$

Clearly, if m_t is not too big, one wants ρ and δ to be near $\rho_{max} \simeq 0.9$ and π, respectively, so as to maximize (39).

Although the large $B_d - \bar{B}_d$ mixing observed argues for the phase δ to be as close to π as possible, the observation of CP violation in the Kaon system requires some non trivial phase δ to exist. In collaboration with Krawczyk, Steger and London [36], I have performed recently a combined fit of x_d and the CP violating parameter ϵ in the Kaon system [10], to determine the allowed range for the parameters in the CKM matrix. Apart from the existing experimental errors on x_d and ϵ, our analysis is uncertain theoretically since one has to estimate the hadronic matrix elements of the $\Delta B = 2$ and $\Delta S = 2$ quark amplitudes, entering

---

[9]This is not quite correct, for large m_t.

[10]ϵ is proportional to $sin\delta$ and so it will vanish if $\delta \to \pi$.

in these quantities. In our work, to be conservative, we assumed a rather broad range of theoretical uncertainty in the calculation of the hadronic matrix elements. Furthermore, we let m_t vary over the full allowed theoretical range, from 40 GeV to 200 GeV. We obtained, in this way, regions in the $\rho - \delta$ plane, allowed by the present experimental results. These regions are in the shape of "moons", which fatten up as one allows the theoretical uncertainty to increase. Our results are shown in Fig. 8.

The lower left-hand figure in Fig. 8 perhaps most closely represents the present "best guesstimate" on the allowed $\rho - \delta$ range. It corresponds to a bag parameter $B_K = 2/3$ in the Kaon system and assumes $100~MeV \leq (f_{B_d}^2 B_{B_d})^{\frac{1}{2}} \leq 200~MeV$ in the B_d system. Furthermore, the top mass is varied over the presently permitted range. Note that if $\rho > 0.3$, δ is indeed rather large: $\delta > 2$. Thus the lower limit on the allowed value of ρ is important. However, it should be pointed out that a minimum value of $\rho > 0.15$ is already demanded by the requirement that $\epsilon \neq 0$, since $\epsilon \sim \rho$. Because of the theoretical uncertainties inherent in calculating x_d and ϵ, the standard model with three generations has still quite a large allowed region in the $\rho - \delta$ plane. A measurement of the top quark mass, however, would do much to reduce this range. This is shown in Fig. 9, where the allowed regions in the $\rho - \delta$ plane are detailed for fixed values of m_t, but otherwise allowing for the full range of the other theoretical uncertainties

The positive result obtained by the NA31 experiment at CERN for the CP violating parameter ϵ' in the Kaon system [24]

$$\frac{\epsilon'}{\epsilon} = (3.5 \pm 1.1) \times 10^{-3} \tag{40}$$

is also consistent with the expectations of the standard model. The importance of this result, however, is that:

1. It shows for the first time that $\Delta S = 1$ CP violating process exist, as predicted by the standard model. The ϵ parameter, measured in K decay, could be due to a purely $\Delta S = 2$, superweak transition [37]. ϵ', however, since it is proportional to the matrix element of H_{weak} between a Kaon and two pions, measures directly the amount of $\Delta S = 1$ CP violation. Thus the result (40) rules out all superweak theories, where CP violation occurs only through the mass matrix.

2. The NA31 result [24] does not further restrict the $\rho - \delta$ plane, because of additional theoretical uncertainties in estimating the hadronic matrix element needed to calculate ϵ'. However, the central range for ρ and δ, as well as the central range of the theoretical calculations of the relevant matrix elements, precisely give the central value in (40). Thus the measurement of ϵ' gives

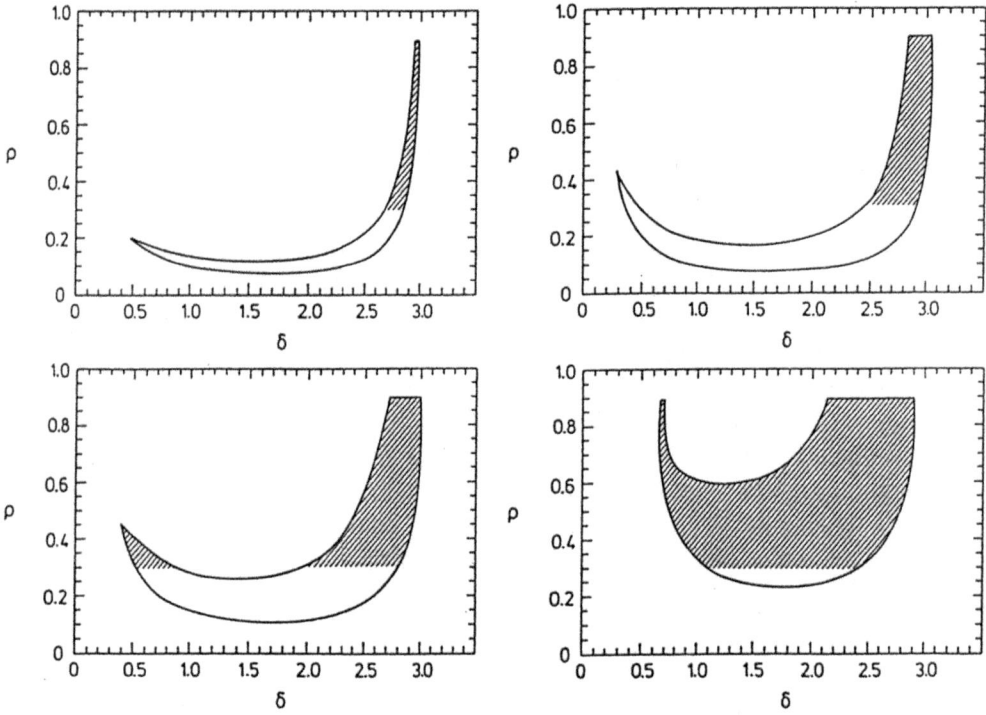

Figure 8: Allowed regions in the $\rho - \delta$ plane for different theoretical assumptions. The crosshatched area lies in the range of Eqs. (32) and (34). For more details, see [36]

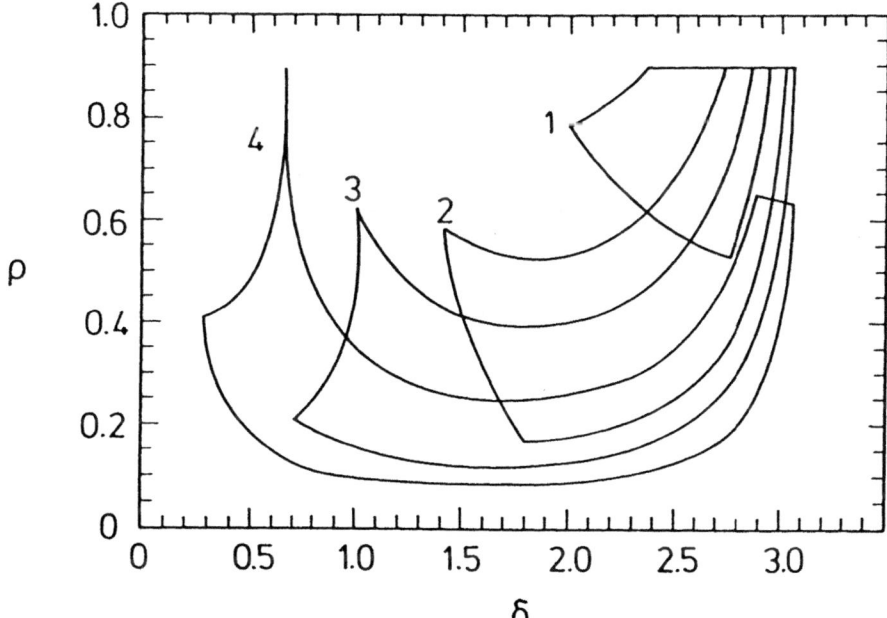

Figure 9: Allowed regions in the $\rho - \delta$ plane for fixed values of m_t, but for the full theoretical uncertainty. Regions 1-4 correspond to $m_t = 60, 90, 120, 180\ GeV$, respectively. From [36]

additional impetus to the notion that the three generation standard model is, essentially, correct.

There is a corollary to the last point which is perhaps worth noting. This concerns the strong CP problem [38], connected with why the effective CP violating parameter $\bar{\Theta} = \Theta + Arg det M$, of the combined electroweak and QCD theory, is so small: $\bar{\Theta} \leq 10^{-9}$. It is possible that $\bar{\Theta}$ vanishes due to some additional high energy symmetry [39]. However, one can also imagine that $\bar{\Theta}$ is tiny because the quark mass matrix is essentially real and CP is broken by the vacuum only [40]. In this latter case, the observed CP violation in the Kaon sector requires another explanation, than that provided by the Cabibbo Kobayashi Maskawa mixing. This necessitates, in general, already a fair bit of fine tuning to get ϵ right [41]. The requirement now to also get a non zero value for ϵ' - and of the right magnitude - make the problem much more difficult. Thus the NA31 result has important implications for alternative models of CP violation

5   The Higgs Enigma

The simplest way to break $SU(2) \times U(1)$ to $U(1)_{em}$ involves a complex scalar doublet Φ and an asymmetric potential

$$V = \lambda[\Phi^{\dagger}\Phi - \frac{1}{2}v^2]^2 \qquad (41)$$

The Higgs boson H is the debris that remains in the theory, after the symmetry breakdown. Because the parameter λ in (41) is unknown, the mass of this excitation is not determined by the theory. There are, however, various theoretical arguments which provide both lower and upper bounds for M_H. Considerations related to the triviality of the $\lambda\Phi^4$ theory [42] have been much discussed in the last years and allow one to set a reliable upper bound for M_H of the order of a TeV [43]. If M_H is above this value, the simple symmetry breaking sector assumed is inconsistent, even as an effective field theory, since excitations appear above the physical cutoff necessary to define the theory. Although the triviality upper bound is theoretically reliable, it is also a long long way away from present experimental probing. Indeed, it will require a machine like the SSC before one can say much about the Higgs, if its mass is near this bound!

The lower bounds on the Higgs mass which exist [44] are less theoretically reliable, since they are based on less firmly established principles (like demanding that radiative effects do not change the nature of the vacuum) and necessitate some constraints on the existing fermion spectrum. The typical value for these bounds, around 10 GeV, furthermore is, or will be soon, accessible experimentally. Thus it is perhaps more worthwhile here to discuss what are the present known experimental bounds on M_H and what are the prospects for improving them, in the near future.

Perhaps the most reliable of the (low mass) bounds on M_H was pointed out by Barbieri and Ericson [45] long ago. It comes from low energy neutron-nucleus scattering, where the presence of a very low mass Higgs alters the shape of the measured angular distribution beyond what is expected, unless

$$\frac{h_{HNN}^2}{M_H^4} \leq 4.3 \times 10^{-10} MeV^{-4} \tag{42}$$

Here h_{HNN} is the Higgs coupling to nucleons. This coupling can be inferred by current algebra methods and is proportional to the pion-nucleon σ-term [46]: $h_{HNN} = \frac{\sigma}{v}$, with $v = (\sqrt{2}G_F)^{-\frac{1}{2}} \simeq 250 \, GeV$. Using the most recent determination of σ [47] ($\sigma \simeq 50$ MeV) yields

$$M_H \geq 3.1 \, MeV \tag{43}$$

There is another low energy bound on Higgs bosons usually quoted, coming from the study of the nuclear deexcitation of the 20.1 MeV state of ^4He [48]. This bound also depends on the value of h_{HNN}. Unfortunately, the excluded region given by this experiment [48]: 3 MeV $\leq M_H \leq$ 14 MeV uses a value for $h_{HNN} \simeq 1.1 \times 10^{-3}$. Using the σ term evaluation above for h_{HNN} gives a much smaller value for this coupling and no bound is set by [48].

There are additional bounds on Higgs bosons which come from direct flavor changing transitions at the quark level ($s \rightarrow dH$; $b \rightarrow sH$), induced by Penguin operators. These transitions, which are shown in Fig. 10, are particularly important if m_t is large. The experimental branching ratio [49] $B(K^+ \rightarrow \pi^+ e^+ e^-) = (2.7 \pm 0.5) \times 10^{-7}$ is below the Penguin prediction for the decay $K^+ \rightarrow \pi^+ H$; $H \rightarrow e^+ e^-$ of Willey [50] which, for $m_t = 45$ GeV, is

$$B(K^+ \rightarrow \pi^+ H; H \rightarrow e^+ e^-) \simeq 7 \times 10^{-6} \tag{44}$$

This excludes Higgs bosons from the region

$$50 MeV \leq M_H \leq 211 MeV \tag{45}$$

where the lower limit is due to an experimental cut in [49] and the upper limit is where the process $H \rightarrow \mu^+ \mu^-$ becomes dominant.

In the case of B mesons, the decay $B \rightarrow HX$ is dominated entirely by the t-quark Penguin and the rate is very large indeed for large m_t [51] [11]

$$B(B \rightarrow HX) \simeq 0.3(\frac{m_t}{M_W})^4[1 - \frac{M_H^2}{m_B^2}]^2 \tag{46}$$

---

[11]The extra factor of $[1 - \frac{M_H^2}{m_B^2}]$ in Eq. (46), comes from the decay amplitude. This additional kinematical suppression was pointed out recently by Bertolini, Borzumati and Masiero [52]

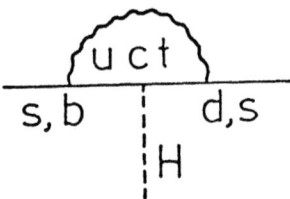

Figure 10: Penguin diagram for Higgs emission.

If $m_t \simeq M_W$, simple charm counting in B decays excludes all $M_H \leq M_B$. For lighter m_t's one must rely on particular exclusive modes and the bounds are more model dependent. In particular, to make use of the CLEO bound on the process $B \to K\mu^+\mu^-$ [54] of roughly 2×10^{-4} to bound M_H, requires an accurate estimate of the relative importance of the $H \to \mu^+\mu^-$ and $H \to \pi^+\pi^-$ modes.

Perhaps the strongest bound to date on Higgs bosons in the GeV range comes from the process $\Upsilon \to H\gamma$. The branching ratio for this process to that for $\Upsilon \to \mu^+\mu^-$ is calculable [55] but, unfortunately, it has very large QCD corrections [56]:

$$\frac{\Gamma(\Upsilon \to H\gamma)}{\Gamma(\Upsilon \to \mu^+\mu^-)} = [\frac{G_F^2 m_b^2}{\sqrt{2}\pi\alpha}(1 - \frac{M_H^2}{M_\Upsilon^2})][\frac{1 - \frac{40}{3\pi}\alpha_s F(\frac{M_H}{M_\Upsilon})}{1 - \frac{16}{3\pi}\alpha_s}] \tag{47}$$

In the above the factor in the first square bracket is the Wilczek result [55], while the second bracket contains the QCD corrections. Note that for light M_H the function F, which is given in detail in [56], is close to unity. Taking this formula at face value CUSB [57] obtains a 90% C.L. bound on M_H of

$$M_H \geq 4.3 \ GeV \tag{48}$$

However, one must worry about the theoretical reliability of this result since the QCD corrections in (47) change the Wilczek value by 60%!

This somewhat somber situation with the present Higgs bounds will change dramatically with the turn on of SLC and LEP. LEP 100 and SLC should be able to discover the Higgs boson if $M_H \leq 45$ GeV, while LEP 200 can explore Higgses up to $M_H \simeq 80$ GeV. At the Z, the most favorable way to look for Higgs bosons appears to be the Bjorken process [58] $Z \rightarrow Hl^+l^-$, which has both a reasonable rate and a good signature. The process proceeds through the HZZ coupling and the presence of the virtual Z produces a dilepton pair which is highly peaked towards large masses. Because of the clean signature, one can hope to see a signal even with only a few events [59], so that with 10^6 Z's one should be able to push the Higgs bound to about 45 GeV (or discover the Higgs!)

At LEP 200 one will look for Higgs bosons in associated production with a Z, where the process $e^+e^- \rightarrow ZH$, for Higgs bosons lighter than 60-70 GeV, has a rate of the order of pb. With an expected integrated luminosity of around 500 pb^{-1} per year, this process could produce of the order of 500 Higgs bosons. However, one will have to be clever to dig out the signal! A promising possibility [60] is to look for events in which the Z decays into neutrinos, which produces rather spectacular events with two unbalanced jets in one hemisphere, plus nothing in the other. Detailed Monte Carlo studies suggest that at LEP 200 a Higgs boson as heavy as 80 GeV will be detectable [60].

By 1995 it appears to me possible that we will know if a Higgs boson exists, with mass less than 80 GeV or so. If no such signal is seen, one will have to wait for the next generation of colliders (SSC, LHC, CLIC) to renew the hunt for the Higgs. Although quite clear strategies to pursue this more distant goal have been mapped out [61], their discussion here would take me too far afield.

I would like to end with a provocative remark. Although finding the Higgs will undoubtedly be an experimental triumph, it may well be a theoretical tragedy. The discovery of a single, light, Higgs will tell us that the asymmetric potential (41) is an adequate description for the symmetry breaking sector of the standard model. This is informative and nice, but it will not help theorists really to understand where this potential comes from! Indeed, if the Higgs is light enough ($M_H \leq 150 - 200$ GeV) the potential (41) is a correct effective description, up to a cutoff lying much above the Planck mass. There is no way to find out, therefore, what the true underlying theory for the symmetry breakdown might be. The enigma of symmetry breaking is perennially cloaked by the discovery of a light Higgs. So, please do not find the Higgs!

6   Conclusions

In 1988 the standard model of the electroweak interactions is in great shape! Let me recapitulate some of the points made, which allow me to be so bullish:

- Although there is still some room for additional families of quarks and leptons, one is getting close to assert that there are only three generations

62

- Top must exist and its mass range is reasonably well bounded between 50 and 200 GeV, with my own personal favorite value being $m_t \sim 80$ GeV.

- One is beginning to check the standard model radiative corrections. Indeed, for light top, these corrections are needed, at the 3σ level, to obtain agreement with experiment.

- Although there have been some exciting developments in flavor physics in the last year, the flavour structure looks still very compatible with the simple Cabibbo Kobayashi Maskawa mixing scheme.

- No sign of Higgs is yet in sight, but this is not a real worry since realistic bounds on M_H are of $O(1\ TeV)$.

This bullish state may or may not persist when the new high luminosity e^+e^- colliders at the Z will come in operation. Clearly, everyone is waiting eagerly for SLC and LEP and their potential for doing precision electroweak physics. Although I am also very curious to see what will happen at SLC and LEP, I would trade this information for the knowledge of what will be said - if anything! - about the standard electroweak model when the University of Bologna will celebrate its millesium!

References

[1] S.L. Glashow, Nucl. Phys. 22 (1961) 579; A. Salam in **Elementary Particle Theory**, ed. N. Svartholm (Almqvist and Wiksell, Stockholm, 1968); S. Weinberg, Phys. Rev. Lett. 19 (1967) 1264

[2] P. Langacker in **Neutrino Physics**, ed. H.V. Klapdor (Springer, Berlin, 1988)

[3] JADE Collaboration, W. Bartel et al., Phys. Lett. 146B (1984) 437

[4] R. Marshall RAL-87-031; T. Greenshaw, private communication to S.L. Wu

[5] S.L. Wu in Proceedings of the 1987 International Symposium on Lepton and Photon Interactions at High Energy, Hamburg, ed. W. Bartel and R. Rückl, Nucl. Phys. B (Proc. Suppl.) 3 (1988) 39

[6] F. Takasaki in Proceedings of the 1987 International Symposium on Lepton and Photon Interactions at High Energy, Hamburg, ed. W. Bartel and R. Rückl, Nucl. Phys. B (Proc. Suppl.) 3 (1988) 17

[7] UA1 Collaboration, C. Albajar et al., Zeit. f. Phys. 37C (1988) 505

[8] G. Altarelli, M. Diemoz, G. Martinelli and P. Nason, CERN-TH 4978/88

[9] P. Nason, S. Dawson and R.K. Ellis , Fermilab preprint 87/222-T

[10] ARGUS Collaboration, H. Albrecht et al., Phys. Lett. B192 (1987) 245

[11] J. Yang, M.S. Turner, G. Steigman, D.N. Schramm and K.A. Olive, Astrophys. J. 281 (1984) 443

[12] ASP Collaboration, C. Hearty et al., Phys. Rev. Lett. 58 (1987) 1711

[13] CELLO Collaboration, H.J. Behrend et al., DESY 88-052

[14] UA1 Collaboration, C. Albajar et al., Phys. Lett. 98B (1987) 271

[15] UA2 Collaboration, R. Ansari et al., Phys. Lett. 186B (1987) 440

[16] P. Colas, D. Denegri and C. Stubenrauch, Saclay preprint DPhPE 88-02

[17] U. Amaldi et al., Phys. Rev. D36 (1987) 1385

[18] A. Sirlin, Phys. Rev. D22 (1980) 971

[19] W. Marciano and A. Sirlin, Phys. Rev. D22 (1980) 2695; D29 (1984) 945

[20] W. Marciano and A. Sirlin, Phys. Rev. D22 (1980) 2695; D29 (1984) 945; M. Consoli, S.Lo Presti and L. Maiani, Nucl. Phys. B223 (1983) 474; Z. Hioki, Prog. Theor. Phys. 68 (1982) 2134; Nucl. Phys. B229 (1983) 284

[21] P. Jenni in Proceedings of the 1987 International Symposium on Lepton and Photon Interactions at High Energy, Hamburg, ed. W. Bartel and R. Rückl, Nucl. Phys. B (Proc. Suppl.) 3 (1988) 341

[22] See for example, G. Altarelli in **Physics at LEP**, ed. J. Ellis and R.D. Peccei, CERN Yellow Report, CERN 86-02

[23] ARGUS Collaboration, H. Albrecht et al., DESY 88-056

[24] NA31 Collaboration H. Burkhardt et al., Phys. Lett. 206B (1988) 169

[25] CLEO Collaboration, A. Jawahery, Talk presented at the XXIV International Conference on High Energy Physics, Munich, August 1988

[26] CLEO Collaboration, D. Kreinek, Talk presented at the XXIV International Conference on High Energy Physics, Munich, August 1988

[27] L. Maiani, Phys. Lett. B62 (1976) 183

[28] L. Wolfenstein, Phys. Rev. Lett. 51 (1983) 1945

[29] G. Altarelli and P. Franzini, Zeit. f. Phys. 37C (1988) 271

[30] D.M. Ritson, in Proceedings of the XXIII International Conference on High Energy Physics, Berkeley, ed. S. Loken (World Scientific, Singapore, 1987)

[31] M. Gilchriese, in Proceedings of the XXIII International Conference on High Energy Physics, Berkeley, ed. S. Loken (World Scientific, Singapore, 1987)

[32] C. Altarelli and P. Franzini, CERN TH 4914/87, to appear in the Proceedings of the Symposium on Present Trends, Concepts and Instruments of Particle Physics in honor of Marcello Conversi's 70th Birthday, Rome 1987

[33] K. Wachs, Thesis, Univ. of Hamburg, Desy grey report (1987)

[34] M. Shifman in Proceedings of the 1987 International Symposium on Lepton and Photon Interactions at High Energy, Hamburg, ed. W. Bartel and R. Rückl, Nucl. Phys. B (Proc. Suppl.) 3 (1988) 289

[35] CLEO Collaboration, A. Bean et al., Phys. Rev. Lett. 50 (1987) 183

[36] P. Krawczyk, D. London, R.D. Peccei and H. Steger, Nucl. Phys. B307 (1988) 19

[37] L. Wolfenstein, Phys. Rev. Lett. 13 (1964) 562

[38] See for example, R.D. Peccei in **CP Violation**, ed. C. Jarlskog (World Scientific, Singapore, 1988)

[39] R.D. Peccei and H.R. Quinn, Phys. Rev. Lett. 38 (1977) 1440; Phys. Rev. D16 (1977) 1791

[40] S. Barr and A. Zee, Phys. Rev. Lett. 55 (1985) 2253

[41] S. Barr, Phys. Rev. D34 (1986) 1567

[42] L. Maiani, G. Parisi and R. Petronzio, Nucl. Phys. B136 (1978) 115; R. Dashen and H. Neuberger, Phys. Rev. Lett. 50 (1983) 1847; D.J.E. Callaway, Nucl. Phys. B233 (1984) 189; M.A. Beg, C. Panagiotakopoulos and A. Sirlin, Phys. Rev. Lett. 52 (1984) 883; M. Lindner, Zeit. Phys. 31C (1986) 295; P. Hasenfratz and J. Nager, Z. Phys. 37C (1988) 477; M. Lüscher and P. Weisz, Nucl. Phys. B290 (1987) 25; B295 (1988) 65

[43] See for example, M. Lüscher and P. Weisz, FSU-SCRI-88-62; J. Kuti, L. Lin and Y. Shen, UCSD/PTH 87-18

[44] R.A. Flores and M. Sher, Ann. Phys. (NY) 148 (1983) 195

[45] R. Barbieri and T.E.O. Ericson. Phys. Lett. B57 (1975) 270

[46] R.D. Peccei, unpublished, T.P. Cheng, IASSNS /HEP-88/22

[47] J. Gasser, H. Leutwyler, M.P. Locher and M.E. Sainio, PSI report PR-88-13

[48] S.J. Freedman et al., Phys. Rev. Lett. 52 (1984) 240

[49] R.J. Cence et al., Phys. Rev. D10 (1974) 776; P. Bloch et al., Phys. Lett. 56B (1975) 201

[50] R.S. Willey, Phys. Lett. 173B (1986) 480

[51] R.S. Willey and H.L. Yu, Phys. Rev. D26 (1982) 3287

[52] S. Bertolini, F. Borzumati and A. Masiero, SISSA 59/88, Nucl. Phys. to be published

[53] W. Schmidt-Parzefall in Proceedings of the 1987 International Symposium on Lepton and Photon Interactions at High Energy, Hamburg. ed. W. Bartel and R. Rückl, Nucl. Phys. B (Proc. Suppl.) 3 (1988) 257

[54] CLEO Collaboration, P. Avery et al., Phys. Lett. 183B (1987) 429

[55] F. Wilczek, Phys. Rev. Lett. 39 (1977) 1304

[56] M.I. Vysotsky, Phys. Lett. 97B (1980) 159; P. Nason, Phys. Lett. 175B (1986) 233

[57] CUSB Collaboration, P. Franzini et al., Phys. Rev. D35 (1987) 2883

[58] J.D. Bjorken in Proceedings of the 1976 SLAC Summer Institute in Particle Physics, ed. M.C. Zipf (SLAC-198, 1977)

[59] H. Baer et al. in **Physics at LEP**, ed. J. Ellis and R.D. Peccei, CERN Yellow report CERN 86-02

[60] S.L. Wu et al. in Proceedings of the ECFA Workshop on LEP 200, Aachen, ed. A. Böhm and W. Hoogland, CERN 87-08

[61] See, for example, I. Hinchliffe in Proceedings of the 1986 Summer Study on the Design and Utilization of the Superconducting Super Collider, ed. R. Donaldson and J. Marx (DPF 1986)

Lithium -7 as a cosmological observable.

BOLOGNA ESO-CERN JUNE 1988

Hubert Reeves   Section d'Astrophysique C.E.N.S. Saclay Gif F - 91191
France

Institut d'Astrophysique de Paris 75014 Paris

Abstract

The present state of the study of the effect of the quark-hadron phase transition on Big-Bang nucleosynthesis (BBN) is reviewed.The following conclusions can be drawn .

Given the uncertainties on the Hubble parameter, the range of Ω b goes from 0.1 to 0.01. This is *appreciably larger* than in the case of a homogeneous density universe. This does not appear to be large enough to allow the baryons to close the universe($\Omega_b < 1$).

The cosmic density of luminous matter (stars and X-ray cluster gas) is $\Omega_L =$ 0.01 whitin a factor of two while the density of clustered matter needed to account for the stability of clusters of galaxy or large scale motions is $\Omega_G = 0.1$ to 0.2.

Thus, within the uncertainties , at one end of the scale the baryonic matter could be entirely luminous (no baryonic dark matter) while at the other end of the scale the clustered matter could be entirely baryonic (no non-baryonic dark matter).

The comparison between the present calculations and the cosmic abundances suggest that the contrast R between the phases is unlikely to be larger than ten. This result is in agreement with the best estimate of the critical temperature of the phase transition $(180 \, \text{MeV} < T_c < 220 \, \text{MeV})$ leading to $R \approx 7$.

Introduction

The importance of lithium-7 abundance for cosmological studies has been known for many years .   (Wagoner, Fowler and Hoyle 1967, Reeves 1971, 1974) . However, until recently,   the interpretation of the data was quite difficult. At least three processes were known to contribute importantly to the observed abundances : Big Bang Nucleosynthesis (BBN), Galactic Cosmic Rays (GCR) , and possibly Stellar Processes (SP). Therefore, as a cosmological indicator , the use of ^{7}Li   was rather limited . As for the isotope ^{3}He , we could only give upper limits to the BBN contribution to the abundance of this isotope , provided that the extent of its galactic and stellar depletion could be properly estimated. And even this

M. Caffo et al. (eds.), Astronomy, Cosmology and Fundamental Physics, 67–78.
© 1989 by Kluwer Academic Publishers.

was quite uncertain.

Furthermore its "message" as a cosmological indicator was not really exciting . Most of the useful information obtained from the success of BBN, (such as the baryonic density of the universe(Reeves 1971) or the number of particle flavors (Yang et al 1984)) came from the abundance of D and ^{4}He . The abundances of ^{7}Li and ^{3}He were compatible with this information but were not telling anything new

The promotion of ^{7}Li to the status of "cosmological observable".

The situation has changed drastically in the recent years. Thanks to a number of new developpments, both observational and theoretical, the isotope lithium-7 has gradually been promoted to the status of a *bona fide* cosmological observable. First because its BBN contribution can now be evaluated more properly . Second because its potential message has been clarified , in relation with baryon inhomogeneities (as stressed, for example , by the anaysis of the physics of the quark-hadron phase transition. (Witten 1984)).

Primordial lithium.

One of the most important event in observational cosmology in recent years has been the discovery of lithium in PopII stars by the Spite (1983 a and b) . Later, a number of other observations have confirmed their data and added a wealth of new measurements (Spite et al 1985) (Hobbs and Duncan 1987) (Rebolo, Molaro and Beckman 1987) . The full data and its significance are best presented in figure 1 (from Cayrel 1986, 1988) . On the abcissa is plotted the ratio of iron to hydrogen, in units of the solar ratio. This parameter is a measure of the importance of stellar nucleosynthesis on a galactic scale at the birth of the corresponding star. The star on the farthest left, for instance, was born when the galactic gas contained less than one part in 3500 of the solar iron abundance. A very primitive star indeed ...

The right part of figure 1 displays , in ordinate, the corresponding magnesium abundance. As expected from typical products of stellar nucleosynthesis, both elements (iron and magnesium) grew together. The case of lithium is completely different (figure on the left) : the abundance of lithium remained almost constant while iron grows from .0003 to 0.1 of the solar value. The message is clear : *the lithium in this range is not mostly produced by stellar processes* . There exist a primordial component which dominates the stellar contribution all through this range.

Observations have shown that this component is mostly made of ^{7}Li : (^{7}Li / ^{6}Li > 10 , Maurice et al 1984). This, however, is not a very tell-taling result since in typical stellar outer layers ^{6}Li is thermonuclearly destroyed one hundred times faster than ^{7}Li.

More interesting result scome from beryllium and boron, two elements which are produced in Galactic Cosmic Rays (Meneguzzi et al 1971) but not in BBN (Wagoner et al 1967). The rate of formation of lithium (both isotopes) is , to better than a factor of two , the same as the rate of formation of boron (both isotopes). It is approximately ten times larger than the rate of formation of beryllium (Reeves and Meyer, 1978, Walker et al 1985). Furthermore, lithium, at all relevant temperature, is destroyed faster than beryllium and boron by stellar processes . Thus the abundance of beryllium gives an estimate of the GCR contributed lithium in a star while the abundance of the boron gives an upper limit to the GCR contribution.

69

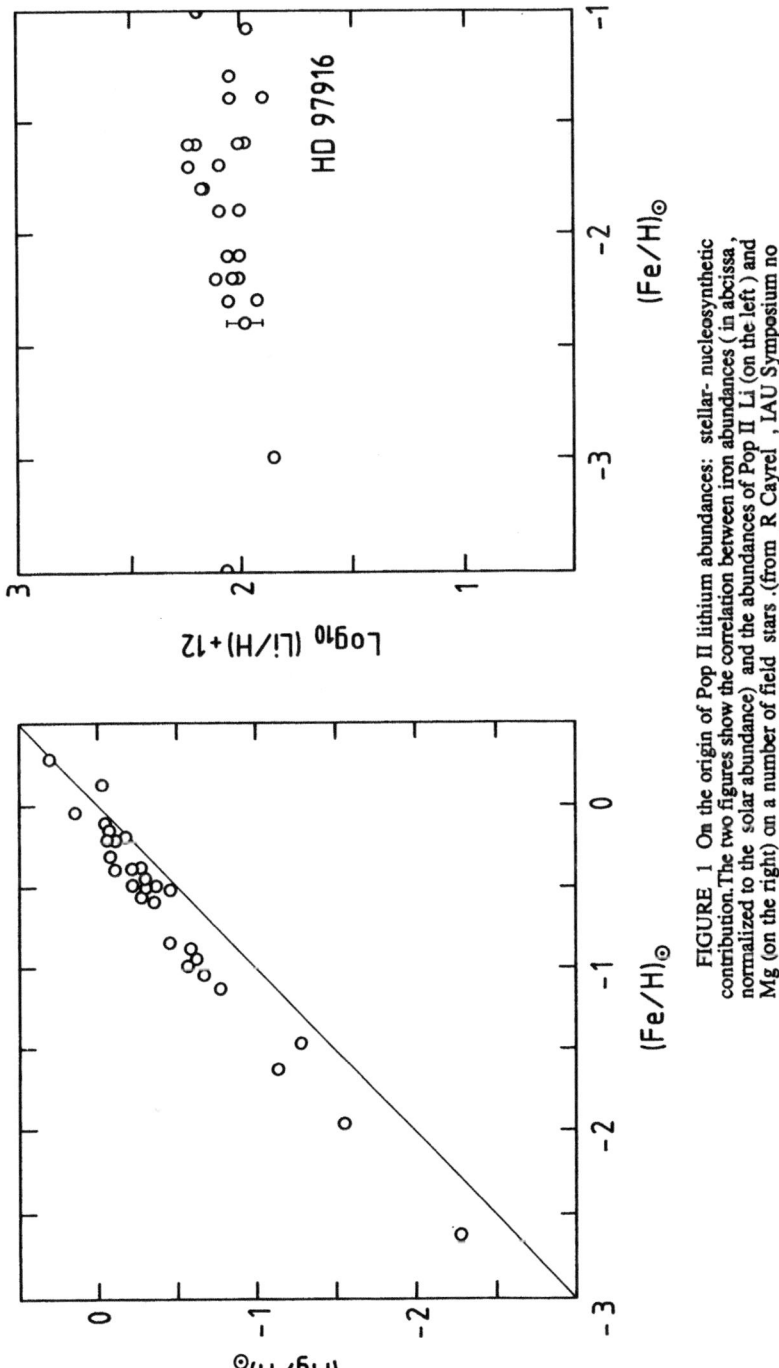

FIGURE 1 On the origin of Pop II lithium abundances: stellar- nucleosynthetic contribution. The two figures show the correlation between iron abundances (in abcissa, normalized to the solar abundance) and the abundances of Pop II Li (on the left) and Mg (on the right) on a number of field stars .(from R Cayrel , IAU Symposium no 1986) . Typically a product of stellar nucleosynthesis, Mg is seen to increase in steps with iron. The Pop II Li abundances appears to be independant of the amount of nucleosynthetic activity, suggesting a primordial origin.

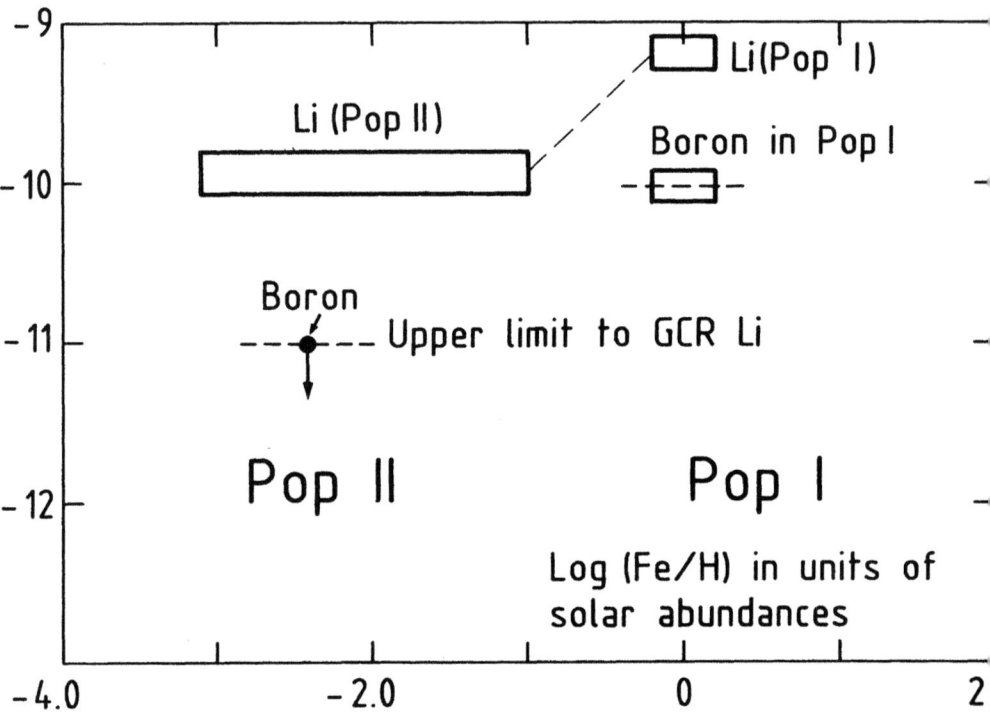

FIGURE 2 On the origin of Pop II lithium abundances: Galactic Cosmic Ray (GCR) contribution. The figure displays the observations of stellar lithium and boron abundances (a pure GCR product) as a function of the iron abundances. For PopII we have only one upper limit of B abundance. The GCR contribution to boron is quite similar to its contribution to lithium. The dashed line shows the upper limit of the GCR contribution to Li in PopII stars.

Beryllium has been recently detected in Pop II stars (Rebolo , Molaro, and Beckman 1988) with hydrogen ratio ^{9}Be / H $= 2$ x 10^{-12}. The corresponding GCR lithium is only one tenth of the observed Pop II abundance.

A search for boron (Molaro 1987) in a Pop II star (HD 140283) has yielded an upper limit of B / Li < 0.04 . The corresponding upper limit to the ^{7}Li isotope is $(7 / H)_{GCR} < 10^{-11}$, at least an order of magnitude smaller than the PopII observations (fig 2) Thus we may conclude that *the primordial component observed in figure 1 is not the result of hypothetical primordial cosmic rays* (Montmerle 1977). The only other process known to us to generate lithium-7 in interesting amount is BBN and we may thus conclude that the lithium in Pop II stars mostly is of cosmological origin.

But , in order to recover the primordial abundance relevant to BBN model calculations ,we must face the question of possible depletion of this element by stellar surface processes.

The mean abundance of lithium in Pop II stars is ^{7}Li / H $= 1.6$ x 10^{-10} with a dispersion of a factor of two (Rebolo et al 1987). This small dispersion is the main argument in favor of the hypothesis that this lithium abundance has not suffered much depletion by processes associated with the stellar surfaces (Michaud 1986) . Following the same logic, it appears reasonable to estimate that *the fractional depletion should not be larger as the observed dispersion.* A thorough study of surface depletion processes by Delyannis et al (1988)has given a similar result . In consequence we estimate an initial value of ^{7}Li / H $= 2.0 + 0.5$ x 10^{-10}.

Vauclair (1987, 1988), studying the effect of rotational mixing on stellar lithium, has argued that the depletion may have been larger and that the primordial value may be the same as the Pop 1 value (^{7}Li / H $= 1.0$ x 10^{-9}) . It remains to be seen if the theory will be able to reproduce the small abundance dispersion displayed in figure 1. More work is being done on this subject.

The lessons from lithium.

The importance of lithium in cosmology stems from two happy circumstances. The first one is the fact that, for homogeneous baryonic density model, the BBN yield of lithium (^{7}Li / H) as a function of baryonic density (ρ_b) shows a deep minimum (a dip) , contrary to the yield of D, ^{3}He and ^{4}He, which are monotonic functions of ρ_b . The second one is the fact that the Pop II observations of lithium correspond to the lithium yield at the bottom of the dip . As a result *this element can be used to test the hypothetical presence of baryonic inhomogeneities at the moment of BBN.*

Imagine that, at the moment of BBN , the cosmic matter was separated into two phases with baryonic densitiy differing by a factor of, say, thirty (due, for instance, to the earlier effect of the quark-hadron phase transition). Then , at least one of theses phases will have a density ρ_b located out of the dip of the lithium yield curve, with consequent overproduction of this element. *Thus the curvature of the lithium yield curve constrains the value of the allowed ρ_b contrast R between the two phases at the moment of BBN* . In case of cosmological models with a distribution of ρ_b , the same limit applies to the width of the ρ_b density distribution.

The real situation is further complicated by the possibility of neutron diffusion between the phases , also affecting the BBN yields (proton diffusion is highly reduced by electromagnetic effects) . The abundance of ^{4}He is particularly sensitive to the presence of neutron.

The quark-hadron phase transition.

The physics of the quark-hadron phase transition (or transitions, since there is a chiral transition and a confinement transition) is presently the object of intense studies (Iso et al 1986) (Satz 1985,1987) (Leutwyler, 1988) etc). Many of the parameters of the transitions are still poorly known despite the vigourous effort being made in QCD calculations on networks.

The relevant parameters , as far as BBN is concerned, are the following. First: the order of the transitions . There are strong indications that in the baryonic density range of the BBN, the transitions are *first order*, leading to nucleation and to bubbles of high density matter in a low density background (or the inverse).

Second : the critical temperature T_c of the transitions. Again in the density range of cosmological interest, the transitions appear to occur at the same T_c and to be simultaneous. QCD calculations give a range of 150 MeV $< T_c$ < 250 MeV . Recent chiral perturbation calculations have been published which quote a narrower range of 180 MeV $< T_c$ < 220 MeV (Gasser and Leutwyler , 1987, 1988)

The density contrast R between the high and low baryonic density phases can be computed, assuming chemical potential equilibrium between the two phases (Sale and Matthews 1986) (Applegate and Hogan 1985) (Applegate, Hogan and Sherrer 1987), (Alcock et al 1987) (Fuller et al 1987) (Kapusta and Olive 1988) . The result depends strongly on the value of the assumed critical temperature. At low T_c , the computed value of R is larger than ten , decreasing gradually at higher T_c . Thus, through the value of R , the lithium abundance is related to the value of T_c . In reality the gradual hadronization of the quark sea leads to a distribution of baryonic densities.

As the universe cools from T_c , at approximately 20 μsec, to one MeV, at one second, the contrast R between the phases is maintained . The neutron to proton ratio (n / p) , governed by weak processses, is the same in all phases . It is given by the Boltzmann formula of mass-action. Below one MeV the weak processes are no more in thermal equilibrium . The neutrons diffuse from high density phases into low density phases, changing both their density and their n / p ratio.

The extent of neutron diffusion is a function of both the fractional volume in each phases and of the mean distance between the high density blobs. As recently shown by Korki-Suonio et al (1988) and Terasawa and Sato (1988,) one must also take into account the diffusion of neutrons back in the high density phases during nucleosynthesis, due to the gradual depletion of neutrons by nuclear processes.

In view of the present difficulty in producing a realistic calculations of all these factors, it appears reasonable to draw, in the appropriate parameter space, the regions corresponding to the possible outcome of these calculations. The two parameters are the baryonic density ρ_b and the n / p ratio of each phases. In the following diagrams ρ_b is plotted vertically, while n / p , plotted horizontally, is given in units of the n / p ratio of the standard (density homogeneous) BBN.

In figure 3, 4 and 5 are plotted the isoabundance curves of D, ^{4}He and ^{7}Li resulting from BBN in this parameter plane(^{3}He is not very interesting).

The lithium dip (fig 5) is transformed in a trench, running from lower left to

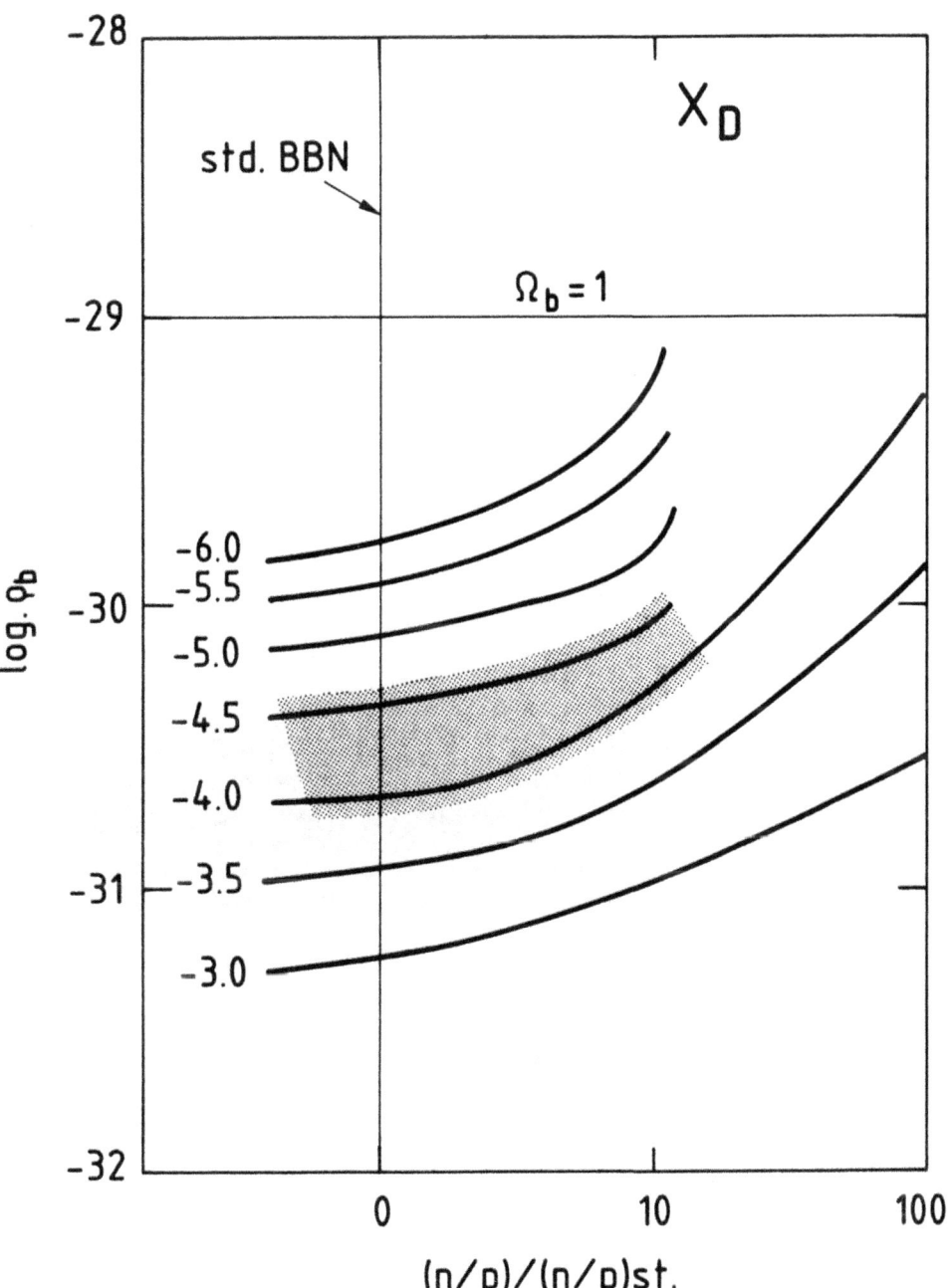

FIGURE 3 Isoabundance curves of D resulting from BBN . The two parameters are the baryonic density ρ_b and the n / p ratio (in units of the n / p ratio of the standard density homogeneous BBN).

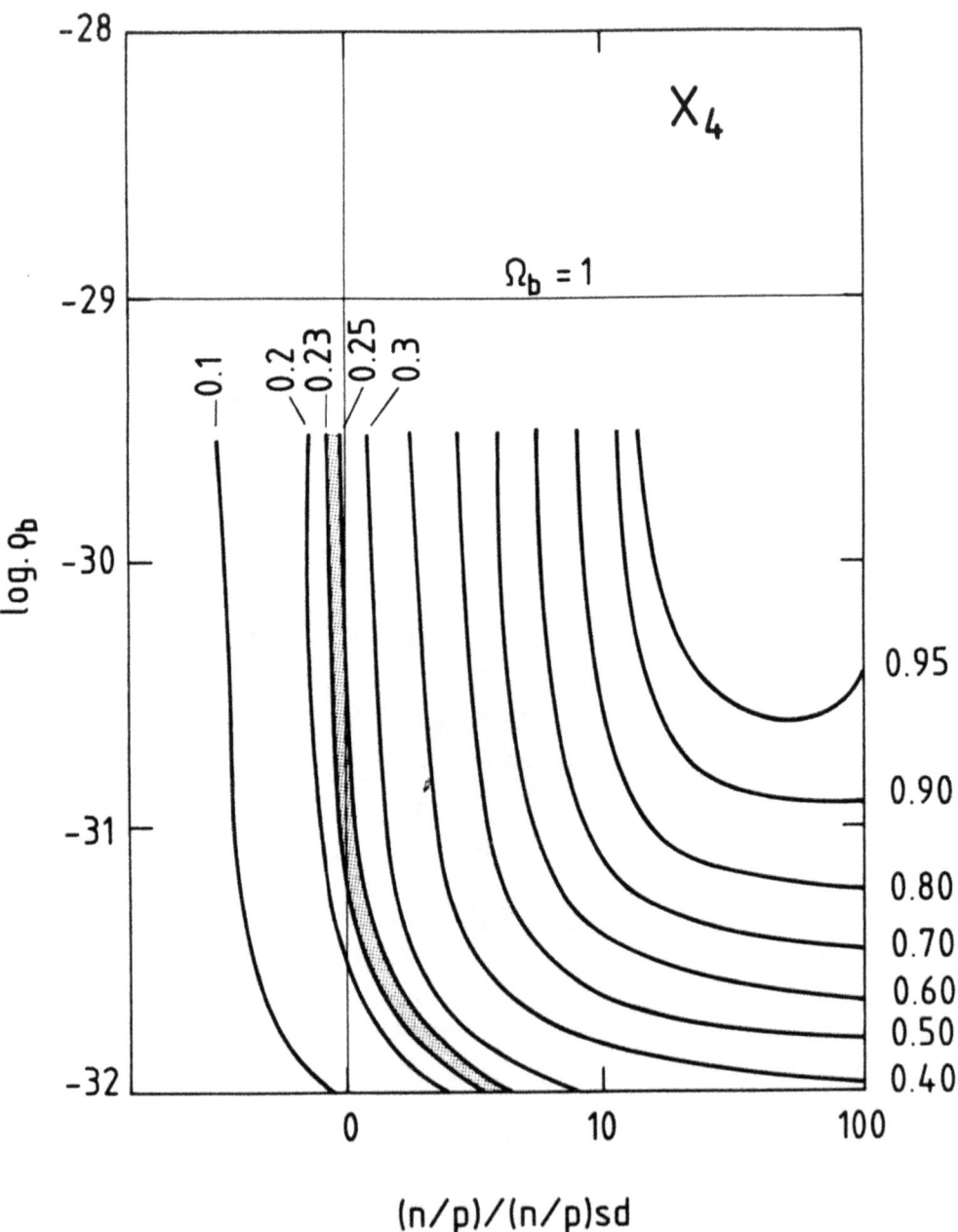

FIGURE 4 Isoabundance curves of ^{4}He resulting from BBN . The two parameters are the baryonic density ρ_b and the n /.p ratio (in units of the n / p ratio of the standard density homogeneous BBN).

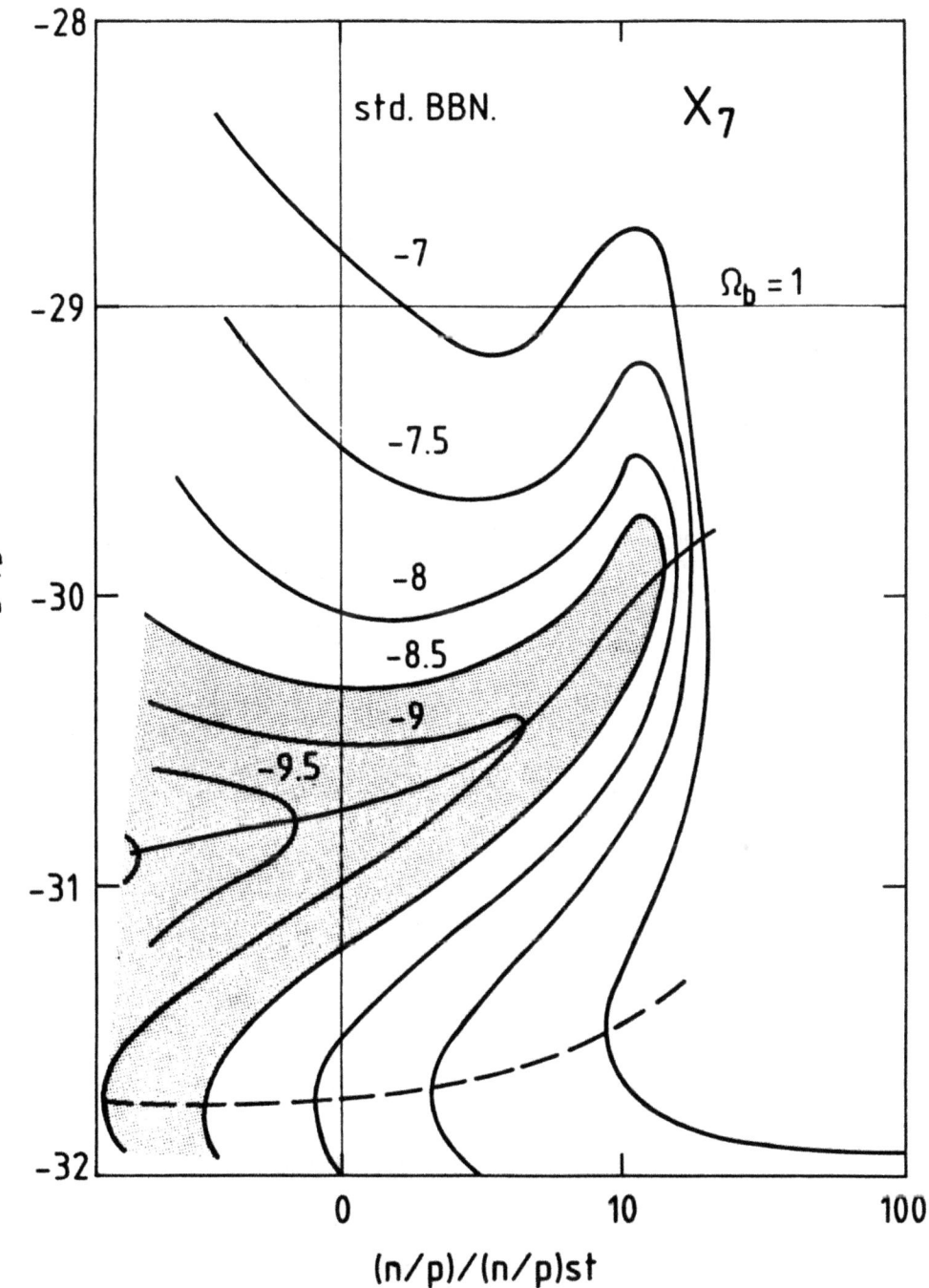

FIGURE 5 Isoabundance curves of ^{7}Li resulting from BBN . The two parameters
are the baryonic density ρ_b and the n / p ratio (in units of the n / p ratio of the standard
density homogeneous BBN).

upper right. The area compatible with the primordial lithium abundance discussed before is hatched. .

As expected , the abundance of ^{4}He depends very sensitively on the n/ p ratio (fig 4) .The primordial abundance of this isotope is between 0.23 and 0.25 , a thin strip running almost vertically in the diagram.

The case of D is complicated by the possible important depletion of this isotope during galactic life . The primordial value of D / H should be between 10^{-5} and a few x 10^{-4}.

The effect of the quark-hadron transition on BBN can be described pictorially using this parameter space (Reeves 1987,Reeves et al 1988). Consider first the effect of the creation of density inhomogeneities with contrast R in a simplified two-phase model (neglecting temporarily the effect of neutron diffusion). In the (ρ_b , n / p) diagram , the two phases are represented by two points at a distance R , one above and one below the mean baryonic density r_b(mean). The position of the points is determined by the fractional volume in each phase.

The neutron diffusion will move the high density point to the left (n / p smaller than in the case of the standard BBN) and the low density point to the right. Complete neutron homogeneization would bring the points all the way to the curved line running almost diagonally in the figure. Back diffusion of the neutrons essentially bring the points back toward the original no-diffusion vertical line .

For a given value of R, all the points, corresponding to all possible values of the fractional volume and of partial neutron diffusion , will be located inside a butterfly-shaped area. Selecting another value of R simply changes the span of the wings.

Results and conclusions.

For each pairs of points, the yield of each cosmological element is obtained from the figures and summed up with an appropriate normalization.

As expected from the width of the lithium trench, *an upper limit of approximately ten can be given to the density contrast R* (as a rule of the thumb, the butterfly must be small enough to fit in the lithium trench). This, in turn , yields *a lower limit to the value of the critical temperature $T_c > 150$ MeV,* in agreement with the QCD or chiral perturbation calculations. This agreement can be consider as another sucessful test of the Big Bang model.

It may be a long time before we get definite results on the effect of the quark-hadron phase transitions on the formation rate of the cosmological nuclides. Nevertheless recents calculations , especially those taking into account the effect of neutron back-diffusion during nucleosynthesis (Terasawa and Sato 1988, Kurki-Suonio et al (1988) already give us the general trends.

We may expect the final results to be quite comparable with the results of homogeneous density calculations . Our present ignorance of the exact values of many relevant parameters of the Q-H transition can be assimilated to corresponding uncertainties on the final results . These uncertainties are likely to decrease as more detailed studies of the transition become available.

As of today the situation can be estimated in the following statements.

1) Given the uncertainties on the Hubble parameter, the range of Ω_b goes from 0.1 to 0.01. This is *appreciably larger* than in the case of a homogeneous density universe. This does not appear to be large enough to allow the baryons to close the universe($\Omega_b < 1$).

The cosmic density of luminous matter (stars and X-ray cluster gas) is $\Omega_L = 0.01$ whitin a factor of two while the density of clustered matter needed to account for the stability of clusters of galaxy or large scale motions is $\Omega_G = 0.1$ to 0.2.

Thus, within the uncertainties , at one end of the scale the baryonic matter could be entirely luminous (no baryonic dark matter) while at the other end of the scale the clustered matter could be entirely baryonic (no non-baryonic dark matter).

The comparison between the present calculations and the cosmic abundances suggest that the contrast between the phases is unlikely to be larger than ten. This result is in agreement with the best estimate of the critical temperature of the phase transition $(180 \text{ MeV} < T_c < 220 \text{ MeV})$ leading to $R \approx 7$.

Bibliography

Alcock, C.R., Fuller, G.M., and Mathews, G.J., Astrophys. J. **320**, 439, 1987

Applegate, J.H., and Hogan, C. 1985 Phys . Rev. **D31** 3037.

Applegate. J.H., Hogan, C.,and Sherrer R.J. 1987a Phys Rev. , **D35** , 1151.

Cayrel. R., I.A.U Symposium no 126 Cambridge 1986.

Cayrel . R., Proceedings of the Alpbach Summer school 1988.

Delyannis, C., Demarque, P., Kawaler, S., Krauss, L., and Romanelli, P. preprint, (1988).

Fuller ,G.M., Mathews G.J., and Alcock, C.R. Phys. ReV . **D37** 1380 1988

Gasser , J and Leutwyler, H., Light quarks at low temperature. Phys Lett B **184, 83** , 1988.

Gasser , J and Leutwyler, H., Thermodynamics of chiral symmetry . Phys Lett B **188,** 477 , 1987.

Hobbs, L.M., and Duncan, D.K., Ap.J. **317, 796**, 1987. (Li in halo stars)

Iso, K., Kodama, H., and Sato, K., Phys. Lett. 337, **169B,** 1986.

Kapusta,J.I., and Olive K.A. 1988 preprint

Kurki-Suonio, H., Matzner , M.M., Centrella, J.M., Rothman, T., and Wilson, J.R., , preprint submitted to Physical Review D 1988 preprint

Leutwyler, H., QCD: low temperature expansion and finite -size effects. Proceedings of the Seillac Conference to appear in April 88.

Maurice, E., Spite., F., and Spite , M., Astron. Astrophys**132, 278** , 1984 (Li-7 / Li -6 ratio in old stars.)

Meneguzzi,M., Audouze, J., and Reeves , H., 1971, Astr. Ap., **15**, 337.

Michaud, G., Ap. J. 302, 650, 1986

Molaro. P. Astr. Ap. 183, 394 1987 (boron in Pop II)

Montmerle 1977 thesis Université de Paris.

Rebolo, R., Beckman,.J., and Molaro, P., Astron, Astrophys. 172 L17 1987 (Li in G 64 -12)

Rebolo, R,, Molaro, P., Abia, C., and Beckman, J.E., Astron. Astrophys., 193, 193-201, 1988. (Be-9 in Pop II)

Reeves, 1971, p . 256 , American Physical Society Meeting , Porto Rico Dec 1971

Reeves . H., Ann Rev Astron Astrophys. **12**,437, 1974

Reeves, H. Varenna School "Confrontations between Observations and Theories in Cosmology" July 1987

Reeves, H., Delbourgo-Salvador, P , Audouze, J. , and Salatti, P. to appear in

Walker , T.P., Mathews, G.J. and Viola, V.E. 1985, Ap.J., **299**,, 745.

Wagoner, R. V., Fowler, W. A., and Hoyle, F., Ap.J. <u>148</u>, 3, 1967.

Witten E., 1984 Phys. Rev. **D30** 272

Yang , J., Turner, M.S., Steigman, G., Schramm, D.N., and Olive, K, 1984, Ap.J. , **281,** 493

RECENT RESULTS FROM THE e^+e^- COLLIDERS TRISTAN AND CESR

Giancarlo Moneti

Syracuse University, Syracuse, New York, U.S.A.

1. Preliminary Results from the Highest Energy Run at TRISTAN

Thanks to a seminar that Kevin Sparks gave in Syracuse a few days ago, I relate to you some preliminary results of the AMY collaboration from data taken at the highest e^+e^- center-of-mass energy (56 GeV) recently attained by the TRISTAN collider at KEK, Japan. Having collected 6 pb^{-1}, they looked at $\gamma\gamma$, e^+e^-, $\mu^+\mu^-$ and $\tau^+\tau^-$ final states. The cross sections and forward-backward asymmetries are all in agreement with the electroweak model with $M_Z = 92.5\ GeV$ for the boson mass and a lower limit $\Lambda_\pm = 70\ GeV$ for the electromagnetic form factors.

The big questions are, of course, "is the top quark there?" or "is a fourth family bottom quark b' there?". The cross section for annihilation into hadrons is three standard deviation below the one expected above threshold for open top production, but it is compatible with production of an additional charge 1/3 b' quark. AMY however looked also for the excess of low thrust events expected when running just above threshold for production of new heavy hadrons. They found none and set the following lower limits (90% CL) for the masses: $m_{b'} > 25.8\ GeV$ and $m_t > 27.7\ GeV$.

2. Status of Measurements of the K-M matrix elements V_{ub}, V_{cb}

It is very appropriate for this conference to review the status of our knowledge of the small elements of the Kobayashi-Maskawa flavour-mixing matrix. On one hand the mixing of generations of quarks and leptons is one of the few clues we have to understand the reason why more than one generation exists at all, and this is one of the most fundamental problems of today's physics; on the other hand the small matrix elements V_{ub} and V_{td} are the ones that carry the phase that, according to the 'standard model' of elementary particles, is the cause of violation of CP symmetry. This, in turn, in the 'standard model' of cosmology, is the cause of the excess of nucleons and electrons over their antiparticles to make the world the way it is.

$|V_{ub}|$ is measured indirectly by combining the measurement of $|V_{cb}|$ with the measurement of the ratio $|V_{ub}/V_{cb}|$. We get an estimate of $|V_{cb}|$ by measuring the semileptonic decay branching ratio B_{sl} of B mesons and their average lifetime τ_B. In fact there are very good reasons to believe that the 'spectator model' is accurately valid for the semileptonic dacays, so that we can write for the semileptonic B meson decay rate:

$$\Gamma_{sl}^B = \frac{B_{sl}}{\tau_B} = \Gamma_0 \eta_{QCD} \left[0.45|V_{cb}|^2 + |V_{ub}|^2\right] \qquad where: \qquad \Gamma_0 = \frac{G_F^2 m_b^5}{192\pi^3},$$

0.45 is a phase-space factors and η_{QCD} is a calculated QCD correction factor.

79

M. Caffo et al. (eds.), Astronomy, Cosmology and Fundamental Physics, 79–82.

Combining the world averages[1] $B_{sl} \equiv B([B_u \text{ or } B_d] \to X\ell\bar{\nu}) = 0.114\pm0.005$ and $\tau_B = 1.18\pm0.14$ ps (this is an average over an unknown mixture of B_d, B_u, B_s and bottom baryons) and neglecting $|V_{ub}|^2$ with respect to $0.45|V_{cb}|^2$, we get an estimate: $|V_{cb}| = 0.05\pm0.01$.

The lowest upper limit for $|V_{ub}/V_{cb}|$ is still the one derived by CLEO, CUSB and ARGUS from the excess of leptons beyond the end point for $b \to c$ decay in the semileptonic B decay.[1] There is however uncertainty in modelling the quark-antiquark bound state: the upper limit varies between 0.10 and 0.17 depending on the model chosen.

$|V_{ub}/V_{cb}|$ can more reliably be related to the branching ratio for B decay into charmless two-body final states.[2] CLEO has preliminary results from a recent run at the $\Upsilon(4S)$ that produced $\sim$ 270,000 $B^\pm$ and 210,000 B^0 decays, about 3.5 times the previous CLEO and ARGUS samples. No charmless decays where found and some of the upper limits for the respective branching ratios are listed in the following table.

Final State	No. of events 90% CL limits	Detection Efficiency	Branch. Ratio 90% CL limit	Theory[2]
$\pi^+\pi^-$	< 8.9	0.45	0.9×10^{-4}	$21 \times 10^{-4}|V_{ub}/V_{cb}|^2$
$\rho^0\pi^-$	< 8.9	0.22	1.5×10^{-4}	$22 \times 10^{-4}|V_{ub}/V_{cb}|^2$
$p\bar{p}$	< 3.9	0.45	0.4×10^{-4}	
$\Delta^0\bar{p}$	< 7.3	0.083	3.3×10^{-4}	
$\Delta^{++}\bar{p}$	< 8.3	0.25	1.3×10^{-4}	
$\Delta^{++}\bar{\Delta}^{--}$	< 3.2	0.12	1.3×10^{-4}	

From the $\pi^-\pi^+$ decay mode we can deduct $|V_{ub}/V_{cb}| \leq 0.21$ (90% CL), still inferior to the limit obtained from the end point of the lepton spectrum from semileptonic B deacays.

The proof that $V_{ub} \neq 0$ would come of course from actually detecting a decay mode in charmless final state. ARGUS[3] recently showed evidence for B decay into $p\bar{p}\pi^+(\pi^-)$ with the p and $\bar{p}$ going in approximately opposite directions and found the following branching ratios for this specific kinematical configuration: $B(B^+ \to p\bar{p}\pi^+) = (5.2\pm1.4\pm1.9) \times 10^{-4}$ and $B(B^- \to p\bar{p}\pi^+\pi^-) = (6.0\pm2.0\pm2.2) \times 10^{-4}$. CLEO analized both the 1985 sample of B's (comparable to ARGUS') and the just mentioned 1987 sample that is not only richer but has also a better momentum and ionization density resolution. Preliminary results, using the same selections used in the ARGUS analysis show no signal and give an upper limit (at 90% CL) of 3.6×10^{-4} for the sum of the two decay branching fractions just mentioned, in clear disagreement with the ARGUS result. Even relaxing the requirement that the p and $\bar{p}$ be approximately anticolinear or substituting other requirements to reduce the combinatorial background, no signal was found in the these two final states. It seems that more data will be needed to find a non-zero value of V_{ub}.

3. Status of $B^0\overline{B}^0$ Mixing

The phenomenon of 'mixing', i.e. the spontaneous transformation of a particle into its own antiparticle, until recently had been observed only in the $K^0\overline{K}^0$ system. Besides being a rare and peculiar phenomenon, it provides, in the case of the $B^0\overline{B}^0$ system, information on the K-M matrix elements V_{td} (and V_{ts}) and on the mass m_t of the top quark. The transition $B^0 \leftrightarrow \overline{B}^0$ is due to a 'box diagram' with two W boson and two quark internal lines. The heaviest quark (presumably the top quark) dominates over the exchange of lighter quarks. The parameter that determines the amount of mixing is given by the expression:

$$x_q \equiv \frac{\Delta M}{\Gamma} = \frac{32\pi}{3} \frac{|V_{tb}V_{tq}^*|^2}{|V_{cb}|^2} B_B f_B^2 \eta_2 \frac{m_t^2}{m_b^4} \Phi\left(\frac{m_t}{M_W}\right)$$

where the unknowns are the K-M matrix element V_{tq} (with $q = d$ or s, V_{tb} is expected to be very close to 1) and the mass of the top m_t. $\Phi(a)$ is a known analytical function and the other factors can be more or less accurately estimated, a reasonable choice being, e.g., $B_B f_B^2 \eta_2 = (0.10 \; GeV)^2$ and $m_B = 5.0 \; GeV$.

One can determine x_q by producing $B\overline{B}$ pairs and measuring how often they end up as two B's or two $\overline{B}$'s. When the $B\overline{B}$ system is produced in a pure $J^P = 1^-$ state, as in the case of the reaction $e^+e^- \to \Upsilon(4S) \to B\overline{B}$, one finds:

$$y_q \equiv \frac{N(B_q B_q) + N(\overline{B}_q\overline{B}_q)}{N(B_q\overline{B}_q)} = \frac{x_q^2}{2 - x_q^2}.$$

One way of measuring y_q is to observe the numbers of same sign and of opposite sign lepton pairs produced in the semileptonic decay of the two B mesons. Until recently evidence of $B^0\overline{B}^0$ mixing was available only from the higher energy colliders PEP and Sp$\overline{p}$S where B^0's are an unknown mixture of B_d's and B_s's. Last year, however, the ARGUS collaboration observed for the first time clear evidence of $B_d\overline{B}_d$ mixing[4] and found $y_d = 0.21\pm0.08$, larger than theoretically expected.

CLEO has recently analyzed the richer sample already mentioned and finds the following preliminary result: $y_d = 0.182\pm0.055\pm0.056$.[5] The fairly large $B_d\overline{B}_d$ mixing is thus confirmed and that implies a very large $B_s\overline{B}_s$ mixing because very reliable theoretical estimates predict $x_s/x_d > 4.8$.[6] It is thus very important to produce $B_s\overline{B}_s$ in quantities large enough to measure their mixing. CLEO collected 110 pb^{-1} at the $\Upsilon(5S)$ that should decay into $B_s\overline{B}_s$, besides than into $B_d\overline{B}_d$ and $B_u\overline{B}_u$. This sample hopefully will detect some decay modes of the B_s and allow to plan a longer exposure to measure $B_s\overline{B}_s$ mixing with CLEO II.

4. Future of B Physics

The most immediate and high expectations for better measurements on B physics come from Cornell. CLEO II is about to be installed and the plans are to start running in March 1989. It will be substantially superior to CLEO because of the CsI electromagnetic shower detector, inside a new superconducting coil, that will cover 95% of the solid angle and have an r.m.s. energy resolution of 1.7% at 5 GeV, increasing to 4% at 100 MeV; the position resolution will range from 5 to 10 mm in the same energy range. This shower detector will have high detection efficiency for π^0's, thus greatly improving the ability to reconstruct B final states. The higher magnetic field (1.5 T) will improve the already excellent momentum resolution in the tracking chambers (already in use) to $\delta p = \sqrt{(0.005)^2 + (0.0015p(GeV))^2}$. An additional 'micro vertex detector' drift chamber with an inner radius of 3.5 cm should allow selection of charm particle decay vertices, thus suppressing the combinatorial background in reconstructing B decay final states.

During the two month run at the $\Upsilon(5S)$ at the beginning of this year, CESR delivered an average luminosity of 18 pb^{-1}/week. After taking into account CLEO inefficiencies, shutdowns and the need for taking data at a center-of-mass energy just below the threshold for $B\bar{B}$ production in order to be able to perform continuum background subtraction, one can still estimate that a million B's per year will be generated in CLEO II when running at the $\Upsilon(4S)$ resonance. Furthermore, the CESR staff has a luminosity upgrade plan that should increase the effective luminosity by a factor between 3 and 10 over the next few years. If this improvement will be successful CLEO II may get tantalizing close to measuring CP violation effects in the bottom sector.

REFERENCES

1. For reviews and original references, see B. Gittelman and S. Stone, in "High Energy Electron Positron Physics", ed. by A. Ali and P. Soding, World Scientific, Singapore, (1988); E.H. Thorndike and R.A. Poling, *Physics Rep.* **157** (1988).

2. M. Bauer and B. Stech, *Zeit. f. Phys.* **C34** (1987) 103.

3. W. Schmidt-Parzefal, Proc. of the 1987 Intern. Symposium on Lepton and Photon Interactions, Hamburg, 1987 and H. Albrecht et al., DESY preprint 88-056 (May 1988).

4. H. Albrecht et al., *Phys. Lett.* **192B** (1987) 245.

5. This numerical result became available a few weeks after the conference, it is included here for completeness.

6. See e.g. G. Altarelli, CERN-TH. 4896/87, unpublished.

THE LARGE-SCALE DISTRIBUTION OF GALAXIES

Margaret J. Geller
Harvard-Smithsonian Center for Astrophysics
60 Garden St.
Cambridge, MA 02138 USA

ABSTRACT: Six strips of the CfA redshift survey extension are now complete. The data continue to support a picture in which galaxies are on thin sheets which nearly surround vast low density voids. In this and similar surveys the largest structures are comparable with the extent of the survey. Voids like the one in Boötes are a common feature of the large-scale distribution of galaxies. The observed structure presents a serious challenge for models.

Within the context of the observed structure in redshift space, we discuss the issue of "fair samples" of the galaxy distribution. We examine statistical measures of the galaxy distribution including the two-point correlation functions. We comment on limits on large-scale flows and on the distribution of groups and clusters of galaxies in the survey.

I. INTRODUCTION

Current observational constraints on the origin and evolution of large-scale structure in the universe yield a puzzling picture. Dynamical studies of individual galaxies, groups, and clusters show that at least 90% (if $\Omega \simeq 0.1$) of the matter in the universe is dark. However, constraints on the relative distribution of dark and light-emitting matter are weak, particularly on large scales.

Maps of the large-scale distribution of galaxies reveal coherent structures comparable in scale with the extent of the surveys. Structures on a scale of $100h^{-1}$ Mpc are a common feature of the distribution of galaxies and direct limits are nonexistent on larger structures. Discovery of much larger structures could challenge the fundamental homogeneity and isotropy assumption of the standard cosmological model. Faith in this assumption is based largely on the remarkable uniformity of the microwave background which contrasts sharply with the lumpy galaxy distri-

M. Caffo et al. (eds.), Astronomy, Cosmology and Fundamental Physics, 83–103.
© 1989 by Kluwer Academic Publishers.

bution. This contrast is a strong constraint on models for galaxy and large-scale structure formation.

The marriage of particle physics and cosmology underlies many of the models which partially account for the salient features of the observations. The results of N-body simulations of the gravitational amplification of primordial density fluctuations depend upon the properties of the seemingly ubiquitous dark matter. Particle physics provides a variety of candidate particles for the collisionless dark matter. The only one of these candidates *known* to exist is the neutrino — and its mass may not be in a cosmologically interesting range.

Models based on collisionless cold dark matter (the axion is a serious contender) are now favored as an explanation for large-scale structure. One troublesome feature of the models which resemble the data is that they require an assumption that galaxies are biased tracers of the matter distribution. The prescription for biased galaxy formation does not yet have a firm physical basis. These models are consistent with the observational constraints on the distribution of dark matter and with the limits on fluctuations in the microwave background. They also succeed in producing the appropriate small and intermediate scale galaxy clustering. However, they fail to produce the strikingly coherent very large-scale observed structures. In particular, the frequency of 5000 km s^{-1} voids appears to be too low in the models.

Redshift surveys are not yet extensive enough to yield constraints on the *typical* properties of the galaxy distribution on scales $\gtrsim$ 10 h^{-1} Mpc. This lack of a fair sample complicates comparisons of the data with models. Larger, deeper surveys require a significant amount of time on 4-m class telescopes. Limited access to such telescopes may be the major obstacle to progress in this field.

Here I outline the observational situation along with some of the important issues raised by the data. Section II is a discussion of redshift surveys. Section III is a review of the recent results of the Center for Astrophysics survey. Implications of the survey and prospects for the future are the subjects of Sections IV and V.

II. THE DESIGN OF REDSHIFT SURVEYS

One of the many goals of redshift surveys is a test of the assumption of large-scale homogeneity and isotropy. Of course, these observations provide a direct test only of the distribution of light-emitting matter. However, as discussed in Section IVc, supplementing redshift surveys with distance measurements may provide interesting constraints on the dark matter distribution. Over the past few years, each new observational approach to redshift surveys has uncovered unexpectedly large structures. The 21-cm surveys by Giovanelli and Haynes (1985) and Giovanelli *et al.* (1986) defined the Perseus-Pisces chain. Optical redshift surveys in deep probes uncovered the void in Boötes (Kirshner, Oemler, Schechter, and Shectman 1981, 1986;

KOSS hereafter). The completion of the first slice in the Center for Astrophysics redshift survey extension (de Lapparent, Geller, and Huchra 1986; LGH) suggested that galaxies are on thin, sharply defined surfaces which surround (or nearly surround) vast voids. This sequence of recent discoveries proves the power of redshift surveys. However, the volumes completely surveyed are at best comparable with the scale of the largest known inhomogeneities (Sections III and IV).

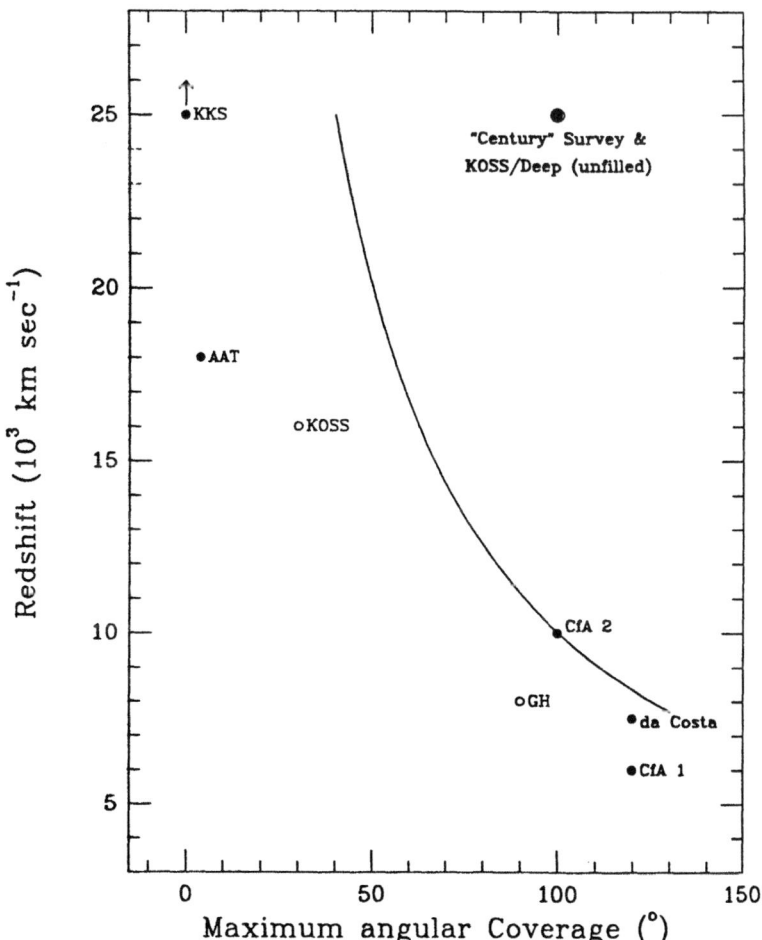

Figure 1: Depth and maximum angular extent of some redshift surveys. The curve is $\theta D = 175h^{-1}$ Mpc.

The sensitivity of redshift surveys to the largest high contrast structures is determined by the depth and maximum angular extent (not necessarily by the solid angle) of the survey. The largest structures which can be well-defined by a survey have a size $\sim D\theta$ where D is the effective depth of the survey and θ is the maximum angle subtended by the sample. In other words, a long thin (but not too thin) strip across the sky is a better configuration for sensitivity to large structures than a "square" patch which subtends the same solid angle.

Figure 1 is a summary of some of the existing redshift surveys. The curve is $\theta D = 175h^{-1}$ Mpc ($H_o = 100$ h km s^{-1} Mpc^{-1}; I take h = 1 wherever the direct dependence on h is unspecified). The •s are surveys which are complete to a particular magnitude limit over the entire region; the os are incomplete surveys.

In Figure 1 the envelope is roughly a line of constant effort. There is a clear message about the parameters of a survey designed to limit structures on larger scales. They must lie above the envelope to the upper right. At least two surveys in this region of the plot are in progress. One of these is a survey of a $1° \times 100°$ strip at a delination of $\delta = 29.5°$ to a limiting magnitude $m_{B(0)} = 17.5$ (effective depth $z \simeq 0.1$). This survey will be complete to the limiting magnitude over the entire region (Geller et al. 1988). The second survey to about the same depth will cover $\sim 25\%$ of a ~ 1000 deg^2 strip in the southern hemisphere (Kirshner et al. 1988).

Several other groups are undertaking large redshift surveys. These include a southern survey (da Costa et al. 1988), the extension of the CfA survey (Huchra et al. 1992), and deeper probes which cover smaller solid angles (KOSS 1986; Peterson et al. 1986; Koo, Kron and Szalay 1986).

III. THE LARGE-SCALE DISTRIBUTION OF GALAXIES

In 1978, Joêveer, Einasto, and Tago (1978) suggested that the large-scale distribution of galaxies has a "cellular" pattern in which rich clusters are connected by "filamentary" structures. The data at that time were incomplete and could only hint at such structure. The discovery of the void in Boötes (KOSS) and the 21-cm survey of the Pisces-Perseus chain (Haynes and Giovanelli 1986; Giovanelli et al. 1986) soon lent support to this picture. However, these first surveys gave no clear message about the frequency of the structures.

It is becoming increasingly clear that large-scale features in the galaxy distribution are ubiquitous. The deep surveys of Koo, Kron, and Szalay (1986) indicate that voids are common even at high redshift. Because these surveys are one-dimensional, the constraints on the sizes of the voids are poor. The AAT surveys also reveal voids along with thin structures perpendicular to the line-of-sight (Peterson et al. 1986). The continuing Arecibo survey delineates nearby voids and appears to support the interpretation of the Pisces-Perseus chain as a filamentary structure.

The extension of the CfA redshift survey, the primary subject of this section, indicates that bright galaxies are distributed on thin *sheets* — two-dimensional structures— which surround (or nearly surround) vast voids. Recent completion of a southern hemisphere survey (da Costa *et al.* 1988) gives some support to this picture although the survey is not sufficiently deep or dense to be directly comparable with the recent CfA results. The 21-cm data (Giovanelli and Haynes 1986) also reveal sheet-like structures in the Perseus-Pisces region. The message of all surveys is now clear: large structures are a *common* feature of all surveys big enough to contain them.

The goal of the Center for Astrophysics redshift survey extension is to measure redshifts for all galaxies in a merge of the Zwicky *et al.* (1961-1968) and Nilson (1973) catalogs which have $m_{B(0)} \leq 15.5$ and $| b_{II} | \geq 40$ °. There will be about 15,000 galaxies in the complete survey; more than 9,000 of these already have measured redshifts. About 2,500 of the galaxies with measured redshifts lie in three "slices" where the survey is now complete: (1) a slice with $8^h \leq \alpha \leq 17^h$ and $26.5° \leq \delta < 32.5°$ (LGH), (2) a slice with $8^h \leq \alpha \leq 17^h$ and $32.5° \leq \delta < 38.5°$, and (3) a slice with $8^h \leq \alpha \leq 17^h$ and $38.5° \leq \delta < 44.5°$. In the southern Galactic hemisphere 3 slices covering $6° \leq \delta > 12°$ and $18° \leq \delta < 30°$ with $0^h \leq \alpha \leq 4^h$ and $20^h \leq \alpha \leq 24^h$ are also complete. More than 60% of the redshift were measured with the 1.5-meter telescope at Mt. Hopkins. The mean external error in the redshift measurements is ~30 km s^{-1} .

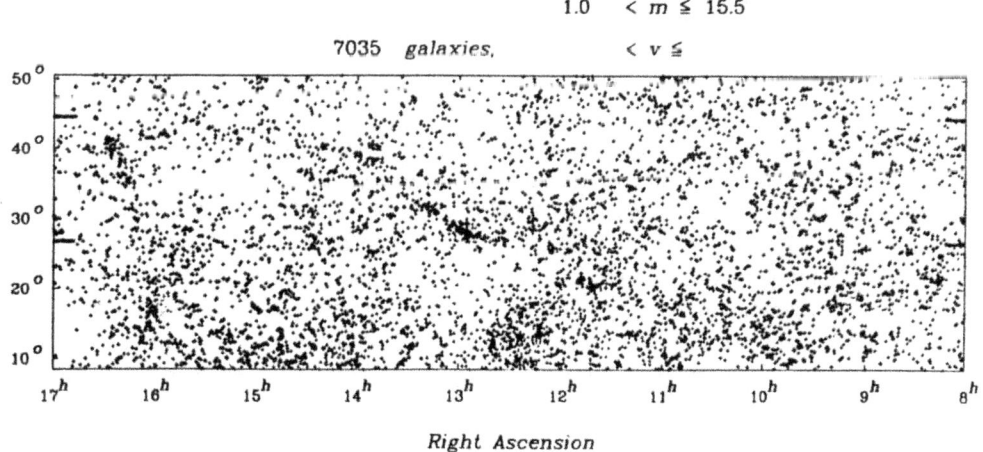

Figure 2: Positions of galaxies in the Zwicky-Nilson merge with $m_{B(0)} \leq$ 15.5. The bold ticks indiciate the limits of the complete strips.

Figure 2 shows the positions of 7031 galaxies in the Zwicky-Nilson merge which have $m_{B(0)} \leq 15.5$ and $8^h \leq \alpha \leq 17^h$ and $8.5° \leq \delta < 50.5°$. The grid is Cartesian in α and δ. The deficiency of galaxies west of 9^h and east of 16^h is caused by Galactic obscuration. The bold ticks indicate the limits of the three complete strips of the survey. The Coma cluster is the dense knot at $\alpha = 13^h$, $\delta = 30°$.

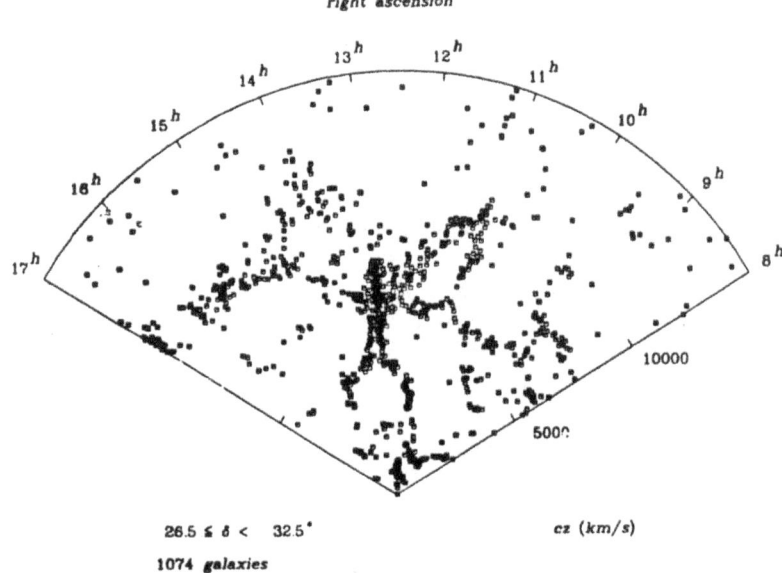

Figure 3: Observed velocity versus right ascension for the survey strip centered at $\delta = 29.5°$. The strip is 6° wide in declination. Only galaxies with velocities $\leq 15,000$ km s^{-1} are shown.

Figure 3 is a plot of the observed velocity versus right ascension for the strip centered at 29.5° (LGH): the strip is 6° wide in declination. The plot includes only the 1067 galaxies with redshifts $\leq 15,000$ km s^{-1}. A galaxy with the characteristic luminosity M* ($\simeq$ -19.4) is at 10,000 km s^{-1} in this survey. Nearly every galaxy in this slice lies in an extended thin structure. The boundaries of the low density regions are remarkably sharp. Several of these voids are surrounded by thin structures in which the inter-galaxy separation is small compared with the radius of the empty region. The edges of some of the largest structures may be outside the right ascension limits of the survey. The only pronounced velocity finger in this slice is the Coma cluster at $\sim 13^h$.

This first slice alone demonstrates that the thin structures in the distribution of galaxies are cuts through two-dimensional sheets, *not* one-dimensional filaments. If the ~ 150 Mpc long structure which extends across the entire survey (from 9^h to 16^h between 7,000 km s^{-1} and 10,000 km s^{-1}) is a filament, a thin linear structure should be visible on the sky. This statement is particularly strong because the structure lies near the survey limit. The required filamentary structure is absent from Figure 2. Because structure on the sky can be caused by patchy obscuration and/or by inhomogeneities in the galaxy catalog, structure on the sky cannot provide complete proof (or disproof) of the filamentary nature of a structure in redshift space. A second argument against the filamentary nature of the structures in Figure 3 is that several thin, elongated structures lie in this single survey slice: the intersection of a slice with a three-dimensional network of filaments is *a priori* unlikely to be a two-dimensional network of filaments. Of course, we could have been lucky (or unlucky).

A geometric structure in which thin sheets surround or nearly surround voids accounts for the data. Examples include "bubble-like" and "sponge-like" geometries. I will use the word "bubble" to convey the image of a structure dominated by thin sheets and holes. Note that the "bubbles" are not necessarily round. In this picture the 150 Mpc "filament" is made up of portions of adjacent "bubbles" and the richest clusters like Coma lie in the interstitial regions (where several "bubbles" come together).

In this interpretation of the data, we assume that the maps are similar in redshift space and in real physical space. Both the cold dark matter (White *et al.* 1987) and the adiabatic models (Melott 1987) suggest that this assumption is reasonable (but see Kaiser 1987).

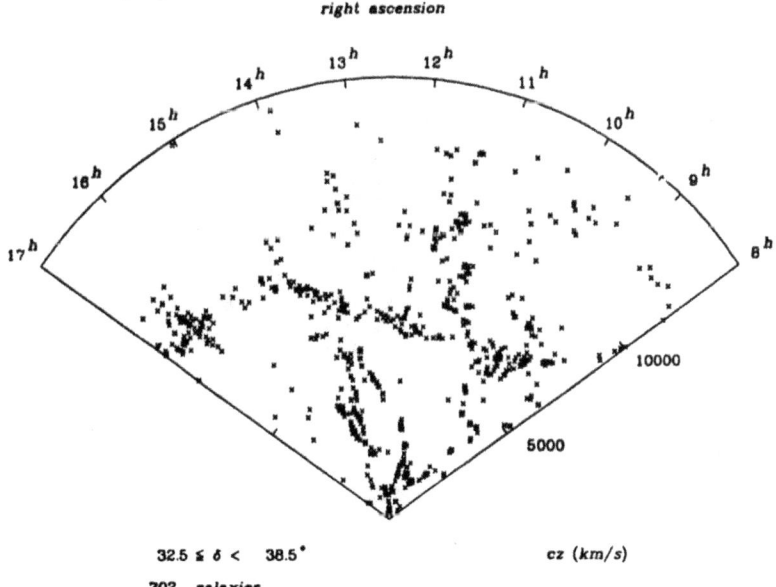

Figure 4: Same as Figure 3 for the slice centered at $\delta = 35.5°$.

Maps of the adjacent slices support the *qualitative* picture suggested by the first slice. Figure 4 shows the slice centered at 35.5°, just to the north of the slice in Figure 3. Once again the galaxies are in thin structures. Furthermore these structures are natural extensions of the structures in Figure 3; the structures are highly correlated in the two slices. The two closed structures at $\sim 11^h$ (9000 km s^{-1} $\lesssim$ cz $\lesssim$ 11,000 km s^{-1}) and at 14^h (7000 km s^{-1} $\lesssim$ cz $\lesssim$ 11,000 km s^{-1}) are not so clearly delineated in Figure 4 as in Figure 3. Sampling of these structures is probably affected by variations in the limiting magnitude of the galaxy catalog.

Figure 5 shows the cone diagram for the two slices taken together. The ⊔s are the data in Figure 3; the ×s are the data in Figure 4. The structures in the two slices are obviously highly correlated. The distribution remains remarkably inhomogeneous with low density voids outlined by thin structures. Because the surfaces are curved or inclined relative to the plane of the slice, the structures are thicker here than in Figures 3 and 4. The largest low density region in the survey is located between 13^h20^m and 17^h with 4000 $\lesssim$ cz $\lesssim$ 9000 km s^{-1} . The diameter of this void is $\sim$ 5000 km s^{-1} or 50 Mpc in the absence of large-scale flows. The underdensity in the region ($\lesssim$ 20% of the mean) and the scale of the structure are comparable with the corresponding parameters for the void in Boötes.

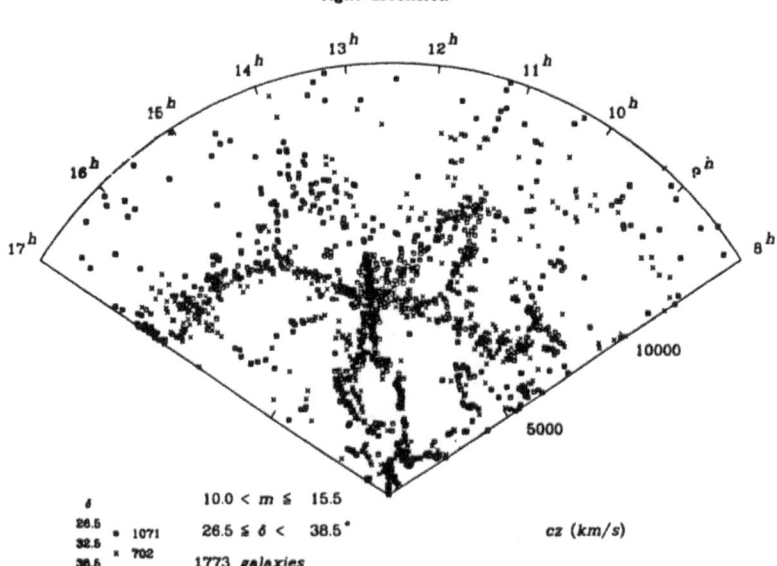

Figure 5: Observed velocity versus right ascension for the 12°-wide strip (the sum of the slices in Figures 3 and 4) centered at $\delta = 32.5°$.

Note that voids in the galaxy distribution are *not empty*: they are regions which are underdense relative to the global mean. The galaxies inside the large void have normal properties and their infra-red Tully-Fisher distances put them at the relative distances indicated by their redshifts (Geller *et al.* 1988). These galaxies may form a tenuous structure which would not be detected in a sparse survey like the KOSS survey of Boötes.

Figure 6 shows the cone diagram for the third slice centered at $\delta = 41.5°$. At first glance this slice gives a somewhat different visual impression from the first two. The reason for the difference is that some of the surfaces lie in this slice; in particular, a portion of the structure surrounding the largest void is nearly in the plane of this slice and appears diffuse. Comparison of this slice with Figure 5 continues to support the large-scale coherence of the structures.

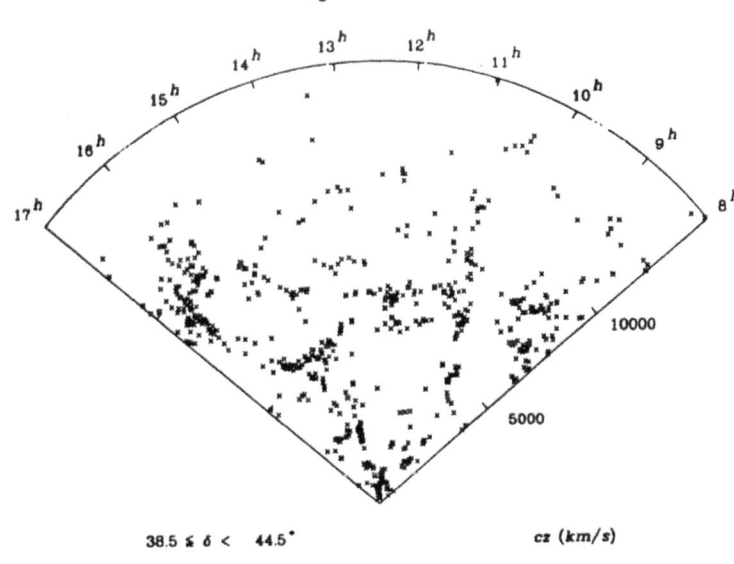

Figure 6: Same as Figure 3 for the slice centered at $\delta = 41.5°$.

The significance of structure in a survey can be evaluated by comparing the observations with "galaxies" randomly distributed in the same volume. Figure 7b shows 2483 points distributed in an 18° slice. Like the data, this simulation is magnitude limited. Figure 7a shows the data. The difference is obvious. If the data were overlaid on the simulation, the number of random points inside a particular void would yield a rough measure of the significance of the structure. More than a hundred points would fall inside the largest of the voids.

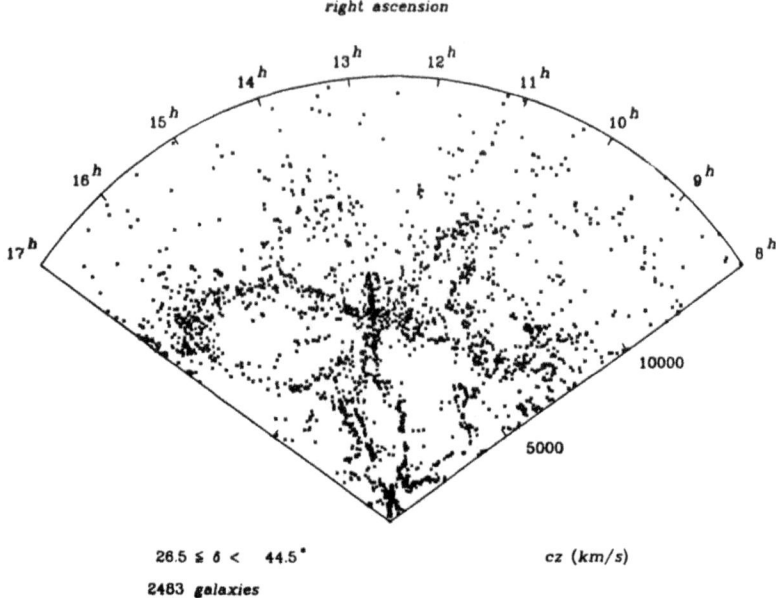

right ascension

26.5 ≤ δ < 44.5°

2483 galaxies

cz (km/s)

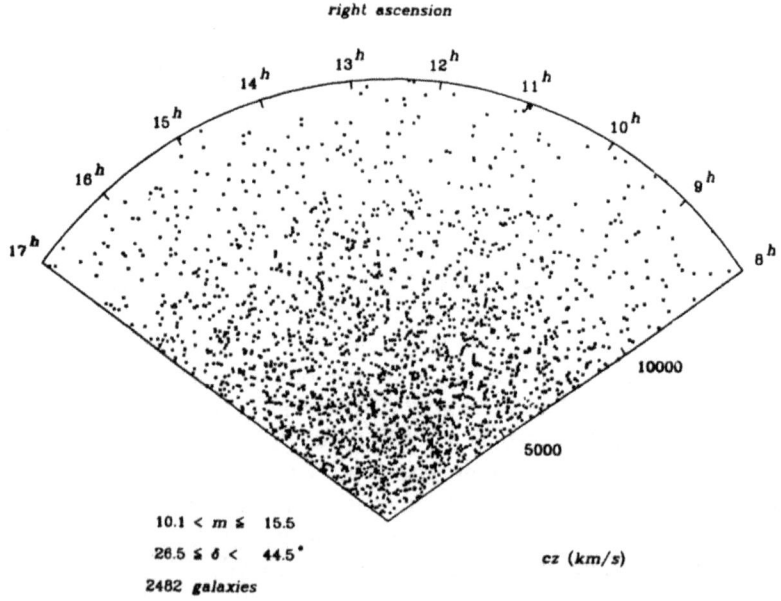

right ascension

10.1 < m ≤ 15.5

26.5 ≤ δ < 44.5°

2482 galaxies

cz (km/s)

Figure 7a: The observed distribution of galaxies in the 18°-wide slice centered at 35.5° (this slice is the sum of the slices in Figures 2.3. and 5). **7b:** A magnitude limited sample of 2483 randomly distributed points (Courtesy of V. de Lapparent).

IV. IMPLICATIONS OF LARGE-SCALE REDSHIFT SURVEYS

a). "Fair" Samples

An important message of these surveys is that the largest inhomogeneities are comparable with the size of the sample. None of the existing redshift surveys are thus large enough to be "fair".

Perhaps even more sobering is that the largest inhomogeneities we detect are the largest we *could* detect within the limits set by the extent of the survey — we have few, if any, reliable direct limits on larger structures in the distribution of light-emitting matter. The size of the inhomogeneities relative to the volume of the surveys may underlie unexplained variations in traditional statistics of the galaxy distribution like the luminosity function (Schechter 1976; KOSS 1983; Bean *et al.* 1983; Davis and Huchra 1982) and the two-point correlation function at large scale (Groth and Peebles 1977; Davis and Peebles 1983; Kirshner, Oemler, and Schechter 1979; Shanks *et al.* 1983). When the inhomogeneities are large compared with the sample volume, mean quantities are not well-defined.

The two strip surveys to $z \simeq 0.1$ (Geller *et al.* 1988; KOSS 1988) discussed at the end of Section I could provide 10% measurements of the mean galaxy density. Of course, if they uncover yet larger structures with high contrast, even larger surveys will be necessary to meet the limit. We do not yet know how large a survey is necessary for a "fair" sample.

b). Correlation Functions

The domination of the sample by large-scale coherent structures and the related 25% uncertainty in the mean density imply that the two-point correlation function is more poorly constrained than previously thought. Figure 8 shows the two-point correlation function $\xi(s)$ where

$$s = \frac{(V_i^2 + V_j^2 - 2V_iV_j\cos\theta_{ij})^{1/2}}{H_o} \tag{1}$$

for the 12° slice with declination between 26.5° and 38.5° (Figure 5). Here V_i and V_j are the velocities of two galaxies separated by θ_{ij} on the sky and H_o is the Hubble constant. We make no correction for the r.m.s. pairwise peculiar velocities of $\lesssim 350$ km s^{-1} (Davis and Peebles 1983; de Lapparent, Geller, and Huchra 1988). The calculation of this correlation function is not seriously affected by the presence of the Coma cluster or of other more poorly sampled clusters in the sample.

Because of the large-scale coherent structures in the sample, the weighting scheme for the calculation of the correlation function *does* affect the result substantially. The calculation of the correlation function and the weighting schemes follow prescriptions in Davis and Peebles (1983). It is probable that none of these schemes are unbiased. The subscript 1 indicates that the member of the pair is unweighted; the subscript ϕ indicates that the member is weighted by $\phi(V)^{-1}$. For these calculations the parameters of the luminosity function are $M^* = -19.15$, $\alpha = -1.2$, and $\phi^* \simeq 0.025 \ \mathrm{Mpc}^{-3} \ \mathrm{mag}^{-1}$ (de Lapparent, Geller, and Huchra 1988).

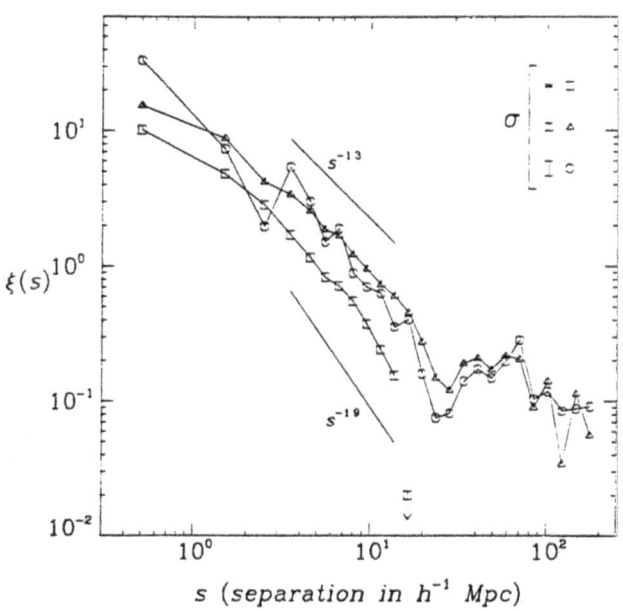

Figure 8: The two–point correlation function $\xi(s)$ for the sample in Figure 5. The symbols denote $\xi_{11}(\sqcup)$, $\xi_{\phi\phi}$ ($\triangle$), and $\xi_{1\phi}$ (○). Note that the amplitude varies by a factor of two among these estimators.

The differences among the estimators in Figure 8 are symptoms of the lack of a fair sample. In the absence of biases introduced by large-scale inhomogeneities, all the estimators should yield the same result to within the statistical noise shown by the error bars at the left of the Figure. The elongated structure at $\sim 10,000 \ \mathrm{km \ s}^{-1}$ increases the amplitude of the correlation function for estimates weighted inversely with $\phi(V)$ relative to the unweighted estimate. Why? This coherent structure contains about a thousand galaxies, nearly half of the sample. The selection function $\phi(10,000 \ \mathrm{km \ s}^{-1}) = 0.01 \ \phi \ (1000 \ \mathrm{km \ s}^{-1})$. The distribution appears more highly correlated when this structure is more heavily weighted. (Note that in the CfA survey to $m_{B(0)} = 14.5$, the opposite effect occurs because the most dense structure, the Local Supercluster, is nearby; see de Lapparent, Geller, and Huchra 1988).

We fit the data in the range $3.5 \leq s \leq 9.5h^{-1}$ Mpc to the standard power law form

$$\xi(s) = \left(\frac{s_o}{s}\right)^\gamma \tag{2}$$

and find that $1.3 \leq \gamma \leq 1.9$ and $5 \leq s_o \leq 12$ h^{-1} Mpc. On scales larger than 20 h^{-1} Mpc the correlation function is indeterminate because its amplitude is comparable with the uncertainty in the mean density. For an average of the estimators in Figure 12, we obtain

$$\gamma = 1.5 \tag{3}$$

and

$$s_o = 7.5h^{-1} \text{ Mpc.} \tag{4}$$

These values agree with the one obtained by Davis and Peebles (1983) for the 14.5 CfA sample (see their Figure 1).

Calculation of the correlation function $\xi(r_p, \pi)$ (Davis and Peebles 1983) where

$$r_p = \frac{V_i + V_j}{H_o} \tan\frac{\theta_{ij}}{2} \tag{5}$$

and

$$\pi = V_i - V_j \tag{6}$$

is a method of examining the distortions in redshift space caused by peculiar velocities. The solid contours in Figure 9 show $\xi(r_p, \pi)$ for the data in Figure 5. All galaxies within 3° of the center of the Coma cluster have been removed from the sample. The weighting scheme is the same as for ξ_{11}. The dashed curves are the expectation for undistorted Hubble flow. The intercept of the $\xi = 1$ contour with the abscissa gives r_{po} and the mean radius of the $\xi = 1$ contour is s_o. The values of these quantities are underestimated relative to calculations weighted by the inverse selection function, ϕ^{-1}. Note that the elongation along the π axis occurs only for $r_p \lesssim 2h^{-1}$ Mpc. The elongation for the $\xi = 1$ contour implies an r.m.s. pairwise peculiar velocity of $\sim$450±150 km s^{-1} in agreement with analyses of other catalogs (Davis and Peebles 1983; Bean et al. 1983). The r.m.s. distortions are also consistent with expectations based on an analysis of groups of galaxies in the 12° slice (Ramella, Geller, and Huchra 1988). In fact, removal of the groups from the survey removes the distortion of the $\xi(r_p, \pi)$ contours (Ramella, Geller, and Huchra 1988).

96

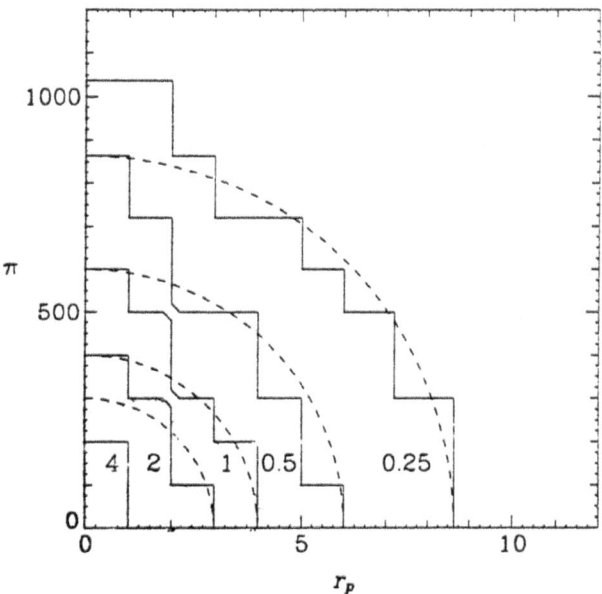

Figure 9: Contours of constant $\xi(r_p, \pi)$ (solid lines) in the plane defined by the projected separation perpendicular to the line-of-sight, r_p (in h^{-1} Mpc), and parallel to the line-of-sight, π (in km s^{-1}). The dashed lines are the expected pattern for pure Hubble flow.

Inclusion of the Coma cluster in the sample makes a substantial change in $\xi(r_p, \pi)$; the distortion along the π direction then extends to $\sim 10 h^{-1}$ Mpc (as in the analysis of the 14.5 sample which is influenced by both Virgo and Coma — see Davis and Peebles (1983)). These distortions are introduced by the virialized cores of the rich clusters, not by large-scale coherent departures from uniform Hubble flow.

c). Coherent Flows

The issue of large-scale coherent flows is of profound importance for the interpretation of redshift survey observations. Lynden-Bell *et al.* (1988) claim a ~ 600

km s^{-1} flow toward a mass concentration in the Centaurus region which they dub "The Great Attractor". Dressler (1988) recently completed a redshift survey in the region and claims that the survey supports the existence of a mass concentration of the required amplitude. However, the interpretation of these observations is complicated by the difficulty of calculating the galaxy density contrast when the mean density is poorly known. The structure in Dressler's survey is qualitatively similar to the structures in the CfA survey slices. Aaronson et al. (1988) have now made infra-red Tully-Fisher distance measurements for galaxies in the region of the Great Attractor as well as in the northern hemisphere. They confirm that there is an anomaly in the region, but the interpretation of the data is not straightforward. One of the problems is the observation that the Centaurus cluster is double along the line-of-sight. Another is the treatment of observational biases (e.g. the Malmquist (1922) bias) in samples where the inhomogeneities are large compared with the extent of the sample.

Because of the limited amount of data and the difficulty of the measurements, the relationship between large-scale coherent flows and the structure in redshift surveys remains a wide open question. However, coherent flows should, for example, be associated with large low density regions. If the matter density inside a void is low compared with the average surroundings (i.e. if the galaxy density contrast is a measure of the matter density contrast), the voids expand relative to the average cosmological flow and the structures should appear elongated in redshift space. For an isolated adiabatic void in an $\Omega = 1$ universe, the outward peculiar velocity $v_{pec} \simeq 0.2 - 0.3 \, v_H$ where v_H is the radius of the void in redshift space. For lower Ω the peculiar velocities are smaller (see Ostriker 1986). The effect of interaction between adjacent shells on peculiar velocities has not been calculated. In a large enough sample containing many voids the intrinsic spatial geometry of the voids averages out and any net elongation could be interpreted as a residual expansion.

The measurement of distances to galaxies in the structures offers a direct probe for large scale flows associated with voids. These flows are a possible discriminant among theoretical models. In the biased cold dark matter models where the matter density contrast is much smaller than the galaxy density contrast, the outflow velocities should be small. The galaxies on the edges of the voids did not move across the low density regions to their current positions; they merely lit up there. Because many spirals lie in the extended sheets, the infra-red Tully-Fisher technique (Aaronson, Huchra, and Mould 1979) can be used to obtain limits at the few hundred kilometers per second level on scales of fifty megaparsecs, within the theoretically predicted range (Geller et al. 1988). The current limit on peculiar velocities on the large shell in Figure 5 is $v_{pec}/v_H \lesssim 0.2$.

d). Comparison of the Data with Models

The net elongation of voids is one of several statistics which may be useful for comparing the data with models. In the absence of "fair" samples, one can still examine the properties of individual structures. In discussing these structures a "void" is a region where the density is less than the global average and the contrast is below some well-defined threshold; analogously, the "sheets" are regions above a threshold. Both the voids and the sheets can be characterized quantitatively.

The frequent mention of the "size" of the Boötes void is a demonstration of the power of a measure of the scale of the "largest" observed structure. The spectrum of void sizes is an important test of models; the small-scale end is a constraint on hot dark matter models (Zel'dovich 1970; Doroshkevich et al. 1980; Centrella and Melott 1983; Centrella et al. 1988) and the large-scale end is most demanding for cold dark matter models (Davis et al. 1985; White et al. 1987) and for the explosive models (Ostriker and Cowie 1981; Ikeuchi 1981; Saarinen, Dekel, and Carr 1986). Determination of the distribution of sizes of voids requires samples much larger than those currently available.

The thickness, coherence, and filling factor of the sheets provide further constraints. The FWHM of the sheets is $\lesssim 500$ km s^{-1} (de Lapparent, Geller, and Huchra 1988a). The thickness as a function of orientation with respect to the line-of-sight restricts physical models. If the sheets were collapsing pancakes, we would expect them to the thinner when they are perpendicular to the line-of-sight than when they are parallel to it. On the other hand, any internal velocity dispersion will make the sheets appear thicker when they are perpendicular to the line-of-sight.

The fraction of the survey volume filled by the coherent structures in the distribution of individual galaxies can be calculated by appropriately binning and smoothing the data. The galaxies fill $\lesssim 20\%$ of the volume and the typical separation of galaxies in the sheets is $3h^{-1}$ Mpc at the survey depth D to which M^* galaxies are included. Remarkably this density is comparable with the surface density of the structures in the deep probes discussed of Koo, Kron, and Szalay (1986).

The "uniformity" of the sheets may provide a constraint on Ω (see Peebles 1986). If $\Omega = 1$ and the distribution of galaxies marks the distribution of matter, it is unlikely that a smooth shell can persist for a Hubble time; gravity causes the galaxies to clump up and "fingers" in redshift space should be apparent. If the actual matter density contrast in the sheets is small and the voids are full of nearly uniformly distributed dark matter (with Ω close to 1), the structures could still be in the linear regime. If, on the other hand, $\Omega = 0.2$ or less (as indicated by the dynamical estimates and by analysis of the abundance of the light elements), the structure could set in early on and then just stretch with the universal expansion.

e). Groups, Clusters, and Large-Scale Structure

The extension of systems of galaxies along the line-of-sight ("fingers" in redshift space) produces distrotions of a map in redshift space relative to one in actual position space. The distortions are obvious for rich systems like the Coma cluster which have a velocity dispersion large compare with the typical FWHM of the extended sheets in the survey (Figures 2 - 7a). The distrotions caused by typical groups of galaxies with a line-of-sight velocity dispersion $\sigma_v = 200$ km s^{-1} are more subtle.

Systems of galaxies — from groups to rich clusters — in the redshift survey can be identified with a single objective algorithm (Ramella, Geller, and Huchra 1988; Huchra and Geller 1982). In the resulting catalog, the derived physical properties of "groups" overlap the properties of systems identified by Abell as "rich clusters".

Figure 10 shows the location of groups and clusters in the slice of Figure 5. Systems of galaxies clearly trace out the same structures defined by the galaxy distribution as a whole. We have not yet calculated the correlation function for the group catalog, but it will be interesting to compare the results with those for rich clusters. The relationship between individual galaxies and systems of galaxies as tracers of the large-scale metter distribution will probably only be understood when complete redshift surveys are extensive enough to include a large number of "rich" systems.

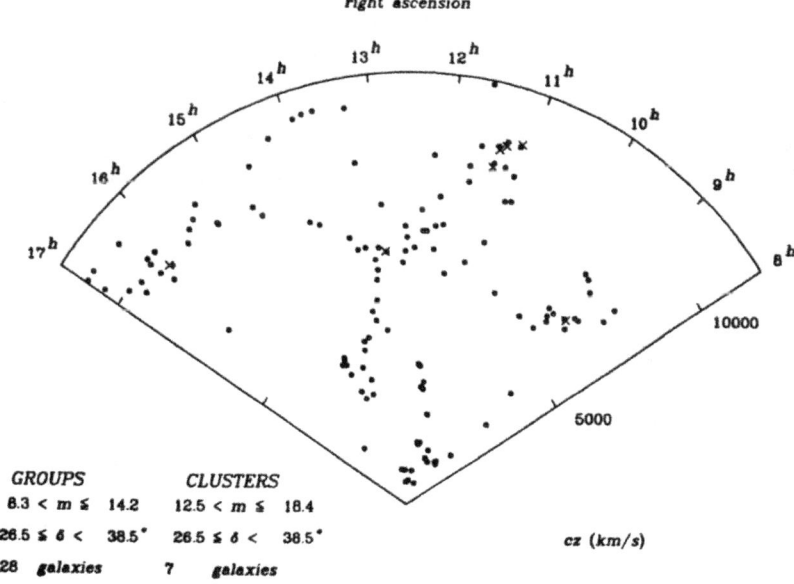

Figure 10: The distribution of centers of groups (∘) and Abell clusters (×) of galaxies in the 12°-wide slice displayed in Figure 5.

VI. DISCUSSION

Theories for the formation of large-scale structure must explain the observed structure in the distribution of galaxies. Is the observed structure consistent with models based on gravitational amplification of random-phase density fluctuations? White *et al.* (1987) and Centrella *et al.* (1988) make quantitative comparisons of the cold and hot dark matter models, respectively, with the available data. The models are $\Omega = 1$ universes. Both the cold and hot dark matter models can successfully reproduce some of the salient features of the data. Both produce large voids and some thin coherent structures. White *et al.* (1987) argue that a standard cold dark matter model with biased galaxy formation produces a distribution of galaxies which is hard to distinguish from the data in Figures 5–9 (see their Figure 10). One quantitative comparison is based on the void in Boötes. Sampling according to the procedure followed by KOSS, White *et al.* find that 3/25 simulations contain a void as large or larger. It is not clear whether the simulations can meet the challenge posed by the increasing number of surveys with more dense sampling than the Boötes survey. There are now five surveys large enough to contain ~ 5000 km s^{-1} voids — all of them do.

Detailed comparison of the data with the models is complicated by the limited number of structures in the samples. However, calculation of statistics like the filling factor could be telling. Another aspect of the models which appears to differ from the data is the relationship between the distribution of rich clusters and the distribution of individual galaxies.

The sharpness of the structures in Figures 5–9 and the possibility that they imply non-random phase initial conditions motivate consideration of alternatives to the standard gravitational models for large-scale structure formation. One suggestion (Ostriker and Cowie 1981; Ikeuchi 1981) is that explosions promote or amplify galaxy formation. Although the origin of such explosions is pure speculation, they can produce voids with radii in the range 5 - 20 h^{-1} Mpc without violating the constraints on these models imposed by the smoothness of the microwave background (Vishniac and Ostriker 1986). Explosions or the decay of superconducting strings (Ostriker, Thompson, and Witten 1986) during the epoch of galaxy formation releasing 10^{61} - 10^{62} (or more) ergs would produce shells with radii of 5 - 6 h^{-1} Mpc or more. Coalescence of these structures is responsible for the production of the larger voids. Detailed investigation of these models would be useful for identifying statistical measures which can differentiate them from the more standard cold and hot dark matter models (see e.g. Ostriker and Strassler 1988). The hydrodynamic models have the appeal that they provide a natural explanation for sharp-edges structures. They may also provide a natural "biasing" mechanism.

Comparisons between models and the data are clearly limited by the absence of "fair" samples. Further progress in mapping out the large-scale structure of the universe requires deeper and more reliable photometric catalogs. Systematic variations in the magnitudes from one region to another in a single catalog (not to mention variations from one catalog to another) almost surely compromise detailed analyses of the properties of the structure. The advent of large format CCD's should go some way toward solving the problem by enabling digital sky surveys.

Another issue which may be clarified by larger redshift surveys is the relationship between individual galaxies and clusters of galaxies as tracers of the large-scale matter distribution. Selection of systems of galaxies from complete redshift surveys rather than from the distribution projected on the sky is clearly more closely related to the underlying physics. When the CfA survey is complete to the limiting magnitude, the sample of rich systems should be large enough to sort out the observational biases.

I thank John Huchra for many years of exciting scientific collaborations and for his comments on a previous version of these lectures. I thank Valérie de Lapparent, Marc Postman, Robert McMahan , and Massimo Ramella for help with some of the figures. This research was supported in part by the Smithsonian Institution and by NASA grant NAGW-201.

REFERENCES

Aaronson, M., Huchra, J.P., and Mould, J. 1979, *Ap.J.*, **229**, 1.

Aaronson, M., Bothun, G.D., Mould, J., *et al.* , in preparation.

Bean, A.J., Efstathiou, G., Ellis, R.S., Peterson, B.A., and Shanks, T. 1983, *M.N.R.A.S.*, **205**, 605.

Centrella, J., Gallagher, J.S., Melott, A.S., and Bushouse, H.A. 1988, preprint.

Centrella, J., and Melott, A.S. 1983, *NAture*, **305**, 196.

da Costa, L.N., Pellegrini, P.S., Sargent, W.L.W., Tonry, J., Davis, M., Meiksin, A., and Latham, D.W. 1988, *Ap.J.*, **327**, 544.

Davis, M. Efstathiou, G., Frenk, C. and White, S.D.M. 1985, *Ap.J.*, **292**, 371.

Davis, M. and Huchra, J.P. 1982, *Ap.J.*, **254**, 437.

Davis, M. and Peebles, P.J.E. 1983, *Ap.J.*, **267**, 465.

Doroshkevich, A.G., Kotok, E.V., Novikov,I.D., Polyudiv, A.N., Shandarin, S.F., and Sigov, Yu.S. 1980, *M.N.R.A.S.*, **192**, 321.

Dressler, A. 1988, *Ap.J.*, **329**, 519.

Einasto, J., Joeveer, M., and Saar, E. 1980, *M.N.R.A.S.*, **193**, 353.

Geller, M.J. *et al.* 1988, in preparation.

Giovanelli, R. and Haynes, M.P. 1985, *A.J.*, **90**, 2445.

Giovanelli, R. and Haynes, M.P. 1986, *Ap. J.*, **292**, 404.

Giovanelli, R., Haynes, M.P., Myers, S.T., and Roth, J. 1986, *A.J.*, **92**, 250.

Groth, E.J. and Peebles, P.J.E. 1977, *Ap.J.*, **217**, 385.

Haynes, M.P. and Giovanelli, R. 1986, *Ap. J. (Letters)*, **306**, L55.

102

Huchra *et al.* 1992, in preparation.

Huchra, J.P. and Geller, M.J. 1982, *Ap.J*, **257**, 423.

Ikeuchi, S. 1981 *Publ. Astr. Soc. Japan*, **33**, 211.

Joeveer, M., Einasto, J., and Tago, E. 1978. *M.N.R.A.S.*. **185**. 357.

Kaiser, N. 1987, *M.N.R.A.S.*, **227**, 1.

Kirshner, R.P., Oemler, A., and Schechter, P. 1979, *A.J.*, **84**, 951.

Kirshner, R.P., Oemler, A. Jr., Schechter, P.L., and Shectman, S.A. 1981, *Ap.J. (Letters)*, **248**, L57.

Kirshner, R.P., Oemler, A., Schechter, P.L., and Shectman, S.A. 1983, *A.J.*, **88**, 1285

Kirshner, R.P., Oemler, A. Jr., Schechter, P.L., and Shectman, S.A. 198, *Ap.J.*, **314**, 493.

Kirshner, R.P., Oemler, A., Schechter, P, and Shectman, S. 1988, private communication.

Koo, D. Kron, R., and Szalay, A. 1986 in *13th Texas Symposium on Relativistic Astrophysics*, M. Ulmer, ed. (World Scientific: Singapore).

de Lapparent, V., Geller, M.J., and Huchra, J.P. 1986, *Ap.J. (Letters)*, **202**, L1.

de Lapparent, V., Geller, M.J., and Huchra, J.P. 1988, *Ap.J.*, **332**, 44.

Lynden-Bell, D., Faber, S.M., Burstein, D., Davies, R.L., Dressler, A., Terlevich, R.J., and Wegner, G. 1988, *Ap.J.*. **326**, 19.

Malmquist, K.G. 1922, *Ark. Mat. Astron. Fys.*, **16**, No. 23.

Nilson, P. 1973 *Uppsala General Catalog of Galaxies, Uppsala Astr. Obs. Ann.*, **6**.

Ostriker, J.P. 1986, in *Galaxy Distances and Deviations from the Hubble Flow*, B. Tully, ed.

Ostriker, J.P. and Cowie, L.L. 1981, *Ap.J. (Letters*, **243**, L127.

Ostriker, J.P. and Strassler, M. 1988, preprint.

Ostriker, J.P. ,Thompson, C., and Witten, E. 1986, *Phys. Rev. Lett. B*, **180**, 231.

Peterson, B.A., Ellis, R.S., Efstathiou, G., Shanks, T., Bean, A.J., Fong, R. and Zen-Long, Z. 1986, *M.N.R.A.S.*, **221**, 233.

Ramella, M., Geller, M.J., and Huchra, J.P. 1988, in preparation.

Saarinen, S., Dekel, A., and Carr, B.J. 1987, *Nature*, **325**, 598.

Schechter, P.L. 1976, *Ap.J.*, **203**, 297

Shanks, T., Bean, A.J., Efstathiou, G., Ellis, R.S., Foong, R., and Peterson, B.A. 1983, *Ap.J.*, **274**, 529.

Vishniac, E.T. and Ostriker, J.P. 1985,*Societa Italiana di Fisica*, **1**, 157.

White, S.D.M., Frenk, C.S., Davis, M., and Efstathiou, G., 1987, *Ap. J.*, **313**, 505.

White, S.D.M., Tully, R.B., and Davis, M. 1988, preprint.

Zeldovich, Ya. B. 1970, *Astron. Astrophys.*, **5**, 84.

Zwicky, F.,Herzog, W., Wild, P, Karpowicz, M. and Kowal, C. 1961-1968, *Catalog of Galaxies and of Clusters of Galaxies*, (Pasadena: California Institute of Technology)

DISCUSSION.

Seitter: I am pleased to add another item to your list of surveys, the MRSP (Muenster Redshift Project). It has so far contributed 24000 new redshifts. Though obtained from low- dispersion objective prism spectra with a mean error of $dz = 0.008$, the MRSP goes sufficiently deep, $z \simeq 0.3$, to make these quite tolerable percentage errors. While the CFA is of great importance in near space at 10–100 Mpc, the MRSP covers the 100–1000 Mpc range. The CFA measures smaller features with high resolution, the MRSP is able to discover larger structures at lower resolution. It is capable of adding with each Schmidt telescope plate measured, another 6000 redshifts in one week of reduction time with fully automated procedures.

Geller: One drawback of this survey is that the error in the redshifts is comparable with the scale of the structure we see. In fact, it is larger than the typical width of the sheets in redshift space.

Brandenberger: How does your interpretation of the survey fit in with the statistical analyses of other surveys which show a genus curve deviation from that in a Gaussian model opposite to that obtained in a toy model of bubbles?

Geller: First I should say that "bubbles" were not meant to be a technical description of the data. The word is meant to call up an image of a structure in which there are thin sheets of galaxies which surround or nearly surround voids. The stucture could be sponge-like or it could be bubble-like. In my opinion, descrimination between these topologies is beyond the power of the current data (and is certainly beyond the power of the data to which the tests you mention have been applied).

Watson: Are there any morphological differences between galaxies in the voids and the galaxies on the "bubble" surfaces?

Geller: There are, of course, very few galaxies in the voids. Thus any conclusion has rather low statistical confidence. Our sample so far reveals no significant differences.

Many have argued that low surface brightness galaxies should be located preferentially in lower density regions. There have been anumber of analyses with conflicting results. It is my impression that the analyses of Binggeli, Bothun and his collaborators, and Thuan and his collaborators are the most convincing; these workers conclude that there is no evidence that low surface brightness galaxies have a distribution very different from that for high surface brightness objects.

Finally, we have typed the galaxies in our first slice centered at $\delta = 29.5°$. We see the normal morphology-density relation. The core of Coma is full of early-type galaxies (as is already well-known). However, both early- and late-type galaxies trace the large-scale pattern we see in Figure 3.

THE COSMIC MICROWAVE BACKGROUND RADIATION AND THE DOG IN THE NIGHT

R. B. Partridge
Haverford College
Haverford, PA   19041
U.S.A.

ABSTRACT.   Recent observational results on both the spectrum and the angular distribution of the cosmic microwave background radiation are reviewed, with special emphasis on null results or upper limits.

—++—

"Is there any other point to which you would wish to draw my
 attention?"
"To the curious incident of the dog in the nighttime."
"The dog did nothing in the nighttime."
"That was the curious incident," remarked Sherlock Holmes.

"The Silver Blaze", by Sir Arthur Conan Doyle

—++—

For all of us, perhaps particularly for those of us whose nation is only one quarter as old as the University of Bologna, it is an honor to help celebrate the 900th anniversary of Alma Mater Studiorum. Floreat!

My task in these proceedings is to review the cosmic microwave background, and I will concentrate primarily on null experiments, particularly upper limits on the spectral distortions and the anisotropies in that radiation.

To emphasize the special nature of such measurements, let me begin with the solution of a well-known case of Sherlock Holmes, a few lines of which are quoted above. The crucial clue in the solution of this mystery was the absence of an event; the dog did not bark. It was in effect a null experiment—the value of the incident lay in what did not happen. Holmes correctly deduced that because the dog did not bark, the intruder must have been its owner.

105

M. Caffo et al. (eds.), Astronomy, Cosmology and Fundamental Physics, 105–130.
© *1989 by Kluwer Academic Publishers.*

1.   NULL EXPERIMENTS: BENEFITS AND PITFALLS

The value of null experiments in physics, as opposed to detective stories, is well known, starting with the crucial experiment (or gedanken experiment) in which Galileo dropped objects from the campanile in Pisa to show that motion in a gravitational field depends on time and distance, but not mass.   That experiment in its refined version by Baron Eötvös is the experimental cornerstone of General Relativity.   There are of course other prominent examples of null experiments--the Hughes-Drever experiment, various experiments on $1/R^2$ laws, and more recently proton-decay experiments.   In each case, we have learned fundamental things about the properties of the Universe from the <u>absence</u> of detectable signals.   The same, I will claim, is true of the cosmic microwave background radiation (CBR):   its main contribution to astrophysics, cosmology and even particle physics lies in what we <u>do</u> <u>not</u> see.   We do not see large perturbations in the spectrum, or a large quadrupole moment, or large amplitude fluctuations on smaller angular scales.

Before presenting and discussing such experimental results, I would like to say a bit more about the nature and peculiarities of null experiments.   Let us start with Galileo on the campanile in Pisa, as shown in the first figure.   Incidentally, this experiment (if it was in fact conducted) would have taken place some 500 years after the foundation of this city's university.

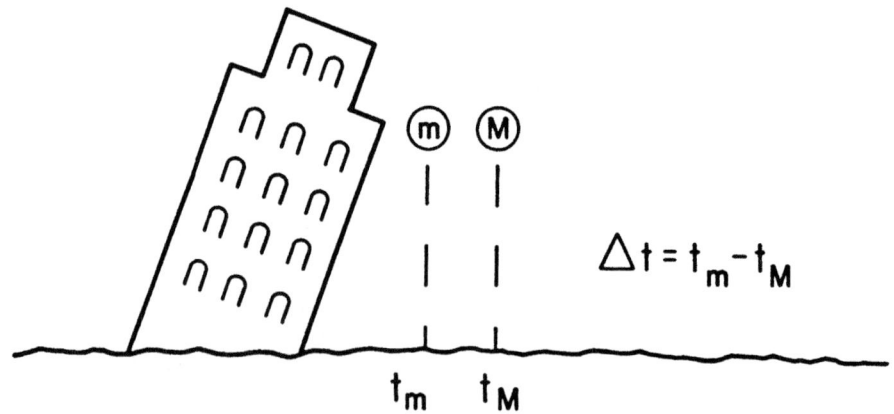

Fig. 1.   A cartoon of Galileo's experiment (or gedanken experiment) at Pisa, testing the hypothesis that unequal masses fall at the same rate, so $\Delta t = 0$.

If in fact the motion is independent of the mass of the object dropped, the difference in the arrival times of the two masses, Δt, should be 0.   But of course any realistic experiment, including this

one if it was ever performed, will have both statistical and
systematic errors present. In the first class one might include the
effects of small wind currents, timing errors and the unsteadiness of
Galileo's hands; in the second class one might include the overall
effect of air resistance. Given imperfect data, how to proceed? Let
us begin by framing Galileo's test in more contemporary statistical
language (essentially a likelihood ratio; Boynton and Partridge,
1973). We frame two hypotheses::

H_1 is the hypothesis that $\Delta t = 0$.

H_2 is the hypothesis that $\Delta t > 0$, that is that lighter masses
fall slower.

We may then define a critical region X for the test between H_1
and H_2 by requiring that for $\Delta t < X$ we are confident at some level
(often taken as 95%) that we will not reject H_2 if it is true. This
is generally referred to as a 95% confidence upper limit on Δt. We
operate by assuming H_2 (that is, $\Delta t > 0$), then using the observational
results to set an upper limit on Δt (for an application to CBR
observations, see Boynton and Partridge, 1973). If we proceed this
way, we leave only a small probability (generally taken as 5% or less)
of making what statisticians call a Type I error (rejecting H_2 if it
is true). That is, there is a probability of 5% or less that the true
value of Δt exceeds our upper limit. Those setting out to make null
experiments, of course, try to make the upper limit on their
observable--and hence on the critical region X--as small as possible.
One obvious way to do so is to minimize statistical error. However X
will also be smaller for data with a given level of statistical error
if the measured departures Δt happen to be small. Likewise, two
experiments which establish the same critical region, and thus the
same upper limit on Δt, can have quite different statistical errors,
as the sketch (Fig. 2) indicates. Which of these two experimental
results is "better"? The value of the critical region alone cannot
tell us, since it is the same in both cases. As a consequence, the
upper limits on Δt would also be the same. I suspect, however, that
the experimentalists in the audience would be a lot happier with the
results shown in sketch A of Figure 2. Clearly we need some other
criterion to measure the quality of a null experiment. That other
measure is the degree to which we can avoid Type II errors, which
involve rejecting the alternate hypothesis H_1 when it is true (see
Lasenby and Davies, 1983). Let us return to our historical analog,
Galileo on the tower of Pisa. If Galileo had actually performed this
experiment honestly, he might have found that his measured Δt was
greater than 0 because of air resistance. Rejecting H_1 would
nevertheless have been an error. To see how likely it is that we are
making such a mistake in a null experiment, we must calculate the
power of the statistical test, which is equal to 1 minus the
probability of Type II error. If the power is low, we have a poor
test even if X and the resulting upper limit on Δt is small. That is
shown schematically in the lower part of Fig. 2. In the case of set
A, the probability of rejecting H_1 (cross-hatched area under dashed
curve) is small and the power is large; in the case of set B, the
cross-hatched area is much larger, so the power of the test is weaker.

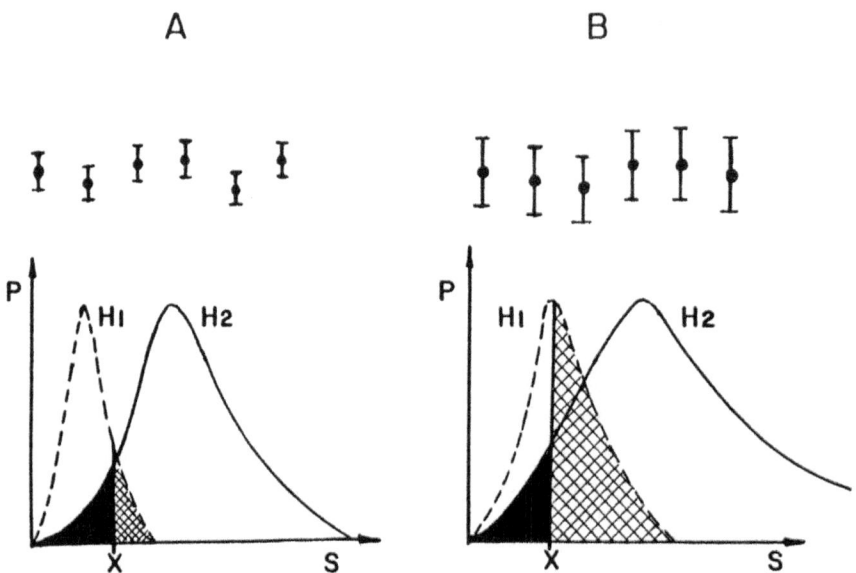

Fig. 2. (Top) Cartoon showing the hypothetical results of two experimental measurements. Both are supposed to have the same value for X, the critical region for the test between H_1 and H_2. Which is a more convincing result? (Bottom) Plot of statistic S (e.g. a χ^2) used to test between H_1 and H_2, for the two sets of experimental results. In both, X is the same and the shaded area is 5%. Thus both produce the same upper limit on Δt. Where the two differ is in their power. For A, the power is high--the probability of Type II error (cross-hatched area) is small. For B, the power is low.

I am sure that much of this is familiar to most of you, but I must say that I was a late convert to this way of thinking, and, as we shall see, the power of various tests used in CBR studies has often been neglected, for instance in one of my papers (1980). Hence it may be that a word of caution is in order for the theorists who make use of the observations I am about to describe. Both the observers and the theorists are now looking much more carefully at the statistical analyses of CBR observations, and I look forward to much more detailed presentations than I can give here from Kaiser and Lasenby (1989), Boughn and Cottingham (1989) and as part of the paper by Readhead et al (1989).

2. INTRODUCTION TO THE CBR

Let me now turn to the real topic of my talk, the cosmic background radiation. It is now generally recognized as radiation left over from the initial hot, dense phase in the history of the Universe (Dicke et al, 1965).* Since its discovery 24 years ago by Penzias and Wilson (1965), we have determined that the spectrum of the radiation is a good match to a Planckian curve with T_0 = 2.75 K (Smoot et al, 1985; Crane et al, 1986; Johnson and Wilkinson, 1987) and that the radiation is isotropic to about one part in a thousand over a range of angular scales from 4" to 180°. It is the apparent lack of spectral perturbations or anisotropies that I will be concentrating on today--the dog that did not bark in the nighttime.

3. ISOTROPY

Radiation left over from the Big Bang is expected to be more or less isotropically distributed. This may be seen by considering the origin of the photons we detect in our measurements of the CBR (Fig. 3). They come from a surface of last scattering, at some fixed redshift z_s, which is isotropically distributed about any observer in the Universe. Indeed the isotropy of the radiation was an important early test of the Big Bang hypothesis. There are, however, three ways in which anisotropies in the radiation may be introduced.
    1. Motion of the observer. The Doppler effect causes an anisotropy of amplitude $\Delta T/T$ = v/c for small values of v. The anisotropy is a pure dipole, written T_1.
    2. Anisotropic expansion (equivalent to a direction-dependent value of Hubble's constant H_0). For most cosmological models, the major multipole component will be a quadrupole moment, T_2. The dipole moment, if present, will be smaller; and there may be some additional low order multipole components.
    3. Inhomogeneities in the distribution of matter on the surface of last scattering (Silk, 1968) or in the intervening matter (Sachs and Wolfe, 1967). These will produce anisotropies on scales of arcseconds to degrees.

*See the contribution of Fred Hoyle for a different view. Much of what I have to say about the CBR is independent of the model assumed for the origin of the radiation.

110

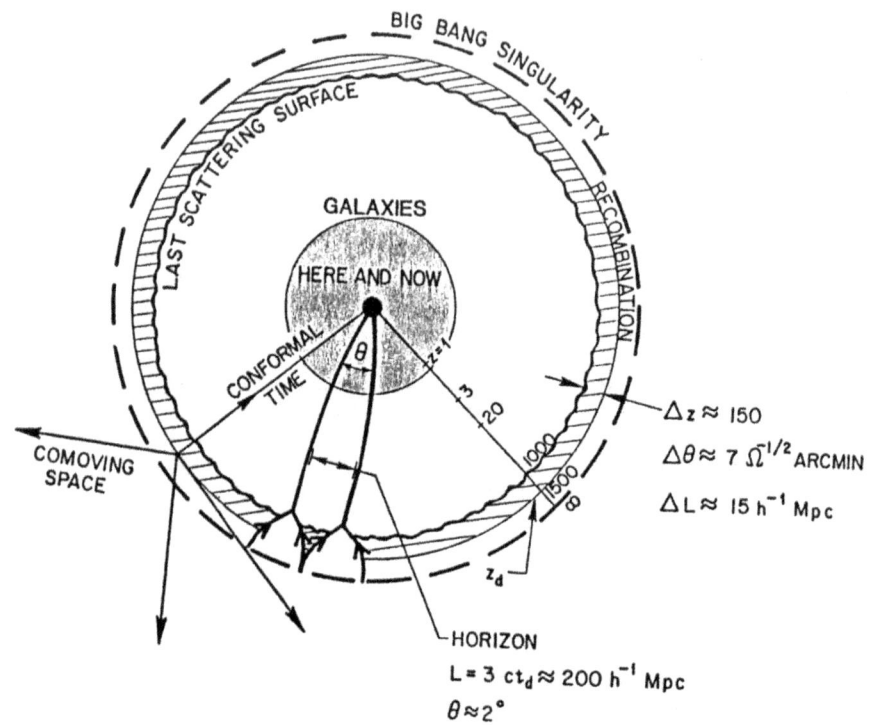

Fig. 3. The origin of CBR photons on the surface of last scattering. Here, it is assumed that the epoch of last scattering is the epoch of recombination of the primeval plasma, with $z_s \sim 1000$. Note that the scale of the causal horizon at this epoch is $\sim 2°$. From Kaiser and Silk (1986).

3.1. The Dipole and Quadrupole Moments

Several groups (Fabbri et al, 1980a; Fixsen et al, 1983; Lubin et al, 1983) have made careful measurements of the dipole moment of the radiation. The most recent work is that of the Moscow group led by Dr. Strukov (Strukov and Skulachev, 1984; Strukov et al, 1987; Klypin et al, 1987). This was a satellite experiment, carried out in such a way as to minimize the radiation of the earth into the side- and back-lobes of their antenna.

The results of some of the recent measurements of the dipole moment are shown in Table 1, and it is clear that there is reasonably good agreement with an amplitude of the dipole $T_1/T_0 \simeq 1.2 \times 10^{-3}$.

These recent experiments show only upper limits for the quadrupole moment T_2. In particular, the Soviet work sets an upper limit of $\leq 3 \times 10^{-5}$ on T_2/T_0. This tight upper limit constrains models for anisotropic expansion (Barrow et al, 1983), or for that matter large-scale magnetic fields (Thorne, 1967). It also implies that the

Table 1 Results (expressed in thermodynamic temperature) of recent measurements of the large-scale distribution of the CBR.

Group	Berkeley	MIT/UBC	Moscow	Princeton
Reference	Lubin et al (1985)	Halpern et al (1985)	Strukov et al (1987) Klypin et al (1987)	Fixsen et al (1983)
Wavelength, mm	3	1-3	8	12
Vehicle	balloon	balloon	satellite	balloon
Detector	heterodyne	bolometric	heterodyne	heterodyne
Dipole amplitude, T_1, mK	3.4 ± 0.2	3.0 ± 0.5	3.16 ± 0.12	3.1 ± 0.2
Direction of solar motion, R.A. and Dec.	$11.2^h, -6°$	(similar)	$11^h3 \pm .16, -7.5° \pm 2.5°$	$11h, -10°$
Limit on quadrupole moment, T_2, mK	0.4^*	n.a.	$\lesssim 0.08^{**}$	≤ 0.19

*Calculated from Table I of Lubin et al by the present author, by taking the quadrature sum of the measured coefficients $Q_1 \cdots Q_5$.

**Taking the more conservative, model-independent upper limit.

much larger dipole moment is not due to anisotropic expansion, in which case we would have expected $T_2 > T_1$. It is also easy to show that the dipole moment T_1 is not due to anisotropically distributed radio sources (Partridge and Lahav, 1988). That leaves the usually accepted explanation of the dipole moment, that it is due to the motion of the observer. If we interpret T_1 this way, and correct the measured value for the motion of the Galaxy within our local group, we may easily derive the velocity vector of our local group of galaxies. This result has proven to be an important benchmark for those studying large-scale velocity streaming in the Universe, a topic addressed here by other speakers including Margaret Geller and Donald Lynden-Bell. Here I will mention only that the amplitude of the velocity is rather high, ~ 600 Km/sec. As other speakers point out, large-scale velocities this high are discomforting for those who favor cold dark matter models for the origin of structure in the Universe.

3.2. Searches for Small-Scale Anisotropies

Let us turn now to the most precise, and perhaps most useful, null experiments on the CBR, searches for anisotropies in the background on angular scales from 90° all the way down to several arcseconds. Such fluctuations may be produced by density inhomogeneities on the surface of last scattering referred to earlier (Silk, 1968), or by inhomogeneities in the matter lying between that surface and the observer (see Sachs and Wolfe, 1967; Sunyaev and Zel'dovich, 1972).

The amplitude of these temperature fluctuations depends on both the redshift of the surface of last scattering z_S, and on various cosmological parameters. If z_S ~ 1000, corresponding to the epoch when the primeval plasma first combines to produce neutral atoms, we have predictions like those in fig. 4 (Bond and Efstathiou, 1984). Note both the dependence on $h = H_0/100$ and Ω, and the general decrease in amplitude at $\theta \lesssim 10'$. The curves shown assume the existence of dark matter in addition to baryonic matter with $\Omega_{bar} \approx 0.06$; pure-baryon models predict values of $\Delta T/T$ about one order of magnitude larger than those shown.

Recently, interest has grown in the possibility that the surface of last scattering may be at much lower redshifts. The shift occurs because explosive, or at least energetic, phenomena associated with galaxy formation reionize the material contents of the Universe and thus reintroduce Thomson scattering of the CBR photons (Ikeuchi, 1981; Ostriker and Cowie, 1981). The same explosive or energetic processes also introduce fluctuations into the CBR (see Hogan, 1980, 1984; Ostriker and Vishniac, 1986); so also will the superconducting cosmic strings being considered by Ostriker et al (1987). These new, astrophysical phenomena complicate the interpretation of the observations. We need to consider the effects on different angular scales. On large angular scales $\gtrsim 5°$, "primordial" fluctuations induced at z ~ 1000 can survive. On smaller angular scales, the primordial fluctuations can be damped out by rescattering in the reionized matter at lower redshifts. It is worth noting that one of

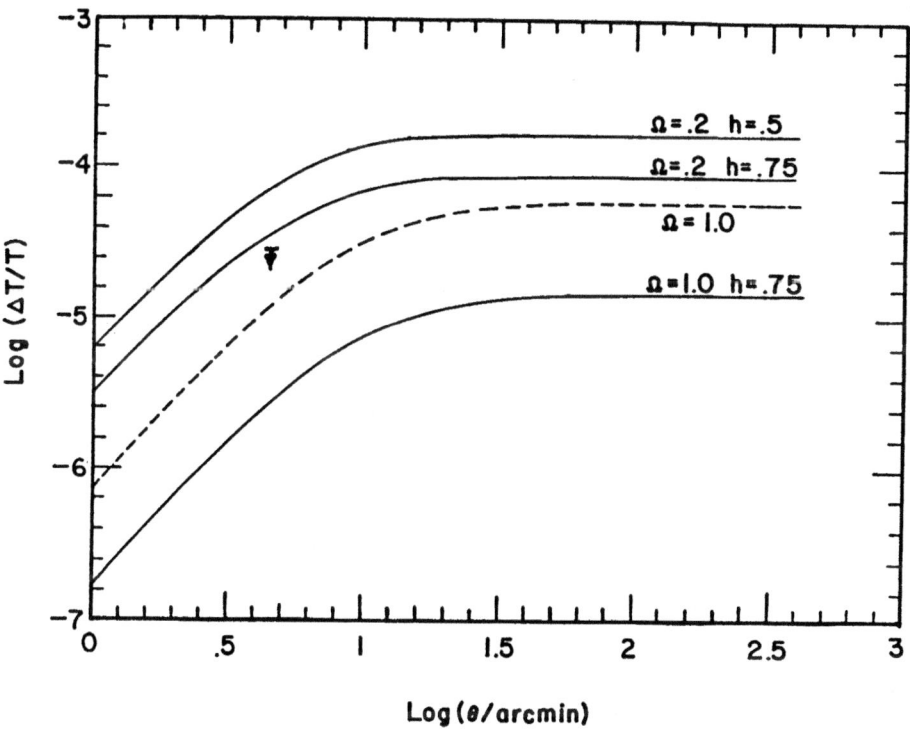

Fig. 4. Predicted values of $\Delta T/T$ for models including dark
matter, for various values of the cosmological parameters
$h = H_0/100$ and Ω (adapted from Bond and Efstathiou, 1984).
The single dashed line is for a hot dark matter model; the
solid lines are for cold dark matter models. The
observational upper limit is the result of Uson and
Wilkinson (1984b) discussed in the text. Here, $z_s \approx 1000$.

the few constraints we might have on particle physics in the early
Universe is the amplitude of density perturbations produced, which in
turn set up temperature fluctuations. One could imagine a scenario in
which rather large density perturbations are allowed, with the
consequent primordial temperature fluctuations subsequently erased by
explosive phenomena at lower redshifts. Only measurements at angular
scales $\gtrsim 5°$ could then provide information about the initial
anisotropy in the CBR, and hence about the physics of the early
Universe.

The most interesting outcome of this new scenario for
cosmologists and radio astronomers, if not for particle physicists, is
the possibility that explosive phenomena will introduce new varieties
of anisotropies into the background radiation (see Hogan, 1980, 1984;
Ostriker and Vishniac, 1986). There is even the possibility that the

fluctuations introduced at lower redshifts could be wavelength dependent (Hogan, 1980). Thus upper limits on, or measurements of, anisotropies in the CBR on scales of a few degrees or less may permit us to say something about astrophysical processes in the "desert years" between redshifts of about 1000 and 3.

After these introductory remarks, let me now turn to the most recent observations, with stress on the most stringent limits. Three ranges of angular scale are represented:--

1.) $1' \lesssim \theta \lesssim 10'$, resulting from measurements with conventional, filled-aperture radio telescopes.

2.) $\theta \gtrsim 1°$, measurements made with specially designed instruments.

3.) $\theta < 1'$, measurements requiring the use of arrays of telescopes and aperture synthesis.

3.2.1. Arcminute scales. Two decades of effort (e.g., Conklin and Bracewell, 1967; Pariiskii et al, 1977; Partridge, 1980; Uson and Wilkinson, 1984a, 1984b; Readhead et al, 1989) have produced the tightest limits on $\Delta T/T$ on these angular scales.

The most fully documented and published work is that of Uson and Wilkinson (1984a), measurements made at $\lambda = 1.5$ cm carried out at the 40-meter telescope of the U. S. National Radio Astronomy Observatory in West Virginia. Differential measurements of the sky at 12 positions near the North Celestial Pole were made (Fig. 5); from these, Uson and Wilkinson derive 95% confidence limits on a scale of 4.5 of $\Delta T/T \lesssim 2.1 \times 10^{-5}$ (1984b). This, however, is one of those cases in which the scatter of the measured points is unusually low (the χ^2 of the data is 7-8 for 11 degrees of freedom). It is that fact which results in a particularly low value for the upper limit on $\Delta T/T$ because the critical region X is small (see Wilkinson, 1988). As a further illustration of this same point, let us compare the 95% confidence level on ΔT with the standard error of each of the individual data points: the ratio is $\Delta T/\sigma \sim 0.06/.25 \approx .25$. We may now compare that quantity with the same ratio found from other searches in which the data are presented in enough detail to allow the calculation to be made (Partridge, 1980; Fabbri et al, 1982; Mandolesi et al, 1986; Davies et al, 1987). Typically we find $\Delta T/\sigma \approx \frac{1}{2}-1$. These considerations suggest the possibility of Type II error, and a calculation of the power of the test used by Uson and Wilkinson shows that to be true: the power is only 10-20%. A more conservative analysis of their important results would give values of $4-5 \times 10^{-5}$ as the 95% confidence upper limit on $\Delta T/T$.

The most recent work on arcminute scales, also at $\lambda = 1.5$ cm, is by Lawrence, Readhead and their colleagues at Cal Tech. Their work has not yet been published, but I understand that $\Delta T/T$ is less than $1-2 \times 10^{-5}$ on a scale of 7'. They have calculated the power of the test on which this 95% confidence limit is based; it is 0.73, so the possibility of Type II error is low. Finally, somewhat more stringent limits may be possible once their observations are corrected for the flux from radio sources which happen to fall in the circumpolar arc

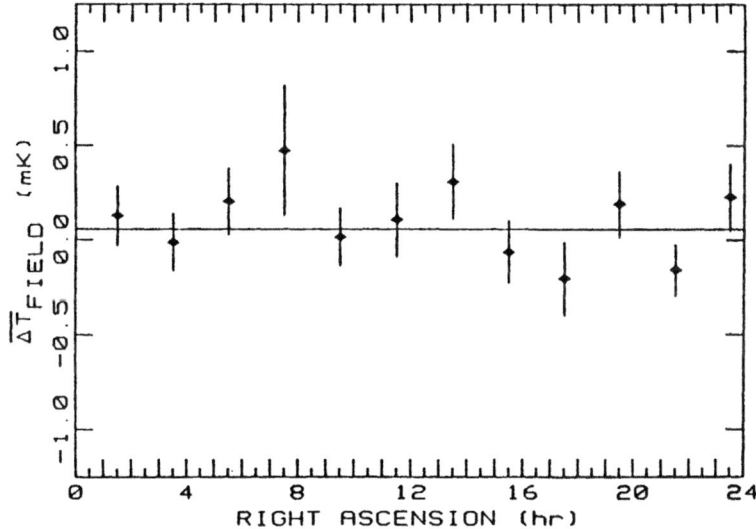

Fig. 5. Measurements of Uson and Wilkinson (1984b)
discussed in the text. These results currently provide the
most stringent upper limits on $\Delta T/T$ that have been published.

they studied (C. Lawrence, private communication). The observations
needed to make such corrections are now underway.

3.2.2. <u>Degree scales</u>. Sensitive searches for CBR fluctuations on
what I shall call intermediate angular scales were pioneered by
Melchiorri and his group (see, e.g., Fabbri et al, 1980b). These
measurements were made in the submillimeter region and thus may be
influenced by the far ir flux discussed below (Matsumoto et al, 1988).
More recent work by Mandolesi et al (1986) at 3 cm was plagued by
atmospheric noise and produced upper limits on $\Delta T/T$ of several $\times 10^{-4}$.
The observations of Pariiskii and his colleagues (see Pariiskii et al
1977; Berlin et al, 1983) require corrections for the effect of radio
sources and some aspects of the statistical analysis (Lasenby, 1981)
and give roughly comparable results.
In any case, these observations have been superseded by recent
work from a high altitude, dry site by Davies and his colleagues
(1987), at λ = 3 cm. Their instrument was designed to minimize some
sources of systematic error which had troubled some other attempts to
measure large-scale isotropy in the CBR (e.g., Mandolesi et al, 1986);
in particular, they combined the radio astronomical techniques of beam
switching and drift scans, and employed a stable, high sensitivity
receiver. The paper reports the detection of sky brightness
fluctuations on an angular scale of 8° at a level corresponding to
$\Delta T/T$ = 3.7$\times 10^{-5}$ if the sky fluctuation is ascribed entirely to the

CBR. Note, however, that this result depends on a rather small number
of independent samples of the sky--see Fig. 3 of their <u>Nature</u> paper
(1987).* There is also the possibility that some part of the observed
fluctuation may be due to patchy emission from our own Galaxy; one way
to check this possibility is to make observations of the same part of
the sky at a longer wavelength and make use of the fact that most
sources of radio emission in the Galaxy have a strong wavelength
dependence, $T (\lambda) \propto \lambda^{\alpha}$, with $\alpha = 2.0-2.7$. These supporting
measurements are under way by Davies and his collaborators;
preliminary reports from the Jodrell Bank group suggest that galactic
emission is <u>not</u> responsible for the observed fluctuations at
$\lambda = 3$ cm. Finally, allow me to mention a concern I have about the
side-lobes (secondary diffraction maxima) of the apparatus they
employed; radiation from bright celestial sources (the Sun, Moon or
Galactic plane) into side-lobes can produce anomalous signals. The
use of a plane reflecting surface in the near field of the horn
antennas they used increases the possibility of side-lobe problems.
It is by no means clear that radiation into the side-lobes can produce
the level of fluctuation Davies et al report, corresponding to
$\Delta T/T = 3.7 \times 10^{-5}$, but it is a question that I hope the observers will
address further. Off-axis response of the apparatus needs to be
checked to $10^{-5}-10^{-6}$ of the on-axis response in order to eliminate the
possibility of solar effects. If the results of this experiment are
confirmed by the work underway, they will provide the first
substantiated evidence for fluctuations in the microwave background.
and it is interesting that the angular scale is large enough so that
it is "primordial" fluctuations that are being probed.

3.2.3. <u>Subarcminute scales</u>. The most recently announced results
concern fluctuations on subarcminute scales. Diffraction limits most
conventional radio telescopes to angular scales > 1′ (but see the
paper by Kreysa here). To reach scales well below 1′ in searches for
fluctuations in the CBR requires the use of extended arrays of
telescopes and the process of aperture synthesis (see, e.g., Verschuur
and Kellermann, 1974). This technique has been used by two groups
working at the Very Large Array (VLA) operated by the U. S. National
Radio Astronomy Observatory in New Mexico (Fomalont et al, 1984 and
Fomalont et al, 1988; and Knoke et al, 1984 and Martin and Partridge,
1988). The observations reported to date were made at a wavelength of
6 cm, the wavelength at which the most sensitive receivers were
available. As shown in Table 2, angular scales of 6" to 60" were
investigated. As is also clear from the table, the sensitivity of
this technique is not yet comparable to the sensitivity of searches
using conventional, filled-aperture radio telescopes (see Partridge,
1988, for details).

*Additional, unpublished observations have been made on adjacent
 strips of the sky. Davies (private communication) reports
 consistency between adjacent strips, showing that the structure on
 the sky is real.

Reference	Angular Scale	Value of, or Upper Limit on $\Delta T/T \times 10^{-4}$
Knoke et al	6"	< 32
(1984)	12"	< 17
	18"	< 12
Martin and	18"–80"	1.7 ± 0.5
Partridge (1988)	36"–160"	1.3 ± 0.2
Fomalont	12"	< 8.5
et al (1988)	18"	< 1.2
	30"	< 0.8
	60"	< 0.6

Table 2. Upper limits on or tentative measurements of $\Delta T/T$ fluctuations on subarcminute scales. All are VLA measurements at $\lambda = 6$ cm.

Perhaps the most intriguing aspect of the results displayed in Table 2 is the apparent qualitative disagreement on whether real CBR fluctuations are present or not on scales of 18"–60". Is the dog barking faintly or not? Allow me to discuss these two sets of measurements in a bit more detail.*

Both groups begin by making a synthesized map of a region of the sky, such as the one shown in Figure 6. The area mapped is chosen to be much larger than the primary diffraction maximum of the individual antennas of the array (called the "primary beam" of the instrument). Thus the instrument is insensitive to radiation from sources at the edge of the map. Indeed that is clear from Figure 6; the radio sources in the field are visible only towards the center of the map. Likewise, any fluctuations in the CBR intensity will be detectable only near the center of the map where the primary beam response is large. On the other hand, instrumental and atmospheric noise will be distributed evenly over the entire map (see Knoke et al, 1984 or Fomalont et al, 1984). A comparison of the variances of the measured flux densities first at the edge of a map and then at its center thus provides a measurement of real fluctuations in the sky; these will produce extra variance at the center. It is important to note that both groups working at the VLA do find such excess variance at the centers of their maps, and that the observational results are in quantitative agreement, even though the two groups looked at different regions of the sky, for different intervals of time, and using different map-making techniques. The disagreement in our final conclusions must have a different explanation.

*This detailed discussion was not included in the talk given at the symposium in Bologna.

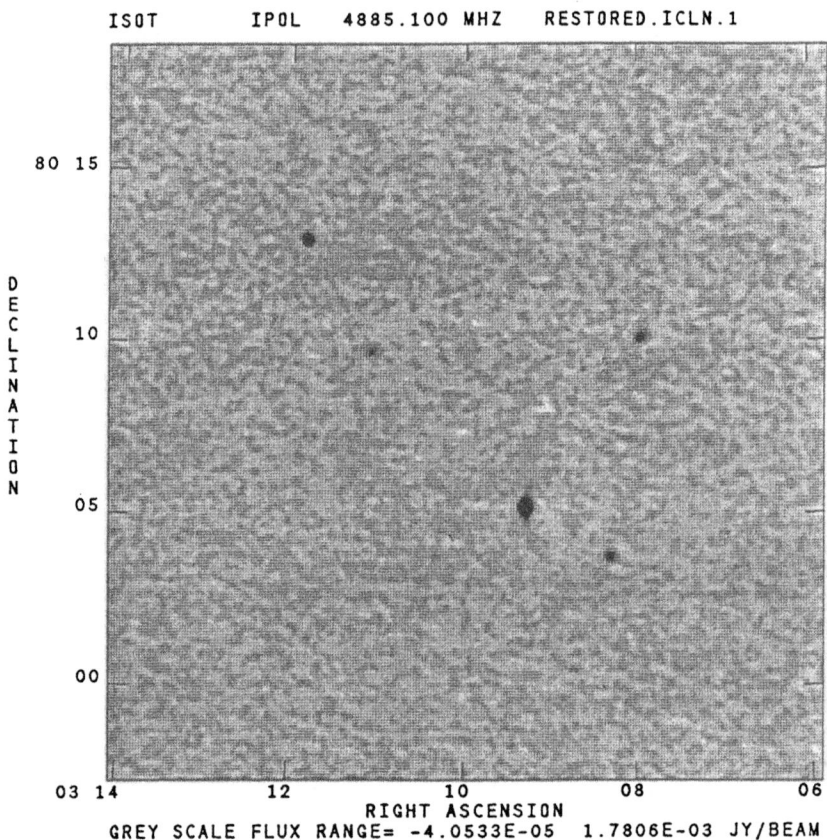

ISOT   IPOL   4885.100 MHZ   RESTORED.ICLN.1

GREY SCALE FLUX RANGE= -4.0533E-05   1.7806E-03 JY/BEAM

Fig. 6. A 6 cm aperture synthesis map of a region of the
sky showing both radio sources (concentrated at the map
center, and instrumental noise (the uniformly distributed
"granularity"). The primary beam has radius 4.5´.

The question is whether this excess variance may be ascribed
entirely to the presence of radio sources (such as those apparent in
Figure 6 and fainter ones) or not. The two groups have treated this
issue differently, and reach opposite conclusions.
In our most recent work (Martin and Partridge, 1988), my
colleague and I began by excising from the map rectangular regions
around the visible sources (see Figure 7). We then divided the
remainder of the map into square cells, and made histograms of the
number of cells with various values of the flux density, S, for both
the center and the edge of the map. The histogram for the outer part

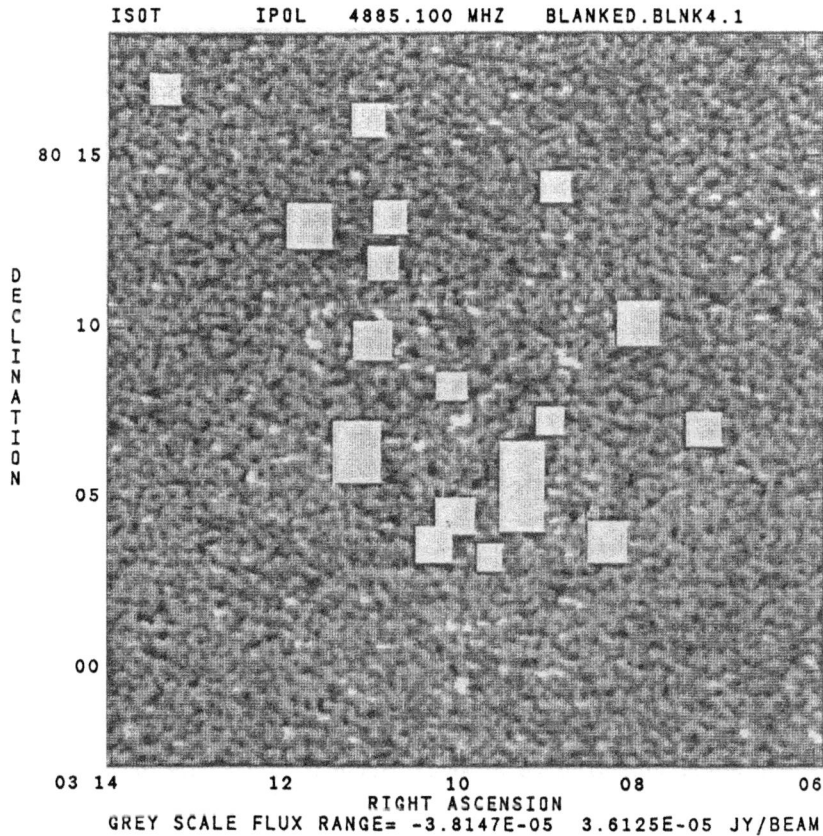

```
ISOT       IPOL     4885.100 MHZ    BLANKED.BLNK4.1
```

GREY SCALE FLUX RANGE= -3.8147E-05   3.6125E-05 JY/BEAM

Fig. 7. Same as fig. 6, but with the bright sources excised. The dynamic range of the plot has been altered also.

of the map was essentially Gaussian, with the scatter due to instrumental and atmospheric noise. The histogram for the central region (Fig. 8), however, displayed a long tail of positive values of the flux density--which we assume was due to the presence of weak radio sources, too faint to have been individually detected in the map. These sources, of course, contributed to the width of the histogram as well. Our next step was to model the contribution of these sources and to subtract it from the histogram. We did so by extrapolating counts of radio sources made at 20 cm (e.g., Mitchell and Condon, 1985) to our observing wavelength of 6.3 cm, and by extrapolating counts made at 6.3 cm (e.g., Fomalont et al, 1984b, and Donnelly et al, 1987) to lower flux levels. Once these extrapolations

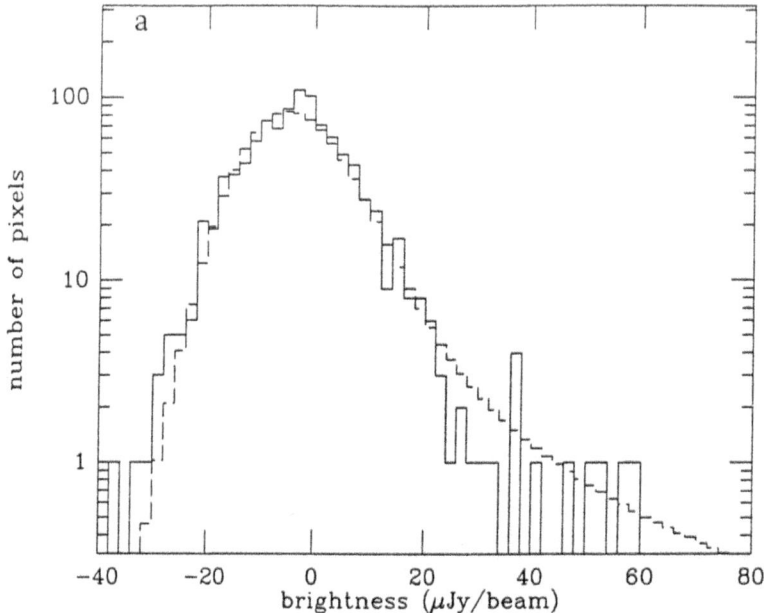

Fig. 8. Frequency distribution of observed flux densities for the central part of the map shown in fig. 6 (solid lines), compared to a model (dashed lines) including instrument noise and flux from faint radio sources (Martin and Partridge, 1988).

gave us the number of faint sources as a function of their flux density, N(S), we were able to calculate how much they added to the positive side of the histogram. We found that the combination of the faint source contributions and instrumental noise was not able to account fully for the variance we saw at the center of the map. To account for all the excess sky variance at the map center would have required roughly twice as many radio sources as our extrapolated counts indicated. I must emphasize, however, that the source counts we employed were extrapolations from counts made at other frequencies and in other parts of the sky. In addition, we did not, by this method, take the side-lobes of weak sources into account (for the 18" resolution map, the side-lobe level was $\lesssim 1.5\%$ or $\lesssim 0.6$ μJy). Note that the side-lobe response can be both positive and negative in an aperture synthesis map.

One additional explanation for the excess sky variance at the center of the map is the presence of fluctuations in the CBR at a level of $\Delta T/T = 1.7 \pm 0.5 \times 10^{-4}$, as shown in Table 2. We checked some other possible sources of systematic and instrumental noise (see Martin and Partridge, 1988, for details), without finding a convincing explanation of the additional sky variance. We thus concluded that we had tentative evidence for the existence of fluctuations in the CBR at the ~ 10^{-4} level.

Now let us consider in outline the approach taken to the same problem by Fomalont, Kellermann and their colleagues in their 1988 paper. They also formed a histogram of the flux densities at the center of the map, and subtracted from it the instrumental variance (as measured by the variance at the edge of the map, plus a small correction). They then modeled the counts of faint sources using a 2 parameter fit:

$$ n(S) = k \left(\frac{S}{20\mu \ Jy} \right)^{\gamma} n_e , $$

where n_e is the Euclidean source count = $90S^{-2.5}$ $ster^{-1}$ Jy^{-1}.
Monte Carlo calculations were then made to scatter sources picked from such a distribution in an idealized "map", and the resulting "map" was convolved with the beam of the VLA. At this stage, this group made the assumption that "the integrated flux density of [each] convolved point source was zero." This is true in an ideal aperture synthesis map before the 'clean' operation has been applied, not necessarily afterwards. In order to ensure that the integrated flux density of each source was zero, a negative bias or offset had to be built into the histogram of source contributions. (Note that this procedure was not used in our work.) The resulting histogram was then further convolved with instrument noise. Finally, values of k and γ were chosen in such a way as to make the model fit the observations with no additional variance left over. In other words, Fomalont et al effectively defined $\Delta T/T = 0$ at an angular scale of 18" by assuming that the radio sources alone were responsible for all the excess fluctuation observed at the center of the map. Given this approach, it is not surprising that the upper limits on $\Delta T/T$ on other angular scales are also small. A crucial question is whether the values adopted for the parameters k and γ are consistent with actual 6 cm source counts at larger fluxes (> 40 µJy); they are consistent. In addition, there is evidence (Fomalont et al, 1988) that regions of 6 cm fluxes between ~ 10 µJy and ~ 30 µJy coincide with faint optical objects, suggesting that they are faint sources.

An obvious way to check the two sets of results is to compare the values Fomalont et al found for k and γ from a fit to their data with the values we found by extrapolating source counts. They are not inconsistent. Thus the simplest possible explanation--that we underestimated source counts for $S \leq 40$ µJy or that Fomalont et al overestimated them--does not provide an adequate resolution of our differences. Another possible source of the disagreement is their assumption of a negative bias in the source flux distribution, or our neglect of the contribution of positive and negative side-lobes to the map flux.

Clearly, further work is needed, both on the analysis of the data in hand and on further and more sensitive observations. Each group now plans to subject data from the other group to its analaysis program. In addition, with Craig Hogan, I have made one preliminary investigation of the role played by faint sources. One can lessen the effect of weak radio sources by observing at a higher frequency: expressed in temperature terms, the spectrum of typical radio sources is $T(\nu) \propto \nu^{-2.7}$. Hogan and I therefore remapped a small area of the sky at 2 cm wavelength, again at the VLA. The receivers at this frequency are less sensitive, so we have only marginal results:

$$\Delta T/T = 2 \pm 1 \times 10^{-4} \text{ at } \theta = 18'' \ (\lambda = 2 \text{ cm}).$$

This result is consistent with our earlier, positive, 6 cm results--but also essentially consistent with the upper limit of 1.2×10^{-4} set by Fomalont et al. Within a year, our two groups will combine forces to use the VLA with its newly installed low-noise receivers at 4 cm to make one more concerted attempt to see if fluctuations on angular scales of 10"-60" are really there. In the meantime, let me repeat earlier warnings about the pitfalls of null experiments. The cautious theorist may want to regard these VLA measurements on subarcminute scales as upper limits only, and should be aware that all the results appearing in Table 2 have built into them corrections for the effect of weak sources, corrections which themselves are uncertain (the whole question of the effect of weak radio sources on searches for fluctuations in the CBR is under review by Franceschini and his colleagues, 1988).

3.3. Upper Limits on Anisotropies--Summary

What have we learned from two decades of observational attempts to find anisotropies in the CBR? In particular, what can we learn from the absence of detectable fluctuations?

If the redshift of the surface of last scattering is at ~ 1000, we know that pure baryon models are essentially excluded; dark matter of some form is required. Even cosmological models including some form of dark matter are just consistent with the results (especially those on arcminute scales). A little breathing space is available if we include bias (e.g., Bardeen, 1986) in our model for galaxy formation. For fuller analyses of the consequences of the observations I have described, readers should consult reviews by colleagues more familiar with the theory (e.g., Kaiser and Silk, 1986; Bond and Efstathiou, 1987).

To what degree can we escape the observational constraints by shifting the redshift of the surface of last scattering to much lower values? As noted above, primordial fluctuations on scales $\lesssim 5°$ will be erased, but new fluctuations will be imprinted. The most detailed work to date (Vishniac, 1987) suggests values of $\Delta T/T$ quite similar to the amplitudes of the primordial fluctuations; in other words, these models too are just consistent with the observations.

There is real power in the continued silence of the dog in the nighttime.

4.   THE SPECTRUM OF THE CBR

A series of measurements using various techniques and conducted over
the past four or five years has established the temperature of the CBR
(T_0 = 2.75 ± 0.04 K) over a wavelength range of ~ 0.1-12 cm, that is
over a wavelength range of more than 100. The value of T_0 itself is
an important cosmological parameter; for instance it is an important
ingredient in nucleosynthesis calculations, which in turn may be used
to set tight upper limits on both the baryon density in the Universe,
ρ_b ≤ 0.06 ρ_c (see Reeves here), and the number of neutrino species
(see Turner here).

I wish to concentrate, however, on what we do <u>not</u> see; in
particular, we do not see substantial evidence for departures from a
pure blackbody spectrum. At least we do not see such evidence in the

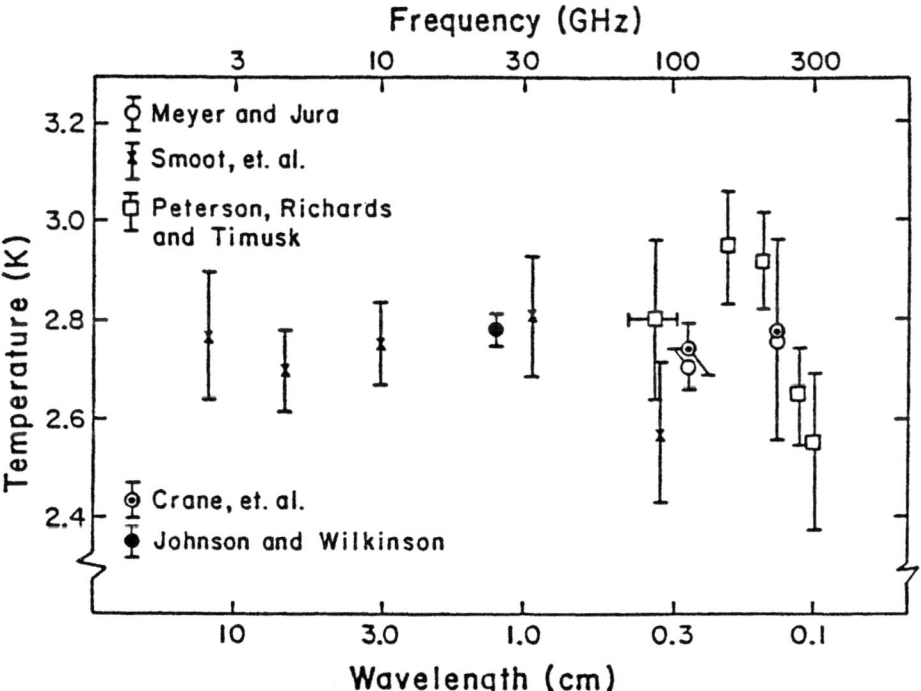

Fig. 9.   Some recent long-wavelength measurements of the
spectrum of the CBR (from Johnson and Wilkinson, 1987).

wavelength range 0.1-12 cm.* A quick glance at the points shown in
Figure 9 and their error bars suggests that the scatter in the data is

*For one measurement of particular importance, see the paper here by
  Philippe Crane and his colleagues (and Crane et al. 1986).

in reasonable agreement with the experimental errors, so that there is no blatant Type II error present. The observations allow no more than a 5-6% distortion in the spectrum in this wavelength range (Smoot et al, 1985). As is well known, the absence of distortion sets interesting constraints on a variety of energy-generating mechanisms which can operate in the redshift range $z \simeq 10^8-10$ (see reviews by Illarionov and Sunyaev, 1975 and Danese and De Zotti, 1977).

In the last 6 months the situation has become much more complicated and frankly much more interesting. Perhaps the dog did bark after all. The major event was the careful rocket experiment of Matsumoto, Richards and their colleagues (1988). This submillimeter observation confirmed the earlier work of Gush (1981) and established the existence of extra flux at wavelengths beyond the peak of a 2.75 K Planck spectrum, particularly at 500-700μ. The experimental results are shown in Figure 10. It is of course possible that the excess

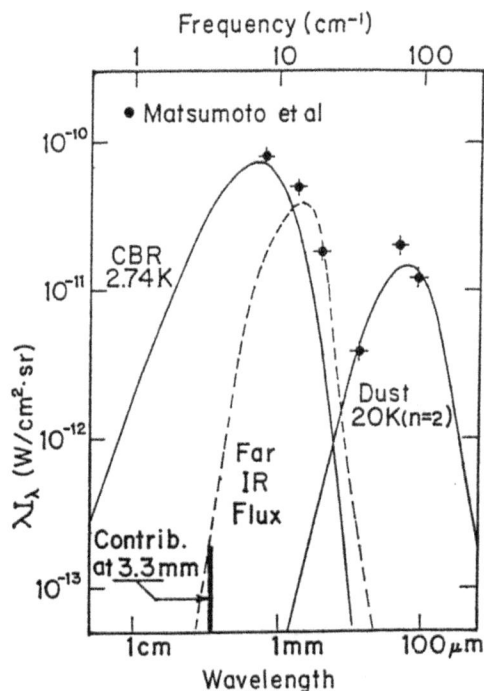

Fig. 10. The CBR, galactic dust emission, and the new far ir flux reported by Matsumoto et al (1988). The solid bar indicates the contribution of this flux to the total background at 3.3 mm.

submillimeter flux is an experimental artifact, but the Berkeley-Nagoya group has been careful to control or rule out a series of possible instrumental effects, including residual atmospheric emission, outgassing from the rocket, etc. Other possible artifacts, like icing up of the antennas, also seem not to produce the kinds of signals seen. In any case, the group plans to fly a somewhat similar experiment about 9 months from now (that is, in early 1989).

Let us accept this as real evidence for excess submillimeter or far ir flux. There are two possible classes of explanations for the observations. One is that they represent a true distortion in the CBR spectrum, and the other is that the flux is a new cosmological background, quite separate from the cosmic microwave background.

How might the CBR spectrum be distorted? One possibility is inverse Compton scattering of the CBR photons by hot intergalactic gas (the Sunyaev-Zel'dovich effect, 1972). This process preserves photon number, but increases the photon energy. The result is a distortion of the type shown in Figure 11, in which the temperature is lowered in the Rayleigh-Jeans region, and increased in the Wien region. The

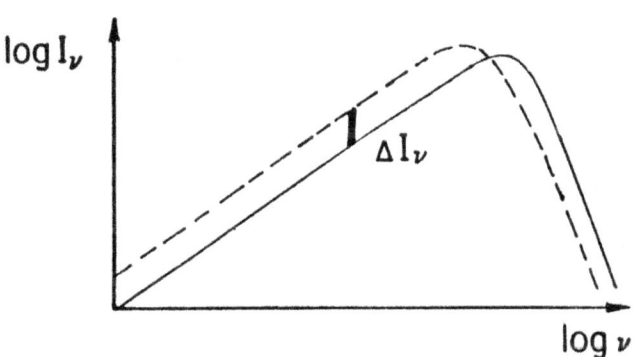

Fig. 11. The effect of inverse Compton scattering (the Sunyaev-Zel'dovich [1972] effect).

magnitude of this effect is parameterized by the quantity introduced by Sunyaev and Zel'dovich,

$$ y = \int_{t_h}^{t_o} c\sigma n_e(t) \left[\frac{kT_e(t)}{m_e c^2} \right] dt \, , $$

where σ is the Thompson scattering cross-section, n_e the electron number density, and t_h the epoch of reheating. To fit the observations of Matsumoto et al, a value of the parameter y defined above of 0.03 is needed. If y takes on this value (Hayakawa et al, 1987; Smoot et al, 1988), the measured CBR temperature in the Rayleigh-Jeans region should be below 2.6 K. The weight of the observational evidence I presented above suggests that T_0 is more like 2.7-2.8 in this region. Very recently, however, work by Kogut et al (1988) suggests a somewhat lower value for T_0 at $\lambda = 3.0$ cm. They find T = 2.61 ± 0.06 K at this wavelength. However, I have some concerns about their calculation of the atmospheric temperature, which must be subtracted from their zenith measurements to give the CBR temperature. If my concerns are justified, the value of Kogut et al might be raised by 20-50 mK, bringing it into better agreement with other measurements in the Rayleigh-Jeans region of the spectrum.

Another possible means of adding a distortion to the microwave background radiation is the radiative decay of light, weakly interacting particles (Kawasaki and Sato, 1987). Particles of mass a few eV and lifetimes of the order of 10^{13} sec could produce the observed spectral distortion; on the other hand, there are indications that radiatively decaying particles with these parameters are excluded by the stability of stars late in their lifetimes.

That leaves us with what I consider a most interesting possibility, namely that the observed far infrared flux is a new cosmological background: the emission from warm interstellar dust in high redshift galaxies. Just such a model has been suggested by Hayakawa et al (1987) and looked at in some detail by Hogan and Bond (1987). The spectrum can be made to fit if we assume a temperature of something like 3.7 K for the dust, an optical depth of order 0.2, and an emissivity law which varies as λ^{-2} (or possibly λ^{-1}).

It is also interesting to note that limits on the large and small-scale anisotropy of the CBR at millimeter wavelengths may be used to set constraints on the anisotropy in this new submillimeter flux. As Figure 10 shows, even at 3.3 mm, the new ir background makes a small contribution to the total measured background. If this far infrared radiation were itself too anisotropic, it would contribute more anisotropy than is observed in the total background at 3.3 mm. Using this approach, Lahav and I (1988) have shown that the maximum allowable dipole anisotropy in the far infrared background is ~ 20%, and the maximum allowed quadrupole moment is 2%. Even more stringent limits may be set by a more sophisticated treatment of the CBR dipole measurements (M. Halpern, in preparation). In any case, the high degree of isotropy inferred from the CBR upper limits suggests that the far infrared background cannot be from local sources, such as the Galaxy. A very similar argument has been carried out for measurements of the small-scale anisotropy by Hogan and Bond (1987). They show that some of the small-scale measurements I have discussed in §3.2 require that the scale of the sources contributing to the far infrared background be less than ~ 3-6 Mpc. Obviously, this new work is so important that many groups will be rushing to perform on the far ir background the kind of null experiments that have already been performed for the CBR itself.

5. ACKNOWLEDGMENTS

This has been a splendid symposium in an equally splendid setting, and I would like to thank all the organizers for making such fine arrangements. Any broad review of this sort owes much to the many colleagues whose work is described. In particular, I am grateful to

Steve Boughn, Rod Davies, Ed Fomalont, Craig Hogan, Ken Kellermann, Buddy Martin and Dave Wilkinson for careful readings of some or all of the draft of this review. Any errors or biases, however, are my own. This review draws on material from earlier reviews I have prepared for Reports on Progress in Physics (Partridge, 1988) and for the 20th Yamada Conference. Much of my own work reported here was supported by grants from the Extragalactic Astronomy Program of the U.S. National Science Foundation.

6. AN HISTORICAL AFTERWORD

Let me conclude by noting that in the past 1% of the age of the University of Bologna, we have extended and sharpened our measurements of the spectrum of the background radiation by roughly a factor of 3, we have measured the dipole moment in the CBR to an accuracy of several percent, and we have lowered limits on anisotropies in the CBR over a wide range of angular scales by a factor of 100 or more. Less parochially, in the past 10% of the age of this university all of modern physics, including both particle physics and cosmology, has emerged. Ninety years ago, the atom was the most elementary of particles, the cosmos was assumed to be static, and there was even confusion as to whether the Galaxy was the Universe or not. Who in 1898 would have dreamed physics and astronomy would come this far, let alone conceived of the directions in which we have come? Who knows what wonders will await us in the next 9 or 90 years? It is even more staggering to think ahead a full 900 years, when the age of this University will have doubled. Think back instead to the time when the University was only half its present age. Nicholas Copernicus was beginning to write his great book De Revolutionibus which began all of modern astronomy. Could he have imagined Grand Unification or CERN or primordial nucleosynthesis or 8-meter telescopes or the VLA?

References

Bardeen, J. 1986, in Inner Space/Outer Space, E. W. Kolb, M. S. Turner, D. Lindley, K. Olive, and D. Seckel, eds., University of Chicago Press.
Barrow, J. D., Juszkiewicz, R., and Sonoda, D. H. 1983, Nature, 305, 397.
Berlin, A. B., Bulaenko, E. V., Vitkovsky, V. K., Kononov, V. K., Pariiskii, Yu. N., and Petrov, Z. E. 1983, in I.A.U. Symposium 104, G. Abell and G. Chincarini, eds., Reidel Publ. Co., Dordrecht, Holland.
Bond, J. R. and Efstathiou, G. 1984, Ap. J. (Letters), 285, L45.
Bond, J. R., and Efstathiou, G. 1987, Mon. Not. Roy. Astr. Soc., 226, 655.
Boughn, S. P., and Cottingham, D. C. 1989, in preparation.
Boynton, P. E., and Partridge, R. B. 1973, Ap. J., 181, 243.
Conklin, E. K., and Bracewell, R. N. 1967, Nature, 216, 777.

Crane, P., Hegyi, D. J., Mandolesi, N. and Danks, A. C. 1986, Ap. J., 309, 822.

Danese, L. and De Zotti, G. 1977, Riv. Nuovo Cimento, 7, 277.

Davies, R. D., Lasenby, A. N., Watson, R. A., Daintree, E. J., Hopkins, J., Beckman, J., Sanchez-Almeida, J., and Rebolo, R., 1987, Nature, 326, 462.

Dicke, R. H., Peebles, P. J. E., Roll, P. G., and Wilkinson, D. T., 1965, Ap. J., 142, 414.

Donnelly, R. H., Partridge, R. B., and Windhorst, R. A. 1987, Ap. J., 321, 94.

Fabbri, R., Guidi, I., Melchiorri, F. and Natale, V. 1980a, Phys. Rev. Letters, 44, 1563.

Fabbri, R., Melchiorri, B., Melchiorri, F., Natale, V., Caderni, N., Shivanandan, K. 1980b, Phys. Rev. D21, 2095.

Fabbri, R., Guidi, I., Melchiorri, F. and Natale, V. 1982 in Proceedings of the Second Marcel Grossmann Meeting on General Relativity, R. Ruffini, ed., North Holland Publ. Co.

Fixsen, D. J., Cheng, E. S., and Wilkinson, D. T. 1983, Phys. Rev. Letters, 50, 620.

Fomalont, E. B., Kellermann, K. I., Anderson, M. C., Weistrop, D., Wall, J. V., Windhorst, R. A., and Kristian, J. A. 1988, Ap. J., submitted.

Fomalont, E. B., Kellermann, K. I., and Wall, J. V. 1984, Ap. J. (Letters). 277, L23.

Fomalont, E. B., Kellermann, K. I., Wall, J. V., and Weistrop, D. 1984, Science, 225, 23.

Franceschini, A., Toffolati, L., Danese, L., and De Zotti, G. 1988, in preparation.

Gush, H. P. 1981, Phys. Rev. Lett., 47, 795.

Hayakawa, S., Matsumoto, T., Matsuo, H., Murakami, H., Sato, S., Lange, A. E., and Richards, P. L. 1987, Publ. Astron. Soc. Japan, 39, 941.

Hogan, C. J. 1980, Mon. Not. Roy. Astr. Soc., 192, 891.

Hogan, C. J. 1984, Ap. J. (Lett.), 284, L1.

Hogan, C. J., and Bond, J. R. 1987, in The Post-Recombination Universe, proceedings of a NATO conference in Cambridge.

Ikeuchi, S. 1981, Publ. Astron. Soc. Japan, 33, 211.

Illarionov, A. F. and Sunyaev, R. A. 1975, Sov. A. J., 18, 691.

Johnson, D. G., and Wilkinson, D. T. 1987, Ap. J. (Letters), 313, L1.

Kaiser, N., and Lasenby, A. N. 1989, in preparation.

Kaiser, N., and Silk, J. 1986, Nature, 324, 529.

Kawasaki, M., and Sato, K. 1987, in Phase Transitions in the Universe and Formation of Hierarchical Structure, K. Sato et al, eds., Univ. of Tokyo Press.

Klypin, A. A., Sazhin, M. V., Strukov, I. A., and Skulachev, D. P. 1987, Soviet Astron. Letters, 13, 104.

Knoke, J. E., Partridge, R. B., Ratner, M. I., and Shapiro, I. I. 1984, Ap. J., 284, 479.

Kogut, A., Bersanelli, M., De Amici, G., Friedman, S. D., Griffith, M., Grossan, B., Levin, S., Smoot, G. F., and Witebsky, C. 1988, Ap. J., 325, 1.

Lasenby, A. N. 1981, Ph.D. Thesis, University of Manchester.

Lasenby, A. N., and Davies, R. D. 1983, Mon. Not. Roy. Astr. Soc., 203, 1137.

Lubin, P. M., Epstein, G. L., and Smoot, G. F. 1983, Phys. Rev. Letters, 50, 616.

Lubin, P. M., and Villela, T. 1985, in The Cosmic Background Radiation and Fundamental Physics, F. Melchiorri, ed., Editrice Compositori, Bologna.

Mandolesi, N. et al 1986, Nature, 319, 751.

Martin, H. M. and Partridge, R. B. 1988, Ap. J., 324, 794.

Matsumoto, T., Hayakawa, S., Matsuo, H., Murakami, H., Sato, S., Lange, A. E., and Richards, P. L. 1988, Ap. J., 329, .

Mitchell, K. J., and Condon, J. J, 1985, A. J., 90, 1957.

Ostriker, J. P., and Cowie, L. L. 1981, Ap. J. (Letters), 243, L127.

Ostriker, J. P. and Vishniac, E. T. 1986, Ap. J. (Letters), 306, L51.

Ostriker, J. P., Thompson, C., and Witten, E. 1987, Phys. Lett., B180, 231.

Pariiskii, Yu. N., Petrov, Z. E., and Cherkov, L. N. 1977, Sov. Astron. Letters., 3, 263.

Partridge, R. B. 1980, Ap. J., 235, 681.

Partridge, R. B. 1988, Rep. Prog. Phys., 51, 647.

Partridge, R. B., and Lahav. O. 1988, submitted to Mon. Not. Roy. Astr. Soc.

Penzias, A. A. and Wilson, R. W. 1965, Ap. J., 142, 419.

Readhead, A. C. S., Lawrence, C. R., Myers, S. T., Sargent, W. L. W., Hardebeck, H. E., and Moffet, A. T. 1989, submitted to Ap. J.

Sachs, R. K., and Wolfe, A. M. 1967, Ap. J., 147, 73.

Silk, J., 1968, Ap. J., 151, 459.

Smoot, G. F., De Amici, G., Friedman, S., Witebsky, C., Sironi, G., Bonelli, G., Mandolesi, N., Cortiglioni, S., Morigi, G., Partridge, R. B., Danese, L., and De Zotti, G. 1985, Ap. J. (Letters), 291, L23.

Smoot, G. F., Levin, S. M., Witebsky, C., De Amici, G., and Rephaeli, Y. 1988, preprint.

Strukov, I. A. and Skulachev, D. P. 1984, Sov. Astron. Letters, 10, 1.

Strukov, I. A., Skulachev, D. P., Boyarskii, M. N., and Tkachev, A. N. 1987, Soviet Astron. Letters, 13, 65.

Sunyaev, R. A. and Zel'dovich, Ya. B. 1972, Comments Astrophys. Space Sci., 4, 173.

Thorne, K. S. 1967, Ap. J., 148, 51.

Uson, J. M., and Wilkinson, D. T. 1984a, Ap. J., 283, 471.

Uson, J. M., and Wilkinson, D. T. 1984b, Nature, 312, 427.

Verschuur, G. L., and Kellermann, K. I. 1974, Galactic and Extragalactic Radio Astronomy, Springer-Verlag, Heidelberg.

Vishniac, E. T. 1987, Ap. J., 322, 597.

Wilkinson, D. T. 1988, in IAU Symposium 130, Reidel Publ. Co., Dordrecht, Netherlands.

DISCUSSION.

Borner: Is there anything known about polarization of the MWB ?

Partridge: Yes, at both large scales (work by Lubin in the early 1980's) and at sca $\leq$ 1' (my work with Nowakovski, Nature 1988). As it happens, there are no dir measurements of linear polarization at arc-min scales, though large polarization wou have been seen by observers like Uson and Wilkinson, because most radio antenna fee are polarized. I think it likely that the polarized component DT/T is $< 10^{-4}$ for measured scales.

Reeves: Please discuss the state of the Sunyaev-Zel'dovich effect and the Kogut dat:

Partridge: The Sunyaev-Zel'dovich distortion can nicely explain the observed far frared excess (see Hayakawa et al. 1987), but there is a problem in the Rayleigh-Tay region. A S-Z distorted spectrum which fits the data of Matsumoto et al. and is cc sistent with CN value at 2.64 mm would give $T_o \simeq 2.6$ K. This value is inconsiste with most measurements at less than 3 mm. A very recent measurement by Kogut al. (1988), however, gives $T_o = (2.61 \pm 0.06)$ K. Results in that paper suggest to me t possibility that the atmospheric contribution has been overestimated; if so, the value T_o should be raised (making it agree with other results better). Thus the discrepan between the observation and S-Z spectrum may remain (Hayakawa et al.).

Predazzi: We have just heard that large voids exist and that the background radiati is very smooth. Are there any obvious relationships between these two issues?

Partridge and Geller: The smoothness of the microwave background and the existen of structure on all scales so far surveyed is a major unresolved problem of cosmolo Note, however, that the microwave background maps the distribution of matter at ve early epochs, before the amplitude of density perturbations had time to grow.

QUARK DECONFINEMENT AND J/ψ SUPPRESSION IN NUCLEAR COLLISIONS

H. SATZ

Fakultät für Physik
Universität Bielefeld
D-4800 Bielefeld, F.R. Germany
and
Physics Department
Brookhaven National Laboratory
Upton, NY 11973, USA

ABSTRACT: We present the physical basis for J/ψ suppression as a signal for quark deconfinement, investigate the relation between colour screening and the binding of heavy quarks, and discuss the momentum dependence of the effect. The results of these considerations are then compared to the presently available experimental results from heavy ion collisions at CERN.

1. J/ψ Suppression as Signal for Quark Deconfinement

Quantum chromodynamics predicts that at sufficiently high density, strongly interacting matter will undergo a phase transition from a hadronic phase, with mesons and baryons as basic constituents, to a plasma of deconfined quarks and gluons. This transition is clearly seen in lattice QCD studies[1] of the energy density ϵ; the behaviour observed for a system of vanishing baryon number density, as a function of the temperature T, is schematically illustrated in fig. 1. At low temperature, we find values consistent with those of an ideal pion gas,

$$\epsilon_\pi/T^4 = \pi^2/10 \simeq 1, \tag{1.1}$$

while for high T, ϵ approaches the form of an ideal quark-gluon plasma,

$$\epsilon_Q/T^4 = 37\pi^2/30 \simeq 12; \tag{1.2}$$

we have here included two quark flavours (u and d). The change-over between the two regimes is quite sudden; for sufficiently light quarks, it appears to be a first order phase transition.[2] The critical temperature is found to be about 200 MeV; this implies that energy density values of 2 - 2.5 GeV/fm^3 are necessary for deconfinement – values more than ten times greater than those of normal nuclear matter.

Is it possible to create such conditions in the laboratory? While always of intrinsic interest to cosmology and astrophysics, the study of critical phenomena in strongly

M. Caffo et al. (eds.), Astronomy, Cosmology and Fundamental Physics, 131–152.

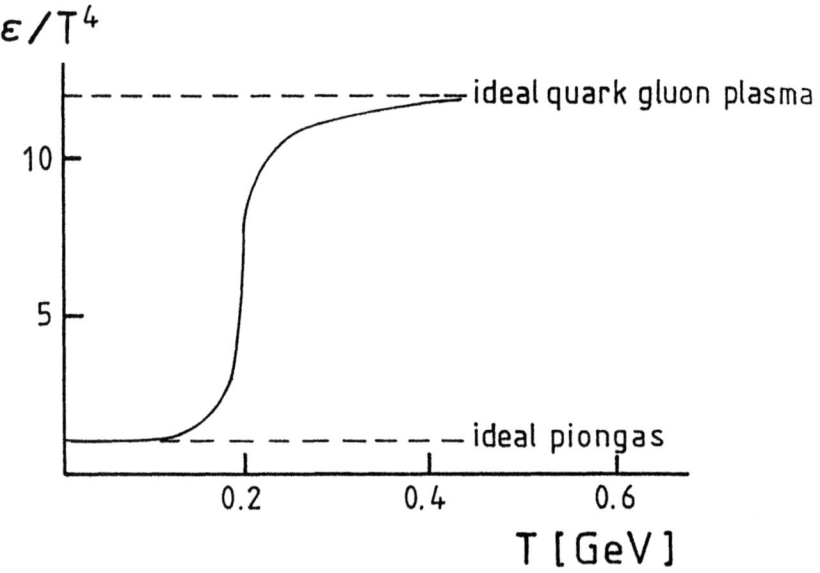

Fig.1: Schematic behaviour of the energy density ϵ of strongly interacting matter.

interacting matter received a tremendous boost by the start of high energy heavy ion experiments at CERN and Brookhaven. We expect these experiments to provide us with the tools to study strong interaction thermodynamics, and we hope that they will produce a medium dense enough for quark deconfinement.

Let us assume then that the head-on collision of two sufficiently heavy nuclei produces a bubble of strongly interacting matter - even though it may be small, short-lived and not fully in equilibrium. How can we ever tell if within this bubble, for a certain length of time, something like quark deconfinement took place? Once the emitted secondaries of the collision reach the counters, spectrometers and calorimeters of the experiment, any primordial state has disappeared. The little bang of nuclear collisions is like its big brother in cosmology: when we can see anything, it's over.

In this paper, we want to discuss and elaborate the proposal[3] that a study of the J/ψ peak in the spectrum of lepton pairs emitted in nuclear collisions can in fact provide the necessary information. We shall first have a look at the general idea, and then consider in some detail a number of different aspects of the deconfinement test. Finally, we shall see what conclusions the experimental results[4] allow so far.

The physical picture leading to J/ψ suppression as test for quark deconfinement in nuclear collisions is quite simple. Within a deconfining medium, quarks cannot bind to form hadrons. If a nuclear collision produces such a medium, then this subsequently expands, cools off, and after passing the confinement point T_c, it hadronises: now the quarks and antiquarks combine to form mesons and baryons. Heavy quark-antiquark pairs ($c\bar{c}$,$b\bar{b}$) are produced by hard, pre-thermal interactions at a very early stage in the collision. In a confining medium, such as the physical vacuum, they subsequently undergo soft hadronic interactions to bind to charmonium ($J/\psi, \psi', ...$) or bottonium ($\Upsilon, \Upsilon', ...$) states, respectively. In a deconfining medium, this binding process is not possible, so the $c\bar{c}$(or the $b\bar{b}$) just "fly apart". At hadronization, the presence of additional thermal c or b quarks is strongly suppressed: $exp(-m_c/T_c) \simeq 0.6 \times 10^{-3}$, with $m_c \simeq 1.5$ GeV, $T_c \simeq 0.2$ GeV. Hence at this stage, a heavy quark cannot find another heavy partner; it instead has to combine with a common light quark to make an open charm or open beauty meson, such as D ($c\bar{u}$, etc.) or B ($b\bar{u}$, etc.). Plasma formation thus leads to enhanced open charm or beauty production, at the expense of charmonium or bottonium states. Since in a normal hadronic interaction, the production of open charm is much more abundant than that of J/ψ's, such an enhancement is not likely to be seen. The resulting reduction of J/ψ formation should, however, be clearly observable and hence provide us with a test for quark deconfinement in nuclear collisions. To make this test unambiguous, we must of course make sure that there is no other way to obtain such an effect.

To illustrate what we expect to see in nuclear collision studies, let us first consider the dilepton spectrum observed in proton-proton interactions (fig. 2), for $M \geq 2.5$ GeV; here M denotes the mass of the $\mu^+\mu^-$ (or e^+e^-) pair. The Drell-Yan continuum, falling as M^{-3}, is due to electromagnetic quark-antiquark annihilation into massive virtual photons, which decay into lepton pairs. Above this continuum we have the very pronounced J/ψ signal, followed by a weaker ψ' peak; these are both due to (non-perturbative) strong interactions. In a deconfining medium, the Drell-Yan continuum should remain functionally unchanged, while $c\bar{c}$ binding becomes impossible. Of course there will always also be non-plasma events – collisions near the surface of the interaction region, etc. – so that in the overall data, the J/ψ peak should become reduced in size, but not completely removed. Matching the functionally identical Drell-Yan continua of the dilepton spectrum from pp interactions and of that from nuclear collisions, we therefore expect to find a behaviour as shown in fig. 2 – if there is indeed deconfinement in the nuclear interaction. We emphasize that it is not necessary to know the relative rate normalization of pp and nuclear collisions; the functional invariance of the Drell-Yan continuum provides an intrinsic scale for comparison. The test thus consists in comparing the corresponding signal-to-continuum ratios.

We have here considered a comparison of pp and AB data, denoting by A and B the mass numbers of the colliding nuclei. The basic idea, however, is simply to compare processes without deconfinement to those in which we suspect plasma formation. We can thus also compare pA with AB collisions, or low multiplicity peripheral with high multiplicity central AB interactions. Since we argued that in a deconfining medium the c and $\bar{c}$ fly apart, we can even compare J/ψ's of high momentum to those of low momentum: sufficiently fast $c\bar{c}$ pairs will still be close enough together to bind into a J/ψ after leaving the plasma bubble; they will thus not feel the deconfinement effect[9].

Let us now turn to some questions that come up when we consider in more detail the scenario just presented.

2. Colour Screening and Heavy Quark Binding

Deconfinement can be thought of as a consequence of colour charge screening in dense matter[10], transforming the vacuum potential $V_0(r)$ into one of much shorter range,

$$V_0(r) \rightarrow V(r) = V_0(r) e^{-r/r_D}. \tag{2.1}$$

When the screening length r_D becomes much smaller than the radius r_B of a given bound state, then this state will dissolve into its constituents. This leads us to expect deconfinement when r_D falls below the size of the common hadrons. Bound states of heavy quarks, however, tend to have much smaller binding radii and hence would "survive" beyond the deconfinement point.

This raises two questions: For what values of r_D do the typical charmonium and bottonium states dissolve? At what plasma temperatures (or densities) are these r_D values obtained?

The first question was recently addressed in the framework of charmonium potential theory[11]. For each heavy quark bound state α, we can define a dissociation energy $E_{dis}^{\alpha}(r_D)$; it measures how far the bound state mass $M^{\alpha}(r_D)$ lies below the large

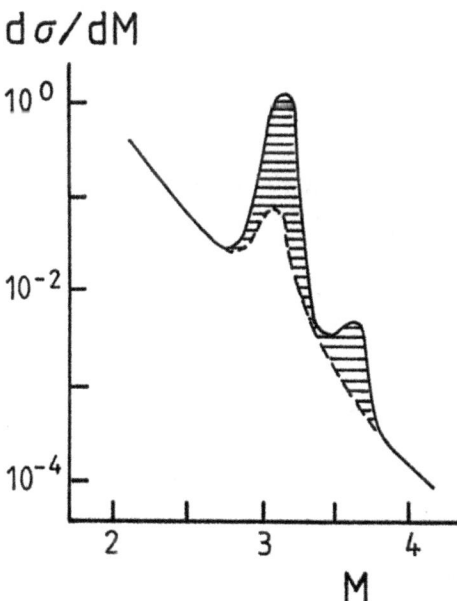

Fig.2: Schematic behaviour of the lepton pair spectrum as function of the pair mass M. The solid line shows the proton-proton form, the dashed line the predicted form of the nucleus-nucleus spectrum, shifted to match the p-p continuum. The shaded area indicates the amount of J/ψ suppression in nuclear collisions.

distance limit of the potential,

$$E_{dis}^{\alpha}(r_D) = V(r_D; r = \infty) - M^{\alpha}(r_D), \qquad (2.2)$$

In the absence of screening $(r_D = \infty)$, there is a strict confinement with $V(r_D = \infty, r = \infty) = \infty$; hence no dissociation is possible. Once screening sets in, however, $V(r_D; r = \infty)$ becomes finite, because the exponential screening factor overcomes the linear rise in r of the confining potential. This makes $E_{dis}^{\alpha}(r_D)$ finite as well. The calculated dependence of $E_{dis}^{\alpha}(r_D)$ on r_D is shown in figs. 3 and 4, for the most important charmonium and bottonium states, respectively. The screening lengths r_D^{dis} necessary to dissociate these states are listed in table 1.

We now want to know at what temperatures these r_D values are attained. This question has been studied extensively in the framework of lattice QCD.[12-16] One there measures the correlation function $C(r, T)$ between two static colour sources separated by a distance r. This function in turn gives rise to the interquark potential $V(r, T)$

$$C(r, T) \sim e^{-V(r,T)/T}, \qquad (2.3)$$

where T is the temperature of the medium containing the sources. Parametrizing $V(r, T)$ in a Debye-screened $1/r$ form[18],

$$V(r, T) \sim (1/r)\, e^{-r/r_D(T)}, \qquad (2.4)$$

we obtain the wanted $r_D(T)$. Studies of this type have been carried out for static quark-antiquark pairs in purely gluonic media[12-15] as well as for media containing light quarks[16]. The overall picture emerging at this time is summarized in fig. 5, together with some of the bound states listed in table 1. We see that all charmonium states except the J/ψ will essentially melt at T_c; this is not surprising, since the higher $c\bar{c}$ excitations have the size of the usual hadrons[19]. The J/ψ itself disappears for $T/T_c \simeq 1.3$, while the decoupling of the Υ requires a much higher temperature $(T/T_c \gtrsim 3)$.

Since about 40 % of the observed J/ψ events in pp interactions come from the decay of an intermediate χ_c,[20] these results imply that plasma formation even with T only slightly above T_c should already have a noticeable effect. For $T/T_c \gtrsim 1.3$, all charmonium states should have melted.

Finally we should note that we have here only considered the deconfinement of static sources due to colour charge screening. Dynamic effects – such as collisions of light quarks with the $c\bar{c}$ system – could certainly bring down further the decoupling temperature for charmonium states[21-23]. Magnetic binding effects[24] as well as different screening behaviour for hidden and open charm states[25] could also still lead to modifications.

State	r_D^{dis} (fm)
J/ψ	0.29
ψ'	0.56
χ_c	0.59
Υ	0.13
Υ'	0.30
χ_b	0.36

Table   1: Screening lengths for the dissociation of heavy quark bound states, from Ref. 11.

	$P_T < 1$ GeV	all P_T	$P_T > 1$ GeV
$S(E_T > 68)/S(E_T < 38)$	0.53±0.11	0.63±0.08	0.72±0.14
$S(54 < E_T < 68)/S(E_T < 38)$	0.66±0.12	0.79±0.10	1.11±0.21
$S(38 < E_T < 54)/S(E_T < 38)$	0.75±0.14	0.85±0.11	0.93±0.16

Table   2: Ratios of the signal-to-background ratios $S(E_T) \equiv [(J/\psi)/cont.]$, for different P_T, from Ref. 4.

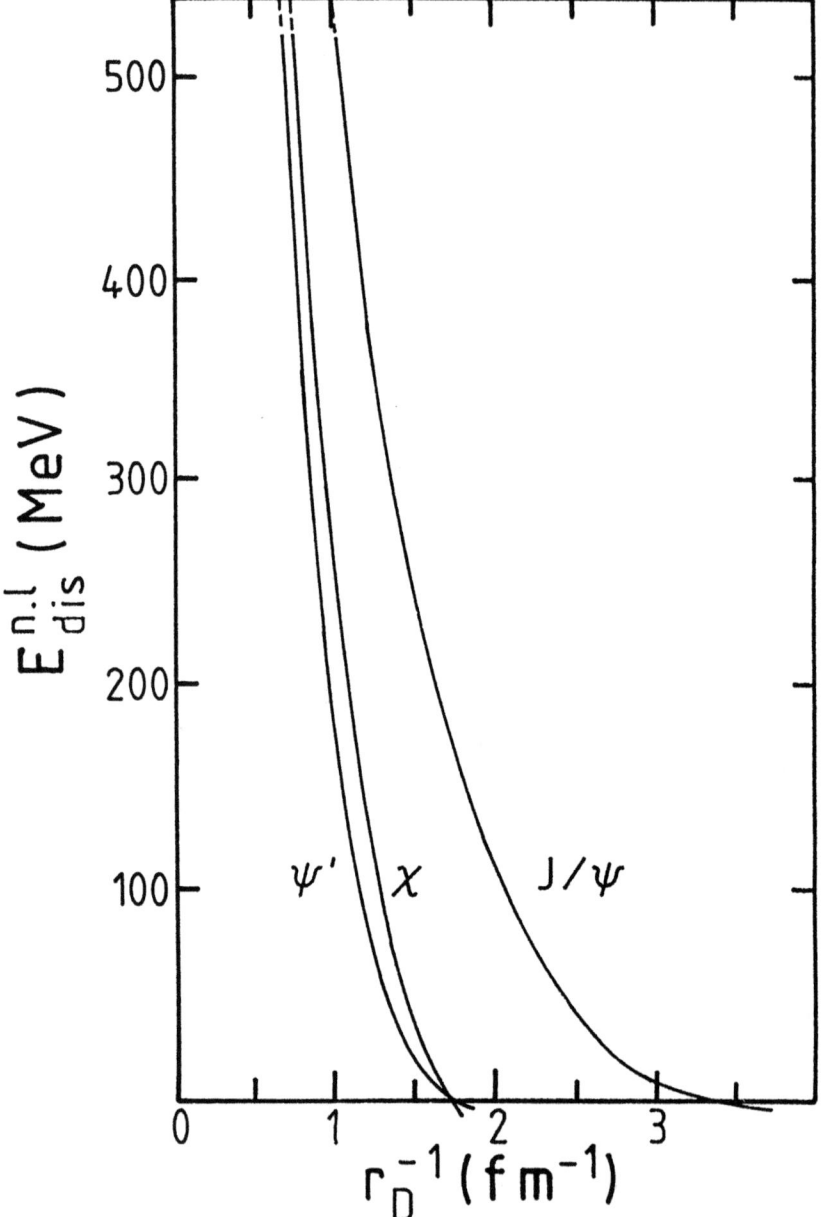

Fig.3: Dissociation energies E_{dis} for charmonium states, as function of the screening length r_D of the medium. From Ref. 11.

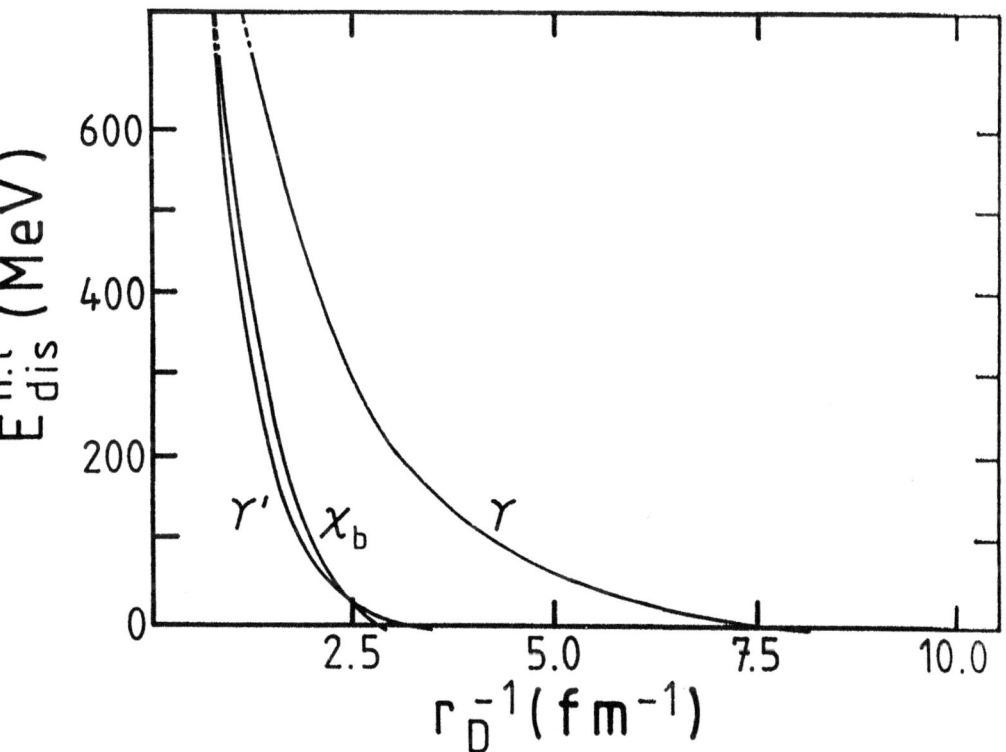

Fig.4: Dissociation energies E_{dis} for bottonium states, as function of the screening length r_D of the medium. From Ref. 11.

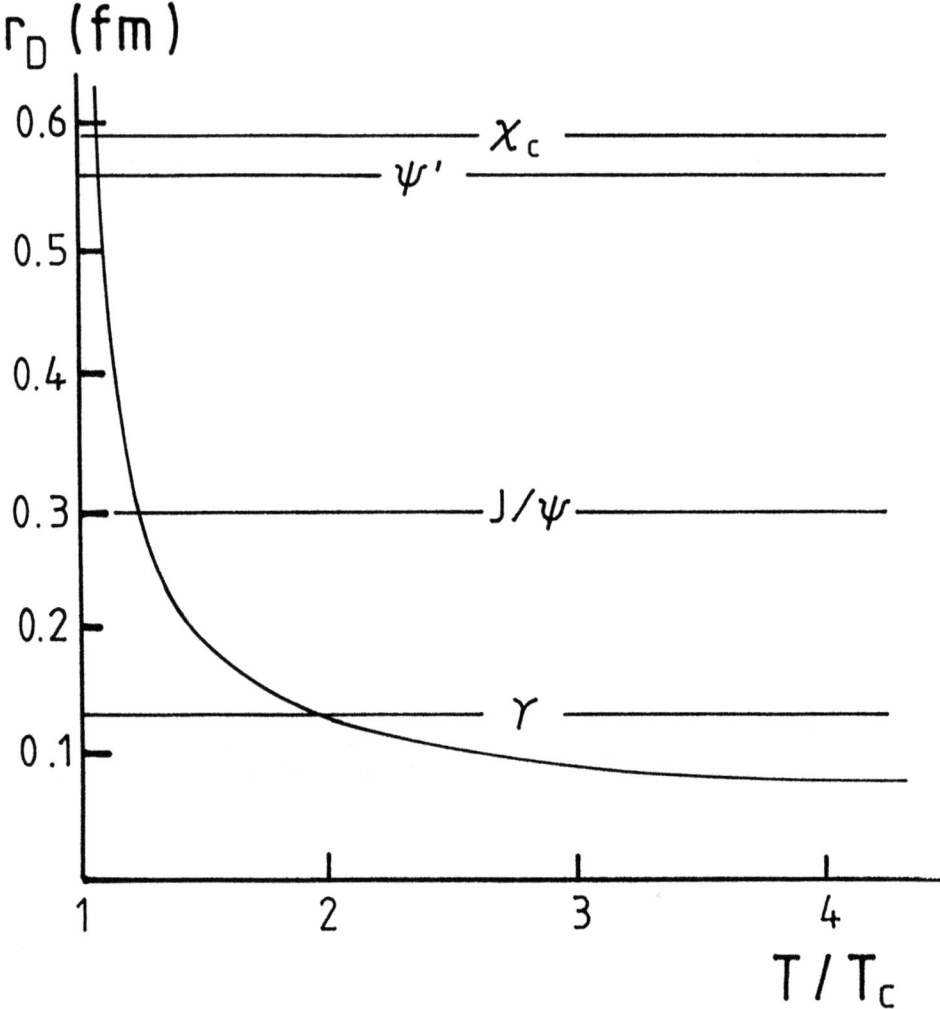

Fig.5: Screening lengths r_D as function of the temperature, together with the dissociation values of some heavy quark bound states. From Ref. 11, Ref. 13 and Ref. 16.

3. The Momentum Dependence of J/ψ Suppression

We had seen that the large separation of c and $\bar{c}$ at hadronisation are crucial in obtaining J/ψ suppression by plasma formation. Any deconfined system we can produce in a nuclear collision will, however, clearly have both a finite spatial size and a finite life-time. For a very fast $c\bar{c}$ pair (relative to the plasma rest-frame) it is therefore possible to escape from the medium before the c and the $\bar{c}$ have separated enough to prevent binding[9]. Let us look at this effect a little more closely; it has recently been studied by a number of authors[26-33].

First we want to know how long it takes for a $c\bar{c}$ pair, produced in a hard collision at some point in space*, to become a J/ψ. The radius of a J/ψ in vacuum is from charmonium spectroscopy[19] known to be $r_{J/\psi} \simeq 0.45$ fm; the momentum of each quark in the J/ψ is[27] $p_c \simeq 0.67$ GeV. The time τ_0, which the $c\bar{c}$ pair would need in its own rest system to separate in free flight a distance $r_{J/\psi}$ is

$$\tau_0 = m_c r_{J/\psi}/p_c \simeq 0.89 \text{ fm} \tag{3.1}$$

using $m_c = 1.32$ GeV[19]. We can consider τ_0 to be the J/ψ formation time, if the $c\bar{c}$ separation is not affected by the binding forces or by the medium. If we assume that at the time of the hard $c\bar{c}$ production the entire $M_{J/\psi} - 2m_c$ goes into kinetic energy, then this would increase p_c and decrease τ_0 by about 10 %. The presence of a medium or more elaborate formation models[28] could, on the other hand, increase the formation time.

In the overall center of mass, which we take to be the cms of the plasma as well, the time τ_0 becomes

$$t_0 = \tau_0 \left(1 + \vec{P}^2/M_{J/\psi}^2\right)^{1/2}, \tag{3.2}$$

where $\vec{P}$ denotes the momentum of the $c\bar{c}$ pair. During the time t_0, the pair will have moved a distance

$$r_0 = \tau_0 \left(|\vec{P}|/M_{J/\psi}\right) \tag{3.3}$$

away from its formation point, measured again in the overall cms. Let us now denote by t_{plasma} and r_{plasma} the life-time and the radius of the plasma bubble, respectively. If either

$$t_0 \geq t_{plasma} \tag{3.4}$$

or

$$r_0 \geq r_{plasma} \tag{3.5}$$

then the $c\bar{c}$ pair considered can still bind to form a J/ψ. In the first case, the plasma has cooled down to hadronize before the c and $\bar{c}$ are far enough apart to prevent binding; in the second, they are out of the plasma bubble before they have separated enough.

---

* We neglect here the space-time interval needed for the hard collision process; it is estimated[9] to be about $1/M_{J/\psi} \simeq 0.06$ fm.

The argumentation just given can be applied directly to the momentum dependence of J/ψ suppression in the transverse direction (orthogonal to the nuclear collision axis). In the longitudinal direction, however, the expected boost invariance will at least at high collision energies lead to a range of moving frames which contain overlapping plasma bubbles; an escape of the $c\bar{c}$ pair is thus not so easy. For details, see Ref. 28).

To obtain the actual behaviour thus predicted for the P_T dependence of J/ψ suppression, it is of course necessary to take into account correctly the geometry of the plasma region and its expansion to the hadronization stage, as well as modifications due to intermediate χ_c production.[26-33] A summary is given in Ref. 33. To illustrate the effect, we normalize the resonance-to-continuum ratio for a process with possible plasma formation to that of a "normal" process not affected by deconfinement (such as the corresponding quantity from p-p collisions) and denote the resulting ratio of ratios by R; $R = 1$ then means no suppression. The behaviour resulting from the finite space-time extension of the plasma bubble produced in nuclear collisions is schematically shown in fig. 6. The transverse momentum limit above which there is no more suppression could in principle be determined either by Eq. (3.4) or by Eq. (3.5); in present experiments, the short plasma life-time seems to make the restriction (10) the relevant one.[26-33]

4. Experimental Results

The experimental study of lepton pairs in the J/ψ mass region was started at CERN in the fall of 1986 by the NA 38 collaboration (Annecy–CERN–Clermont-Ferrand–Ecole Polytechnique/Palaiseau–Lisbon–Lyon–Orsay–Strasbourg–Valencia). In a first run, they studied $\mu^+\mu^-$ pairs produced by oxygen nuclei at 200 GeV/nucleon incident on stationary uranium and copper targets. Including all pair masses, they obtained about 2.5×10^6 oxygen-uranium and 8×10^5 oxygen-copper events. In addition, the reaction proton-uranium for 200 GeV protons was studied in the same set-up. For these data, a first analysis has been published[4]. In a second run in 1987, the experiment was repeated with sulphur at 200 GeV/nucleon incident on uranium. Also further proton-uranium data were taken. These measurements, containing a similar number of dimuon events, are now being evaluated; some preliminary results have just been presented[5-7].

This paper is not meant to give a quantitative analysis of the available data; we only want to outline what at this time seem to be the outstanding features.

In the experiment, the mass and the momentum of the muon pairs is measured; in addition, the associated (neutral) energy E_T emitted in the plane orthogonal to the beam axis is determined by calorimeters. Since this transverse energy provides a measure of the associated hadron multiplicity, high E_T means central collisions with many hadronic secondaries, low E_T more peripheral events of lower multiplicity. Extremely peripheral ("glancing") collisions are excluded simply because they lead to a low dimuon rate.

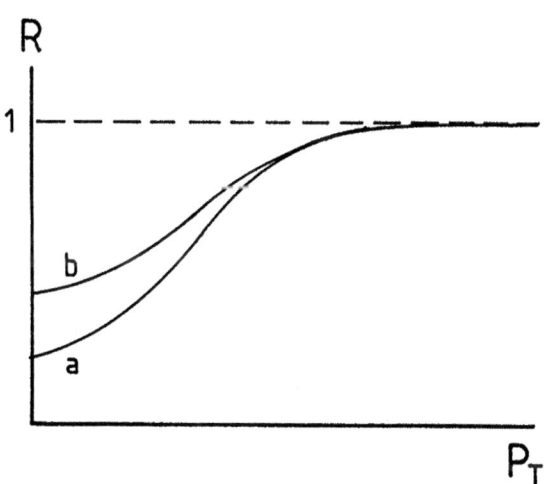

Fig.6:   Schematic behaviour of the J/ψ suppression ratio R as function of P_T, for $P_L = 0$ (a) and for all P_L (b). From Ref. 27.

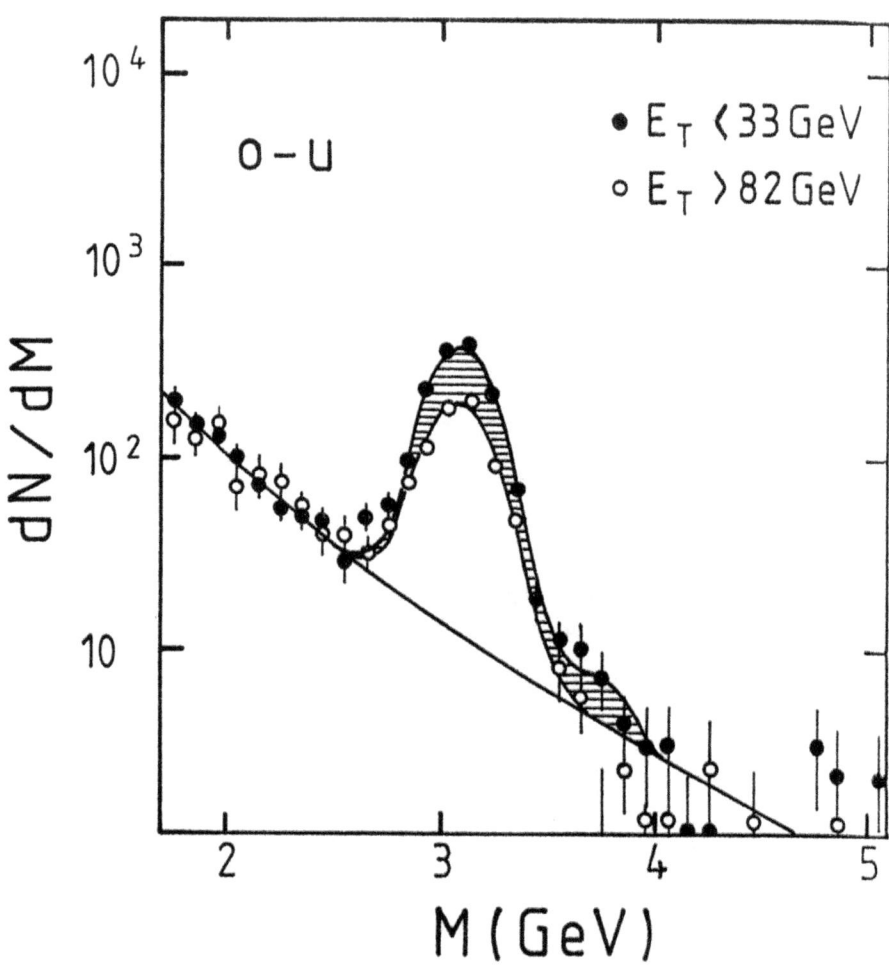

Fig.7: The dilepton spectrum from oxygen-uranium collisions for $E_T < 33$ GeV and with $E_T > 82$ GeV, with superimposed backgrounds. The shaded area indicates the amount of J/ψ suppression. From Ref. 6.

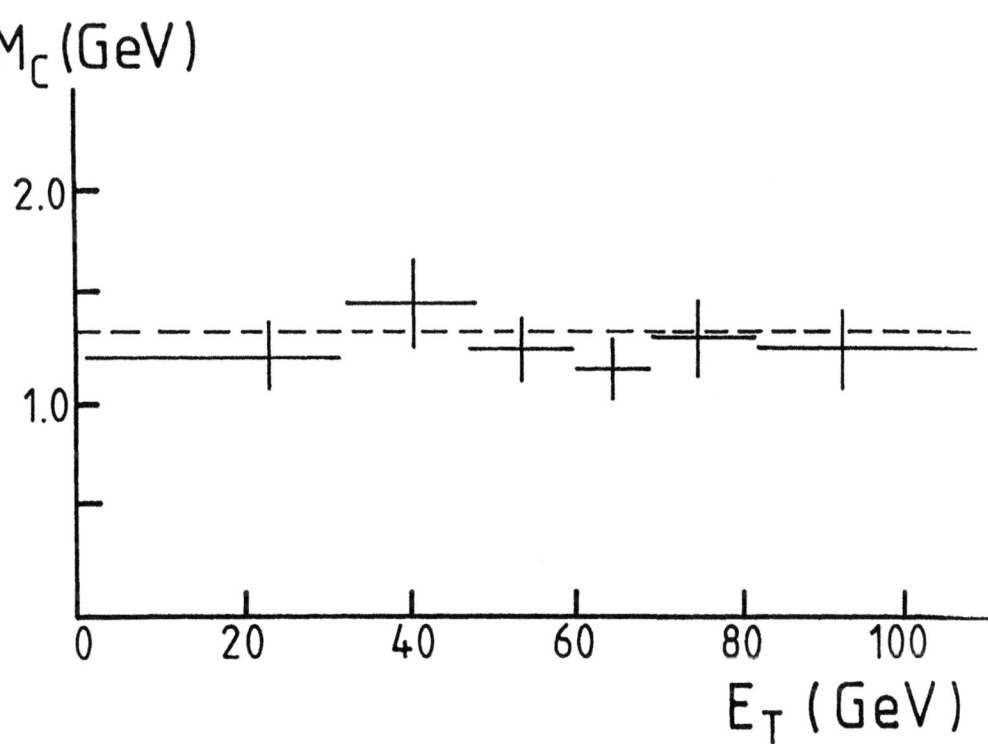

Fig.8: The Drell–Yan fit parameter M_c as function of E_T, for oxygen-uranium data. From Ref. 6. The dashed line shows the corresponding value from fits of proton-proton data.

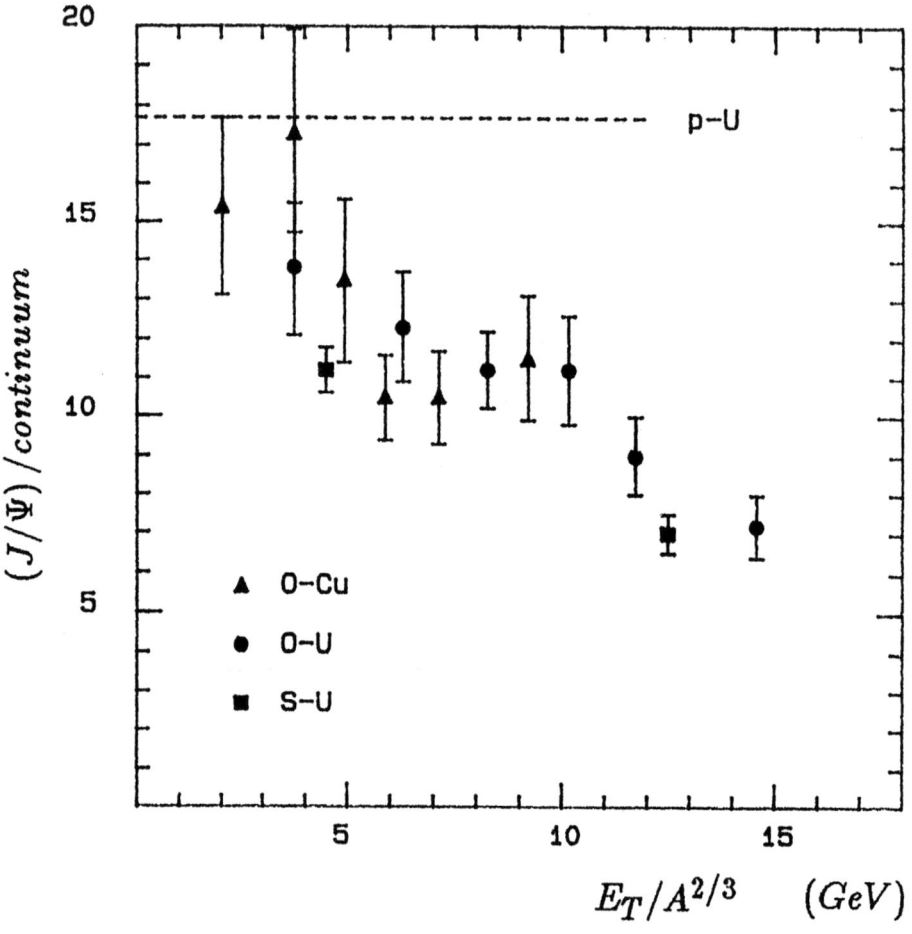

Fig.9: The E_T dependence of J/ψ suppression in oxygen-copper, oxygen-uranium and sulphur-uranium collisions; from Ref. 4, Ref. 5, and Ref. 6. The dashed line indicates the overall J/ψ-to-continuum ratio in proton-uranium collisions, from Ref. 7.

In fig. 7, the muon pair spectra from oxygen-uranium collisions[6] for $E_T < 33$ GeV and $E_T > 82$ GeV are superimposed by matching the fitted Drell-Yan continua. We note that at high E_T, the J/ψ signal has become considerably weaker; it has decreased by the shaded area. Any evidence for the ψ' has disappeared at high E_T. These results contain all the data for the dimuon mass interval shown – i.e., they are integrated over the transverse and longitudinal momenta of the pairs. Using a Drell-Yan fit,

$$(dN/dM) \sim M^{-3} e^{-M/M_c}, \quad M_c = \text{const.}, \tag{4.1}$$

one can determine the functional form of the continuum for different values of associated transverse energy. It is found to remain essentially independent of E_T; the value $M_c \simeq 1.33$ GeV, which is obtained from corresponding proton-proton studies[8], provides a good fit to all E_T bins studied, as seen in fig. 8. Relative to this continuum, the J/ψ signal decreases as E_T increases: when we go from the $E_T < 33$ GeV to the $E_T > 82$ GeV sample, the resonance-to-continuum ratio decreases by about 50 % .

First results on oxygen-copper and sulphur-uranium indicate a very similar pattern[5,6]. Since the projectile volumes are different, we need a suitable variable for a direct comparison of the three cases. In fig. 9, we plot the dependence of the resonance-to-continuum ratio as function of $E_T/A^{2/3}$; this should give us approximately the same transverse energy per unit volume in each case. Indeed we find a rather universal decrease of about 50 % between the lowest and the highest values of $E_T/A^{2/3}$ measured.

Preliminary data[7] from proton-uranium collisions, on the other hand, do not show any significant E_T dependence. The overall value of the resonance-to-continuum ratio for this case is indicated in fig. 9; as expected, the nucleus-nucleus curve converges to this value in the low E_T limit.

Let us now turn to the P_T dependence of the J/ψ suppression. In table 2, we show the ratio of ratios, high E_T to low E_T resonance-to-continuum, for $P_T < 1$ GeV, $P_T > 1$ GeV, and all P_T, from oxygen uranium collisions[4]. There is a noticeable trend towards less suppression at higher P_T. Future analyses of the data will hopefully consider this in smaller P_T steps. What seems to be emerging already now, however, is that the ratio high E_T to low E_T events for the continuum (pairs of mass $1.6 \leq M \leq 5.1$ GeV, excluding the J/ψ region) is essentially P_T-independent. On the other hand, the same ratio in the J/ψ region is strongly P_T-dependent. It is smallest at $P_T = 0$, then increases, and for $P_T \geq 3$ GeV has reached the value of the corresponding continuum ratio - i.e., the suppression present at low P_T has dissapeared. In fig. 10 we show this behaviour[4], suitably normalized[27].

In contrast to this, the ratio of high E_T to low E_T events for the J/ψ region in proton-uranium collisions does not show any P_T dependence[4,6].

The data so far available can certainly only give us a first idea of what may be happening. With this caveat, we summarize the results so far obtained from nucleus-nucleus collisions:

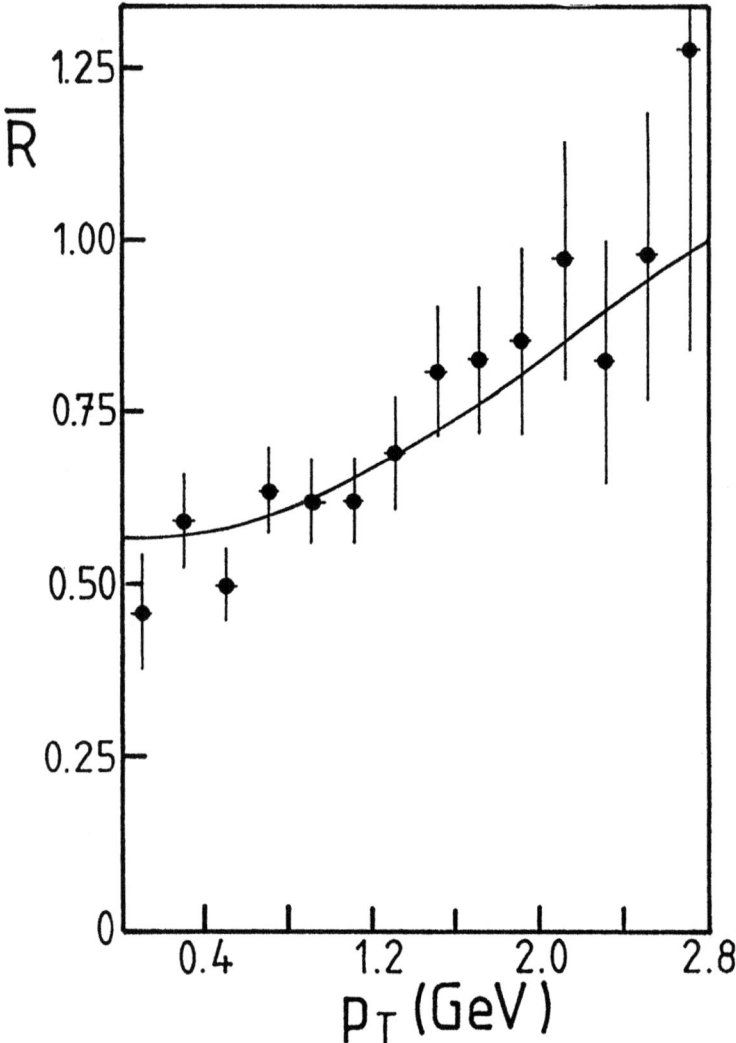

Fig.10: The P_T dependence of J/ψ suppression in oxygen-uranium collisions; here $\bar{R}$ denotes the ratio of the $E_T > 68$ GeV and the $E_T < 38$ GeV results, normalised to unity in the case of no suppression. From Ref. 27.

- for the continuum, the functional form in M is approximately E_T and P_T independent and agrees with that found in proton-proton collisions;
- the J/ψ signal relative to this continuum is quite strongly dependent on both E_T and P_T;
- with increasing E_T, the J/ψ peak above the continuum is significantly reduced, by about 50 % at the largest E_T values measured so far;
- the J/ψ suppression is strongest at low P_T ($\sim 50 - 60\%$) and disappears for $P_T \geq$ 3 GeV;
- in contrast to the behaviour found in nuclear collisions, J/ψ production by proton-uranium collisions has so far not shown any noticeable dependence on either E_T or P_T.

These observations are in accord with what we would expect for quark deconfinement in nuclear collisions. Could this be energetically possible? In Ref. 6, the highest E_T bin ($E_T \geq 82$ GeV) of the oxygen-uranium data corresponds to an overall transverse energy (neutral plus charged secondaries) of about 150 GeV. With the NA 38 set-up, this is spread over about two units in rapidity, so that we have $E_T^{tot} \simeq 75$ GeV per unit rapidity. With a longitudinal formation length $l_0 \simeq 1$ fm and with

$$\epsilon = E_T^{tot} / \pi \left(1.15 A^{1/3} \right)^2 l_0, \qquad (4.2)$$

this leads to an energy density of at least $\epsilon \simeq 2.8$ GeV/fm^3. As mentioned above, lattice calculations predict deconfinement at energy densities of 2 - 2.5 GeV/fm^3; it is thus theoretically conceivable that quark matter formation could occur in the NA 38 experiment.

5. Concluding Remarks

Before we can consider the observed J/ψ suppression as evidence for quark deconfinement, we must obviously check if there are other mechnisms which could give rise to such an effect. This question has therefore attracted much interest, once the experimental results became known, and by now a number of alternative suppresion schemes has appeared[34-43]. A detailed discussion of this work is beyond the scope of the present survey; we refer the interested reader to Ref. 44. Here we note only that conventional absorption schemes do not suffice, but that a combination of initial state gluon scattering with final state absorption in extremely dense hadronic matter can in fact also account for the present data[41-43]. Whether global colour screening (deconfinement) or local scattering (absorption) is the decisive mechanism for J/ψ suppression can only be answered by further studies with heavier nuclei. Nevertheless, we can certainly conclude that the experimental results obtained so far are both interesting and encouraging for our study of strongly interacting matter at high density.

REFERENCES

1. See e.g. H. Satz, Ann. Rev. Nucl. Part. Sci. <u>35</u> (1985)245.

2. See e.g. F. Karsch, Z. Phys. C - Particles and Fields <u>38</u> (1988)147.

3. T. Matsui and H. Satz, Phys. Lett. <u>178B</u> (1986)416.

4. M. C. Abreu et al. (NA 38), Z. Phys. C - Particles and Fields <u>38</u> (1988)117. (All E_T values quoted in this reference have to multiplied by 1.35).

5. A. Romana (NA 38), report at the International Conference on Physics and Astrophysics of Quark-Gluon Plasma, Bombay/India, 8.-12.2.1988.

6. P. Bordalo (NA 38), report at the Rencontre de Moriond, Les Arcs/France, 14.-18.3.1988.

7. L. Kluberg (NA 38), report at the Discussion Meeting on Lead Beam Experiments at the SPS, CERN, 28.-29.4.1988.

8. N. S. Craigie, Phys. Rep. <u>47</u> (1978)1.

9. F. Karsch and R. Petronzio, Phys. Lett. <u>193B</u> (1987)105.

10. See e.g. H. Satz, Nucl. Phys. <u>A418</u> (1984)447c.

11. F. Karsch, M. T. Mehr and H. Satz, Z. Phys. C - Particles and Fields <u>37</u> (1988)617.

12. T. DeGrand and C. DeTar, Phys. Rev. <u>D34</u> (1986)2469.

13. K. Kanaya and H. Satz, Phys. Rev. <u>D34</u> (1986)3193.

14. J. Engels et al., Nucl. Phys. <u>B280</u> (1987)577.

15. N. Attig et al., "Polyakov Loop Correlations in Landau Gauge and the Heavy Quark Potential", CERN Preprint CERN-TH.4955/88 (January 1988); Phys. Lett. B, in press.

16. R. V. Gavai et al., Phys. Lett. <u>205B</u> (1988)295.

17. J. C. Gale and J. Kapusta, Phys. Lett. <u>198B</u> (1987)89.

18. J. Kapusta and T. Toimela, "Friedel Oscillations in Relativistic QED and QCD", Minnesota Preprint (February 1988).

19. S. Jacobs, M. G. Olsson and C. Suchyta III, Phys. Rev. <u>D33</u> (1986)3338.

20. Y. Lemoigne et al., Phys. Lett. <u>113B</u> (1982)509.

21. G. Röpcke, D. Blaschke and H. Schulz, Phys. Lett. <u>202B</u> (1988)479.

22. G. Röpcke, D. Blaschke and H. Schulz, "Dissociation Kinetics and Momentum Dependent J/ψ Suppression in a Quark-Gluon Plasma", Preprint Dresden/Rossendorf 1988.

23. B. Svetitsky, Phys. Rev. <u>D37</u> (1988)2484.

24. T. H. Hansson, Su H. Lee and I. Zahed, Phys. Rev. <u>D37</u> (1988)2672.

25. R. Vogt and A. Jackson, "The Charmonium and D Mesons at Finite Temperature", Stony Brook Preprint (July 1987).

26. J. Blaizot and J. Y. Ollitrault, Phys. Lett. <u>199B</u> (1987)499.

27. F. Karsch and R. Petronzio, Z. Phys. C - Particles and Fields <u>37</u> (1988)627.

28. P. V. Ruuskanen and H. Satz, Z. Phys. C - Particles and Fields <u>37</u> (1988)623.

29. T. Matsui, Z. Phys. C - Particles and Fields <u>38</u> (1988)245.

30. M.-C. Chu and T. Matsui, Phys. Rev. <u>D37</u> (1988)1851.

31. J. Kapusta, Phys. Rev. <u>D36</u> (1987)2857.

32. J. Cleymans and R. L. Thews, "Simple Model for the Suppression of J/ψ' s in Relativistic Heavy Ion Collisions", Arizona Preprint (January 1988).

33. F. Karsch, "Momentum Dependence of J/ψSuppression", CERN Preprint TH 4946/88 (January 1988).

34. J. Ftáčnik, P. Lichard and J.Pišut, "An Alternative Mechanism of J/ψ Suppression in Heavy Ion Collisions", Bratislava Preprint (1988).

35. C. Gerschel and J. Hüfner, "The Contribution of the Final State Interactions to the Suppression of the J/ψ Meson Produced in High-Energy Nucleus-Nucleus Collisions", Orsay Preprint (February 1988).

36. S. Gavin, M. Gyulassy and A. Jackson, "Hadronic J/ψ Suppression in Ultrarelativistic Nuclear Collisions", Berkeley Preprint LBL-24790 (February 1988).

37. R. Vogt et al., "J/ψ Interactions with Hot Hadronic Matter", Stony Brook Preprint (February 1988).

38. A. Capella et al., "Nuclear Effects in J/ψ Suppression", CERN Preprint CERN TH.4974/88 (February 1988).

39. S. Raha and B. Sinha, Phys. Lett. <u>198B</u> (1987)543, and "Quark-Gluon Plasma Diagnostics and J/ψ Suppression in Nuclear Collisions", Saha Institute Preprint Calcutta (1988).

40. R. V. Gavai and S. Gupta, "J/ψ Production in the Central Region of PP and AA Collisions", Tata Institute Preprint TIFR/TH/88-18 (1988).

41. S. Gavin and M. Gyulassy, "Tranverse Momentum Dependence of J/ψ Production in Nuclear Collisions", LBL-Preprint LBL-25663, August 1988.

42. J. P. Blaizot and J. Y. Ollitrault, "The P_T Dependence of J/ψ Production in Hadron-Nucleus and Nucleus-Nucleus Collisions", BNL Preprint, August 1988.

43. J. Hüfner,Y. Kurihara and H. J. Pirner, "Gluon Multiple Scattering and the Transverse Momentum Dependence of J/ψProduction in Nucleus- Nucleus Collisions", Heidelberg Preprint, August 1988.

44. J.-P. Blaizot, Report at the International Conference on Ultra-Relativistic Nucleus-Nucleus Collisions, Quark Matter '88, Lenox, MA, USA, September 1988 (to appear in Nuclear Physics).

152

DISCUSSION.

Reeves: Can you use the E_t and P_t dependence of J/ψ suppression to estimate the critical temperature?

Satz: The amount of suppression as function of E_t should provide a measure of the temperature of the system, while P_t gives information about the plasma life-time.

Derado: Why other experiments do not see any indication of deconfinement, or why NA38 does not see any indication of open charm increase in the J/ψ region?

Satz: It is difficult to answer to the first part of the question, because of the different conditions of the specific experiments. For the second part of the question, even in pp open charm production is so high that it is impossible in present conditions to observe the difference between E_t small and E_t large in the NA38 experiment.

Predazzi: Although many effects are expected to reduce its relevance, would you not expect a similar reduction effect to be at work in the ϕ region?

Satz: That is not clear, because: 1) the ϕ background is due to soft hadronic processes, so that its A dependence is not clear, in contrast to the Drell-Yan background of the J/ψ; 2) the suppression of thermal s-quarks at hadronisation is only 0.1, in contrast to 0.0001 for thermal c-quarks; hence a recombination is much more possible for the ϕ; 3) the formation of $s\bar{s}$ pairs is not as well understood as that of $c\bar{c}$ pairs; thermal equilibrium could at high T even enhance $s\bar{s}$ formation; this has even led to ϕ enhancement by plasma formation. In summary: for the J/ψ, we have a perturbative $c\bar{c}$ formation process on top of a perturbative background; hence only the binding of $c\bar{c}$'s to make J/ψ's is affected by plasma formation. In the ϕ, $s\bar{s}$ formation, the binding and the background probably are affected. That makes things much more complicated.

A BRIEF STATUS REPORT ON THE SLAC LINEAR COLLIDER (SLC)*

Gerson Goldhaber
Physics Department, University of California and
Lawrence Berkeley Laboratory†
Berkeley, California 94720 USA

ABSTRACT. Some aspects of SLC operation and running conditions in 1988 are discussed.

1. The SLC

The SLC, shown in Fig. 1, operated with 3 bunches - 2 electron bunches (separated by 59 ns) and one positron bunch[1]. All bunches are stored in North (electron) and South (positron) damping rings. The positron bunch and the first electron bunch are accelerated to the full LINAC energy. The second electron bunch is accelerated to 33 GeV and is then deflected by a kicker magnet to the positron target to supply the next positron bunch. The positrons are then collected and brought back to the front end of the LINAC to be accelerated.

The e^+ and e^- beams are extracted from the damping rings at the front end of the SLC and accelerated, one behind the other, in the LINAC. At the end of the LINAC they are deflected into their respective arcs and brought around into the final focus region whose function it is to generate tiny spots at the interaction point. (Design specifications are 2μ x 2μ).

The energy upgrade of the LINAC readily achieved 50 GeV beams. The machine is routinely run with 46 GeV $e^\pm$, corresponding to an estimate of the Z° mass of 92 GeV/c². In order to achieve

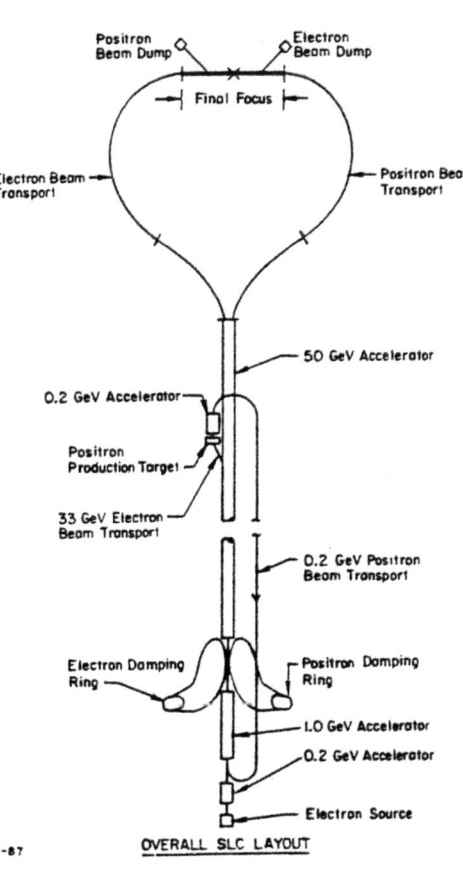

Figure 1. SLC Layout.

153

M. Caffo et al. (eds.), Astronomy, Cosmology and Fundamental Physics, 153–157.

small spots at theinteraction point, the damping rings are required to provide the LINAC with very low-emittance beams. Emittances measured at the exit of the damping rings are now within design specification for both beams. At this time emittance growth occurs in the LINAC so that typically the emittance of the beams is about $\lesssim$ twice nominal as they enter the arcs. This problem is being worked on at present.

With the present setup of the arcs and final focus, small beam sizes are achieved routinely for both e^+ and e^-, $3\mu \times 5\mu$ ($4\mu \times 4\mu$) being the best achieved for $e^-(e^+)$ beams.

2. Luminosity

The luminosity at the interaction point (IP) can be written as follows:

$$L = 1.5 \times 10^{27} \text{ cm}^{-2}\text{s}^{-1} \frac{f}{30\text{Hz}} \frac{N_-}{10^{10}} \frac{N_+}{10^{10}} \frac{4\mu}{\sigma_x} \frac{4\mu}{\sigma_y} \tag{1}$$

where f is the collision frequency, N_- and N_+ are the number of electrons and positrons per bunch at the IP, respectively, and σ_x and σ_y are the transverse beam sizes at the IP for the larger of the two beams. The achieved values for these parameters are summarized in Table I. The first column lists the best achieved value, the second column lists typical values for recent colliding-beam runs, and the third column lists possible values expected for 1989. The rate of Z° production at the peak of the Z° resonance can be written in terms of the luminosity as follows:

$$Z^\circ \text{ rate} \approx \frac{4}{\text{day}} \frac{L}{1.5 \times 10^{27} \text{ cm}^{-2}\text{s}^{-1}} \varepsilon \tag{2}$$

where ε is the average efficiency for colliding beams.

The entire machine was recently operating at a repetition rate of 30 Hz with parts of the LINAC running at 60 Hz in preparation for an increase to that frequency. One of the major difficulties so far has been a low value of ε (typically 1-4%). During the present SLC shut down a considerable effort is directed at improvement of the reliability of critical components. This should result in considerably higher values of ε.

TABLE I. SLC parameters at the interaction point for colliding beams

Parameter	Best Achieved Values	Typical Value for Recent Running	Possible Values for 1989
f	30 Hz	30 Hz	60 - 120 Hz
N_-	1.5×10^{10}	0.9×10^{10}	$2.0 - 3.0 \times 10^{10}$
N_+	0.9×10^{10}	0.5×10^{10}	$1.5 - 2.0 \times 10^{10}$
$\sigma_x \times \sigma_y$:			
electron beam	$3\mu \times 5\mu$	$6\mu \times 6\mu$	$3\mu \times 3\mu$
positron beam	$4\mu \times 4\mu$	$8\mu \times 8\mu$	$4\mu \times 4\mu$

3. Wire Scanner

The beam size is measured by scanning the beam across a thin carbon fiber, "wire" which can be "flipped-up" into the path of the e^+ and e^- beams at the collision point. Each beam is measured separately on both a horizontal and vertical wire[2]. The wire flipper has three wire diameter sizes - 4μ, 7μ and 28μ. Usually we use one of the two smaller diameter wires. As the beam is scanned across the wire the current in the wire is digitized allowing one to extract a beam profile. Thus by allowing for the contribution from the wire the beam size in both x and y is obtained (Fig. 2a). At larger $e^\pm$ beam currents $\gtrsim 5 \times 10^9$ the wire signal becomes non-linear and bremsstrahlung from the intercepted beam is used instead of the current in the wire. The bremsstrahlung photons are detected upstream in a monitor which generates Čerenkov light after the photons traverse a converter (Fig. 2b). Wire scans are done under automatic software control, the magnets are capable of $\leq 1\mu$ size scan steps. Fits are performed on the data providing the beam sizes. Both beams can be measured in less than a minute with good reproducibility.

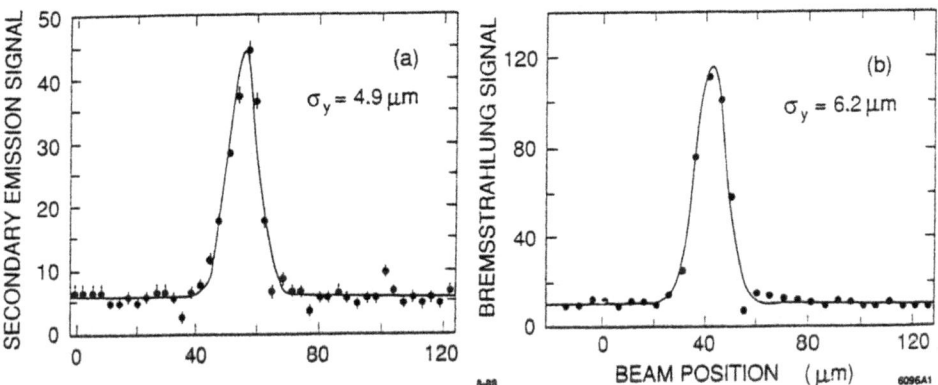

Figure 2. Typical electron beam profiles for (a) the secondary emission signal and (b) the bremsstrahlung signal.

4. Beam-Beam Deflection

Once the two beams are small they are brought into collision. To do this we utilize the deflection of one beam by the other as they pass by each other[3]. One beam is kept at a fixed x, y location while the other is scanned across it. The deflection of this beam is monitored by beam position monitors about 3 meters downstream of the collision point. The deflection is mapped out as a function of the position of the moving beam. The point of zero deflection gives the maximum overlap of the beams. For typical beam parameters, the maximum deflections measured are 50-100 μradians. This procedure is under computer control and takes about twenty seconds. The process can thus be repeated at regular intervals to ensure that the beams remain in collision. Figure 3 shows an example of this deflection.

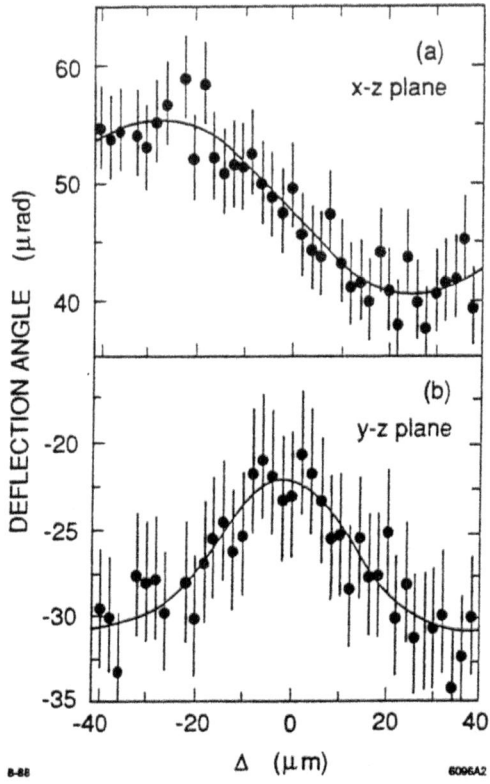

Figure 3. Example of the deflection angle in the (a) x - z plane (horizontal) and (b) y - z plane (vertical) for a positron beam scanned across an electron beam in the x direction. The beams are offset by approximately 10μ in the y direction.

5. Backgrounds

Machine backgrounds have been studied by the MARK II group. The detector is both triggered using a random coincidence with the beam-crossing signal and a physics trigger which is designed to capture almost 100% of all visible Z° decays. The physics trigger used to study the backgrounds has two main pieces, a) small angle luminosity monitors (for Bhabha scattering) and b) a Z° trigger.

In addition to the detector elements, the radiation levels in monitors strung along the e^+ and e^- final focus beamlines are also recorded. In this way background sources can be isolated by studying correlations. Two major background sources have been observed: a) electromagnetic debris resulting from $e^\pm$ striking the beampipe and synchrotron masks close to the MARK II and b) muons which are produced when tails of the $e^\pm$ beams impinge on final focus, fixed aperture machine protection masks and movable final focus scrapers needed to trim halo from the beams. In most cases the muons are produced 30-60 meters from the detector, are deflected by magnetic elements on the beamline and enter the MARK II roughly parallel to the beam axis. The number of muons hitting the detector varies from 5 to 100 depending on beam conditions.

During a current SLC shut down both additional collimators (near the beam switchyard) and torroidal iron magnets (placed around the beam pipe near the muon sources) are being installed. These new additions should reduce the muon background substantially.

6. Recent Achievements

6.1. BNS OR LANDAU DAMPING[4])
This involves introducing a momentum spread in the early part of the acceleration in the LINAC and removing it in the latter part. This results in reducing resonant emittance growth due to the action of the head on the tail of the beam. The technique resulted in reducing sensitivity to launch conditions into the LINAC, yielding gains in intensities of about a factor of two in both e^- and e^+ beams.

6.2. FAST FEEDBACK LOOP
The introduction of a fast feedback loop (pulse to pulse corrections) has helped improve beam stability in the LINAC. Slower loops controlled by the mainframe VAX were replaced by microcomputer based loops to provide positional and energy stability of the beam entering the arcs.

6.3 EXTRACTION LINE ENERGY SPECTROMETERS
Two spectrometers were introduced, one in each of the two extraction lines. The purpose is to measure both beam energies on a pulse to pulse basis. The method is to measure the relative position of two synchrotron light strips on a fluorescent screen, before and after deflection in the spectrometer. These devices are very successful and will allow significant $M(Z^\circ)$ and $\Gamma(Z^\circ)$ measurements. So far they achieved beam energy measurements with an absolute accuracy of $\pm$ 40 MeV per beam and after a careful survey a $\pm$ 15 MeV accuracy is expected.

6.4 TWO PHOTON EVENT
One candidate of the type $e^+e^- \rightarrow e^+e^-\ e^+e^-$ was observed in the MARK II detector for a total luminosity corresponding to about 0.5 Z°. This event - as well as beam gas events observed - indicates clearly that even with present background conditions Z° events could be readily observed. When the SLC running starts again in early 1989 background conditions and luminosity should be considerable improved.

References

1. SLAC Linear Collider Conceptual Design Report, SLAC-Report-229, 1980.
2. R. Fulton *et al.*, SLAC-PUB-4605, 1988, submitted to *Nucl. Instr. and Meth.*
3. P. Bambade, SLAC-CN-303, 1985.
4. V.E Balakin, A.V. Novokhatsky and V.P. Smirnov, XIIth Int. Conf. on High Energy Accelerators, Fermilab, 1983 (p. 119).

* Invited talk at International Symposium on Modern Cosmology in Retrospect in associated with the third ESO-CERN Symposium, Bologna, Italy. The information presented is updated to September 30, 1988.

† This work was supported by the Director, Office of Energy Research, Office of High Energy and Nuclear Physics, Division of High Energy Physics, of the U.S. Department of Energy under Contract No. DE-AC03-76SF00098.

THE LARGE SCALE STRUCTURE OF THE UNIVERSE

Nicola Vittorio
Dipartimento di Fisica, Universita' dell'Aquila
P.le dell'Annunziata 1, 67100 L'Aquila,Italy

Abstract

We discuss the constraints that three different class of observations can set on various model for the formation of the large scale structure of the Universe. These observations involve measurements of galaxy peculiar velocity and gravity, and the cosmic microwave background angular distribution. We use as a guide line a biased, cold dark matter scenario, and we compare it with an open, baryon dominated universe. The first model fails by a large margin in reproducing large scale, large amplitude bulk motions, while the second performs, at least in this context, much better and naturally predicts large scale flows. The magnitude of the dipole anisotropy of the galaxy distribution seems to constitute a rather weak test, unable to distinguish between models which differ for the amount of large scale power. On the contrary, the direction of this dipole provides a more sensitive and discriminant test. Finally, we discuss the large scale pattern of the cosmic microwave background and present results of Monte Carlo simulations of the microwave sky, which seem to be the only way for properly comparing theory and observations.

1. Introduction

It is usually assumed that the pattern outlined in the sky by galaxies, clusters and superclusters of galaxies is determined by the gravitational amplification of primordial density fluctuations. The statistical properties of these fluctuations, at some arbitrary but very early epoch, constitute the initial conditions for the origin of the large scale structure of the Universe. Once that a cosmological model is specified, e.g., by fixing the density parameter Ω_0 and the Hubble constant H_0 (we will assume hereafter a vanishing cosmological constant), the initial conditions are fully specified by fixing the nature (adiabatic or isocurvature), the amplitude and the spectrum of these density fluctuations. Being an expanding system, the Universe retains memory, at least on large scales, of the initial conditions. Since different initial conditions imply very different outcomes, observations of the large scale structure provide a unique tool for constraining cosmological scenarios and models for the early universe.

The possible choices for the initial conditions are considerably reduced in the framework of the inflationary scenario, which up to now represents the greatest input from particle physics on cosmology (see Blau and Guth, 1987 and references therein). Inflation resolves in a natural way, e.g., the flatness problem of the standard Big-Bang cosmology: the density parameter is today indistinguishable from unity, as a consequence of the early accelerated expansion phase. But, most of all, it is suggested

159

M. Caffo et al. (eds.), Astronomy, Cosmology and Fundamental Physics, 159–180.
© *1989 by Kluwer Academic Publishers.*

that microphysical processes during inflation may be responsible of the generation of density fluctuations. These fluctuations are then expected: i) to have a scale invariant power spectrum, as a consequence of the self similar de Sitter expansion during the inflationary era; ii) to be gaussian distributed, as the quantum fluctuations of the scalar field; iii) to correspond to fluctuations in the total energy density and, then, to represent real wrinkles in space–time (adiabatic fluctuations). The still missing prediction, from the early universe physics, involves the amplitude of these fluctuations. Having density fluctuations of the right amplitude to form structure seems to require a fine tuning of the potential associated with the scalar field responsible for inflation.

In the past few years several scenarios have been proposed for the formation and evolution of the large scale structure. All these models assume that the universe is flat, accordingly to the theoretical prejudice motivated by inflation, and are dominated by massive weakly interacting particles, for not violating primordial nucleosynthesis constraint on the abundance of baryons in the universe (see Boesgard and Steigmann, 1987, and references therein). On this line, two main models emerged, which reproposed the two alternative views for the fomation of the large scale structure present at the end of the 70's (see, e.g., Peebles, 1981; Shandarin, 1981): hierarchical clustering (or bottom–up scenario) vs. fragmentation (or top–down scenario). Both the models are fully specified and, then, potentially falsifiable.

After the claimed evidence for a non zero electronic neutrino rest mass (Lubimov et al., 1980), several authors examined the possibility that massive light ($\sim 30eV$) neutrinos may dominate the dynamics of the universe today. In this case, (see, e.g., Bond and Szalay, 1988, and references therein) a top–down scenario is expected, since neutrino free streaming erases density fluctuations on small scales: galaxies are then expected to form after pancake collapse, via fragmentation. This model, after the first enthusiasms, was criticized on several grounds (Kaiser, 1983; White, Frenk, and Davies, 1983), and some variations of this scenario were proposed, where, e.g., unstable neutrinos decay through a non radiative channel (e.g., Doroshkevich *et al.* , 1987, and references therein).

The second model assumes that the mass of the Universe is dominated by weakly interacting particles such as axions or photinos or gravitinos (see, e.g., Blumenthal et al., 1984). In this case the free streaming length is vanishingly small, and a bottom up scenario is expected where galaxies form via hierarchical clustering of previously formed subgalactic units. This scenario seems to provide the most efficient way for forming galaxies. However, the flatness of the model can be reconciled with the dynamical estimates of the density parameter only by assuming that galaxies form in rare, very dense regions, i.e., around the maxima of the density field (Bardeen et al., 1986). Using luminous galaxies to evaluate the density parameter would underestimate the true value of Ω_0. Additional, less clustered material is expected to dominate the gravitational field over large scales. The amplitude of the density perturbation is fixed by requiring that the rms mass fluctuation averaged in a sphere of radius $8\,h^{-1}$ Mpc equals $b^{-1} = 0.4$, where b is the biasing parameter (hereafter h=$H_0\,\mathrm{km\,s^{-1}Mpc^{-1}}$). This constitutes the so–called biased cold dark matter (CDM) scenario, which performs extremely well on the small, non linear scales, but it seems to have problems, as we will see, on larger scales.

As we still do not have *the* theory for the formation of the large scale structure, we should keep in mind an alternative view, which has been recently reproposed by Peebles (1987). The universe is open as observed, its dynamics is presently determined by baryonic material only, and its large sale structure formed out of primordial isocurvature density fluctuations. Isocurvature fluctuations, unlike the adiabatic ones, imply local inhomogeneities only in some species, in this context in the baryons only. This model has non–standard assumptions from the inflationary point of view, also if the possibility of having either open universes and/or isocurvature primordial fluctuations has been recently discussed in the framework of inflation (see, e.g., Peebles, 1988). We will consider a model with $\Omega_0 = 0.4$ and a primordial spectral index n=-1. This model is unbiased, and the normalization is done as in the CDM model, but with b=1. We will refer to this model as the PIB (Peebles Isocurvature Baryonic) model.

All these models must be checked against observations. From the observational point of view, most of the information we have on the large scale structure is provided by galaxies. Here the approach is twofold. On one hand, galaxy redshift surveys and number counts map the distribution of luminous matter. On the other hand, measurements of galaxy peculiar velocities reveal the dark matter distribution. In fact, the observed distorsions of the Hubble flow are due to local departures from homogeneity, which are determined, if dominant, by the dark matter. The two distributions, of dark and luminous matter, a priori are not expected to coincide: this is e.g., the case in the biased CDM model. Both the mentionned approaches probe the present local universe ($R \lesssim 100\,\mathrm{h}^{-1}$ Mpc). Density fluctuations at earlier epochs, ($\sim 700{,}000$ years after the Bang) are constrained by observations of the angular structure of the cosmic microwave background (CMB).

In this article we review the present status of the biased CDM and PIB models, stressing the observations and the statistical tools that can be used for testing different scenarios. The peculiar velocity and acceleration fields are discussed in Sect.2 and 3, respectively. In Sect.4, we will discuss the CBR large scale pattern and the predicted angular anisotropy. Finally, in Sect.5, a brief summary.

2. Peculiar Velocity

In a slight inhomogeneous universe, a random placed observer acquires, in a Hubble time, a peculiar velocity relative to the comoving frame (Peebles, 1980):

$$\mathbf{v}_{CMB}(\mathbf{x}) = \frac{1}{3}\Omega_0^{0.6}\mathbf{g}(\mathbf{x}) \qquad (1)$$

Here Ω_0 is the density parameter, $\mathbf{x}$ the observer comoving coordinate vector, measured in $\mathrm{km\,s}^{-1}$, and the dipole $\mathbf{g}$ is defined as follows:

$$\mathbf{g}(\mathbf{x}) = \frac{3}{4\pi}\int \delta(\mathbf{y})\frac{\mathbf{y}-\mathbf{x}}{|\mathbf{y}-\mathbf{x}|^3}d^3y \qquad (2)$$

where $\delta(\mathbf{x}) = \rho(\mathbf{x})/\rho_b - 1$ is the fractional mass density contrast, relative to the background density ρ_b.

The simplest statistical quantity to study the peculiar velocity field is the bulk velocity, or center of mass velocity of a large volume relative to the CMB:

$$\mathbf{v}_{bulk}(\mathbf{x}; R) = \int d^3y \ \mathbf{v}_{CMB}(\mathbf{y})\Phi(|\mathbf{y} - \mathbf{x}|) \tag{3}$$

where the function $\Phi(s)$, suitably normalized, describes the region over which the average is taken. If density fluctuations form, at least on large scales, a random Gaussian field, their statistics is fully determined by their power spectrum. In particular, the amplitudes of the bulk velocities are expected to have a Maxwellian distribution, with a dispersion:

$$\sigma_v^2(R) = \frac{\Omega_0^{1.2}}{2\pi^2} \int_0^\infty P(k)W^2(kR)dk \tag{4}$$

For a randomly placed observer, at the 95% confidence level, the bulk flow on scale R has an amplitude $< 1.6 \ \sigma_v(R)$. In Eq.(4), $W(kR)$ is the Fourier transform of $\Phi(s)$ and, for simplicity, we consider a spherically simmetric region and a gaussian weighting scheme. This implies $W(kR) = \exp(-k^2R^2/2)$, which acts as a low pass filter, suppressing the small $(kR \to \infty)$ wavelength perturbations and providing us with informations about the density inhomogeneities on scales larger than the sample.

In Fig.1a we show $\sigma_v(R)$, predicted in the biased CDM scenario. Large bulk flows are not expected in this model (Vittorio and Silk, 1985; Vittorio, Juszkiewicz, and Davis, 1987; Bond, 1987). The choice of R in Eq.(4), appropriate to model recent observations, has been a subject of controversy. The depth of the Rubin-Ford sample of spiral galaxies (Rubin $et\ al.$ 1976), re-observed by Collins $et\ al.$ (1986), is $\sim 65\,h^{-1}$ Mpc. The Dressler $et\ al.$ (1987) elliptical galaxy sample extends to similar distances, and, if volume-limited, it would imply $R \sim 40\,h^{-1}$ Mpc. However, Kaiser (1988) has argued that because of the method used by Dressler $et\ al.$ to reduce the data, the effective R of their elliptical sample is reduced to $15\,h^{-1}$ Mpc. Górski and Hoffman (1988) suggest that Kaiser's formalism mixes a priori and a posteriori probabilities, and that when $W(kR)$ is constructed in a self-consistent, a priori fashion, the effective depth increases to $R = 25\,h^{-1}$ Mpc. In any case, for any $R > 15\,h^{-1}$ Mpc, a very conservative lower limit, the model fails, as recently stressed by Bertschinger and Juszkiewicz (1988), in reproducing the observed velocities of $970 \pm 300\,{\rm km\,s^{-1}}$ (Collins $et\ al.$, 1986) and $599 \pm 104\,{\rm km\,s^{-1}}$ (Dressler $et\ al.$, 1987), confirming earlier analysis.

The difficulty of the model in reproducing the Dressler $et\ al.$ result is even more severe from another point of view. The bulk velocities of two different concentric, spherically simmetric samples, are not indipendent. The conditional probability

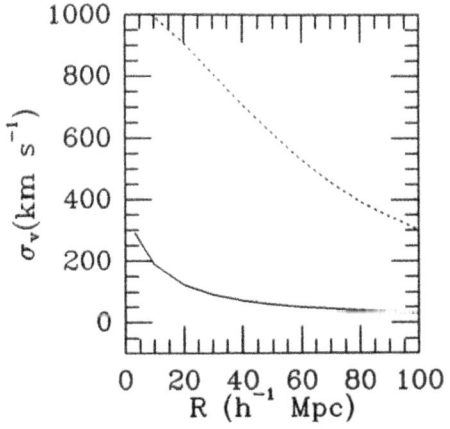 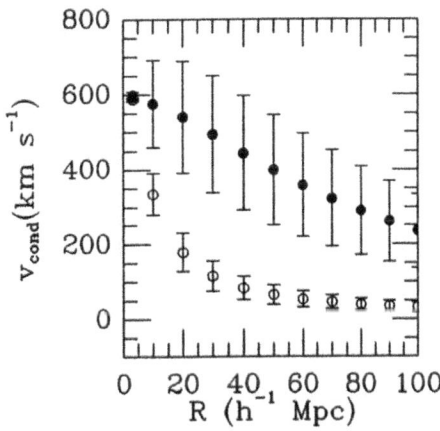

Fig.1: Rms bulk velocities (left panel) predicted in CDM (continuous line) and PIB (dotted line) models vs. the sample depth R, defined as in the text. The average bulk velocity, V_{cond} (right panel), under the condition of having a Local Group peculiar velocity of $600 \, km \, s^{-1}$. Open and filled circles refer to CDM and PIB models, respectively. Error bars are the dispersions around the mean.

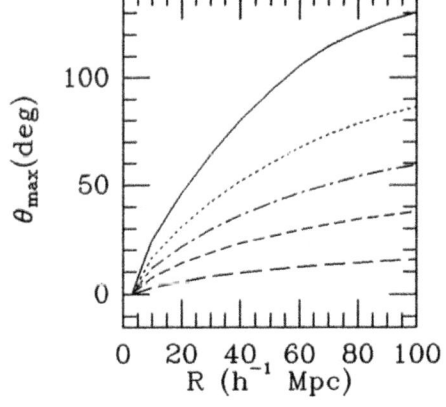 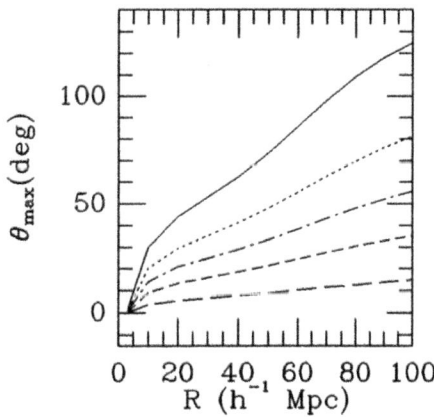

Fig.2: Confidence levels on the maximum misalignment angle expected in CDM (left panel) and PIB (right panel) models between the apex of the CBR dipole anisotropy and the bulk velocity of a sample of depth R: 5% (long dashed line), 25%,50%,75%, 95% (continuous line).

distribution for $\mathbf{B} \equiv \mathbf{v}_{bulk}(\mathbf{x}, R_B)/\sigma_v(R_B)$, given $\mathbf{A} \equiv \mathbf{v}_{bulk}(\mathbf{x}, R_A)/\sigma_v(R_A)$, is (Vittorio, Juszkiewicz, and Davis, 1987):

$$p(\mathbf{B}|\mathbf{A})d^3B = \frac{1}{\sqrt{2\pi}} \frac{1}{(1-\gamma^2)^{3/2}} \exp\left[-\frac{1}{2}\frac{(\mathbf{B}-\gamma\mathbf{A})^2}{\sqrt{1-\gamma^2}}\right] d^3B \qquad (5)$$

where $\gamma \equiv \sigma(R_A, R_B) \left[\sigma_v(R_A)\sigma_v(R_B)\right]^{-1}$ is, in modulus, less than unity, and the correlation between the two bulk velocities is given by:

$$\sigma(R_A, R_B) = \frac{\Omega_0^{1.2}}{2\pi^2} \int_0^\infty P(k)W(kR_A)W(kR_B)dk \qquad (6)$$

The observations of the angular distribution of the CMB have unambigously discovered a dipole anisotropy (see Partridge, this volume), which implies a Local Group peculiar velocity of $600\,\mathrm{km\,s^{-1}}$. Then, let us fix $\mathbf{v}_{bulk}(\mathbf{x}, R_A) = 600\,\mathrm{km\,s^{-1}}$, with $R_A = 3\,h^{-1}$ Mpc. In Fig.1b we show the value $V_{cond}(R_B) = \langle|\mathbf{B}|\rangle\,\sigma_v(R_B)$, obtained by the distribution given in Eq.(5). The corresponding error bars are the dispersions around the mean for any R_B. It is evident that there is no way of reconciling the biased CDM model with flows as large as $600\,\mathrm{km\,s^{-1}}$ on scales $> 15\,h^{-1}$ Mpc.

Obviously, reducing the biasing parameter helps, as the amplitude of density fluctuations scales as b^{-1}. Adopting $b = 1.6$ may help to save the CDM model (Kaiser 1988). However, it should be stressed that the biasing factor is not a parameter one may vary freely to maximize the probability of finding flows, coherent over large scales. N-body simulations of the non linear clustering in a flat CDM model require $b \simeq 2.5$: otherwise, the small scale peculiar velocities and galaxy clustering would be inconsistent with the observations (Davis *et al.* 1985). Besides, there is another reason, which makes lowering b very difficult. All dynamical estimates give $\Omega_0 \simeq 0.3 \pm 0.1$ (Peebles 1986 and references therein). To make these measurements consistent with an $\Omega_0 = 1$ CDM universe, one needs $2 \lesssim b \lesssim 3$ (see, e.g. Dekel and Rees 1987).

Starting from the distribution given in Eq.(5), it is easy to construct the conditional probability distribution $p(\theta|v_A, v_B)$ for the misalignment of two bulk velocities, given their amplituides v_A and v_B (Juszkiewicz, Vittorio, Wyse, and Bardeen, 1988). In Fig.2a we show confidence limits on θ, the angle between $\mathbf{v}_{bulk}(\mathbf{x}, R_A)$ and $\mathbf{v}_{bulk}(\mathbf{x}, R_B)$, expected in a biased CDM scenario. There, $v_A = 600\,\mathrm{km\,s^{-1}}$ and $R_A = 3\,h^{-1}$ Mpc, and $v_B = V_{cond}(R_B)$. A good alignment between two bulk flows is expected only on very small scales, where the motion is highly coherent.

For comparison, in Fig.1a, Fig.1b, and Fig.2b respectively, the PIB model predictions for $\sigma_v(R)$, V_{cond}, and θ are shown. This model has been recently considered by Peebles (1987), in particular for taking into account large amplitude, large scale flows, and, in fact, it definetly performs better than CDM, at least in this context.

After considering the bulk flow solution to their data, Lynden-Bell et al. (1988) proposed what they believe is a better fit for their sample. The velocity field in this model is spherically symmetric about the Great Attractor, a point located $42\,h^{-1}$ Mpc from the Local Group. As the velocity profile around the Great Attractor is related to its overdensity Δ, one can evaluate, in a given theoretical scenario, the abundance of such objects. In fact, the mean number density of high density regions in a random Gaussian field, is (Doroshkevich 1970; Bardeen *et al.* 1986):

$$n_{>\nu} \simeq \frac{(\sigma_1/\sigma_0)^3}{(2\pi)^2 3^{3/2}} (\nu^2 - 1) \exp(-\nu^2/2) \tag{7}$$

where $\nu = \Delta/\sigma_0$ and $\sigma_m^2(r) = (9/2\pi^2 r^2) \int_0^\infty P(k) k^{2m} j_1^2(kr) dk$, $m = 0, 1$. The quantity $\sigma_0^2(r) = \langle \Delta^2 \rangle$ is the variance of the density field smoothed on scale r, i.e., ν is the amplitude of the volume-averaged density contrast, expressed in standard deviations. Here and below $j_0(x)$ and $j_1(x)$ are spherical Bessel functions. Bertschinger and Juszkiewicz (1988) have carefully addressed this point and argued that the number (per Hubble volume) of regions dense enough to produce the observed peculiar velocities, is vanishingly small in a biased CDM model, while is reasonably large for the PIB model. In the biased CDM model, the density contrast, averaged over a sphere centered around the Great Attractor and with the Local Group at the border, represents a 7.3 σ_0 fluctuation, a very rare one in Gaussian statistics. In fact, the mean number of such objects in a biased CDM universe is less than one per 10^5 Hubble volumes. For the PIB model this number is definitely larger: ~ 900 Great Attractors are expected in a Hubble volume.

A more sophisticated statistical tool for studying the peculiar velocity field is given by the velocity correlation function (Peebles, 1987; Kaiser, 1988; Groth and Juszkiewicz, 1988; Davis, Gorski, and Strauss, 1988):

$$\xi_v(r) \equiv \langle \mathbf{v}(r) \cdot \mathbf{v}(0) \rangle = \frac{\Omega_0^{1.2}}{2\pi^2} \int_0^\infty dk \ P(k) \ j_0(kr) \tag{8}$$

As ξ_v can be evaluate directly from the raw data, it provides an interesting alternative test for the models. The advantage is, of course, of using all the informations available and not highly synthetic representations of observations. This is, on the contrary, what happens when one talks about bulk flows or Great Attractors.

In Fig.3 we show ξ_v, evaluated by Groth and Juszkiewicz (1988) from the Dressler *et al.* elliptical sample. Groth and Juszkiewicz (1988) argue that the observed velocity correlation function is significantly different from the CDM prediction, both in shape and amplitude (see Fig.3). The observed velocity field appears to be much more correlated, than expected in the biased CDM model, in particular at large separations. The velocity correlation function predicted by the PIB model (see Fig.3), appear to have more large scale power than required by the observations. However, this could not be a problem. The coherence length for the velocity field would be in this model greater than the sample depth. So, the catalogue would be too shallow to constitute a fair sample and a direct comparisons with the ensemble-averaged ξ_v given in Eq.(8) is not meaningfull.

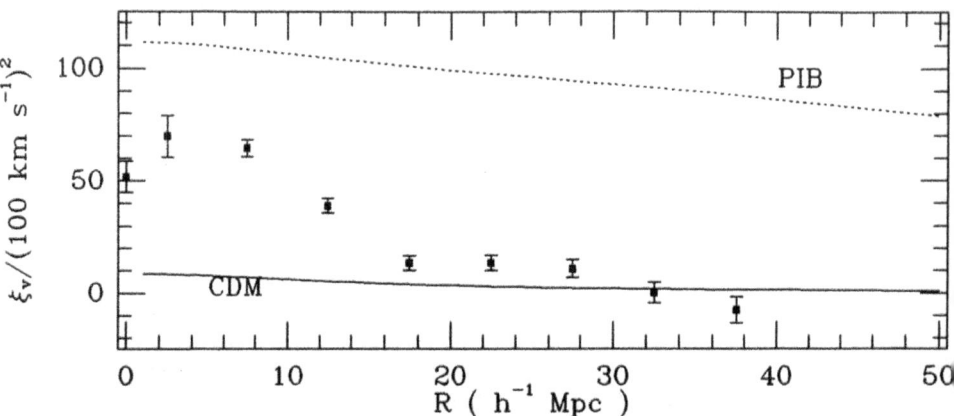

Fig.3: The dot product correlation function, $\xi_v(r)$, as evaluated by Groth and Juszkiewicz (1988) from the elliptical sample (filled squares); the solid and dotted lines represent the biased CDM and PIB model predictions.

3. Peculiar Acceleration

A complementary approach consists in directly estimating, from a given galaxy sample, the dipole g [cf. Eq.(1)]. When sufficient sky coverage is available, this can be done by using either galaxy redshift surveys, to directly map $\delta(\mathbf{x})$ in our vicinity, or the flux–weighted angular dipole of galaxy counts.

Both these approaches have been applied to an infrared, magnitude–limited sample of galaxies observed by IRAS (InfraRed Astronomical Satellite). This sample has the advantage of having selection criteria uniform over all the sky and a very high sky coverage. It has been shown that the distribution of galaxies in the catalogue exhibits a dipole anisotropy, whose apex roughly coincide (to within $\sim 20° - 30°$) with that of the CMB (see, e.g., Strauss and Davis, 1988, and Yahil ,1988, and references therein). A similar approach has also been applied to optical galaxy catalogues (Davis and Huchra 1982; Lahav 1987).

If the IRAS galaxies trace the mass distribution, the flux–weighted angular dipole of the galaxy counts it is equals to

$$g_R(\mathbf{x}) = \frac{3}{4\pi} \int \delta(\mathbf{y}) \; \frac{\mathbf{y} - \mathbf{x}}{|\mathbf{y} - \mathbf{x}|^3} \; \Gamma(|\mathbf{y} - \mathbf{x}|, R) \; d^3y \qquad (9)$$

The functional form of the window function depends upon the luminosity function of the IRAS galaxies. Adopting $\Phi(L) \propto L^{-2}[1 + L/(\beta L_*)]^{-\beta}$ implies $\Gamma(x, R) = [1 + x^2 / (2.4R^2)]^{-2.4}$ (Yahil et al., 1986). The parameter R gives the effective depth of the IRAS at a given flux $S : R = \sqrt{L_*/(4\pi S)}$. In linear theory the dipole $\mathbf{g}_R$ is proportional to the peculiar acceleration , $\mathbf{a}_R$, exerted on the Local Group by all the galaxies nearer than R. The quantity $\mathbf{V}_{obs}(\mathbf{x}, R) = (1/3)\Omega_0^{0.6}\mathbf{g}_R$ is then the velocity acquired by an observer at position $\mathbf{x}$, because of all the inhomogeneities contained in a spherical volume of radius $\sim$ R, centered in $\mathbf{x}$. Clearly, for $R \to \infty$, $\mathbf{g}_R \to \mathbf{g}$, and $\mathbf{V}_{obs}(\mathbf{x}, R) \to \mathbf{v}_{CMB}(\mathbf{x})$ [cf. Eq.(9) and Eq.(2)].

Again, $|V_{obs}|$ is expected to be a random variable, with a Maxwellian distribution, and variance (Juszkiewicz, Vittorio, Wyse, and Bardeen, 1988)

$$\sigma_{obs}^2(R) = \frac{\Omega_0^{1.2}}{2\pi^2} \int_0^\infty dk P(k)\Theta^2(kR) \; ; \tag{10}$$

$$\Theta(kR) = \int_0^\infty ds\, \Gamma(s, R)\, j_1(ks) \; . \tag{11}$$

For a volume limited sample, $\Theta(x) = 1 - j_0(x)$. As $\Theta(0) = 0$ and $\Theta(\infty) = 1$, Θ acts as a high pass filter and supresses the contribution of density perturbations on scales larger than the depth of the sample. In fact, $\mathbf{V}_{obs}$ is determined only by the inhomogeneities contained in the sample.

The considerations discussed up to now are, of course, valid in the context of the linear limit of the gravitational instability theory. We expect this limit to hold on scales $\gtrsim 10\,h^{-1}$ Mpc, i.e., where density fluctuations averaged over large volumes are much less than unity. From the observational point of view, one is forced to assume that the IRAS galaxy traces the mass. Possibly, a comparison of the dipoles obtained for an infrared and optical selected samples could quantify at least the relative bias of the infrared to the optical galaxies. Since the Local Group peculiar velocity relative to the CMB is known, an independent estimate of $\mathbf{g}$ can in principle lead to determine, by using Eq.(1), the density parameter Ω_0. To have a good estimate of $\mathbf{g}$, and then of Ω_0, it is necessary that inhomogeneities on scales larger than the sample are not important. Only in this case, infact $\mathbf{g}_R \to \mathbf{g}$. If there is substantial power at wavelength greater than the sample depth, the dipole moment $g_R = |\mathbf{g}_R|$ is likely to be underestimated, and this will result in an overestimate of Ω_0 [cf. Eq.(1)]. The standard method used to check the absence of large scale, large amplitude density fluctuations involves dividing the sample into a set of nested spherical subsamples. Then, a sequence of dipole moments, $\mathbf{g}_R$, is calculated for the material contained in spheres of increasing radii R. If the variations of g_R remain within the noise, it is usually concluded that the sequence converges and the contribution from material at large distances can be neglected. It has been claimed in the literature that the IRAS data do support such convergence of the peculiar acceleration, which already occurs at distances $\sim 40\,h^{-1}$ Mpc, (see, e.g., Strauss and Davis, 1988). Unfortunately, this procedure is very unreliable. In fact, the criterion of convergence for the cumulative acceleration provides a condition

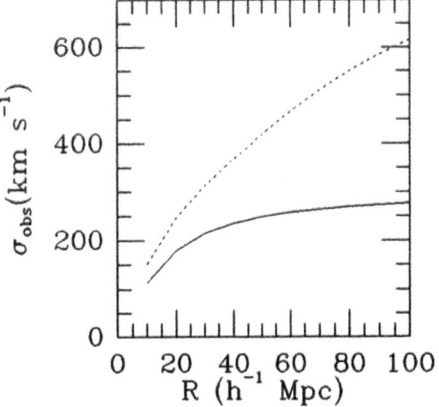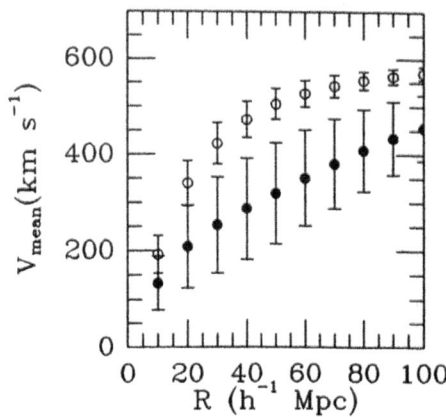

Fig.4: CDM (continuous line) and PIB (dotted line) model predictions for σ_{obs} (left panel) vs. the sample depth, defined as in the text. The mean value, V_{mean} (right panel), of $V_{obs}(\mathbf{x}, R)$ under the condition of having a Local Group peculiar velocity of $600\,\mathrm{km\,s^{-1}}$ is also shown. Open and filled circles refer to CDM and PIB models, respectively. Error bars are the dispersions around the mean.

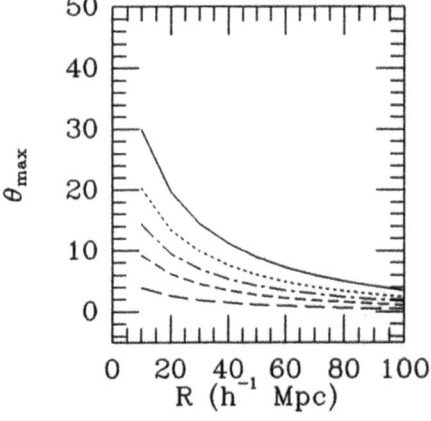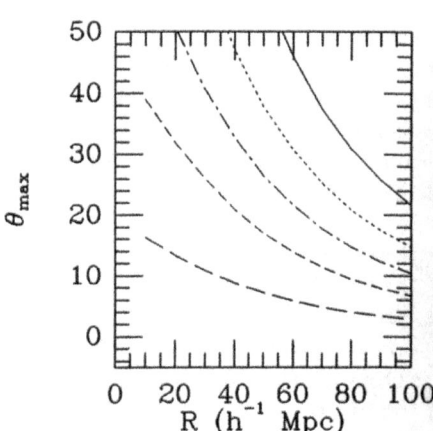

Fig.5: Confidence levels on the maximum misalignment angle expected in the CDM (left panel) and PIB (right panel) models between the apex of the CBR dipole anisotropy and V_{obs} at different confidence levels: 5% (long dashed line), 25%,50%,75%, 95% (continuous line).

which is definitely necessary but not sufficient to exclude the existence of large scale, large amplitude density fluctuations. Consider the case in which $P(k) \propto k^{-1-\epsilon}$, with ϵ small and positive. In this case, $\sigma_{obs}(R) \propto R^{\epsilon/2}$. The low value of the exponent can mimic the apparent convergence of the peculiar acceleration. At the same time, density fluctuations on large scales are so important that the rms bulk velocity of the sample is divergent: $\sigma_v(R) = \infty$ (Vittorio and Juszkiewicz, 1987). Incidentally, if $\epsilon = 0.2$, the slope of the predicted two point correlation function agrees with the observed one: $\xi(r) \propto r^{-1.8}$.

Fig.4a shows $\sigma_{obs}(R)$ for the biased CDM model. Due to the lack of large scale power in the model, $\sigma_{obs}(R)$ genuinely converges at large scales to a finite value. In this scenario the dipole anisotropy is determined on scales $\lesssim 100\,h^{-1}$ Mpc and mostly at distances $\sim 40\,h^{-1}$ Mpc, as observed. However, the rms amplitude of the Local Group peculiar velocity $[\equiv \sigma_{obs}(\infty)]$ is far below the observed 600 km s^{-1}.

One can also evaluate the mean value, V_{mean}, of $|\mathbf{V}_{obs}(\mathbf{x}, R)|$, under the condition that the Local Group velocity is 600 km s^{-1}. This can be done by using the distribution $p(\mathbf{B'}|\mathbf{A'})$, defined in Eq.(5), but with new variables $\mathbf{B'} \equiv \mathbf{V}_{obs}(\mathbf{x}, R_B)/\sigma_{obs}(R_B)$, $\mathbf{A'} \equiv \mathbf{V}_{obs}(\mathbf{x}, R_A)/\sigma_{obs}(R_A)$, $\gamma' \equiv \sigma'(R_A, R_B)[\sigma_{obs}(R_A)\sigma_{obs}(R_B)]^{-1}$ and $\sigma'(R_A, R_B) = (\Omega^{1.2}/2\pi^2)\int_0^\infty P(k)\Theta(kR_A)\Theta(kR_B)dk$. The quantity V_{mean} is shown in Fig.4b, where the error bars are again the dispersion around the mean. The predictions of the biased CDM and PIB models are distinguishable only at the 1 sigma level. This analysis confirms that it is difficult, from estimating g_R, to distinguish among models which differ for the amount of large scale power.

As the acceleration is a vector, we should also discuss its convergence in direction. This is, at least in principle, a discriminant test. Because of Eq.(1), we expects perfect alignment between $\mathbf{v}_{CMB}$ and $\mathbf{g}_R$, provided that the power on scales beyond those probed by the sample is negligible (i.e., $\mathbf{g}_R = \mathbf{g}$). In Fig.5 we present upper limits, at different confidence levels, of the misalignment angle θ between $\mathbf{v}_{CMB}$ and $\mathbf{V}_{obs}(\mathbf{x}, R)$ (Juszkiewicz, Vittorio, Wyse, and Bardeen, 1988). For the biased CDM, it turns out that θ is much smaller than observed ($\sim 20^o \rightarrow 30^o$), as a consequence of lack of fluctuations on large scale. However, this is the misalignement expected in the framework of linear theory only, and it is likely to be increased by shot noise and local non-linear effects (Villumsen and Davis 1986). The spread in θ is wider for the PIB model, where there is more power on large scales: there is $\sim 95\%$ chance of having θ smaller than $\sim 20^o$ for a sample depth $\sim 100\,h^{-1}$ Mpc.

4. CBR Large Scale Pattern

The study of the large angular scale ($\gtrsim 1^o$) CBR temperature distribution is very important for at least two reasons. On one hand, the large sky coverage, provided by balloon and satellite experiments, ensures we observe a significant sample of the sky. On the other hand, the observational upper limits to the CBR temperature anisotropy can be interpreted independently of the presence or absence of late reheating of the intergalactic medium. In fact, the gravitational potential fluctuations responsible for the CBR temperature fluctuations (Sachs and Wolfe, 1967) are independent of the location of the last scattering surface. A detection of large scale CBR temperature fluctuations would be, of course, of fundamental interest, not only because it would confirm the current ideas on the origin and the evolution of the large scale structure of the universe, but also because it would provide a direct measure of the amplitude of the initial density fluctuations in the framework of the linear theory. This, in turn, would constrain both the epoch of galaxy formation and the properties of the dark matter in the universe.

Under the assumption of primordial gaussian density fluctuations, the temperature fluctuations of the microwave sky are expected to be a 2-D random gaussian process, $\Delta(\hat{n})$, where $\hat{n}$ identify a generic direction in the sky. The statistics of the CBR temperature field is completely described by the two point angular correlation function $C(\alpha) \equiv \langle \Delta(\hat{n}_1)\Delta(\hat{n}_2) \rangle$, where $\alpha = \cos^{-1}(\hat{n}_1 \cdot \hat{n}_2)$. It is usual to take into account the finite angular resolution of the antenna by modeling the antenna beam with a gaussian profile with a dispersion σ (or equivalently, FWHM $= 2.35\ \sigma$). Then, the smoothed correlation function is (Scaramella and Vittorio, 1988)

$$C(\alpha, \sigma) = \frac{1}{4\pi} \sum_{l=2}^{\infty} w_l \ (2l+1) \ P_l(cos\alpha) \, exp\left\{ -\left[\left(l + \frac{1}{2} \right) \sigma \right]^2 \right\} \qquad (12)$$

$$w_l = \frac{A}{16} \left(\frac{H_0}{c} \right)^{3+n} \frac{\Gamma[l+(n-1)/2] \ \Gamma[3-n]}{\Gamma[l+(5-n)/2] \ \Gamma[(4-n)/2]} \qquad (13)$$

The effect of the beam, as expected, acts as a low pass filter, which severely attenuates the contribution from harmonics of order $l \gg 1/\sigma$. The behaviour of $C(\alpha, \sigma)$ for different spectral indeces is shown in Fig.6. In Eq.(12) the sum starts form l=2: the monopole component is of course unobservable by difference measurements and the dipole component is dominated by our peculiar motion relative to the CBR. The basic assumption in evaluating the coefficients in Eq.(13) is that the temperature fluctuations are determined by potential fluctuations on the last scattering surface. In Eq.(13), A is the overall amplitude of the density fluctuations and n is the density fluctuation spectral index.

The difficulty in detecting CBR temperature fluctuations has raised the question of devising the best observational strategy. For this goal, it is necessary to calculate for different models the expected pattern of the microwave sky. This can be done by using $C(\alpha, \sigma)$ given in Eq.(12) (Bond and Efstathiou, 1987; Vittorio and Juskiewicz, 1987; Scaramella and Vittorio, 1988). Two quantities are relevant for the observations: the number of upcrossing regions in the sky and their angular dimension. By upcrossing regions one indicates the regions in the sky where the CBR temperature fluctuation is higher than ν times the rms value [i.e., $C^{1/2}(0, \sigma)$]. If these regions are sufficiently abundant and large, one could look for rare but very hot spots in the microwave sky (Sazhin, 1985). On the other hand, beam–switching at an angular scale less than the typical hot spot angular diameter could produce a strong reduction in any detectable anisotropy. As all the measurements are differential, the

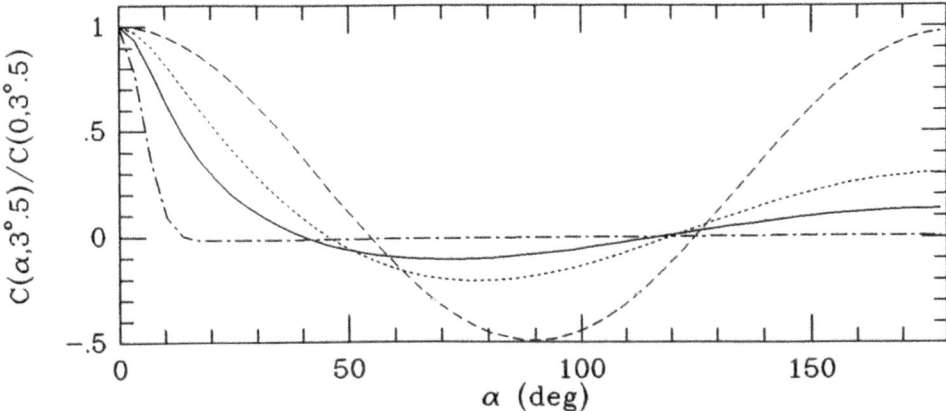

Fig.6: The smoothed autocorrelation function as a function of the angular lag, α, vs the primordial spectral index: $n = 3$ (dashed line); $n = 1$ (continuous line); $n = 0$ (dotted line); $n = -2.9$ (dashed line). Here the antenna beam has $\sigma = 3^{\circ}.5$

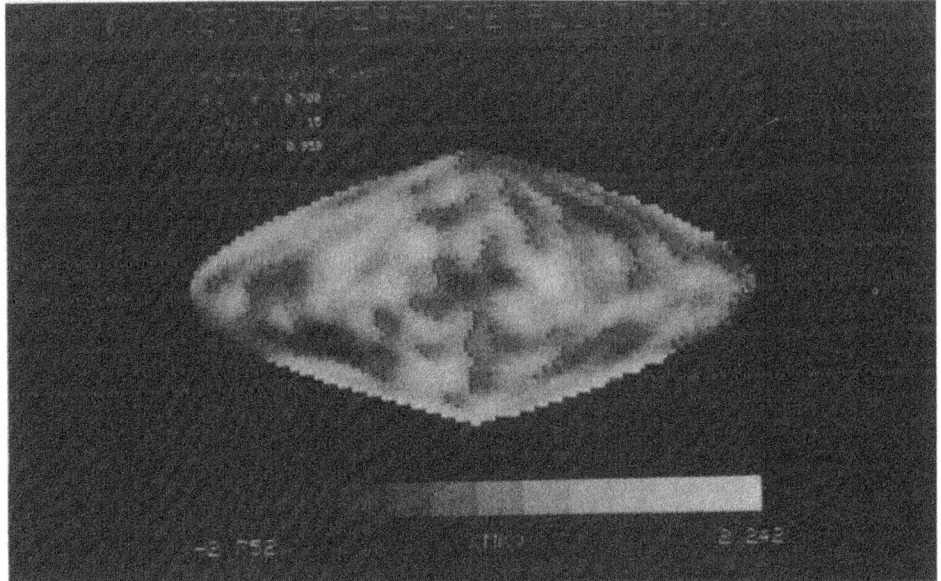

Fig.7: Temperature pattern of the microwave sky on large angular scales. The universe is assumed to be flat, with adiabatic, gaussian distributed, scale invariant density fluctuations. CBR angular anisotropies are determined by the Sachs and Wolfe effect.

knowledge of the typical hot spot angular diameter can at least be a guide in designing the observational configuration.

If the primordial fluctuations have a Zel'dovich spectrum (i.e., n=1), the number of upcrossing regions expected in all the sky has an analytical expression (Vittorio and Juszkiewicz, 1987):

$$N_{>\nu} = \frac{650}{\sigma^2} \frac{\nu \ e^{-\nu^2/2}}{[-ln \ 2\sigma \ + 3.78]} \tag{14}$$

In this formula σ is measured in degrees. As it should have been expected, this number depends only upon the antenna beam size. In fact, this is the only characteristic scale which is introduced observing the otherwise scale invariant temperature distribution. The number of upcrossing regions scales as σ^{-2} for very steep density fluctuation power spectra. This dependence flattens out for negative spectral indeces. In fact, lowering n reduces the small scale (relative to the large scale) power: in this case the finite antenna beam size has pratically no effect. The dependence of $N_{>\nu}$ on the antenna beam size (for $n \gtrsim 0$) implies that hot spots in the CBR temperature distribution appear as unresolved sources: in fact, their number continuously increases on improving the antenna angular resolution.

The expected angular diameter, D, of the upcrossing regions is inversely proportional to their number and depends on the threshold as ν^{-1} (for $\nu \gg 1$). For a Zel'dovich spectrum, the angular diameter of the upcrossing regions is given by (Vittorio and Juszkiewicz, 1987):

$$D = \frac{5.6}{\nu}\sigma[-ln \ 2\sigma + 3.78] \tag{15}$$

Here σ is measured in degrees. Again, as it should have been expected for a scale invariant process, the dimension of the hot spot is determined mainly by the smearing of the antenna beam. For fixed ν, D increases linearly with σ for positive n's, but flattens out to $90°$ (formally for $\nu = 1$) when only the quadrupole is dominant (i.e., $n \lesssim -2$) (Scaramella and Vittorio, 1988). The quantities in Eq.(14) and Eq.(15) are average quantities. In Fig.7 we show a Monte Carlo realization of the microwave sky: CBR temperature anisotropies are generated through the Sachs and Wolf effect (1967) out of primordial, scale invariant, gaussian distributed density fluctuations. Here we assumed an antenna beam $\sigma = 3°$.

To compare theory and observations we have to remember the intrinsic differential nature of the observations: the observable is either $T_1 - T_2$ (in a so-called single subtraction experiment), or $T_0 - (T_1 + T_2)/2$ (in the so-called double subtraction experiment). Here labels 0,1,2 identify the central and outer beams respectively. On intermediate angular scale ($1° < \alpha < 10°$), the rms CBR temperature anisotropy, as it would be measured in a single subtraction experiment, is (Vittorio, Matarrese, and Lucchin, 1987)

$$\frac{\delta T}{T}\Big|^2_{rms}(\alpha,\sigma) = A \ \frac{F^2(\Omega_0^{-1} - 1)}{\Omega_0^2} \ \frac{2}{\pi^{3/2}} \ \frac{\Gamma\left(\frac{3-n}{2}\right)}{\Gamma\left(2 - \frac{n}{2}\right)} \left(\frac{H_0\Omega_0}{2c}\right)^{n+3}\sigma^{1-n}$$

$$\sum_{m=1}^{m=\infty} \frac{(-1)^{(m-1)}}{(m!)^2} \Gamma(\frac{2m+n-1}{2})(\frac{\alpha}{2\sigma})^{2m} \tag{16}$$

The function $F(y) = 2y/[5 + 15y^{-1} + 15\sqrt{1+y}\ y^{-3/2}ln(\sqrt{1+y} - \sqrt{y})]$ takes into account the reduced growth of fluctuations in an open universe (Peebles, 1980) and the constant A is the overall amplitude of the density fluctuations. If n=1, the scale invariance of the spectrum reflects on the fact that the expected CBR anisotropy depends only on the ratio α/σ. In Fig.8a we show the predictions of a biased CDM model for a single and a double subtraction experiment.

Melchiorri et al. (1981) reported a positive detection of CBR temperature fluctuation, commonly considered as an upper limit because of the uncertainties in possible galactic contamination. This far infrared, balloon borne experiment involved a single beam subtraction with $\alpha = 6°$ and $\sigma = 2°.2$. The deduced upper limit is, at the 90% confidence level, $\Delta T/T < 4 \cdot 10^{-5}$. A good fit to the theoretical prediction of a biased (b=2.5) CDM model is given by (Vittorio, Matarrese, and Lucchin, 1988)

$$\frac{\delta T}{T}|_{rms}(6°, 2°.2) \simeq 2.4 \cdot 10^{-5+(0.23\Omega_0-1.15)n}\ \Omega_0^{-1.15}. \tag{17}$$

More recently, Davies et al. (1987) reported also a positive detection of CBR temperature fluctuations. The experiment operated at radio wavelengths and used a double subtraction technique, in order to minimize atmospheric contaminations. The antenna beam size and the beamswitching angle were $\sigma = 3°.5$ and $\alpha = 8°.2$, respectively. The published data refer to a strip of the sky of constant declination $(= 40°)$, and imply a rms CBR temperature fluctuation $C(0,\sigma) = 3.7 \cdot 10^{-5}$, when analysed with the likelihood method. The Davies et al. result is by large above the 1 sigma prediction of the biased CDM model (see Fig.8b) and constitutes a potential problem for this scenario.

The analysis of CBR anisotropy data with the likelihood method (see, e.g., Kaiser and Lasenby, 1987) requires an explicit guess for the functional form of the CBR temperature correlation function. In other words, the amplitude of the desumed CBR anisotropy depends upon the considered model. Davies et al. assumed the gaussian functional form:

$$C(\alpha,\sigma) = C_D(0,\sigma)exp\left\{-\alpha^2/\left[2\left(2\sigma^2 + \theta_c^2\right)\right]\right\},$$

with a sky intrinsic coherence angle of $\theta_c \simeq 4°$. As this is not expected in the standard galaxy formation scenario, the data have been reanalysed using the correlation functions given in Eq.(12) (Vittorio et al., 1988). The results of this reanalysis, pre sented in Fig.9, show that in the framework of the maximum likelihood analysis, white noise or scale–invariant density fluctuation spectra are also consistent with the Davies et al. data, producing best fit values of $C_{0M}^{1/2} = 10^{-4}$ and $C_{0M}^{1/2} = 5.9 \cdot 10^{-5}$, respectively. The likelihood ratio (LR), which provides an a posteriori confidence on a result significantly different from zero, is comparable for the three choices: $LR \sim 10$.

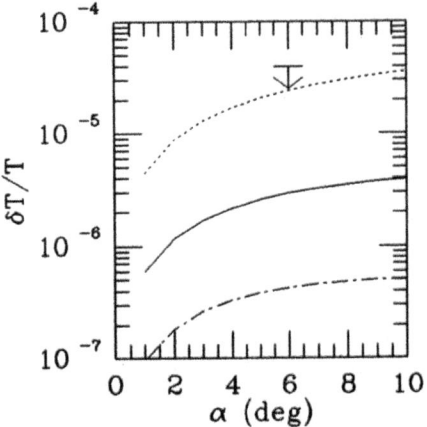 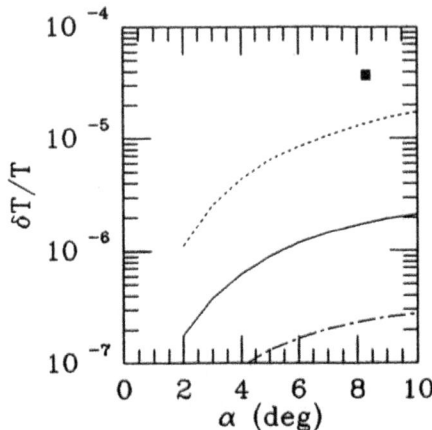

Fig.8: CBR anisotropy expected in a biased (b=2.5) CDM dominated universe for a single sub-traction (left panel) and a double subtraction (right panel) experiment. Dotted, continuous and dotted–dashed lines refer to different primordial spectral indeces: n=0,1 and 2, respectively. The Hubble constant is assumed to be $50 \, \mathrm{km \, s^{-1}/Mpc}$. The arrow refers to the Melchiorri et al. upper limit. The filled square refers to the Davies et al. reported detection.

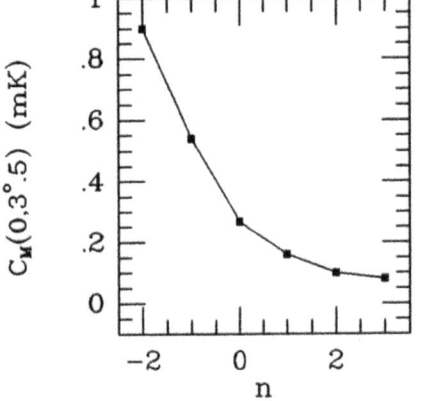 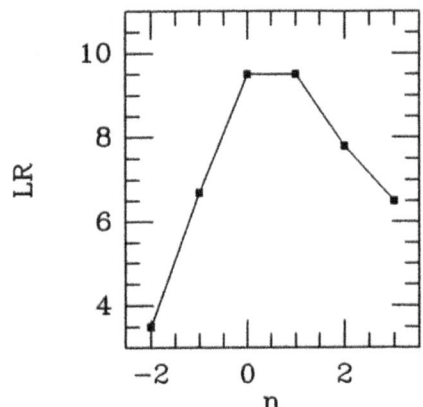

Fig.9: The quantity $C^{1/2}(0, \sigma)$, and the likelihood ratio obtained analysing the Davies et al. data, as a function of the assumed primordial spectral index.

In order to check the stability of these results, we applied a technique which is usually known as 'jackknife' analysis (see, e.g., Efron, 1979). The idea is to create several subsamples by eliminating subsets of data, and to analyze each of them as the original, complete data set. The comparison among the results obtained in this way quantify the contribution of each data point to the result. Since the Davies *et al.* data are heavily oversampled (there are only seven truly independent points in their data set) we eliminate ~ 10 data around any given point. By changing this point, Vittorio *et al.* (1988) built 61 pseudosamples of 60 data each and analysed each of these pseudosamples with the likelihood method. The results of this analysis are shown in Fig.10. No anisotropy is found by neglecting data around right ascension $219°$ (declination $= 40°$). A part from this region, this analysis reveals a fair stability of the derived amplitude for the CBR anisotropy. So, the jackknife analysis suggests that most of the Davies *et al.* (1987) signal is coming from a particular region of the sky (cf. Watson et al., 1988).

One could think that this is region constitutes a hot spot in the microwave sky due to primordial, adiabatic density fluctuations. In order to better investigate this possibility, the best strategy is to perform Monte Carlo simulation of the observations. Also, this seems the only method for fairly comparing theory and observations, since it takes into account all the important experimental effects (such as receiver noise, modulation geometry, beam pattern,etc.) otherwise neglected in a purely theoretical analysis. The only parameter to be fixed in the simulations is the rms amplitude of the CMB temperature field. This value is obtained by ensemble averaging over all the possible realizations of last scattering surfaces, i.e., over all the possible observers. As the CMB angular correlation does not vanish at large angular scales,at least for $n \lesssim 1$ (cf. Fig.6), the microwave sky, on large, does not provide a so–called fair sample. In other words, averaging over our own last scattering surface it is not equivalent to ensemble averaging. This would be eventually the case if $n > 2$, which implies a correlation length $\theta_c < 7°$. So, for scale–invariant intial conditions, a proper comparison of theory and observations must involve Monte Carlo simulations.

One thousands different realizations of the theoretical sky have been generated. Temperature fluctuations are generated, via the the Sachs and Wolf effect, by adiabatic, scale invariant density fluctuations (i.e., n=1). Their rms value is fixed to be $5.9 \cdot 10^{-5}$, as found analysing the Davies et al. data for $n = 1$. Each theoretical sky is sampled as in the real experiment, so a data set of 70 points per strip is generated. The receiver noise is simulated by adding to the theoretical temperature fluctuations a white noise of amplitude (rms) ~ 0.22 mK, the amplitude of the error bars in the Davies et al. data. The result of the simulations indicates that only 25% of the observers would have reported a positive detection, with $LR \gtrsim 10$. According to these observers, the expected amplitude of the CBR temperature fluctuations is 0.22 mK, with a standard deviation of 0.04 mK (see Fig.11). In the assumed theoretical scenario, it is quite unprobable to find temperature fluctuations of the amplitude reported by Davies et al., also if the rms temperature fluctuation (ensemble average) has just that amplitude. This is due to the fact that the signal to noise ratio [rms temperature fluctuation $(= 0.16mK)$ to receiver noise $(0.22mK)$] is only 0.75. A positive detection

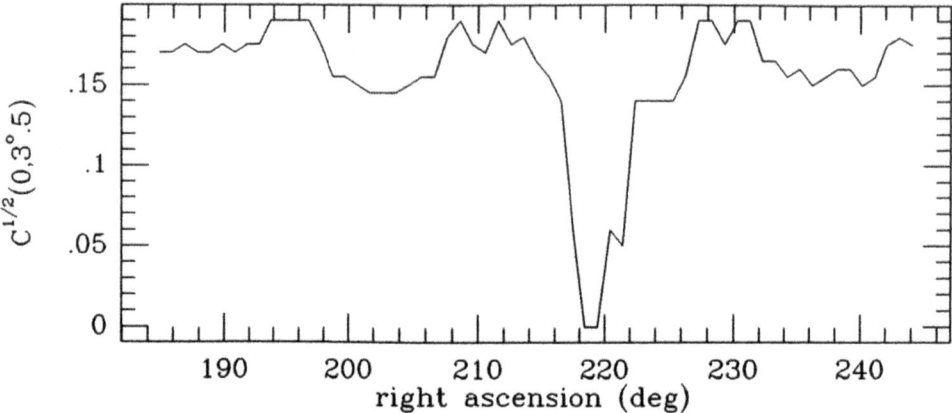

Fig.10: Results of the jack-knife analysis for the Davies et al. data. Eliminating data in the region around $\alpha = 219^\circ$, two null results are obtained.

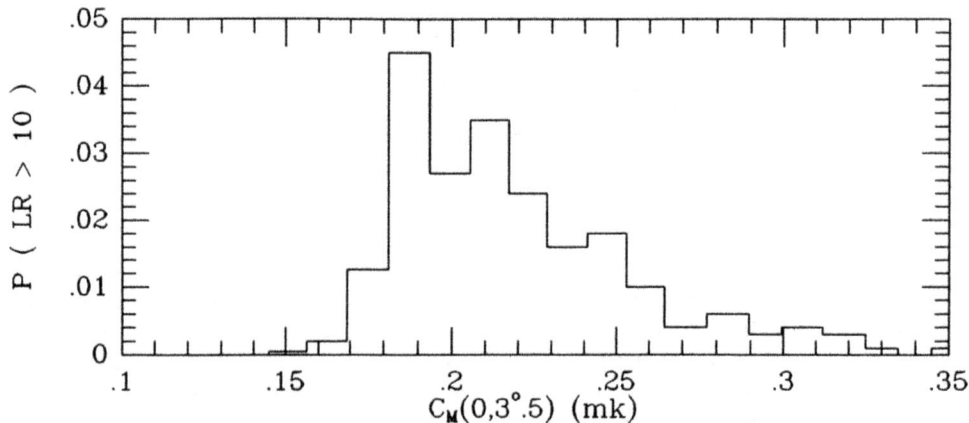

Fig.11: Frequency of the likelihood results of one thousands data set, obtained from the Monte Carlo simulation of the Davies et al. experiment. Here we select only those realizations with a likelihood ratio greater than 10.

of CBR anisotropy is possible only for those rare realizations of last scattering surfaces, allowed from Gaussian statistics, in which temperature fluctuations stick out of the noise. This is why the distribution shown in Fig.11 exhibits a long tail towards high values of the temperature fluctuations. A natural outcome of this analysis is the signal to noise ratio necessary for having a detection at a significant confidence level. The fraction of observers which would report a positive detection would be $\sim 45\%$ and $\sim 95\%$ for signal to noise ratios of 1 and 2, respectively.

If the same level of CBR fluctuations reported by Davies et al. were confirmed from data with smaller error bars, the consistency with a primordial origin due to adiabatic density fluctuations would have a higher confidence. If the simulated experimental noise is reduced to, e.g., $\sigma_{err} = 0.16$ mK (which corresponds to doubling the Davies et al. integration time), 45% of the observers would report a positive detection. For them, the mean value of the CBR temperature anisotropy would be 0.17 mK, with a standard deviation 0.03 mK. A fraction 5% (30%) of the realizations would have $LR \gtrsim 10$ and $C^{1/2}(0, \sigma) \lesssim 0.13$ (0.18) mK.

5. Summary

We presented a discussion of some of the constraints that can be put on galaxy formation scenarios by using astronomical observations of the large scale structure of the universe.

A universe dominated by weakly interacting particles such as photinos or axions, where galaxies form only in the peaks of the density field seems to have problems in reproducing large scale, large amplitude bulk flows. In fact, if $\Omega_0 = 1$ and $b = 2.5$, the rms bulk flow miss the observed value $\sim 600\,\mathrm{km\,s}^{-1}$, reported by Dressler et al. by a large factor. In the same scenario, as found by Bertschinger and Juszkiewicz (1988), the Great Attractor represents a 7 sigma fluctuation. This makes the abundance of these objects very low, less than one per 10^5 Hubble volumes. It must be stressed that these conclusions based on fits to the data (e.g., bulk flow or Great Attractor) are only as strong as the fits themselves. The estimate of the velocity correlation function has the advantage of being derived directly from the observations. The predicted velocity correlation function differs, both in shape and amplitude, from that derived from the elliptical sample (Groth Juszkiewicz, 1988; Kaiser, 1988; Davis, Gorsky, Strauss, 1988). The observations suggest a velocity field much more correlated, than expected in the CDM model. All these analysis confirm the conclusions reached earlier by Bond (1986) and Vittorio et al. (1986): large scale streaming motions are a serious embarassment for the CDM cosmogony.

The PIB model, with $\Omega_0 = 0.4$, naturally leads to large scale coherent flows and inhomogeneities. Both the amplitude of the bulk flow and the abundances of Great Attractor, as estimated by Bertschinger and Juszkiewicz (1988), seem in better agreement with the observations. It remains to be seen if the value of the primordial spectral index (n=-1), derived from phenomenological arguments, can be justified by early universe microphysics (cf. Peebles 1988). The velocity correlation function predicted by the model exhibits a larger amplitude than required by the observations (see, Groth and Juszkiewicz, 1988). However, as already discussed, this could not be a problem, as in this case the catalogue would be too shallow to constitute a fair sample.

We presented predictions for the peculiar acceleration exerted on the Local Group, by the material inside a galaxy sample of depth R. The convergence of the peculiar acceleration, exerted by the material contained in nested subsamples, to a finite value is only a necessary condition for the absence of large scale, large amplitude density inhomogeneities. To confidently assess this absence, we need to independently estimate either the peculiar velocity of the considered sample relative to the comoving frame, or the misalignment angle between the apeces of the dipoles of the CBR temperature and galaxy distribution. This test provides a discriminant probe for the large scale structure. In the framework of the linear theory, in the biased CDM model the misalignment angle (95% confidence level) is much smaller than observed, while in the PIB model is statistically consistent with the observations. As already discussed, this is only the misalignment expected in the linear theory, and an other additional contribution should be expected becuase of small scale non linear clustering (Davis and Villumsen, 1987). The estimates of Ω_0 derived from these data could be uncertain and sensitively depend on the implicitly assumed power spectrum.

In the PIB model, CMB temperature fluctuations are expected to have a coherence length of few degrees, and a standard deviation $\sim 10^{-5}$, still below but close to experimental bounds on angular scales of few degrees (Peebles, 1987b). The biased CDM scenario is definetly below the Melchiorri et al. upper limit on 6°. A reanalysis of the Davies et al. CBR anisotropy data shows that the Davies et al. data are consistent with either white noise or scale invariant primordial density fluctuations and that these are preferred among other scale-free spectra. The amplitude of these fluctuations is ~ 15 times larger than in the biased CDM scenario. The statistical significance of this consistency is low, as shown by Monte-Carlo simulations of the microwave sky and of the specific experiment. Monte Carlo simulations of CBR experiments can also be used to define the best experimental configuration for searching CBR temperature anisotropy. This can be done as a first approach by studying the expected large scale pattern of the CBR and by predicting abundances and angular diameter of the hot spots in the CBR temperature distribution. For example, if $\sigma = 3^\circ.5$, as in the Davies et al. experiment, the expected angular diameter of a 2 sigma (i.e, $\nu \sim 2$) hot spot is $\sim 20^\circ$ for $0 \lesssim n \lesssim 1$, and is less than the Davies et al. beam switching angle only if $n > 1$ and/or $\nu \gtrsim 4$. Numerical simulations, however, have the advantage of easily taking into account detailed effects such as noise (detector, atmosphere, instrumentations, etc), sky coverage, modulation geometry, beam pattern, etc. Therefore, they seem to be extremely promising, from one side, for fairly comparing theory and observations and properly constraining theoretical models, and, from the other side, for assessing the best observational strategy for a positive detection of CBR temperature anisotropies.

Acknowledgments

I am particularly indebted with Ed Groth and Roman Juszkiewicz for permission of using their still unpublished work on the velocity correlation function, reported in Sect.2. The results reported in Sect.3 have been obtained in collaboration with Jim Bardeen, Roman Juszkiewicz, Rosie Wyse: I am grateful for their permission to report on our still unpublished results.

References

Bardeen, J. M., Bond, J. R., Kaiser, N., and Szalay, A. S. 1986, *Ap.J.*, **304**, 15.

Bertschinger, E., and Juszkiewicz, R. 1988, *Ap.J. (Letters)*, in press

Blau, S.K., and Guth, A.H., 1987, *300 Years of Gravitation* eds. S.W.Hawking and W.Israel (Cambridge Un.Press, Cambridge)

Blumenthal, G., Faber, S., Primack, J., and Rees, M. 1984, *Nature*, **301**, 584.

Boesgard, A.M. and Steigman, G. 1985, *Ann.Rev.Astron.Astrophys.*, **23**, 319.

Bond, J. R. 1986, in *Galaxy Distances and Deviations from Universal Hubble Expansion*, ed. B. F. Madore and R. B. Tully (Boston: Reidel), p. 255

Bond, J.R., and Efstathiou, G., 1987, *M.N.R.A.S.*, **226**, 655.

Collins, C. A., Joseph, R. D., and Robertson, N. A. 1986, *Nature*, **320**, 506.

Davies, R. D., Lasenby, A. L., Watson, R. A., Daintree, E. J., Hopkins, J., Beckman, J., Sanchez-Almeida, J., and Rebolo, R. 1987, *Nature*, **326**, 462.

Davis, M., Górski, K., and Strauss, M. 1988, in preparation

Davis, M., Efstathiou G. G, Frenk, C.S. and White S.M.. 1985, *Ap. J.*, **292**, 371.

Davis, M., and Huchra, J. 1982, *Ap.J.*, **254**, 437.

Dekel, A., and Rees, M. J. 1987, *Nature*, **326**, 455.

Doroshkevich, A. G., Klypin, A.A., Kholopov, M.U., 1987, in *The Large Scale Structure of the Universe, IAU Symposium N° 130*, ed. Audouze, J., Pelletan, M. C., and Szalay, A. (Kluwer Academic Publishers)

Doroshkevich, A. G. 1970, *Astrofizika*, **6**, 320.

Dressler, A., Faber, S. M., Burstein, D. , Davies, R. L. , Lynden-Bell, D. , Terlevich, R. J. , and Wegner, G. 1987 , *Ap. J. (Letters)*, **313**, L37.

Górski, K., and Hoffman, Y. 1988, preprint

Groth, E., and Juszkiewicz, R. 1988, in preparation

Juszkiewicz, R., Vittorio, N., Wyse, R. , and Bardeen, J. 1988, preprint

Kaiser, N. 1983, *Ap.J.Lett*, **217**, L17.

Kaiser, N. 1988a, *M.N.R.A.S.*, **231**, 149.

Kaiser, N., 1988b, To appear in *Large Scale Structure and Motions in the Universe* eds. G. Giuricin, F. Mardirossian, M. Mezzetti, M. Ramella. Trieste, Italy, April 1988.

Lahav, O. 1987, *M.N.R.A.S.*, **225**, 213.

Lubimov,V.A., Novikov, E. G. , Nozik , V. Z. , Tretyakov , E. F. , Kosik , V. S. 1980 , *Phys.Letters*, **94B**, 266.

Lynden-Bell, D. , Faber, S. M. , Burstein , D. , Davies , R. L. , Dressler, A. , Terlevich , R. J. , and Wegner, G. 1988, *Ap.J.*, **326**, 19.

Melchiorri, F., Melchiorri, B., Cecarelli, C., and Pietranera, L. 1981, *Ap.J.*, **250**, L1.

Peebles, P. J. E. 1980, *The Large-Scale Structure of the Universe* (Princeton: Princeton University Press)

Peebles, P. J. E. 1981, in *The Origin and evolution of Galaxies* ed. Jones, B.J.T., and Jones, J.E. (Reidel)

Peebles, P. J. E. 1986, *Nature*, **321**, 27.

Peebles, P. J. E. 1987a, *Nature*, **327**, 210.

Peebles, P. J. E. 1987b, *Ap.J.Lett.*, **315**, L73.

Rubin, V. C., Ford, W. K., Thonnard, N., Roberts, M. S., and Graham, J. A. 1976, *A.J.*, **81**, 687.

Sachs, R.W., and Wolfe,A.M. 1967, *Ap.J.*, **147**, 73.

Sazhin, M.V., 1985, *M.N.R.A.S.*, **216**, 25p.

Scaramella, R. and Vittorio, N., 1988, *Ap.J.Lett* in press

Shandarin, S. 1981, in *The Origin and evolution of Galaxies* ed. Jones, B.J.T., and Jones, J.E. (Reidel)

Strauss, M., and Davis, M. 1987, in *Large Scale Motions in the Universe, Proceedings of the Vatican Conference,* ed. Rubin, V. N. (in press)

Villumsen, J., and Strauss, M. 1987 , *Ap.J.*, **322**, 37.

Vittorio, N., Juszkiewicz, R. 1987, in *Nearly Normal Galaxies,* ed. S. M. Faber (New York: Springer-Verlag)

Vittorio, N., and Juszkiewicz, R., 1987, *Ap.J.Lett.*, **314**, L29.

Vittorio, N., Juszkiewicz, R., and Davis, M. 1986, *Nature*, **323**, 132.

Vittorio, N., and Silk, J. 1985, *Ap.J. (Letters)*, **293**, L1.

Vittorio, N., Matarrese, S., and Lucchin, F. 1987, *Ap.J.*, **328**, 69.

Vittorio, N., de Bernardis, P., Masi, S., and Scaramella, R., 1988, *Ap.J.* in press

Yahil, A. 1987, in *Large Scale Motions in the Universe, Proceedings of the Vatican Conference,* ed. Rubin, V. N. (in press)

Watson, R.A., Rebolo, R., Beckman, J.E., Davies, R.D., and Lasenby, A.N., 1988, in *Large Scale Structures and Motions in the Universe* Trieste, Italy, Reidel Publishing Company, Giuricin, G., Mardirossian, F., Mezzetti, M., and Ramella, F. editors

White, S.D.M., Frenk, C.S., Davis, M., 1983, *Ap.J.Lett.*, **274**, L1.

STATUS AND PHYSICS OF THE FERMILAB TEVATRON

Leon M. Lederman
Fermi National Accelerator Laboratory
P. O. Box 500
Batavia, IL   60510

ABSTRACT.   The first superconducting proton synchrotron became operational in 1983.   Since then, extensive fixed target and colliding beam facilities have been added to create the TEVATRON.   Here we describe the fixed target physics program, largely from the major runs of 800 GeV in 1985 and 1987.   The first collider run of p̄p at 1.8 TeV took place from January-May 1987.   The operation and physics results of the collider will also be presented.   Finally, an outline of future improvements will be given.

1.   Introduction

To set the stage for the discussion of Fermilab's TEVATRON, I'd like to refer to Table I which is a chronology of the Laboratory.   We see the 200 GeV machine coming on in 1972 and going to 400 GeV in two years, not by an "improvement" program but just by clever design of accelerator magnets.   R. R. Wilson claims that his main technical contribution to the Fermilab accelerator was in the design of the magnet.

In 1973, R&D on Superconducting accelerator magnets was begun and this culminated in the so-called Energy Saver, the first superconducting synchrotron.   The 6.3 km ring was completely superconducting in May of 1983.   Physics started (slowly) in 1983/84 with the energy increasing to 800 GeV and efficiency gradually improving.   In 1985 a fixed target run of 8.5 months took place and I'll tell you some of these results.   At the end of this run, a partially completed Collider Detector (CDF) was rolled in and the proton-antiproton collider was initiated in a five week run with collisions at 800 GeV x 800 GeV.

Fermilab now alternates fixed target program using 800 GeV protons and colliding beam physics with 900 GeV in each beam making CM collisions at 1.8 TeV.   We will review some of the results.   When one thinks about 1.8 TeV or, soon, 2 TeV recall:

$$2 \text{ TeV} = 3 \text{ ERGS} \tag{1}$$

and for the Astrophysicists, this energy is equivalent to:

181

M. Caffo et al. (eds.), Astronomy, Cosmology and Fundamental Physics, 181–210.
© *1989 by Kluwer Academic Publishers.*

$$2 \times 10^{16} \; ^{\circ}K \text{ or } 10^{-12} \text{sec after the Big Bang} (2)$$

Table II lists the experiments completed in the 1987 fixed target run. Table III lists the approved program now and Table IV lists active proposals. The classifications indicate the scope of the physics topics covered.

2. Fixed Target Program

Figure 1 shows the integrated luminosities in the 1985 and '87 runs. It should be noted that in 1987, some 15 experiments (and test beams) took data and a total of 35,000 high density tapes were written in the 9 month run.

Some impressionistic view of the experimental results and of the power of the superconducting machine can be gleaned from the following listing of pieces of research carried out in the pre-TEVATRON era and experiments having the "TEVATRON Advantage."

BEFORE & AFTER

TOPIC: Sigma Beta Decay

BEFORE	AFTER
World Collection: 400 events	E-715 (1984) 80,000 events

TOPIC: CP Violation in 3 pion decay (η_{o+-})

8 Experiments yielded 1200 events	E-621 (1988) 3.2×10^{6} events

TOPIC: Drell-Yan Dimuons

500 Upsilons with 2% Energy Resolution (E-288)	E-605 (1985) 20,000 Upsilons with 0.2% resolution

TOPIC: CP in the 2-pion Decay (ϵ'/ϵ)

E-617 (1981) 14,000 $K_L \rightarrow 2\pi$	E-731 (1985) 300,000 $K_L \rightarrow 2\pi$

TOPIC: Photoproduction of Charm

Total Sample from e^+e^- Machines is $\leq$ 2000 charm events reconstructed	E-691 (1985) 10,000 reconstructed charm events

TOPIC: High Energy Neutrino Interactions

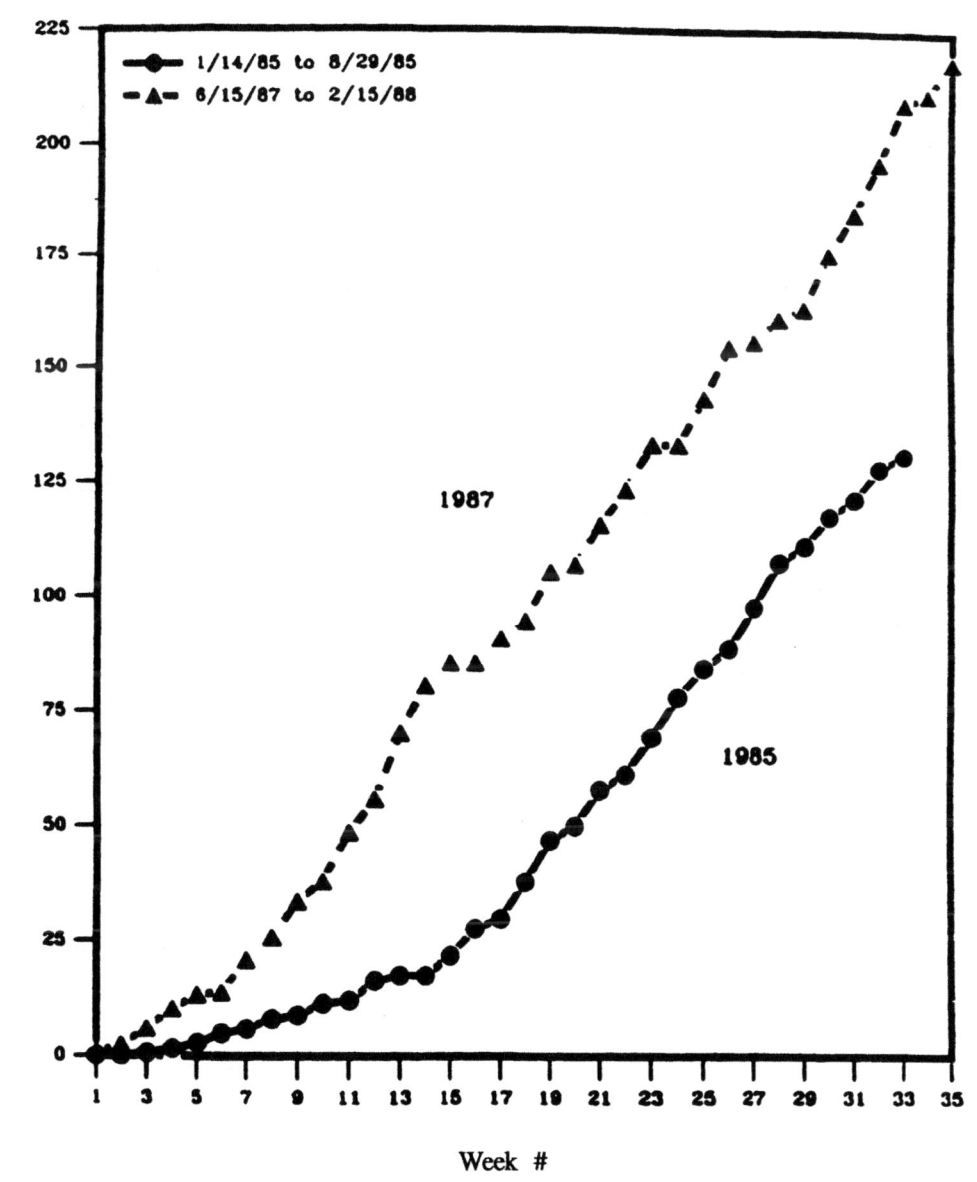

Fig. 1. TEVATRON Fixed-Target Operation: Integrated Intensity at 800 GeV

Events with $E_\nu >$ 300 GeV   200,000 Charge Current events
  Appx   zero   with $E_\nu >$ 300 GeV

What is this advantage?   There are four components:

(1)   The energy of 800 GeV is clearly one.   This gives higher yields, especially of more massive states, and it gives better quality beams.

(2)   The duty factor of the machine is 23 seconds of "slow spill" compared to 57 seconds of repetition rate.   This is to be compared to the pre-TEVATRON duty factor of ~ 1 sec spill every 10-15 seconds (depending on the financial situation).   Thus data taking is four times faster with the TEVATRON.

(3)   The redesigned beam lines are more efficient and of higher quality.

(4)   Finally, the TEVATRON (as do all other machines) profits from the advanced technology in the detector business.   In the particular case of Fermilab, there are the applications, for example, of silicon micro strips, ring imaging Cerenkov counters, on the hardware side, and the ability to process huge amounts of data in a Fermilab-developed parallel microprocessor system (ACP).

These advantages apply to the fixed target research at $\sqrt{s}$ = 42 GeV. As technology, intensity and quality of performance improve, this "programmatic" research will continue to firm up our "standard model" base by the application of precision and variety.   To illustrate this, I'll present a sampling of data from the 1985 and 1987 Fixed Target runs.

To this audience, general and quasi-philosophical comments are more valuable than a detailed review of experimental results.   It is very difficult to admire results without a feeling for the difficulties and the subtle cleverness by which these were overcome.

The scaffolding around the partially obscured monument which will explain all of microscopic physics (as well as the creation and evolution of the universe) is called the Standard Model.   Since this has received a lot of attention in this meeting, I simply remind you that it lists three generations of fermions; each generation has two quarks and two leptons.   This assembly of six quarks and six leptons is replicated in the antiworld.   The forces which cause interactions between fermions are carried by spin zero gauge bosons of which there are 12.   There are symmetries which make the SM more elegant and the Electroweak unification is one indicator of the deeper truth underneath the scaffold.

The Fermilab fixed target program largely addresses open questions in the SM.   Because there are many loose ends and inconsistencies, precise and detailed measurements serve to strengthen this base [to mix metaphors a bit] but the history of physics is rich in cases of "programmatic" research leading to major discoveries.   The Collider, while having a strong programmatic role, is basically aimed at extending the SM beyond its current level of validity which is something like 100 GeV i.e. the mass of the W, Z bosons.

Thus there is much we do not know about the masses, lifetimes, branching ratios and general spectroscopy of mesons and baryons containing the charmed quark. Nature provides us an infinitude of up and down quark states but only in high energy collisions can we find states of $u\bar{s}$, $d\bar{c}$, $u\bar{b}$ or $c\bar{b}$ etc. Charm was discovered in 1975 (13 years ago) and many of the properties were obtained in electron-positron colliders. Only in the past 3 or 4 years have charmed particles been detected clearly in hadron machines. The problems are that charm production is only 10^{-3} of the total inelastic cross-sections. Thus there is a signal to background problem. Several generations of hadron charm experiments failed or merely acquired a few tens of events.

Issues in charm physics are the production mechanisms as probed by hadrons (π, p, k), by photons, by muons and by neutrinos. Precise lifetimes and decay modes of charmed mesons probe the electroweak interactions of the c-quark. Charmonium spectroscopy, long studied by e^+e^- still has open questions related to χ-(pseudoscalar)$_+$ states. The properties of charmed baryons are very little studied in e^+e^- machines.

The situation with b-quark particles is obviously much worse. Although the discovery of the b-quark (via its $b\bar{b}$ "-onium form) was made in a hadron machine, almost all our present knowledge comes from e^+e^- machines, most notably Cornell's CESR and DESY's DORIS. An important exception is the UA1 observation of mixing in the $B\bar{B}$ system.

TEVATRON experiments are relevant to all of these issues. Thus there are many important quantities which the SM must include but which are not measured.

We also note that the top quark is not yet established but this is clearly a Collider problem. The ν_τ must be made in the TEVATRON but experiments designed for its detection have been postponed and are currently inactive.

Another general topic has to do with QCD studies generally related to hard collisions of quarks and gluons and studies which test our understanding of the strong color force between quarks.

Theoretical progress in making more precise QCD calculations has encouraged experimental tests especially where the strong interaction version of radiative corrections are involved. Fixed Target experiments probe these via detection of direct photons (radiated from quarks and gluons), by studying quark-quark scattering (the quarks materialized as either narrow jets of particles or as single very high p_t hadrons).

The electroweak force is studied principally via the CP violation which is built into the SM but whose origin is still not understood. This process, discovered in the decay of neutral K mesons in 1965, has very great significance, also in astrophysics. We exist because of CP violation. Tevatron studies (some beautiful work is also done at CERN) here are extensive.

Briefly, K_1 and K_2 are CP eigenstates in the neutral kaon system with the observable short and long-lived state:

$$K_s = K_1 + \epsilon K_2 \text{ and } K_L = K_2 + \epsilon K_1 \tag{3}$$

$K_2 \rightarrow \pi\pi$ is CP violating.

$$\eta_{+-} = \frac{A(K_L \rightarrow \pi^+\pi^-)}{A(K_s \rightarrow \pi^+\pi^-)} \quad \text{is a complex CP violating} \tag{4}$$

parameter as is

$$\eta_{oo} = \frac{A(K_L \rightarrow \pi^o\pi^o)}{A(K_s \rightarrow \pi^o\pi^o)} \tag{5}$$

Equivalent parameters often used are ϵ and ϵ' where

$$\eta_{+-} = \epsilon + \epsilon'$$
$$\text{and} \quad \eta_{oo} = \epsilon - 2\epsilon' \tag{6}$$

The ϵ parameter is due to mixing and decay of the K-states. The ϵ' parameter is the focus of Tevatron experiments and is entirely due to CP violation in the decay 2π amplitude. A recent CERN result shows $\epsilon' = .003 \pm .001$. If confirmed this would be the first direct observation os CP violation in the decay and for example would be in contradiction with the so-called superweak theory of CP violation. An accurate value of ϵ' would reflect on the deepest effect of SM electroweak processes.

We will briefly describe E-731 and list some of the other CP violation experiments in the fixed target program.

Three Examples of Fixed Target Research

A. Charm Physics.

Here I select photoproduction of charm as represented by E-691 and E-687.

Fermilab's E-691 and E-687 concentrate on the photoproduction of large numbers of charmed quarks. E-691 has collected over 10,000 reconstructed events with a meson or baryon containing the charmed quark. This is a qualitative increase over all previous experiments and is, to date, the largest sample of such data in a single experiment. The data analysis has yielded new and improved numbers on the production dynamics, masses, lifetimes and decays of the charmed particles: $D^o, D^+, D_s^+, \Lambda_c$. Using a spectrometer which has been "battle-tested" in several early experiments, they added a silicon microstrip detector at the front end of the spectrometer with enough spatial resolution to detect secondary vertices as close as 50μ from the primary interaction. The beam had an average energy of $\sim$ 150 GeV photons, tagged as to energy. They recorded 10^8 triggers with about a factor of two in the way of charm enhancement. The silicon data was used off-line with a powerful special purpose processor (ACP) to select

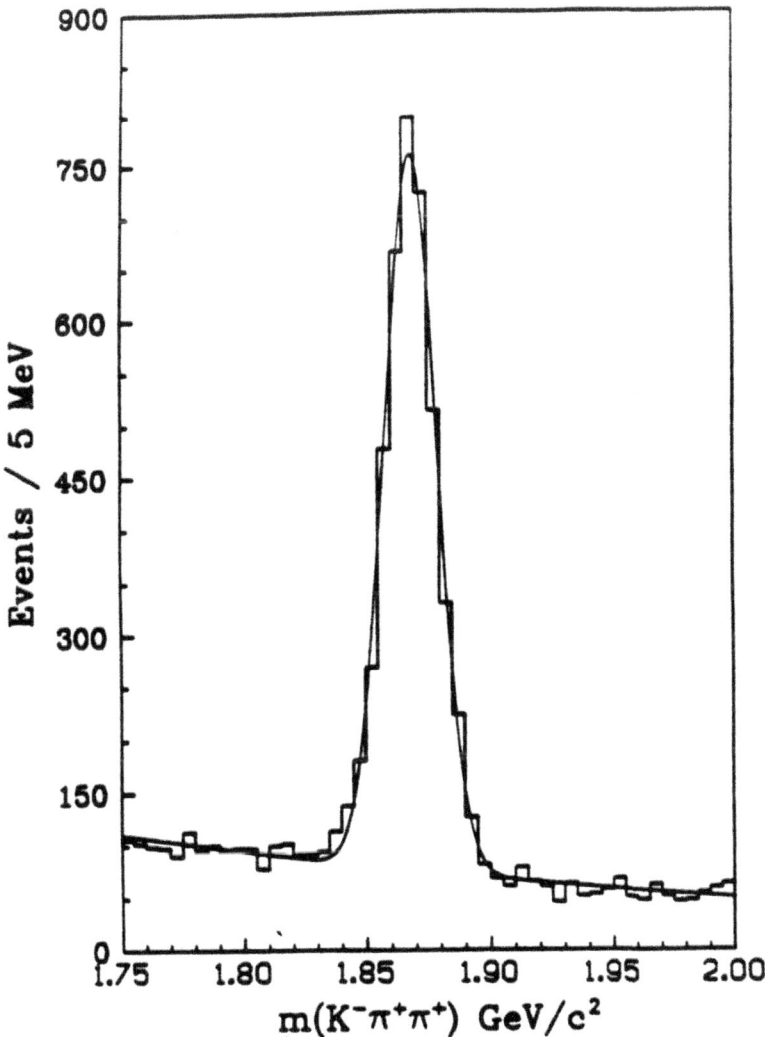

Fig. 2. The $D^+ \to K^- \pi^+ \pi^+$ mass spectrum with the
vertex cuts used for the D^+ π^- analysis
(E-691)

188

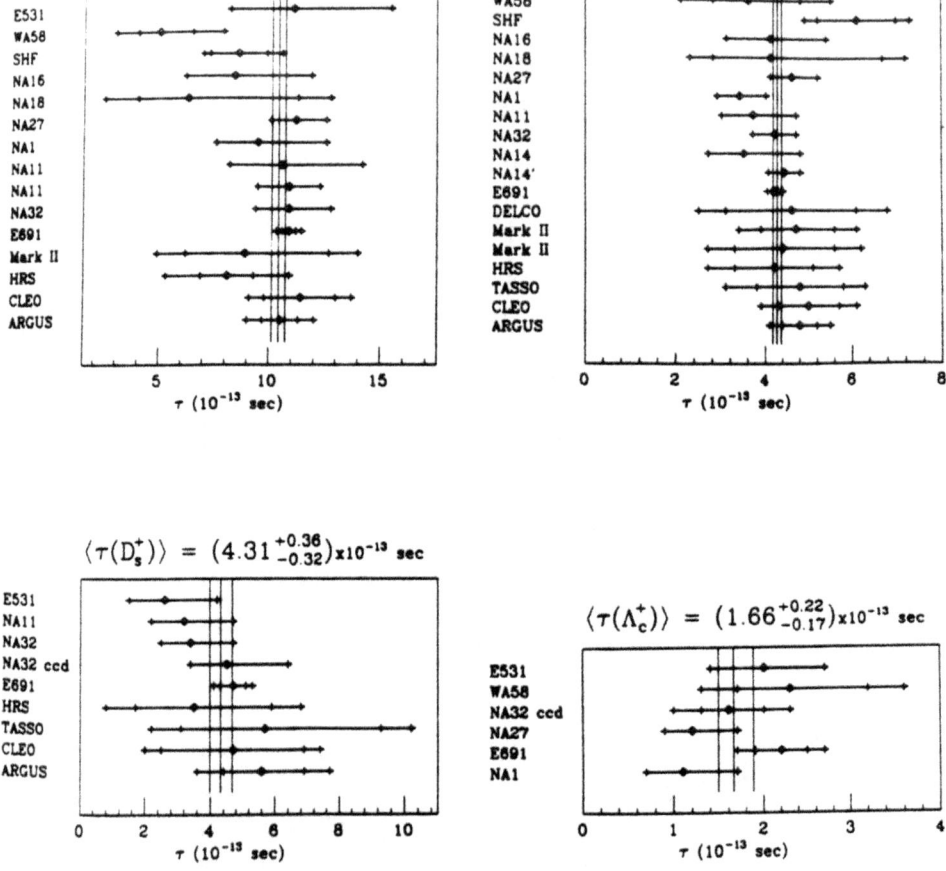

Fig. 3. Comparison of Measurements of Various Charm Particle Lifetimes

events rich in the decay of charmed particles. Fig. 2 illustrates the cleanliness of the sample. Some of the experimental results are presented in Fig. 3.

E-687, using a more advanced detector in a higher energy and more intense beam, is expected to considerably extend these measurements and also provide some data on the photoproduction of B's. Their apparatus is shown in Fig 4. In the 1987 run they collected some 15,000 charm decays.

E-769 extends the study of charm to hadroproduction, using beams of pions, kaons and protons. In the 1987 run, they wrote 500 million events on tape and preliminary analysis indicates that backgrounds under D-meson peaks are no worse than in the photoproduction data.

This brings up a philosophical point: clearly out of 500 million trigger, only a few tens of thousand are charm events. Thus the off-line analysis must select one in 10,000 recorded events. This off-line analysis is carried out (as in E-691) with the ACP device. One is tempted to include such a discrimination in the on-line processing so as to decrease the load at the tape-writing and off-line machinery where E-769 is limiting. Further progress will come from more powerful recording technology or in more elaborate data processing before recording. R&D in both directions is hot and heavy. The stakes are high since the beauty physics in hadron fixed target research is a worthy goal. As with Charm, the rates are very high, far higher than in the e^+e^- colliders but the backgrounds are also high and the needle-in-a-haystack metaphor is very appropriate.

B. CP Violation

Another research topic of great interest has to do with CP violation. E-621 measured η_{+-0} i.e the CP violating decay of K^o to $\pi^+\pi^-\pi^o$. They use two parallel beams, one for K_s one for K_L. In 10% of their 3 million $K\pi3$ events, they find $\eta_{+-o} \cong .002\pm .016$ and a phase angle of -19 ± 84°. They expect the complete analysis to yield an error of ± 003 in η_{+-o}.

E-731 is after the direct CP violating parameter ϵ'/ϵ. The objective is a total error of ± .001. The role of systematic errors in presenting ratios of numbers of decay events is crucial and great efforts were taken in the design of the experiment to minimize these. They obtained over 300,000 $K_L \rightarrow \pi^o\pi^o$ (the limiting precision reaction). Since the publication of a finite ratio by NA31 (CERN) of .0033±.0011 confirmation of the "first evidence of direct CP violation" is extremely important. As a follow-up to E-731, the 1989 run will see E-773, an experiment to measure the phase difference $\eta_{oo} - \eta_{+-}$ to ±1°.

C. Hyperon Physics

As a final example of the variety and quality of TEVATRON fixed target work, there is E-756, an attempt to measure the magnetic moment of the Ω^- hyperon. The research began with an attempt to produce polarized Ω^-'s at 2.5 mr off a proton beam hitting a

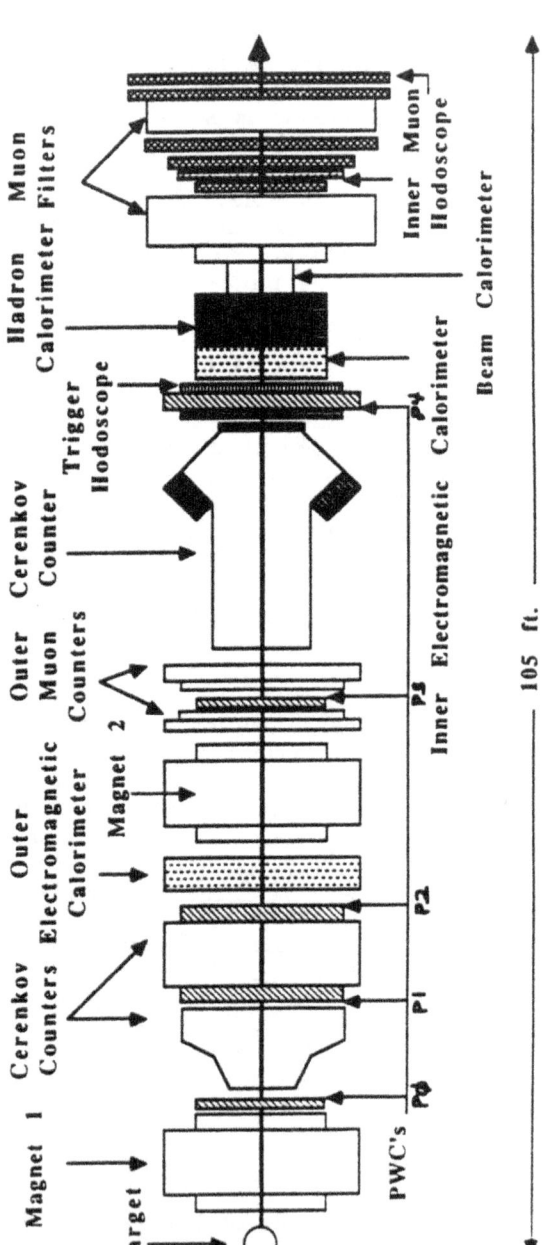

Fig. 4. E-687 Experimental Setup

production target. The Ω^-'s were produced, observed via Λ,K decay modes and found to be unpolarized.

The experimenters then tried "polarization transfer" via a preliminary proton target producing a neutral beam containing polarized lambdas which strikes the production target at zero degrees. Some 20,000 Ω^- were produced and a reasonable polarization determined. Now the magnetic moments of hyperons are calculated from one input as shown in the following table:

TABLE I

Hyperon	μ (expt)	Theory
Λ	- .613 ± .005	Input
Ξ^-	- .69 ± .04	- .49
Ω^-	?	- 1.84

The Ω^- (a three strange quark state) should have $\mu_\Omega = 3\mu_\Lambda$. The Λ^0 state is uds where the single strange quark determines the moment. Some of the beautiful hyperon peaks are shown in Fig. 5.

The E-756 results are $\mu(\Omega^-) = -2.0 \pm .2$ nm

The future of Fixed Target may address many issues: no one has ever seen a ν_τ and $\nu_\tau - \nu_\mu$ oscillations are interesting, especially to astrophysics. CP violation, if it is really directly observable in K decay, may require another round. The intense beams of hyperons seen in E-756 are sources of rare decay modes as well as CP violating reactions etc. The EMC effect in muon scattering and the curious polarization results from CERN in deeply inelastic scattering must be clarified.

Beauty physics is an open frontier and will provide the greatest challenge to technique in the near future. At the TEVATRON, E-771, E-690 and P-789 are dedicated B-physics experiments although many of the charm spectrometers have beauty capability and will constitute the first round of what may well grow into the breakthrough that can, in principle, collect many thousands of reconstructed $B\bar{B}$ pairs. The 1989 run will set the stage for B-physics at the Tevatron fixed target program. The reason, we must emphasize, is that whereas beauty is made in only one out of one million hadron collisions, there are in principle 10^{12} hadronic collisions per second and therefore 10^{11} $B\bar{B}$ pairs per day - the genius that can collect and analyze, say 10^6 of these will be surely richly rewarded.

3. The Collider

A. Introduction

The Fermilab Collider is based upon an antiproton source which collects $\bar{p}$'s from main ring proton collisions at 120 GeV. Collisions

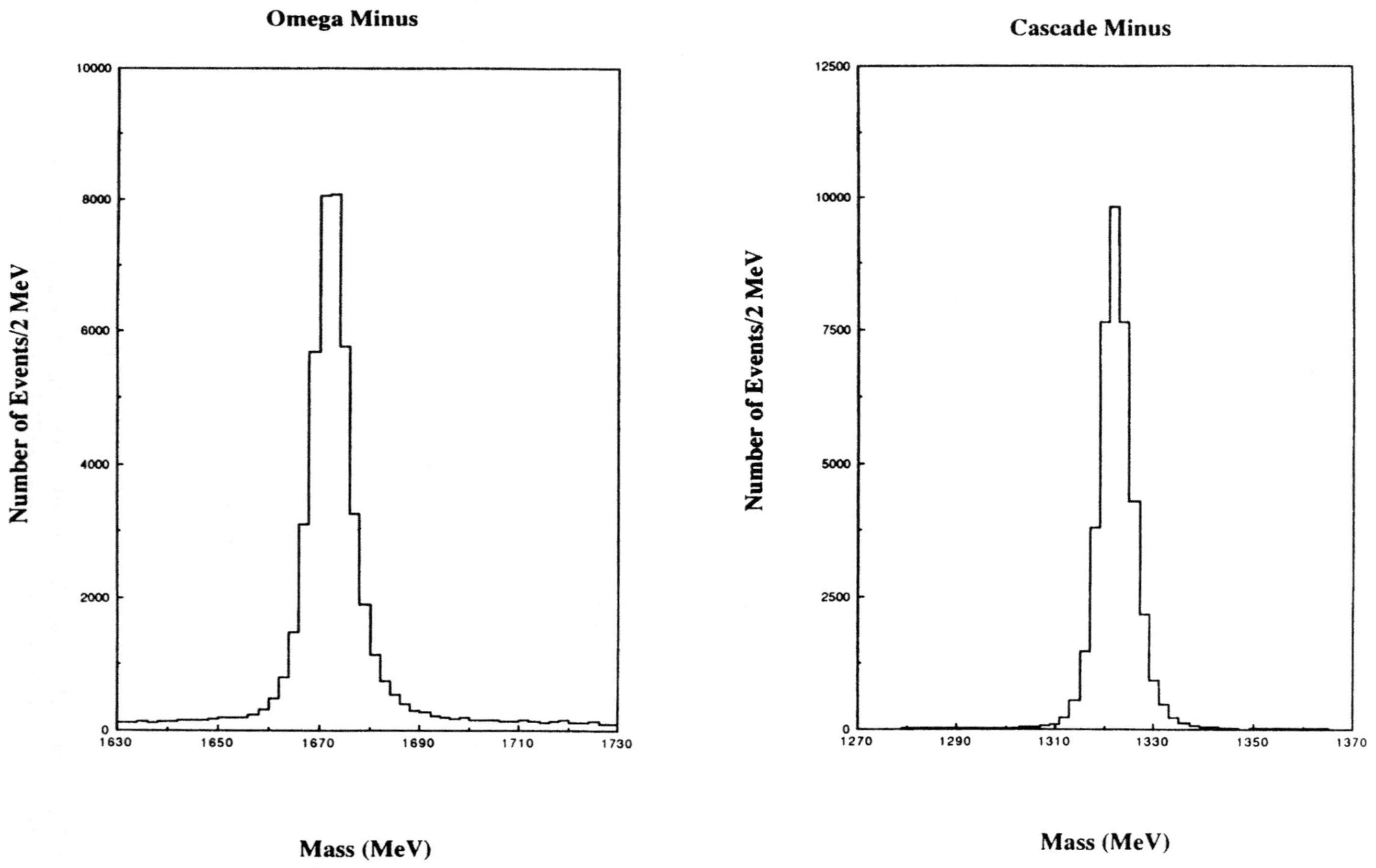

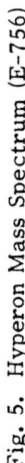

Fig. 5. Hyperon Mass Spectrum (E-756)

take place in the cryogenic TEVATRON ring at a flat-top energy of 900 GeV. Holding superconducting magnets at full field is no problem since no energy is used. The cryo pumping is a great convenience to obtaining a very good vacuum.

Originally only one (B Zero) of the six long straight sections in the TEVATRON was available for a collision region. However it was realized early on that the extraction devices at the D Zero straight section were not needed if we were not extracting beam for fixed target and that D Zero could be a second high luminosity interaction region. Smaller experiments that did not require strong focussing of their beams were eventually squeezed into C Zero and E Zero.

B. The pbar Source

This has been described in many places. Based upon the CERN pioneering experience the Fermilab group added innovations and designed a system that would produce a luminosity of 1×10^{30} cm^{-2} sec^{-1}. It has two concentric rings, each about 500 feet in circumference, designed for particles near 8 GeV. Here $\bar{p}$'s are collected from the 120 GeV target via an axial focussing lithium lens, carrying a current discharge of 600 K amp. The accepting ring is the Debuncher ring. Antiprotons are treated with RF in a variety of ways designed to increase density in phase space. After some milliseconds the $\bar{p}$'s are transferred to the Accumulator ring. About 10^7 $\bar{p}$'s are transferred each 3 seconds or so. In this way about 10^{10} $\bar{p}$'s are collected per hour. In the Accumulator, $\bar{p}$'s are cooled and stacked, waiting for transfer back into the accelerator complex. They are extracted from the accumulator in 6 bunches, each holding somewhat over 10^{10} pbars.

These are captured into the Main Ring RF and accelerated to 150 GeV where they are transferred to the TEVATRON ring. Here they encounter 6 bunches of protons of somewhat more intensity circulating in the opposite sense. The relative diffuseness of both beams makes collisions rare. The two counterrotating beams are then accelerated up to 900 GeV and held there. Powerful superconducting quadrupoles are turned on to squeeze the beam, down to a fraction of a millimeter in cross section.

Now collisions take place.

In January of 1987, the process of commissioning the new machine and its new detector, CDF, began. The record of luminosity improvements in the 1987 data are in Figure 6. The peak luminosity achieved was 1.2×10^{29} cm^{-2} sec^{-1} and a total of 70 nb^{-1} was delivered to the CDF detector. Although it would appear that this first run achieved over 10% of its design, it should be noted that the design had in it a cross section for production of $\bar{p}$'s by protons that was in error by a factor of 2.5. The machines then all gave over 25% of the design performance. [Added note: In the 1988 run, a world's record luminosity of 1.3×10^{30} cm^{-2} sec^{-1} was achieved as of Sept. 8].

194

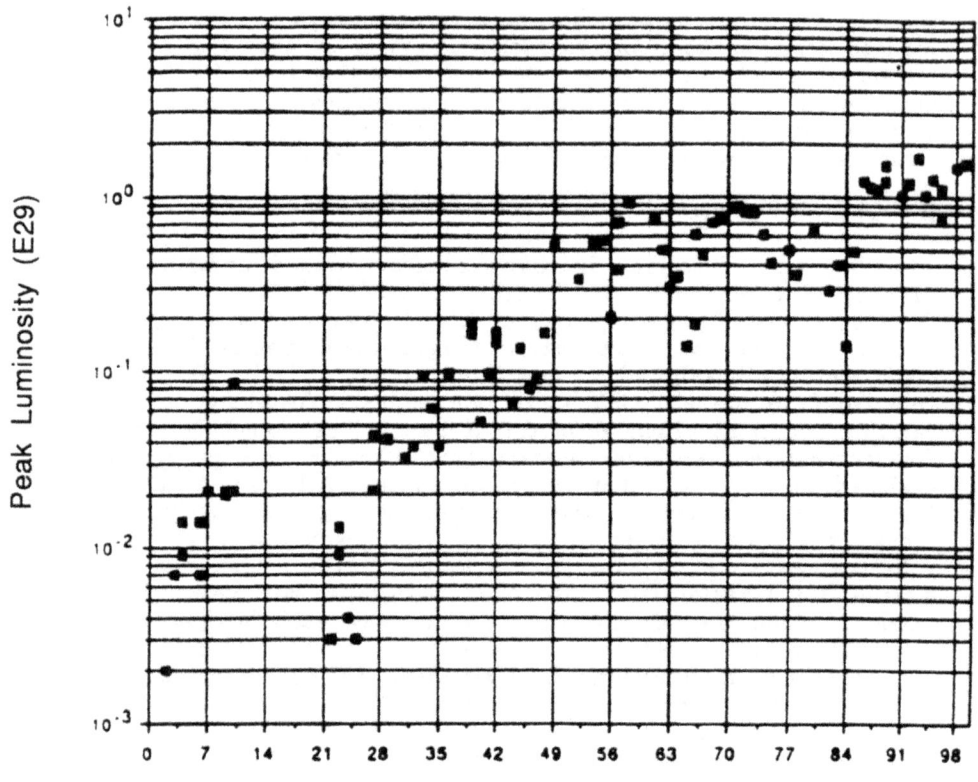

Day (beginning Feb. 1, 1987)

Fig. 6. TEVATRON peak luminosity (10^{29} cm^{-2} sec^{-1}) for February 1 –
May 11, 1987. Each point represents a separate pbarp store.

CDF (Collider Detector at Fermilab)

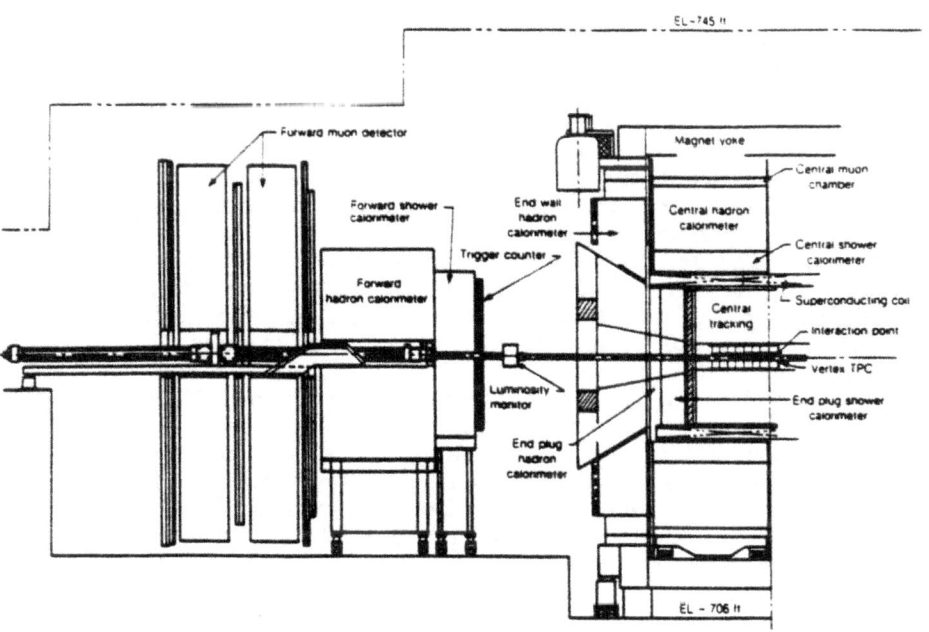

Fig. 7. A Cutaway View of CDF

C.   CDF

This is a collaboration of 3 national labs, 10 universities and scientists from Japan and Italy.   It took about 7 years to build under the leadership of Roy Schwitters (Harvard) and Alvin Tollestrup (FNAL). A cutaway view is shown in Fig. 7.   Since UA1 was started in 1979, CDF has had some of the advantages of UA1 design and construction experience although not the benefits of two or three years of running experience.   Nevertheless, CDF has a much greater granularity (by a factor of 50) as can be seen from unfolding the detector into a "lego" plot of pseudorapidity (essentially the longitudinal coordinator) vs azimuthal angle.   (Fig. 8).

To say that CDF bears great similarity to other collider detectors is to remark that most humans have roughly the same symmetry, numbers of appendages, orifices and protuberances.   Vertex [etection (via a TPC), central tracking, electromagnetic calorimetry, hadron calorimetry and muon detector are the eyes, ears, arms and legs of our metaphor.   The key to the quality of instrument is its performance in action - the trigger, the data acquisition system, the off-line software all depend on the quality of hardware design and fabrication and the anticipation, in the design, of difficulties, backgrounds,, unknowable obstacles, etc.

In Collider Run I, the machine delivered about 70 nb^{-1} to the detector.   These arcane units simply mean that a process that has a cross   section of 1 nb $(10^{-33}$ $cm^2)$ would have generated 70 events.

The detector on-time was such that about 33nb^{-1} of data were written to tape.   The first results from analysis of these data (about 2% of the data accumulated by UA1 and UA2 at CERN) are more an indication of detector performance than new physics although we must recall that the observations are at 1.8 TeV and in fact some greater sensitivity is indeed achieved in the search for supersymmetric particles and for compositeness.

A set of issues appropriate for hadron colliders could be:

1.   Elastic and Diffractive Scattering
2.   Minimum bias events
3.   QCD processes: jet structure
4.   W, Z
5.   Search for new particles.

The CDF group has 20 Ph.D. theses in preparation as a result of the first run [also as a result of an intense desire of students to graduate].   Only the highlights of the first physics results are given here.

1.   Elastic Scattering, etc.

Although CDF has small angle (~ microradian) detection capability, especially for high mass difffraction studies, no results were forthcoming

197

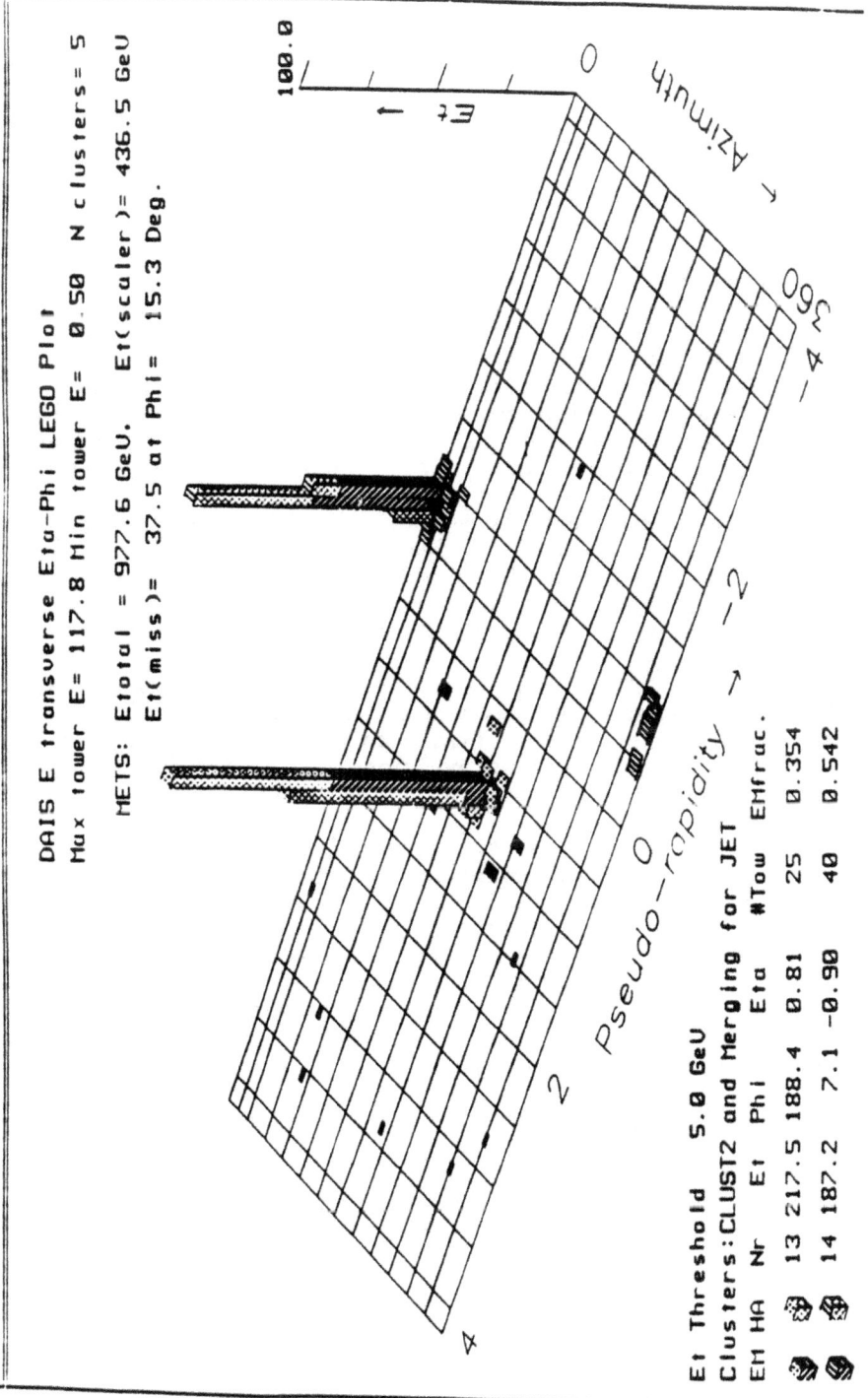

Fig. 8. A Typical 2-Jet Event (pseudorapidity vs azimuthal angle)

from RUN I (see E-710 below). However CDF does measure multiplicities, p_t spectra, particle production (mass identified) and assorted correlations. These analysis should all be ready for the Munich Conference in August.

2. Minimum Bias Events

CDF is making an analysis of minimum bias events but this has not been brought to "conference report" status.

3. Jets and QCD

Fig. 9 shows the test for quark compositeness - the p_t dependence of single jets. CERN results set a limit on the compositeness scale of $\Lambda \leqq 400$ GeV. CDF's similar analysis raises this to ≥ 700 GeV with 95% confidence level. Multijets is a hot subject with some sensitivity to the important strong coupling parameter α_s. CDF is working on the analysis of multijets. Dijet angular distributions are in excellent agreement with QCD predictions.

The issue in jets should be emphasized. These are interpreted as outgoing constituents: quarks and gluons, which materialize, outside the collision volume, as narrow bundles of hadrons. The detector must define a cluster algorithm that includes the "jet," excludes random tracks and corrects for the inevitable errors. The trigger definition is obviously crucial.

The influence of electrons vs hadrons, the influence of cracks, the calibration for neutral particles and a host of other considerations all contribute an inherent uncertainty to the definition and energy resolution. Since all hadronic collisions result, finally, in leptons, photons and jets, the understanding of how to deal with jets is crucial to progress.

Another QCD effect is the observation of "direct" photons i.e. produced in the constituent collision. According to QCD these arise as a result of gluon bremstrahlung. The CDF detector studies the transverse profile of the electromagnetic shower at the "shower max" in order to have maximum sensitivity to the background π^0's. Fig 10 shows CDF results contrasted with brand X.

Direct photons is a fairly subtle signature and its detection at the level seen in CDF (1 event per 10^7 interaction) is greatly encouraging for the future of hadron colliders.

4. W and Z Physics

Here, with 26 W's and some 6 Z's, CDF can only provide a point on the $\sigma(E)$ curve. This indicates W production is 3 times as much as at 630 GeV, a result expected from Drell-Yan type theory.

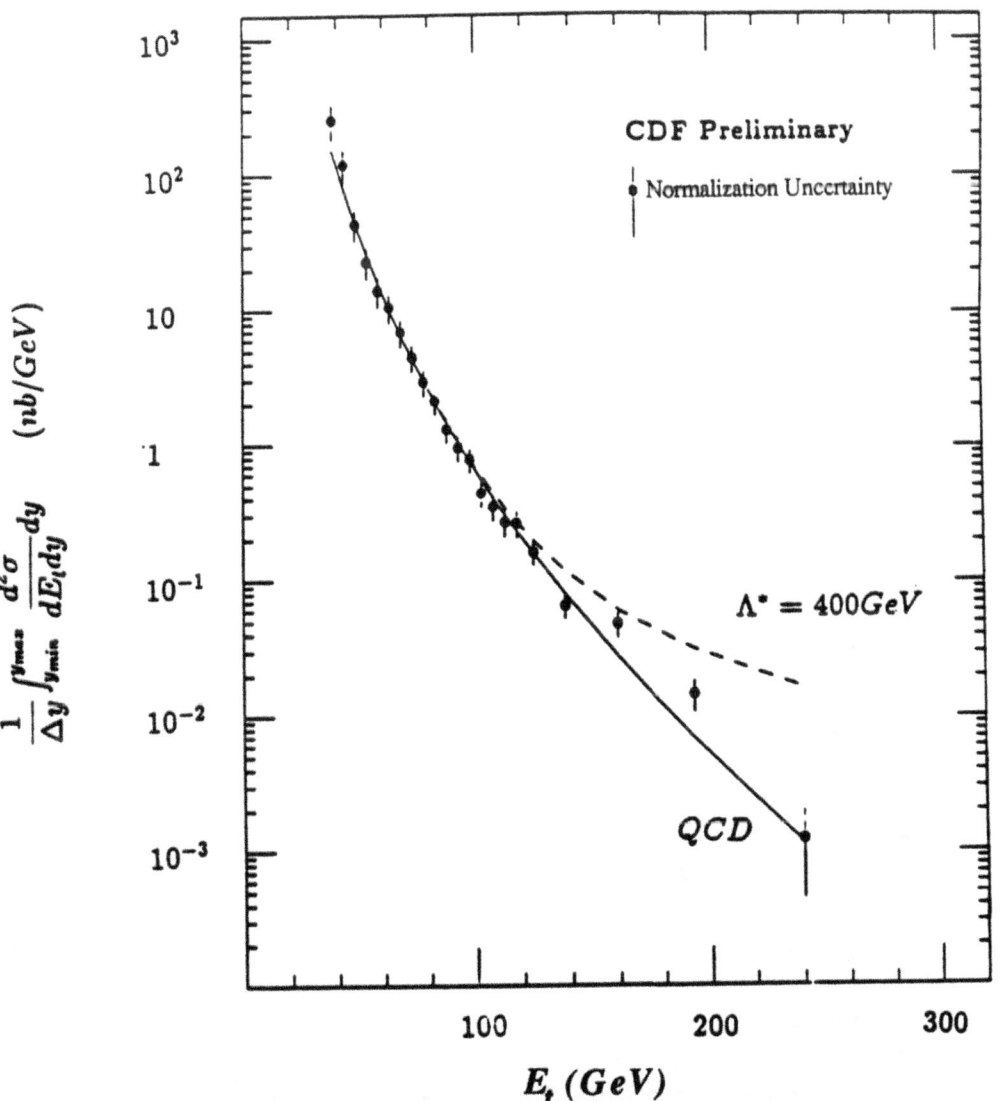

Fig. 9. Inclusive Jet E_t Distribution (CDF)

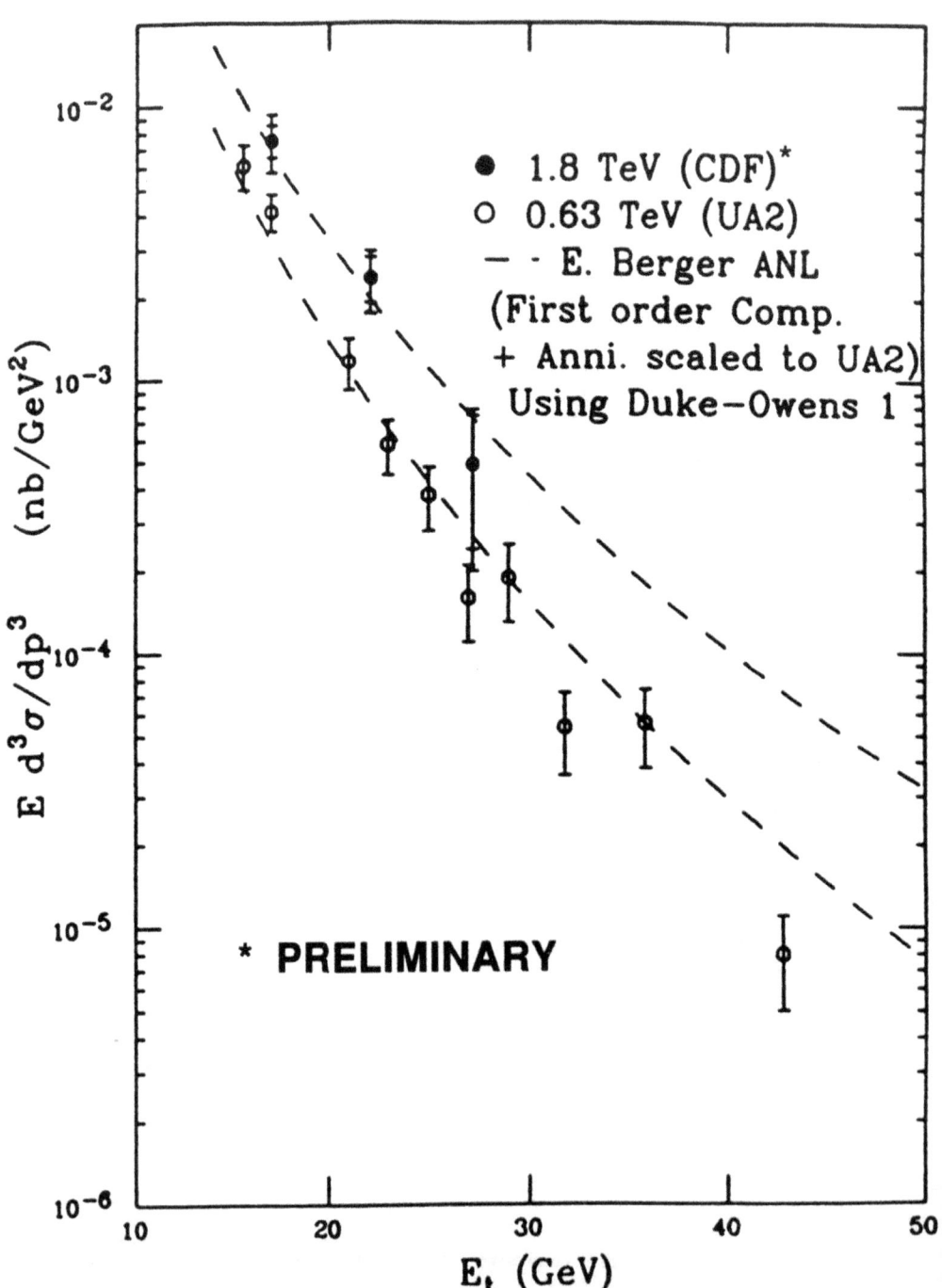

Fig. 10. "Direct" Photon Spectrum

5.   Search for New Particles

Here we should start with the top quark and note that CDF had very little sensitivity in the first run. This is one of the prizes of the <u>next</u> CDF run where one expects 1-2 pb^{-1} and a sensitivity to top of somewhat under 100 GeV of mass (present UA1 limit $M_t > 41$ GeV).

In $\bar{p}p$ collisions, supersymmetric particles can be pair-produced. The cross sections are calculable in QCD. The collider looks for the lowest mass states, i.e. if the mass of the gluino is greater than the mass of the squark, then

$$\bar{p}p \rightarrow \tilde{q}\tilde{\bar{q}} \rightarrow q\bar{q}\,\tilde{\gamma} \tag{7}$$

where $\tilde{\gamma}$ is the lowest mass neutralino (photino?). This reaction gives two jets and missing E_T since γ is assumed to escape. Alternatively:

$$\bar{p}p \rightarrow \tilde{g}\tilde{g} \rightarrow q\bar{q}\,\tilde{\gamma} + q\bar{q}\,\tilde{\gamma} \tag{8}$$

and we have 4 jets and missing E_T. The CDF analysis gives limits which considerably extend the UA1 results: For gluinos, CDF finds Mg > 82 GeV/c and for squarks Mg > 89 GeV. These limits do assume that the lightest "ino," i.e. photino say, is lighter than the squarks or gluinos.

These examples indicate that CDF is working well at least at the level of peak luminositiers of $\sim 10^{29}$ cm^{-2} sec^{-1}. Below, we'll discuss the prospects for Collider Run II.

D.   D0

In 1982 it was recognized that a second interaction region could be constructed if the extraction magnets could easily be removed. In 1987, a D Zero collaboration was formed under Paul Grannis of SUNY, Stony Brook. The design of D0 had CERN experience as an advantage. The decision was to complement CDF by building a non-magnetic detector which concentrates on (1) hermiticity (2) good calorimetric resolution (3) electron and muon identification and coverage. It is a collaboration of 21 institutions including 3 national labs, French, Soviet and Brazilian institutions. The calorimeter is liquid Argon and uranium plates (gee whiz numbers: 15,000 liters of liquid Argon and 300 tons of uranium plates.

The construction schedule calls for a complete detector installed in mid-1990. It will be installed in the D straight section with its own Low β (high luminosity) quadrupole set.

E.   Elastic and Diffraction Scattering E-710

This is a small experiment based upon Roman pots to detect scattered particles within a few mm of the beam out to 124m from the IR. It is installed in the E-straight section where beam is transferred from MR to TEVATRON. Fig. 11 show some of the data obtained by this collaboration in its first physics run.

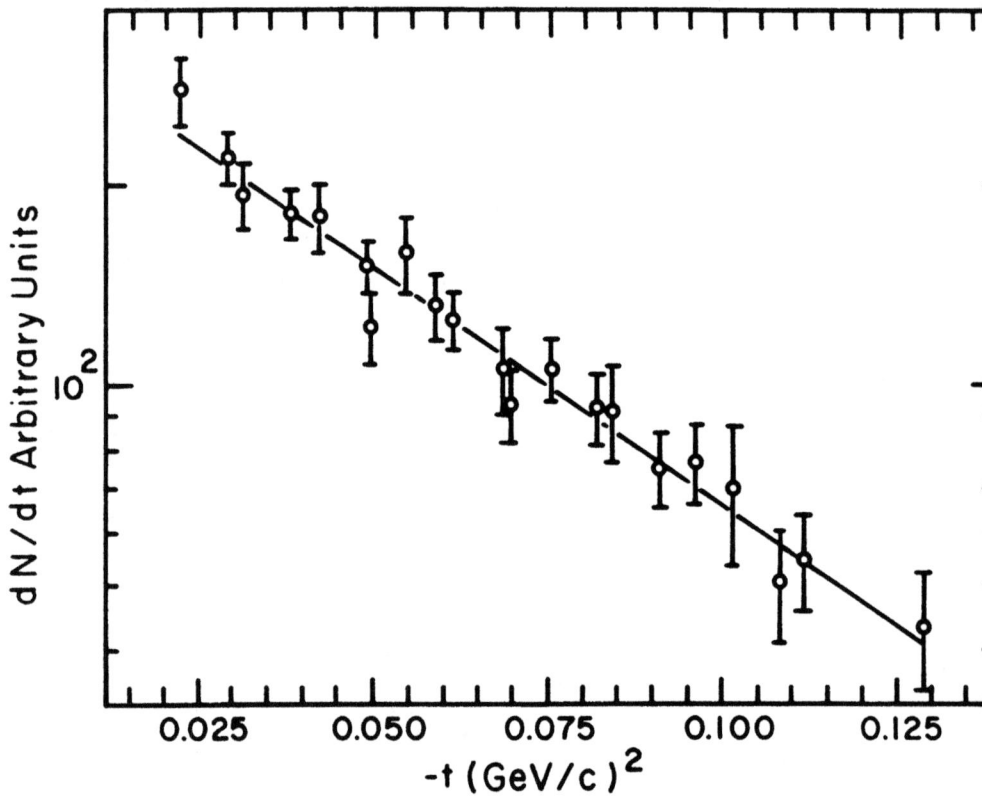

Fig. 11. Elastic scattering distribution obtained in one run in Experiment E-710.

F.   Search for Quark-Gluon Plasma E-735

This research, squeezed into a busy straight section (C Zero) is designed to try to observe signatures of quark/gluon deconfinement in 1.8 TeV collisions.   The relatively simple detector consists of scintillation hodoscopes, wire chambers, time-of-flight particle identification, and a bending magnet - all at 90° to the collision line. The idea is to measure the multiplicity as a function of average P_t which, some theories say, can indicate a change of phase.   Another indicator could be a "burst of strangeness."   Fig. 12 indicates some preliminary data from this group.

G.   Collider Run II

This second collider run will begin in June of 1988 and is scheduled to deliver 1 pb^{-1} to the CDF detector.   Based upon studies that have taken place in the $\bar{p}$ source during the 1987/8 fixed target run, we expect this to take about 8 months after a 2 month commissioning period.   This is 30 x the first run and may well be called the "discovery run."   This should yield ~ 800 W's, 200 Z's, a considerably extended quark jet distribution, sensitivity to top quark (mass < 100 GeV), and the many other discovery objects that proliferate in the theoretical literature.

As noted above, there is a reasonable project that this goal will be exceeded before the end of Calendar 1988.

IV.   The Future

We complete this report with a brief discussion of how the physics program Laboratory will evolve over the next 5 years.

The Laboratory has an ambitious upgrade plan which will bring the Fixed Target energy from 800 GeV to 900 GeV and the collider from 900 to 1000 GeV.   The collider luminosity goal is 5 x 10^{31} cm^{-2} sec^{-1}, a factor of 50 over design while the same improvements will raise the fixed target intensity from 1.7 x 10^{13} ppp to 3-4 x 10^{13} ppp. The procedure for accomplishing these has, at this time, several possibilities.   All options are based upon a first phase which involves many improvements including raising the energy to 1 TeV, extensive modifications of the linac to reach 400 MeV with improved emittance (beam quality), installing very strong focussing quadrupoles to decrease the transverse dimensions of the beams at the collision points (low β quads), a variety of $\bar{p}$ source and cooling improvements and the installation of separators to decrease the influence of one beam or the other.   These "adiabatic" steps should result in a peak luminosity of 5 x 10^{30} cm^{-2} sec^{-1}.

The second phase of the upgrade involves a choice of options whose details will not be described here.   The physics objective however is very clear: in the collider mode, the upgrade will double the mass range of the discovery potential.   This can only be demonstrated

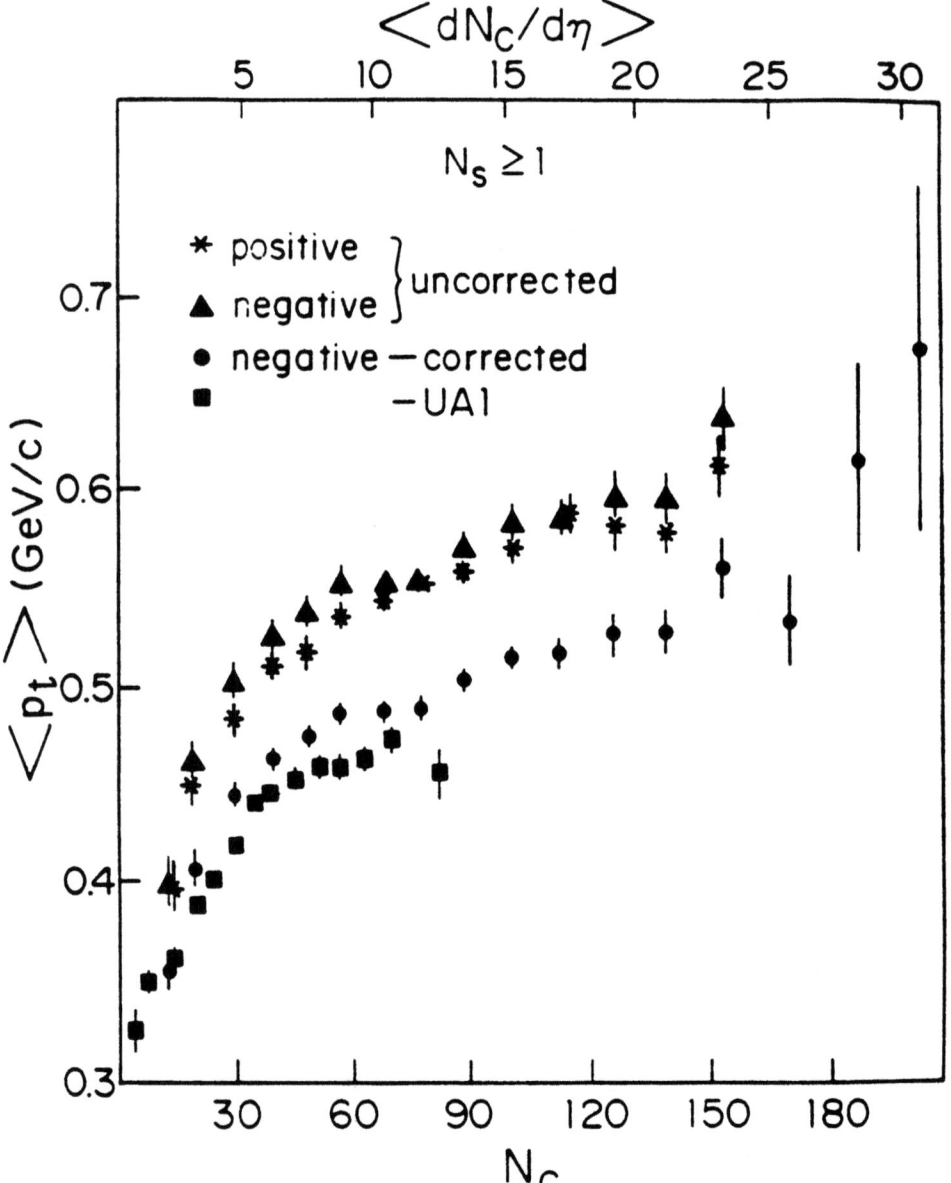

Fig. 12. E-735 – Average P_t vs Number of Charged Particles

by appealing to a variety of theoretical speculations, each of which makes predictions as to the existence of hypothetical particles. Usually nothing is known about these particles, however the theories have the ability to calculate the production cross-sections in hadron-hadron collisions as function of the unknown masses.

In this way one can use these theories as convenient tests of the relative value of changes in accelerator parameters e.g. how many more X-particles do we make if we double the luminosity? If we raise the energy 50%?

This kind of testing for a variety of theories can give a measure of the value of the TEVATRON Upgrade. As an example, we look at new gauge bosons W', Z', heavy quarks e.g. top or 4th generation quark, Q or supersymmetric particles (squarks & gluinos) and technipions.

It is this kind of study which leads to the conclusion that a 2 TeV $\bar{p}p$ collider at 5×10^{31} will essentially double the mass reach for discovery over the original 1×10^{30} cm^{-2} sec^{-1} design.

As a final comment we note that the Fermilab Upgrade assumes SSC will be built in the U.S., will turn on in the 1997 time period (with physics publications a year or two later) and that the U.S. will maintain a vigorous program of lower energy physics in the interim.

The same set of improvements will result in better proton beams for fixed target physics, especially in the domain of B-physics, intense beams of neutral kaons for CP violation etc. All in all, a vigorous future for the TEVATRON.

TABLE I

CHRONOLOGY OF FERMILAB

- 1968   CONSTRUCTION START
- 1972   200 GeV BEAM
- 1973   R&D ON SUPERCONDUCTING MAGNETS
- 1974   400 GeV BEAM
- 1979   "ENERGY SAVER" START
- 1981   PBAR SOURCE (TeV I) START
- 1982   FIXED TARGET (TeV II) START
- 1983   FIRST "SAVER" BEAM
- 1984   800 GeV   F. T. RUN   ($\sim$ 5 MO)
- 1985   800 GeV   F. T. RUN   (8 MO) AND 1.6 TeV COLLIDER TEST
- 1987   900 x 900 COLLIDER I 800 GeV F. T. (9 MO)
- 1988   COLLIDER RUN II   JUNE → ?

TWO MODES

- 800 → 1000 GeV $\sqrt{s}$ = 42 GeV FIXED TARGET
- 900 x 900 → 2 TeV = $\sqrt{s}$ COLLIDER

TABLE II

EXPERIMENTS COMPLETED IN 1987-88 FIXED-TARGET RUN

FIXED-TARGET

ELECTROWEAK

E-632 (MORRISON/   WIDE BAND NEUTRINOS IN THE 15 FT DUBBLE CHAMBER (16/84)
       PETERS)
E-733 (BROCK)   NEUTRINO INTERACTIONS WITH QUAD TRIPLET BEAM (4/26)
E-745 (KITAGAKI)   NEUTRINO PHYSICS WITH QUAD TRIPLET BEAM (10/43)
E-770 (SMITH)   NEUTRINO PHYSICS WITH QUAD TRIPLET BEAM (4/28)

DECAYS AND CP

T-721 (ROSEN)   CP VIOLATION (8/44)
E-731 (WINSTEIN)   MEASUREMENT OF ϵ'/ϵ (5/27)
E-756 (LUK) Ω^- MAGNETIC MOMENT (4/16)

HEAVY QUARKS

E-653 (REAY)   HADRONIC PRODUCTION OF CHARM AND B (19/79)
E-705 (COX)   CHARMONIUM AND DIRECT PHOTON PRODUCTION (8/47)
E-769 (APPEL)   PION AND KAON PRODUCTION OF CHARM (8/25)

HARD COLLISIONS AND QCD

E-711 (LEVINTHAL)   CONSTITUENT SCATTERING (3/23)
E-772 (MOSS)   NUCLEAR ANTIQUARK STRUCTURE FUNCTIONS (9/26)

OTHERS

T-755 (MAJKA/   STREAMER CHAMBER TESTS (2/10)
       SLAUGHTER)
E-776 (BAKER)   NUCLEAR CALIBRATION CROSS SECTIONS (3/7)

NOTE: E = EXPERIMENT; T = TEST. NUMBERS IN PARENTHESES DENOTE TOTAL
       NUMBER OF INSTITUTIONS AND PHYSICISTS, RESPECTIVELY.

TABLE III

CURRENTLY APPROVED FERMILAB EXPERIMENTS

FIXED-TARGET

ELECTROWEAK

E-665	(MONTGOMERY)	MUON SCATTERING WITH HADRON DETECTION (13/79)
E-782	(KITAGAKI)	MUON SCATTERING WITH TOHOKU BUBBLE CHAMBER (7/33)

DECAYS AND CP

E-761	(VOROBYOV)	HYPERON RADIATIVE DECAY (6/16)
E-773	(GOLLIN)	PHASE DIFFERENCE BETWEEN η_{oo} and η_{+-} (4/12)
E-774	(CRISLER)	ELECTRON BEAM DUMP PARTICLE SEARCH (4/7)

HEAVY QUARKS

E-687	(BUTLER)	PHOTOPRODUCTION OF CHARM AND B (8/58)
E-690	(KNAPP)	HADRONIC PRODUCTION OF CHARM AND B (5/21)
E-760	(CESTER)	CHARMONIUM STATES (7/59)
E-771	(COX)	BEAUTY PRODUCTION BY PROTONS (9/68)

HARD COLLISIONS AND QCD

E-672	(ZIEMINSKI)	HIGH P_T JETS AND HIGH MASS DIMUONS (7/28)
E-683	(CORCORAN)	PHOTOPRODUCTION OF JETS (9/33)
E-704	(YOKOSAWA)	EXPERIMENTS WITH A POLARIZED BEAM (16/50)
E-706	(SLATTERY)	DIRECT PHOTON PRODUCTION (9/75)

COLLIDER

E-710	(OREAR/ RUBINSTEIN)	TOTAL CROSS SECTION (6/18)
E-713	(PRICE)	HIGHLY IONIZING PARTICLES (2/3)
E-735	(GUTAY)	SEARCH FOR QUARK GLUON PHASE (7/52)
E-740	(GRANNIS)	D0 DETECTOR (20/124)
E-741/ E-775	(SCHWITTERS/ TOLLESTRUP)	COLLIDER DETECTOR AT FERMILAB (20/247)

OTHERS

E-466	(PORILE)	NUCLEAR FRAGMENTS (3/7)
E-754	(SUN)	CHANNELING TESTS (4/8)
E-778	(EDWARDS)	STUDY OF SSC MAGNET APERTURE CRITERION (5/15)
E-790	(SCIULLI)	ZEUS CALORIMETER MODULE TESTS (7/?)

NOTE: E = EXPERIMENT; T = TEST. NUMBERS IN PARENTHESES DENOTE TOTAL NUMBER OF INSTITUTIONS AND PHYSICISTS, RESPECTIVELY.

TABLE IV

PENDING PROPOSALS

P-682	(UNDERWOOD)	POLARIZED BEAM
P-688	(DITZLER)	POLARIZED BEAM
P-699	(STANEK)	POLARIZED BEAM
P-781	(RUSS)	LARGE-X BARYON SPECTROMETER
P-783	(REAY)	TEVATRON BEAUTY FACTORY
P-784	(LOCKYER)	BOTTOM AT THE COLLIDER
P-785	(BONNER/ PINSKY)	LOW ENERGY ANTIMATTER
P-786	(WILSON)	HEAVY QUARKS WITH MUONS
P-787	(GUTAY)	SEARCH FOR QUARK GLUON PHASE
P-788	(BERNSTEIN)	NEUTRINO OSCILLATIONS WITH NEUTRAL BEAM
P-789	(KAPLAN)	PRODUCTION AND DECAY OF B-QUARK MESONS AND BARYONS
P-791	(APPEL)	HADROPRODUCTION OF BEAUTY AND CHARM PARTICLES

210

DISCUSSION.

Trimble: Can you typically schedule all or most of the approved experiments, or are there 2 or 3 times as many ideas than the two dozens or so programs that you showed are scheduled for 1988 ?

Lederman: We always have a waiting list. We do approve more experiments than we can afford to build and run. The penalty is time. Experiments must wait up to 2-3 years longer than they should because of limited resources. This is a policy decision that waiting is better than not approving. Not all physicists agree.

COSMIC STRINGS AND GALAXY FORMATION: OVERVIEW AND RECENT RESULTS

Robert H. Brandenberger
Department of Physics
Brown University
Providence, R.I. 02912, USA

ABSTRACT. The cosmic string model of galaxy formation is briefly summarized and compared with other theories. Some recent results are presented: the origin of galactic angular momentum via tidal torquing is discussed, and the evolution of the mass function is studied.

1. Cosmic Strings and Galaxy Formation

Cosmic strings[1] are linear topological defects which form during a phase transition in the very early universe - typically about $10^{-35}s$ after the big bang. They arise in many but not all gauge theories in which the original symmetry group is spontaneously broken. Cosmic strings are thus lines of trapped energy with unusual dimensions. Their energy per unit length μ is about 10^{15} tons/cm and their width is about 10^{-22} of the radius of a hydrogen atom. ·

At the time of the phase transition a network of strings forms which can be viewed as a collection of random walks with curvature radius of the order of the horizon. The evolution of the cosmic string network from the time of the phase transition to times of relevance for galaxy formation has been studied both numerically[2] and analytically.[3] The crucial process is loop formation which occurs when infinite strings intersect.[4] The conclusion of these studies is that the cosmic string network approaches a scaling solution with on the order 1 infinite string segment per Hubble volume and a distribution of loops given by

$$n(R,t) = v\, R^{-4} \left(\frac{z(R)}{z(t)} \right)^{-3}. \tag{1}$$

Here R is the radius of the loop $z(t)$ is the redshift at time t, ν is a constant of the order 1 and $n(R,t)dR$ is the number density of loops in the radius interval $[R, R + dR]$.

The basic premise of the cosmic string model of formation of structure is that cosmic string loops form accretion seeds for structures such as galaxies and clusters.

211

M. Caffo et al. (eds.), Astronomy, Cosmology and Fundamental Physics, 211–219.
© *1989 by Kluwer Academic Publishers.*

Gas and dust begins to cluster about the loops at the time t_{eq} of equal matter and radiation when pressure becomes negligible. Small loops seed galaxies, larger loops seed clusters. In more quantitative terms: objects with mean separation d are seeded by loops of radius $R(d)$ where $R(d)$ is chosen such that the number density of loops with radius $\geq R(d)$ equals the observed number density d^{-3}. Up to this point, the analysis is independent of μ and of the dark matter content of the universe. [We shall follow theoretical prejudice and assume that $\Omega = 1$. Hence there must be dark matter - it can be either hot (close to relativistic velocities at t_{eq}) or cold (negligible thermal velocities at t_{eq})]. μ does not change the distribution of loops, it does however crucially influence the mass accreted onto a seed. μ can be determined by demanding that the mass of a cluster agree with observations.[5] The result is $G\mu \sim 10^{-6}$ which translates into a scale of symmetry breaking of about 10^{16} GeV in the underlying gauge theory.

The cosmic string theory of galaxy formation differs significantly from other currently popular models, for instance the "cold dark matter model" (CDM). In the latter the primordial energy density perturbations can be viewed as plane waves of all wavelengths superimposed with random phases. In the cosmic string model the initial density distribution is a collection of point perturbations of various amplitude superimposed on a homogeneous background.

The cosmic string model predicts a two point correlation function of clusters[6] which is independent of any parameters in the model, and the predictions match with observations within the observational error bars. Since $G\mu$ is fixed by demanding that the theory give the correct cluster mass, there are no more free parameters to adjust when studying galaxy formation. The predictions depend only on whether one has hot or cold dark matter. In the following sections I shall summarize some recent work on galaxy formation with cosmic strings, performed in collaboration with E. P. S. Shellard.[7]

2. Angular Momentum from Tidal Torquing

Tidal torquing as the source of angular momentum of galaxies has been studied in the context of the CDM model by many authors (see e.g. Ref. 8). Here I shall show that in the cosmic string model tidal torquing can generate enough angular momentum to explain observations, provided the dark matter is cold. The mechanism is sketched in Fig. 1. If galaxies are elliptical rather than spherical, the gravitational forces between nearest neighbors generate torques and hence angular momenta. In the cosmic string theory the eccentricity ϵ of a galaxy is generated by accretion onto a moving loop[9] and turns out to be about 1. The distance between nearest neighbor galaxies is set by the galaxy correlation function.

The angular momentum $\underline{L}$ is obtained by integrating the torque from t_{eq} to the

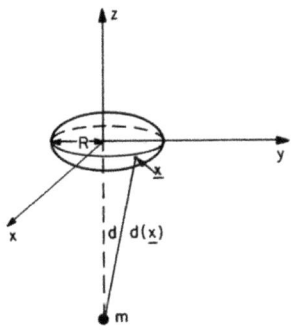

FIG. I

1. Sketch of the tidal torquing mechanism. A galaxy with "radius" R and center at the origin of the coordinate system is subject to a torque produced by its nearest neighbor galaxy with mass m located a distance d along the negative z axis.

present time t_0 :

$$\underline{L} = \int\limits_{t_{eq}}^{t_0} dt' \int d^3\underline{x}\, \underline{x} \cdot \underline{F}(\underline{x}, t') \tag{2}$$

where the spatial integral runs over the virialized volume and the force density $F(\underline{x}, t')$ has amplitude

$$F(\underline{x}, t') = G\, m(t')\rho(\underline{x})d^{-2}(\underline{x}, t') \tag{3}$$

$m(t')$ is the mass of the galaxy exerting the force and $\rho(\underline{x})$ is the density at $\underline{x}$. The resulting force is a quadrupole force which scales as $d^{-3}(t)$. The integral (2) is dominated by the time t_f when all free gas has accreted onto some string loop. For $t > t_f$ the integrand decreases as z^3. Thus, we can roughly evaluate (2) as

$$L \simeq \frac{2}{3}\, G\, t_f\, m(t_f)\, \hat{m}(t_f)\, R^2(t_f)\, d^{-3}(t_f) \tag{4}$$

where $\hat{m}$ and R are mass and virialized radius of the galaxy acquiring angular momentum.

The separation $d(t)$ between nearest neighbor galaxies is $d(t) = d(R)/N_0$ where N_0 is the number of galaxies typically found in a sphere of radius $d(R)$ about a given galaxy. From the observed amplitude of the galaxy correlation function it follows that $N_0 \simeq 3$. If M_{12} is the mass of a galaxy in units of $10^{12} M_\odot$ and $R_{0.1}$ the radius of the halo in units of 0.1 Mpc, then for cosmic strings and cold dark matter we obtain

$$L \sim 10^{73} \ cm^2 \ g \ s^{-1} \ \left(\frac{\epsilon}{2} \ M_{12}^2 \ R_{0.1}^2 \ N_0^3 \right)$$

(5)

For hot dark matter masses and radii are typically one order of magnitude smaller. In addition, most of the mass has accreted in the last expansion time, whereas for cold dark matter t_f occurs at a redshift of about 10. This leads to a further suppression of L :

$$L \sim 10^{67} \ cm^2 \ g \ s^{-1} \ \left(\frac{\epsilon}{2} \ M_{11}^2 \ R_{0.01}^2 \ N_0^3 \right).$$

(6)

We conclude that for cosmic strings and cold dark matter tidal torquing can explain the observed angular momenta of spiral galaxies which range from $10^{73} - 10^{75} cm^2 g s^{-1}$. For hot dark matter the answer is negative and we need a different mechanism for explaining the angular momenta of spiral galaxies (maybe along the lines of the one proposed in Ref. 10).

3. Evolution of the Mass Function

If all cosmic string loops accrete mass independently up to the present time t_0, then the mass function $n(M)$ of galaxies would follow immediately from (1). For the range of R of relevance for galaxies, $n(R) \sim R^{-5/2}$. With cold dark matter the mass $M(R)$ accreted about a seed loop of radius R is proportional to R. Hence

$$n(M) \sim M^{-5/2} ,$$

(7)

much too steep to match observations (which give $n(M) \sim M^{-1.2}$ with substantial uncertainty in the exponent). $n(M)dM$ is the number density of objects in the mass interval $[M, M + dM]$. For hot dark matter, free streaming reduces accretion onto small loops, with the result[11] that $M(R) \sim R^3$ and hence

$$n(M) \sim M^{-3/2}$$

(8)

in rather good agreement with observations. Note that hot dark matter with random phase adiabatic energy density perturbations is ruled out since in this case no galactic scale perturbations form independent of larger structures. Since cosmic string loops are perturbations which survive neutrino free streaming, hot dark matter with cosmic strings yields an interesting cosmological model with several good features such as a reasonable mass function and flat halo velocity rotation curves.[11]

Here we shall take a closer look at galaxy formation with cosmic strings and cold dark matter. We shall show[7] that the actual mass function is quite different than (7) and in much better agreement with observations. In fact, the assumption that all string loops accrete mass independently breaks down. Already at a redshift $z(t_f) \simeq 10$ all mass has accreted onto some loop. Thereafter, competition between loops and merging will be important. Note that for cosmic strings and hot dark matter only about 10% of all matter has virialized at t_0.

We first consider the mass function at redshift $z(t_f)$ in a model with cosmic strings and cold dark matter. We must incorporate the fact that many small loops will be inside the turn-around radius of larger loops and will not seed independent structures. Let $F(R)$ denote the fraction of space inside the turn-around radius of loops with radius $\geq R$. Then the effective mass function of structures is given by

$$\hat{n}(R) = n(R)\left(1 - F(R)\right).\tag{9}$$

In Fig. 2 we plot the resulting mass function at $z(t_f)$. We checked the results by numerical simulations and found agreement within the statistical error bars.

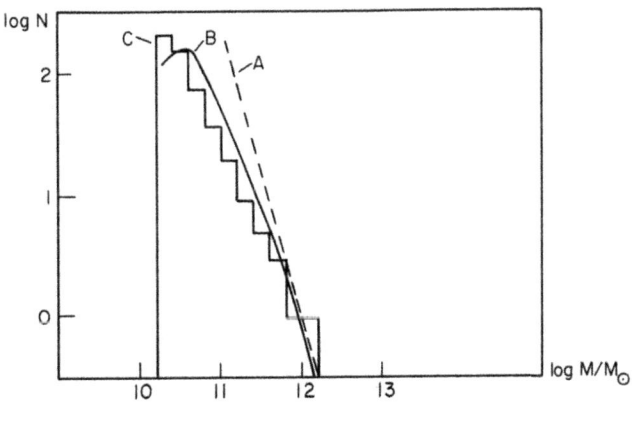

FIG. 2

2. The initial mass function for cosmic strings and cold dark matter (redshift $z \sim 10$). Curve C presents the numerical results, curve B gives the analytical prediction based on Eq. (9), curve A shows the original mass function with a slope $-5/2$.

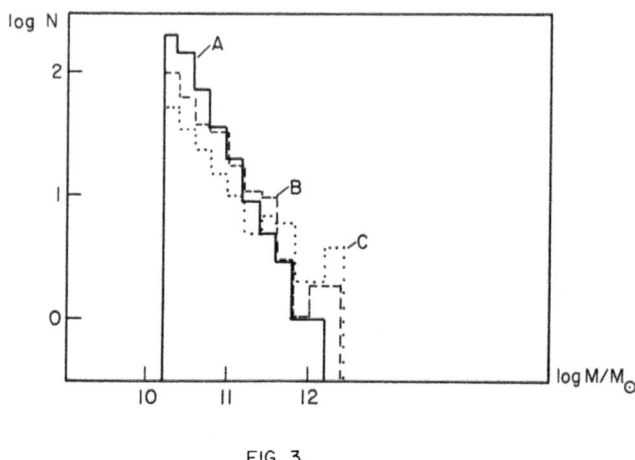

FIG. 3

3. The evolution of the mass function. Curve A is the initial mass function at a redshift of 10. Curve B is the mass function at redshift 2.5 and curve C the mass function at $z = 1.2$.

Next, we studied the evolution of the mass function for $z < 10$ using an N body code developed by E. P. S. Shellard (for more details see Ref. 7). The code is a "sticky particle" code. Two bodies merge in a given time step δt if the minimal distance between the points is smaller than the sum of the virialized radii and if the relative velocity is smaller than the escape velocity from the combined object. In Fig. 3 we plot the time dependence of the mass function and in Fig. 4 we compare the results with observational data. Between redshifts 10 and 1, the slope of $n(M)$ decreases significantly, especially at the low mass end. The final curve does not follow a single power law. The effective slope evaluated at $10^{12} M_\odot$ changes from -1.8 at $z = 10$ to -1.25 at $z = 1$. The final curve is in reasonable agreement with observations.

4. Conclusions

We have presented new results on angular momentum generation and mass function evolution in the cosmic string model. With cold dark matter, tidal torquing can easily explain the observed angular momenta of spiral galaxies, with hot dark matter a different mechanism is required. With cold dark matter, competition between loops and merging are important in analyzing the mass function. We conclude that, in contrast to naive expectations, the mass function agrees fairly well with observations.

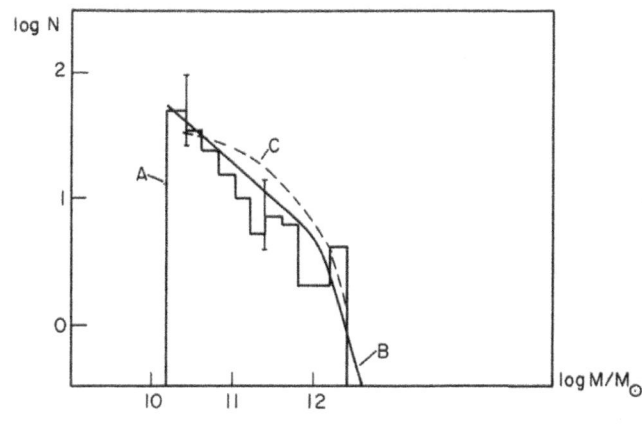

FIG. 4

4. A comparison with observational data. Curve A gives the cosmic string mass function at $z = 1.2$, curve B is the total mass function given by Binggeli (Ref. 12) and curve C are the observational data from Bahcall (Ref. 13). Note that curves B and C can be shifted horizontally by changing the mass to light ratio. Thus, really only the slopes of the three curves should be compared. The error bars for the two points on curve A represent the statistical uncertainties based on two runs.

REFERENCES

1. For reviews see e.g.,
   A. Vilenkin, *Phys. Rep.*, **121** (1985) 263;
   R. Brandenberger, *Int. J. of Modern Physics*, **A2** (1987) 77;
   N. Turok, in *"Astroparticle Physics,"* ed. A. De Rujula and P. Shaver, (World Scientific, Singapore, 1988).

2. A. Albrecht and N. Turok, *Phys. Rev. Lett.* **54** (1985) 1868;
   D. Bennett and F. Bouchet, *Phys. Rev. Lett.* **60** (1988) 257.

3. T. W. B. Kibble, *Nucl. Phys.* **B252** (1985) 277;
   D. Bennett, *Phys. Rev.* **D33** (1986) 872;, *Phys. Rev.* **D34** (1986) 3592.

4. E. P. S. Shellard, *Nucl. Phys.* **B283** (1987) 624.

5. N. Turok and R. Brandenberger, *Phys. Rev.* **D33** (1986) 2175;
   H. Sato, *Prog. Theor. Phys.* **75** (1986) 1342;
   A. Stebbins, *Ap. J. (Lett.)* **303** (1986) L1.

6. N. Turok, *Phys. Rev. Lett.* **55** (1985) 1801.

7. R. Brandenberger and E. P. S. Shellard, "Angular Momentum and Mass Function of Galaxies Seeded by Cosmic Strings," Brown preprint BROWN-HET-663, June 1988.

8. P. J. E. Peebles, *Ap. J.* **155** (1969) 393;
   S. White, *Ap. J.* **286** (1984) 38;
   J. Barnes and G. Efstathiou, *Ap. J.* **319** (1987) 575.

9. E. Bertschinger, *Ap. J.* **316** (1988) 496.

10. W. Zurek, *Phys. Rev. Lett.* **57** (1986) 2326.

11. R. Brandenberger, N. Kaiser, and N. Turok, *Phys. Rev.* **D36** (1987) 2242;
    R. Brandenberger, N. Kaiser, D. Schramm and N. Turok, *Phys. Rev. Lett.* **59** (1987) 2371;
    E. Bertschinger and P. Watts, *Ap. J.*, in press (1988).

12. B. Binggeli, in *"Nearly Normal Galaxies,"* ed. S. Faber, (Springer, New York, 1987).

13. N. Bahcall, *Ann. Rev. Astr. Astrophys.* **15** (1977) 505.

DISCUSSION.

McCrea: Does your model predict the sort of large-scale structure about which we heard this morning? in particular does it predict filaments or sheets of galaxies (or clusters) ?
Brandenberger: The cosmic string theory predicts sheets of galaxies with a mean separation of 25 Mpc, i.e. the size discussed this morning. However, the regions between the sheets are not predicted to be devoid of galaxies. The sheets stem from overdensities behind moving long strings. These are called wakes. If the length of a string segment is t and its transverse velocity v, then the width of the wake will be vt. Wakes formed at t_{eq} will have the most time to accrete matter and will be the most prominent. They will appear as thin sheets of galaxies with a mean separation of 25 Mpc, the comoving horizon at t_{eq}. For more details I refer to a paper by A. Stebbins, S. Veeraraghavan, R. Brandenberger, J.Solk and N. Turok, entitled "Cosmic String Wakes", *Ap. J.* **322**, 1 (1987).

Widrow: Can large voids bounded by sheets of galaxies be created in a cosmic string scenario?
Brandenberger: see answer to the question by McCrea.

Lyth: How seriously do you take the difficulty of reconciling galaxy forming cosmic strings with inflation? I refer to the following problem (D.H. Lyth, *Phys. Lett.* **196B**, 126, 1987). If inflation is not the cause of galaxy formation and cosmic strings are, the energy density during inflation has to be much less than the energy per unit length of the string ($\hbar = c = G = 1$). This is hard to reconcile with the fact that the strings are not formed before inflation.
Brandenberger: It is not easy to construct models which have an inflationary phase followed by cosmic string formation with a symmetry breaking scale of $\sigma = 10^{16}$ Gev. However, there is no no-go theorem . Note that the temperature T_c of the cosmic string phase transition is $\sqrt{\lambda}\sigma$, where λ is a self coupling constant. Thus, by making λ sufficiently small, T_c can be smaller than the reheating temperature. However, this involves fine tuning. I think the cosmic string theory should be studied independently from inflation. After all the foundations of the inflationary universe are not very solid.

Watson: In your model the total angular momentum must be zero. Over what scale of the universe would you expect the angular momentum to be zero?
Brandenberger: The angular momenta of a large number of galaxies will average to zero. The net angular momentum will satisfy the usual causality constraints, i.e. vanishes on scales larger than the horizon. Thus, on scales larger than that of a cluster (the comoving horizon scale at t_{eq}) the net angular momentum will drop sharply.

RESULTS FROM THE FREJUS EXPERIMENT

(Aachen-Palaiseau-Orsay-Saclay-Wuppertal Collaboration)

L. MOSCOSO
DPhPE-CEN-Saclay
91191-Gif-sur-Yvette CEDEX
France

ABSTRACT. A study of neutrino interactions recorded by the FREJUS detector is presented. A detailed comparison with a Monte Carlo calculation based on atmospheric neutrino (anti-neutrino) interactions in the detector shows that the behaviour of the apparatus is well understood. The analysis of the nucleon stability is described and lifetime limits are given for various decay modes with a charged lepton or an anti-neutrino in the final state. Neutrino and muon data have been also analyzed to search for a signal coming from the supernova SN1987A, yielding a negative answer.

The aim of this paper is to present the analysis of data recorded in the FREJUS detector on the three following subjects: atmospheric neutrinos, nucleon decay, and high energy neutrinos from the supernova SN1987A.

The FREJUS nucleon decay detector[1] is a very fine grain tracking calorimeter with $5 \times 5 mm^2$ cells and a calorimetric sampling of 3mm of iron. The total weight is 900 tons with an average density of 2. One thousand vertical planes of flash chambers with alternately horizontal and vertical cells provide two orthogonal views, 113 planes of Geiger tubes are used to trigger the detector. The average radiation length is 7cm and the collision length is 42cm. The rock cover is about 1800m in average for cosmic ray muons reaching the detector.

Data for this analysis were recorded between February 19th, 1984 and March 21st, 1988. Before June 1985, the mass of the detector was gradually increased from 240 to 900 tons. After June 1985, data were accumulated with the full size detector.

All data were classified by a visual scan. A rescanning was carried out with the help of a pattern recognition program.

1. Atmospheric neutrinos

The analysis of the neutrino interactions in the detector follows a double purpose. Firstly, it provides a direct measurement of neutrino fluxes which can be compared to the theoretical predictions based on neutrinos (anti-neutrinos) production by mesons produced by primary cosmic rays interactions in the atmosphere. Secondly, it allows to verify the understanding of the apparatus responses in view of the analysis of the nucleon decay which was the main motivation of this experiment, and to calculate the background level in the different decay modes (see next section).

Recently, the KAMIOKANDE collaboration[2] has reported the observation of an abnormally low number of ν_μ interactions when compared to the predictions. The analysis of the neutrino interactions in the FREJUS detector will be reported and compared to the expectations calculated from a Monte Carlo simulation.

M. Caffo et al. (eds.), Astronomy, Cosmology and Fundamental Physics, 221–230.
© *1989 by Kluwer Academic Publishers.*

In order to remove events produced by neutral hadrons or photons coming from interactions outside the detector and charged particles entering the detector, all events with the interaction vertex located at less than 50cm from the detector edges have been rejected; this reduces the fiducial mass to 554 tons. The total exposure sensitivity is 1.45 ktonxyr and 190 neutrino (anti-neutrino) interactions have been selected. The number of "contained events", for which all the secondary tracks stop inside the detector, is 137.

The selected events have been measured track by track on a graphic display by associating hits recorded in both horizontal and vertical projections. The primary and secondary vertices have also been measured in both views. Electromagnetic showers have been calibrated between 0.25 and 6 GeV with a test detector exposed to an electron beam at DESY. The energy measurement resolution is: $\Delta E/E \approx 12\%$ at 1 GeV. The momenta of stopping muons and pions are determined from the range of their tracks. The estimated error on the momentum is 10 MeV/c for a 500 MeV/c muon perpendicular to the iron plates. All events with an electromagnetic shower converting at less then 5cm from the primary vertex are classified as charged current (CC) v_e interactions.

Events with no visible electron are classified as CC v_μ interactions, if the highest momentum particle doesn't show a visible interaction in the detector. All the other events are called neutral current (NC) v interactions.

To compare the selected events with the predictions, we have performed a Monte Carlo simulation program of neutrino and anti-neutrino interactions at momenta between 200 Mev/c and 20 Gev/c, where the energy dependence of the neutrino fluxes is given by the predictions from ref. 3 at the FREJUS latitude[3].

The neutrino interactions in the detector were simulated by taking into account the quasi-elastic, resonant pion production (dominated by the $\Delta(1232)$) and multi-pion production processes for both charged and neutral currents; nuclear interactions of pions in ^{56}Fe have been included[4]. The simulation of the particle propagation has been performed by taking into account the sampling and the actual geometry of the detector.

The simulated events, which correspond to an exposure of 4 ktonxyr, have been scanned, measured and analyzed in the same way as the data.

Table I gives the numbers of neutrino (anti-neutrino) interactions for different types and containment criteria for real data. Those numbers are compared to the Monte Carlo expectation scaled to a 1.45 ktonxyr sensitivity. The quoted errors are statistical only and do not include an overall systematic error of 10-20% due to flux and cross-sections uncertainties. The agreement between Monte Carlo and data is fair.

Table I

Comparison of the data to the expected neutrino events for the present exposure of 1.45 ktonxyr (quoted errors are statistical only).

	DATA	M.C.	DATA/M.C.
ALL	190	190.7	1.00 ± 0.05
v_e	62	59.1	1.05 ± 0.14
v_μ	109	120.3	0.91 ± 0.09
N.C.	19	11.3	1.68 ± 0.38
CONTAINED	137	144.2	0.95 ± 0.08
v_e	59	54.7	1.08 ± 0.15
v_μ	62	79.0	0.78 ± 0.10
N.C.	16	10.5	1.52 ± 0.38

The visible energy distribution for the selected events is also well described by the theoretical model. This is depicted in figure 1.

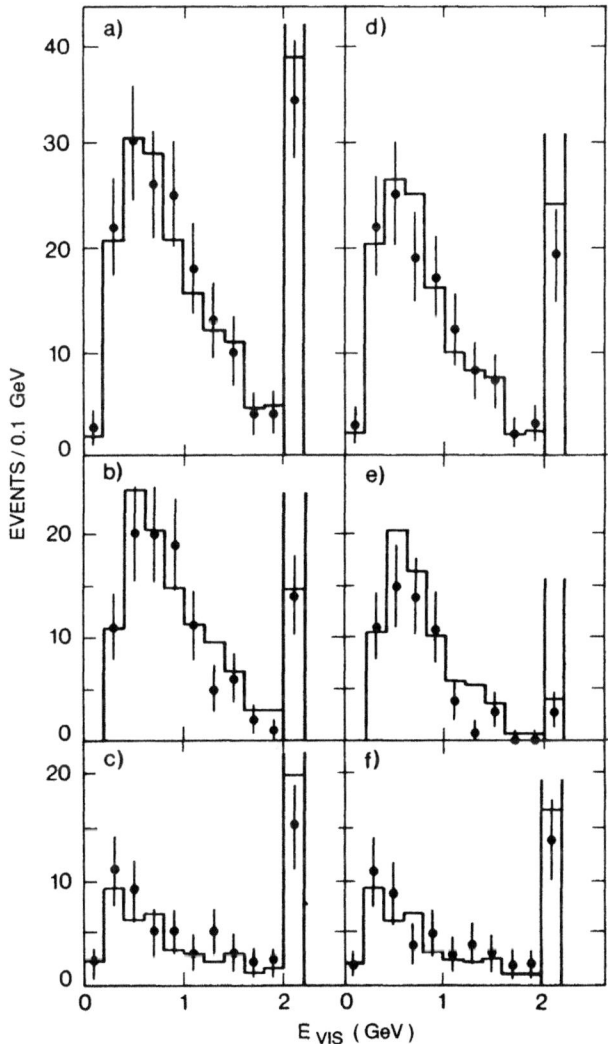

Figure 1. Visible energy distribution for: (a) all events, (b) ν_μ charged current, (c) ν_e charged current, (d) all contained events, (e) ν_μ contained charged current and (f) ν_e contained charged current. The last energy bin sums all events with their visible energy larger than 2 Gev. The histograms show the distributions expected from atmospheric neutrino interactions.

The main feature of the KAMIOKANDE effect is that the number of electron-like single ring events is in good agreement with the predictions of the Monte Carlo calculation and that, on the other hand, the number of muon-like single ring events is 59 ± 7 % of the predicted number.

In the FREJUS experiment, 62 CC $\nu_e (\bar{\nu}_e)$ interactions and 109 CC $\nu_\mu (\bar{\nu}_\mu)$ interactions have been selected and are in good agreement with the corresponding values (59.1 and 120.3) predicted by our

Monte Carlo calculation. In particular, the number of CC ν_μ ($\bar{\nu}_\mu$) interactions is $91 \pm 9\%$ (statistical error) of the predicted number. The ratio of the number N_e of the CC ν_e ($\bar{\nu}_e$) interactions to the number N_μ of the CC ν_μ ($\bar{\nu}_\mu$) interactions is $N_e/N_\mu = .57 \pm .09$ (statistical error), in good agreement with the expected value of .49 calculated from the Monte Carlo simulation. The comparison with the KAMIOKANDE result can be made by quoting $R = (N_e/N_\mu)_{DATA}/(N_e/N_\mu)_{MC}$ for both experiments.

This gives:

$R = 1.16 \pm .18$ for the all events in the FREJUS experiment
$= 1.37 \pm .25$ for the contained events
$= 1.78 \pm .27$ for the KAMIOKANDE experiment.

The FREJUS result is then in agreement with the expected value $R=1$, whereas the KAMIOKANDE result is at 2.9 standard deviations from 1.

2. Nucleon decay

The analysis of the nucleon decay is made by using the data recorded between February 19th, 1984 and November 13th, 1987.

The event selection criteria and measurements are the same that for contained neutrinos. The total fiducial sensitivity is 1.3 ktonxyr.

The analysis of the nucleon decay with a charged anti-lepton in the final state is based on the total energy-momentum conservation. This property cannot be used in the case of the nucleon decay with an anti-neutrino in the final state and cuts have been applied on the data by taking into account that all visible tracks come from one meson decay[5].

The detection efficiency has been calculated by generating nucleon decays in all the analyzed modes, taking into account the nuclear effects (absorption, elastic and inelastic scattering, and charge exchange). All tracks leaving the nucleus were propagated in the detector and all events were scanned, measured, and analyzed in the same way that the real data.

The background has been estimated by using two methods. The first method is to generate by a Monte Carlo calculation a large sample of neutrino interactions in the detector, as described in the previous section. The second method[6] is to use neutrino(anti-neutrino)- aluminum interactions recorded by the Aachen-Padova experiment aat CERN[7]. This experiment has used a spark chamber detector with a granularity similar to that of our detector. The exposure was equivalent to a sensitivity of 75 ktonxyr. Both calculations show that the background levels are in good agreement with the number of events recorded in our experiment.

Table II gives the number of selected events, the detection efficiencies, and the 90% confidence level limits of the nucleon lifetime divided by the branching ratio for the analyzed decay modes.

TABLE II

Detection efficiencies, selected events and 90% confidence level limits on the nucleon lifetime divided by the branching ratio. Results from KAMIOKA and IMB are also given.

Decay mode	Overall efficiency	FREJUS DATA	τ/B (10^{31}Years)	KAMIOKA τ/B (10^{31}Years)	IMB τ/B (10^{31}Years)
$e^+ \pi^\circ$	.31	0	5.1	19	25
$e^+ K^\circ$	.16	0	2.6	12	8.5
$e^+ \eta$	.22	0	3.8	11	20
$e^+ \rho^\circ$	.10	2	0.8	7.8	3.2
$\mu^+ \pi^\circ$	.26	0	4.5	15	8.9
$\mu^+ K^\circ$	.20	0	3.4	8	6.6
$\mu^+ \rho^\circ$	.21	1	2.5	6.8	2.6
$e^+ \pi^-$	.16	0	3.1	8.2	6.6
$e^+ \rho^-$	.20	0	3.9	4.5	2.0
$\mu^+ \pi^-$	.12	0	2.3	6.6	4.5
$\mu^+ \rho^-$	.09	0	1.8	1.5	1.1
$e^+ e^+ e^-$	.70	0	12.		51
$\mu^+ \mu^+ \mu^-$	.60	0	10.		19
$\bar{v} \pi^+$	.16	11	1.0	2.0	
$\bar{v} K^+$	.12	1	1.5	6.4	1.6
$\bar{v} \rho^+$	.15	0	2.4	2.	1.4
$\bar{v} K^{*+}$	.11	0	1.7	1.8	1.8
$\bar{v} \pi^\circ$	.10	1	1.3	6.8	2.0
$\bar{v} K^\circ$	.10	1	1.3	6.1	2.4
$\bar{v} \eta$	.16	0	2.9	7.1	4.4
$\bar{v} \rho^\circ$	.13	4	0.9	0.9	0.4
$\bar{v} \omega$	.11	1	1.3	5.4	1.9
$\bar{v} K^{*\circ}$	.12	0	2.2	1.7	0.9

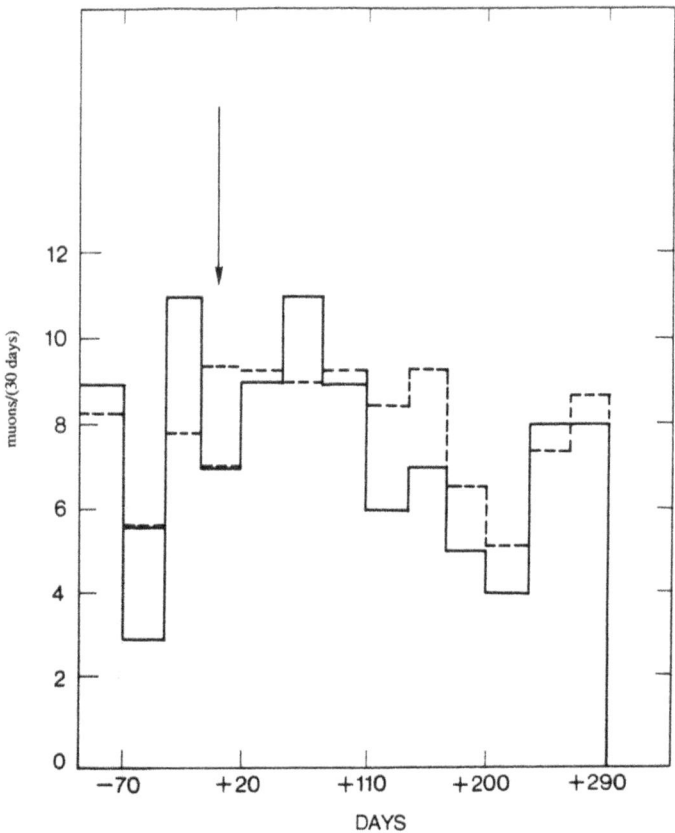

Figure 2. Time distribution of crossing muons in a 2° half angle cone pointing to SN1987A for a period going from 100 days before the SN explosion to 280 days after. Dashed line shows the estimated background distribution. The arrow indicates the SN explosion date.

For the charged current neutrino interactions in the detector, only events with visible energy greater than 1 GeV were used between February 23rd, 1987 and March 21st, 1988. Figure 3 shows the distribution of the Sin δ (where δ is the declination) as a function of the right ascension α for the reconstructed origin of the 15 selected neutrinos. No event has been found found within 30° from the SN direction, when 1.5 are expected.

3. The supernova SN1987A

The possibility of detecting high energy neutrinos produced from SN1987A by interactions of accelerated protons with the ambient gas has been discussed by several authors[8].

The analysis presented in this section is based on the study of through-going muons pointing in the supernova direction, which may be produced by high energy ν_μ($\bar{\nu}_\mu$) charged current interactions in the rock surrounding the detector, and on the study of the charged current interactions in the detector of both ν_e($\bar{\nu}_e$) and ν_μ($\bar{\nu}_\mu$) coming from the SN direction.

The first method is more sensitive to very high energy neutrinos (lower spectral index), but is contaminated by downward going muons produced in the earth atmosphere, since it is impossible to determine the directionality of a crossing muon. The second method offers the advantage to select only upward going neutrinos and to allow an analysis of ν_e($\bar{\nu}_e$) and ν_μ($\bar{\nu}_\mu$). On the other hand, this method is mainly sensitive to high spectral index.

The analysis of crossing muons was made on data recorded between February 23rd and November 30th, 1987. Only muons in a 2° half angle cone pointing to the SN direction were selected. Their total number was 72, when the expected number obtained from an off source calculation was 84. This gives a 90% confidence level upper limit of 10.3 events.

Figure 2 shows the time distribution of the selected muons for a period going from 100 days before the SN explosion to 280 days after. This is compared to the expectation from an off source calculation. No effect is seen. The limit on the luminosity may be estimated by a Monte Carlo calculation which includes the detector lifetime, the geometrical acceptance, the reconstruction efficiency, the angular distribution of the muon direction with respect to the neutrino direction, and the neutrino-nucleon charged current cross section.

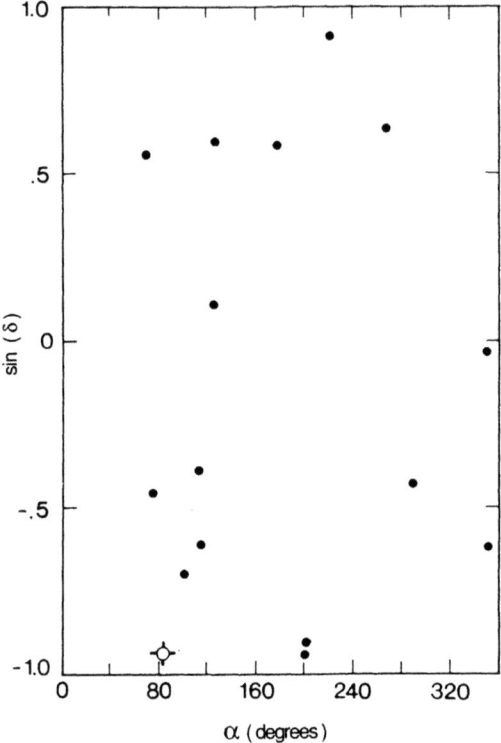

Figure 3. Sky map in equatorial coordinates of the reconstructed source of neutrinos with visible energy larger than 1 Gev. The open point indicates the SN1987A position.

If one assumes an energy dependence of the neutrino-nucleon cross section of the form: $\sigma = \sigma_0 \ln(1+aE)$ with $a = 2m/M_W^2 \approx 3 \cdot 10^{-4}$ Gev^{-1}, where m and M_W are the masses of the proton and of the W, and with $\sigma_0 = 2 \cdot 10^{-35}$cm^2 for neutrinos, 10^{-35}cm^2 for anti-neutrinos.

If the flux spectrum is given by $d\Phi/dE \propto E^{-\gamma}$, it is easy to calculate:

$$L = K \left[\frac{X_T^{2-\gamma}}{(\gamma-2) \int_{X_T}^{\infty} x^{-\gamma} \ln(1+x) dx} \right] \frac{dN}{dt}$$

with $K = 2 \cdot 10^{50}$ erg/s for neutrinos and $K = 10^{50}$ erg/s for anti-neutrinos and with $X_T = aE_T$, where E_T is the threshold energy. The term between brackets is ≈ 1 for $\gamma > 2.5$, it is a decreasing function of γ and diverges for $\gamma = 2$.

As no event has been recorded during the observation period of 392 days, a 90% confidence level upper limit of $dN/dt = 6.8 \cdot 10^{-8}$ s^{-1} is derived. Figure 4 shows the luminosity limits for ν and $\bar{\nu}$ emission from the supernova as a function of the differential spectral index γ, with the assumption that the ν and $\bar{\nu}$ luminosities are equal. The curve (a) in figure 4 is obtained from crossing muons for a threshold energy $E_T = 1$ GeV and is only valid for ν_μ and $\bar{\nu}_\mu$. Curves for other values E'_T of the threshold energy may be calculated by: $L_{lim}(E'_T) = L_{lim}(E_T)(E_T/E'_T)^{\gamma-2}$. The curve (b) is obtained from ν interactions in the detector and is valid for any value of the threshold energy $E_T \geq 1$ GeV and for $\nu_{e,\mu}$ and $\bar{\nu}_{e,\mu}$. The limits for values of the threshold energy $E'_T < 1$ GeV may be calculated by: $L_{lim}(E'_T) = E'^{2-\gamma}_T L_{lim}(E_T \geq 1$ GeV) with E'_T in GeV.

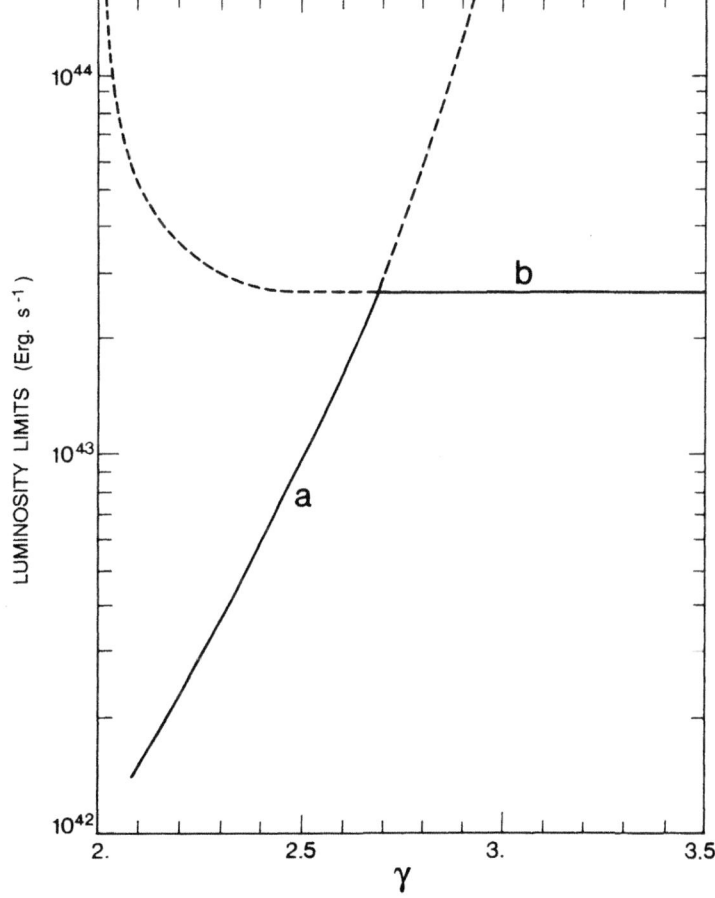

Figure 4. 90% confidence level limits of the luminosity for neutrinos coming from the supernova SN1987A obtained from: (a) crossing muons events and (b) neutrino interactions in the detector.

4. Conclusions

The analysis of neutrino interactions in the FREJUS detector has shown that the visible energy distributions are in good agreement with the predictions based on neutrinos (anti-neutrinos) production by mesons produced by primary cosmic rays interactions in the atmosphere.

Lower limits on nucleon lifetime have been calculated for decay modes with charged leptons and anti-neutrinos in the final state. The limits on τ/B range between $\approx 10^{31}$ and 10^{32} years depending on the decay mode.

A negative search for a signal from the supernova SN1987A, by using neutrino interactions in the detector and crossing muon tracks, yields upper limits on the neutrino luminosities.

5. References

1) Ch. Berger et al., Nucl. Instr. and Methods, **A262**(1987)463.
2) K.S. Hirata et al., Phys. Letters **B205**(1988)416.
3) T.K. Gaisser, 11th Conference on neutrino and Astrophisics. Dortmund 1984.
4) 'Study of atmospheric neutrino interactions with the FREJUS detector' (Frejus collaboration), to be published.
5) Ch. Berger et al., 'Results from the FREJUS experiment for nucleon decay modes into anti-neutrino+meson', submitted to Nuclear Phisics B.
6) W. Kolton, 'Etude du bruit de fond neutrino (anti-neutrino) attendu dans l'expérience du FREJUS sur la durée de vie du nucléon'. PARIS VI thesis, CEA-N-2522.
7) H. Faissner et al., Proceeding of the International Neutrino Conference, Aachen 1976
   H. Faissner et al., Phys. Rev. Letters **41**(1978)213.
8) R.M. Kulsrud et al., Phys. Rev. Letters **28**(1972)636
   V.S. Berezinsky et al., Il Nuovo Cimento **8C**(1985)185
   T.K. Gaisser aand T. Stanev, Phys. Rev. Letters **58**(1987)1695.

SUPERCLUSTERS OF GALAXIES: FRACTAL PROPERTIES

Jaan Einasto
Tartu Astrophysical Observatory
202444 Toravere, Estonia, U.S.S.R.

ABSTRACT. The distribution of galaxies has fractal properties, expressed in self-similarity of filaments and voids. Fractal properties have been observed in the scale interval from a few to several hundred Megaparsecs. The lower limit of the fractal structure is given by dynamical processes in clusters of galaxies, the upper limit cannot be derived from available observational data.

The classical cosmology is based on the following paradigms:
    (i) the Universe is homogeneous except the smallest scales;
    (ii) the Universe is expanding smoothly;
    (iii) the basic constituent of the Universe is ordinary matter;
    (iv) the evolution of the Universe can be described by an expanding Friedmann model.

Modern observations have shown that none of these paradigms holds. Instead, new paradigms have been introduced:
    (i) the Universe is inhomogeneous except the largest scales;
    (ii) large-scale coherent motions exist in the Universe;
    (iii) the basic constituent of the Universe is dark matter;
    (iv) the Friedmann expansion was preceded by an exponential (inflationary) stage.

Large scale motions, problems associated with dark matter and the inflationary cosmology were discussed elsewhere during this conference. This report is devoted to the first new paradigm, the inhomogeneity of the large scale distribution of galaxies.

Already first modern studies of the 3-D distribution of galaxies in space demonstrated that the Universe has a certain structure: in addition to clusters and second order clusters known earlier there exist filaments of galaxies and clusters of galaxies forming together with clusters the basic elements of superclusters. The space between clusters and filaments is devoid of any visible objects, respective voids have diameters from several tens to hundreds of Megaparsecs (Joeveer and Einasto 1978, Tarenghi *et al.* 1978, Tifft and Gregory 1978, Tully and Fisher 1978).

M. Caffo et al. (eds.), Astronomy, Cosmology and Fundamental Physics, 231–234.
© *1989 by Kluwer Academic Publishers.*

Recent deep surveys have confirmed these results (de Lapparent, Geller and Huchra 1986).

The correlation analysis has shown the presence of a puzzle in the large scale structure: the correlation length for galaxies is about 5 h^{-1} Mpc (Davis and Peebles 1983), but for clusters of galaxies 25 h^{-1} Mpc (Klypin and Kopylov 1983, Bahcall and Soneira 1983). Jones and Jones (1985) and Einasto, Klypin and Saar (1986) suggested that this behavior may be due to the fact that the ratio of the volume occupied by systems of galaxies to the total volume of the sample (the filling factor) is decreasing with increasing depth of the samples. This tendency can be describe naturally if we suppose that the distribution of galaxies in space is fractal as suggested by Mandelbrot (1982), Lachieze-Rey (1986), Calzetti et al. (1987) and Ruffini, Song and Taraglio (1988).

Jones et al. (1988) and Klypin et al. (1988) determined the fractal dimension of various samples of galaxies in our vicinity. In superclusters the fractal dimension is about 2 and in intersupercluster region about 1.3 which corresponds to dominating sheet-like and filamentary objects, respectively. Einasto and Einasto (1988, hereafter EE) suggested that the fractal structure of the Universe has some analogy with trees: filaments of different richness correspond to branches of various thickness. They also emphasized that as in all natural objects the fractal description of the Universe should have a lower and an upper limit.

To check this behavior EE used one of the principal properties of fractals, the self-similarity of structures. They determined mean diameters of voids for samples of galaxies of different size (the sidelength of cubic volumes) from 1 h^{-1} to 100 h^{-1} Mpc, and for samples of clusters of galaxies of size from 100 h^{-1} to 250 h^{-1} Mpc. Void diameters were defined as distances between two successive density maxima along rectangular beams put in various directions in sample volumes. To check these results a second method was also applied: the mean radii of empty spheres surrounded by galaxy systems were derived. In both cases constant relative resolution was used, which is determined in the first method by the thickness of beams and in the second one by the grid size of the density matrix. The results obtained with both methods are plotted in Fig. 1.

We see that the mean diameter of voids is proportional to the size of the sample. This indicates that at fixed relative resolution voids and filaments are self-similar. A small systematic difference exists between results obtained with two different methods due to differences in void diameter definition, but the general trend is identical in both cases. There exists a well defined lower limit of void diameters and the fractal structure given by the mean diameter of clusters of galaxies (at smaller scales voids are not defined since respective samples contain only one density maximum if any). The structure and diameter of clusters is determined by dynamical processes, these processes determine the lower limit of the fractal structure. Fractal properties are expressed in the filamentary nature of the distribution of galaxies. These structures are dynamically young and therefore have preserved their primordial form.

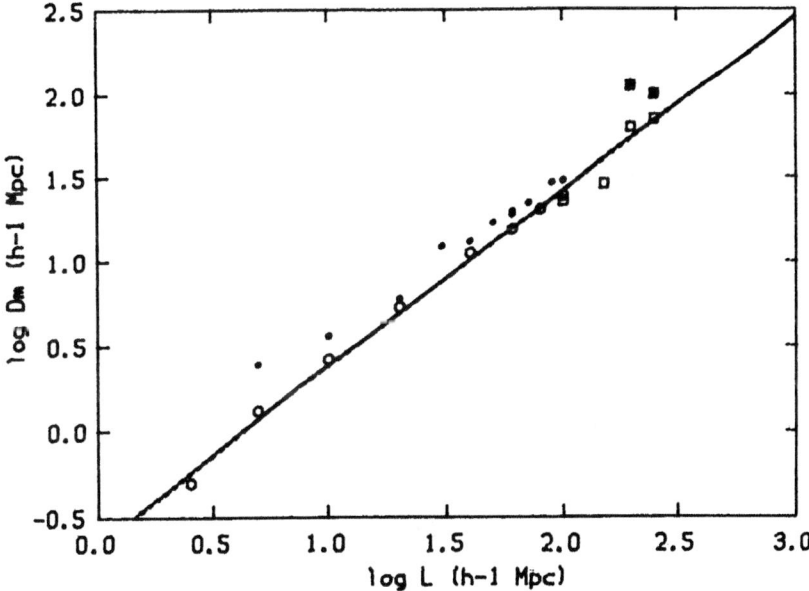

Fig. 1. The mean diameter of voids, D_m, for samples of different size, L, expressed in units h^{-1} Mpc. Galaxy samples are plotted by circles, cluster samples by squares; open and filled symbols denote results obtained by the empty beam and sphere methods, respectively. The regression line, determined from galaxy sample data and empty beam method, is drawn by a solid line; the slope of the line is 1.04 ± 0.05.

Available data do not indicate the presence of an upper limit of the self-similarity of voids. Whether this effect is real or due to the lack of data is still not clear and deeper complete redshift surveys are needed to answer this question.

To check the reality of these results the diameters of voids have been determined for random samples, too. Random samples have also mean diameters of voids proportional to sample sidelength. Detailed properties of voids are, however, completely different. In particular, the filling factor of real and random samples is different, also the dependence of void diameters on the mean density of particles in samples (for details see the original paper by EE).

These results demonstrate that the inhomogeneity of the Universe extends to rather large scales. None of the 3-D galaxy samples available can be considered as a fair sample of the Universe. The same conclusion was reached by visual inspection of new redshift data by Geller (1988). Thus care is needed to use available galaxy samples to probe the Universe and to determine its global properties, in particular parameters of the correlation function.

REFERENCES

Bahcall, N. and Soneira, R., 1983. *Astrophys. J.*, **270**, 20.
Calzetti, D., Einasto, J., Giavalisco, M., Ruffini, R., and Saar, E., 1987. *Astroph. Sp. Sc.*, **137**, 101.
Davis, M. and Peebles, P.J.E., 1983. *Astrophys. J.*, **267**, 465.
de Lapparent, V., Geller, M. and Huchra, J., 1986. *Astrophys. J.*, **302**, L1.
Einasto, J., Klypin, A. A. and Saar, E., 1986. *Mon. Not. R. astr. Soc.*, **219**, 457.
Einasto, M. and Einasto, J., 1988. *Mon. Not. R. astr. Soc.* (in press).
Geller, M., 1988. This conference.
Huchra, J.P., 1988. Redshift compilation.
Joeveer, M. and Einasto, J., 1978. *The Large Scale Structure of the Universe,* eds. M.S. Longair and J. Einasto, Reidel, Dordrecht, p. 241.
Jones, B.J.T. and Jones, J., 1985. Preprint.
Jones, B.J.T., Martinez, V., Saar, E. and Einasto, J., 1988. *Astrophys. J.* (in press).
Klypin, A.A., Einasto, J., Einasto, M., and Saar, E., 1986. *Mon. Not. R. astr. Soc.* (in press).
Klypin, A.A. and Kopylov, A.A., 1983. *Pis'ma v Astr. Zh.*, **9**, 41.
Lachieze-Rey, M., 1986. Preprint.
Mandelbrot, B.B., 1982. *The Fractal Geometry of Nature.* Freeman and Co., San Francisco.
Ruffini, R., Song, D.J., and Taraglio, S., 1988. *Astr. Astroph.*, **190**, 1.
Tarenghi, M., Tifft, W.G., Chincarini, G., Rood, H.J. and Thompson, L.A., 1978. *The Large Scale Structure of the Universe,* eds. M.S. Longair and J. Einasto, Reidel, Dordrecht, p. 263.
Tifft, W.G. and Gregory, S.A., 1978. *The Large Scale Structure of the Universe,* eds. M.S. Longair and J. Einasto, Reidel, Dordrecht, p. 267.
Tully, R.B. and Fisher, J.R. 1978. *The Large Scale Structure of the Universe,* eds. M.S. Longair and J. Einasto, Reidel, Dordrecht, p. 214.

UNDERGROUND PHYSICS

E. Bellotti
Laboratori Nazionali del Gran Sasso - I.N.F.N.
S.S. 17 bis km 18 + 910
67010 - ASSERGI (AQ)
Dipartimento di Fisica - Universita' di Milano
and Sezione I.N.F.N. Milano
Italy

ABSTRACT. Some of the experimental subjects usually investigated in underground laboratories are shortly reviewed; among them double beta decay and solar neutrinos receive more attention. In fact, present experiments on $\beta\beta$ decay and the new thecniques recently developed are very promising and garantee that in a near future very stringent limit on neutrinoless decay and Majoron emission will be reached and two neutrino mode observed for a few nuclei.
Solar neutrino flux measurements is one of the major task of physicists in next years; experiments presently in operation or in construction are described. In a second part of the paper a status report of the Gran Sasso Laboratory is given.

1. INTRODUCTION

It is the tradition of physicists to go in mines or other deep underground places to investigate on very specific problems. Among them, I would mention the many experiments on very energetic muons and muon bundles, the detection of atmospheric neutrinos, the search of solar neutrinos, the investigation on matter stability and on rare decays; finally let me recall the extremely important observation of neutrinos from 1987 Supernova.
    The reason to go underground lays on the fact that large thickness of rock are a very efficient shield against cosmic radiation. Only very energetic muons, neutrinos of any energy and exotic particles (magnetic monopoles, WIMP's etc) can penetrate hundreds meters of rock. Therefore two classes of experiments are usually carried out underground: experiments which, at least in principle, require no background radiation at all and experiments whose aim is to detect the very penetrating particles to investigate on their nature on their origin. In this paper, I would shortly review some of the experimental subjects of both kind. In a second part a "status report" of the Gran Sasso Laboratory will be given.

2. PROTON STABILITY

Enthusiasm for minimal models of grand unified theories has been

235

M. Caffo et al. (eds.), Astronomy, Cosmology and Fundamental Physics, 235–254.

chilled by results obtained on nucleon stability experiments. Limits of 10^{32} years have been obtained for many decay modes (DEG-87). It seems very difficult to improve significantly these limits; very huge and very well instrumented detectors are needed. Projects in this sense are in discussion (a very large - 30000 tons-water Cerenkov detector in Japan and a 3000 tons liquid argon detector at Gran Sasso); however significant improvements are not expected in a short time.

3. DOUBLE BETA DECAY

Recently very interesting results have been obtained on the experimental front as well on the understanding of theoretical problems (see f.i. KOT-87 and KLA-87 and references therein). Most significant experimental results are the observation of the 2 neutrino decay mode of Se (ELL-87), the improvement of limits on neutrinoless decay of Ge, the many new ideas in the field.

Table I reports some of most relevant results on $\beta\beta$ decay; the decay through the emission of a Goldstone boson (majoron) has been also considered.

TABLE I

Some double beta decay rates or limits (years)

Nuclide	2ν $(0^+\to0^+)$	0ν $(0^+\to0^+)$	0ν $(0^+\to2^+)$	majoron	reference
^{76}Ge	$>8\cdot10^{19}$	$>(5-9)10^{23}$	$>2\cdot10^{23}$	$>1.4\cdot10^{21}$	CAL-88
	$>5\cdot10^{20}$	$>3\cdot10^{23}$	$>3.4\cdot10^{22}$	$>1.2\cdot10^{21}$	FIS-87
	-	$6\cdot10^{22}$	$(1.5\pm.5)10^{22}$	-	MOR-87
	-	$3\cdot10^{23}$	$8\cdot10^{22}$	$6\pm2\cdot10^{20}$	AVI-87
^{82}SE	$11^{+.8}_{-.3}10^{20}$	$>1.1\cdot10^{20}$	-	$7.3\cdot10^{20}$	ELL-87
^{136}Xe	-	$>3\cdot10^{21}$	$1.3\cdot10^{21}$	$1.\cdot10^{20}$	KUZ-87
	-	$>2.4\cdot10^{21}$	-	-	BAR-87
	-	$>2\cdot10^{21}$	-	-	FIO-88/1

$2\nu(0^+\to0^+)$ two neutrino decay from g.s. to g.s.

$0\nu(0^+\to0^+)$ neutrinoless decay from g.s. to g.s.

etc.

For a complete review of experimental results on double beta dacay see f.i. (AVI-87, FIO-88/1).

As it results from Table I, the two positive results on majoron emission and on neutrinoless decay on the first excited state obtained by F.T. Avignone et al. and by the Bordeaux-Saragoza Collaboration respectively, have not been confirmed by other experiments.

The use of isotopically enriched samples is now applied also to huge detectors. The ITEP-Eravan group (AVI-87, VAS-87) is operating a Ge detector of ≈ 90 cm^3, enriched to 85% ^{76}Ge.

Realistic proposals to built large array of Ge detectors made with enriched Ge have been put forward by the V.H. Klapdor at Heidelberg, in collaboration with Soviet groups, and by F. Avignone and Coll. in USA.

The Milano group is now operating a multicell proportional chamber filled with natural Xenon at 9 atms.; the chamber will be filled with enriched ^{136}Xe(i.a.65%) before the end of the year; by extrapolating present results (ALE-88), it is expected to reach limits of a few 10^{22} years on the neutrinoless decay and significant limits on majoron emission; the observation of the 2ν mode is not outside of the possibility of this experiments.

Other detectors based on the use of enriched Xenon are in use in Soviet Union (BAR-87 and KUZ-87).

A TPC has been constructed by Caltech and now it is in operation in the Gothard Tunnel (BOE-88).

Liquid Argon and Xenon detectors have been developed by E. Aprile (APR-87) and there are proposals to use them for double beta decay searches.

The use of isotopically enriched materials in large quantity allows us to be confident in significant improvements in the above quoted limits on the neutrinoless decay and on a systematic observation of the 2ν mode.

Very recents results on bolometric devices obtained by a few groups in the world are of extreme interest; this subject has been discussed in a few review papers; see f.i. (SAD-88).

This completely new technique, which use in $\beta\beta$ decay searches was proposed in (FIO-84) developed quite fast in the last year and it merits great attention.

Principle of operation is very simple: consider a pure crystal of an insulator; the heat capacity is given by the Debye low:

$$c = 1.2 \cdot 10^{16} \cdot m/a \cdot (T/\theta)^3 \text{ MeV/K}$$

where m is the mass of the sample (g) and A_3 its atomic weight, T the absolute temperature and $1.2 \cdot 10^{16}$ m/A $(T/\theta)^3$ MeV/K the Debye temperature, which is of the order of one or a few hundreds of K (e.g. it is 640 K for Si). If the crystal is kept at few millikelvin or tens of mK, the energy deposition of keV's or MeV's is sufficient to increase the temperature in a measurable way.

Detectors of various dimensions have been constructed and operated; let me quote a few results.

Mc Cammon et al. (MOS-88), with a crystal of HgCdTe reached an energy resolution of 17 eV on the iron lines; Alessandrello et al. (ALE-88), which are more interested in large mass detectors, cooled

238

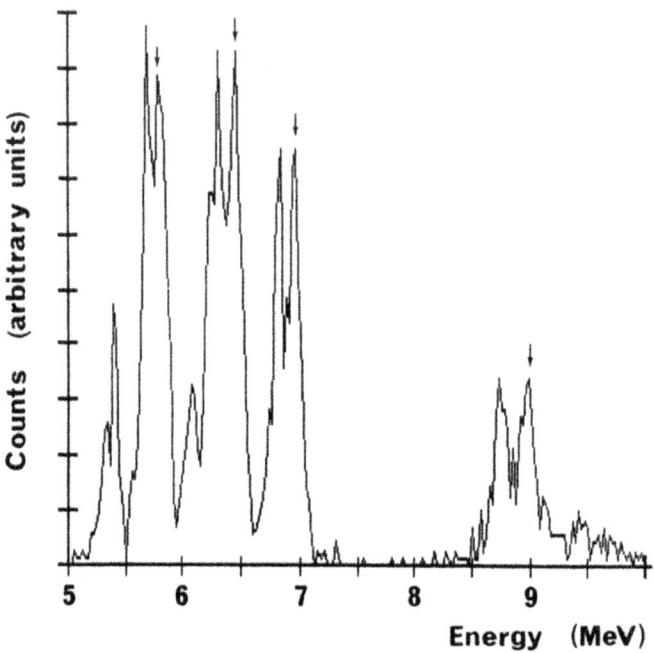

Fig. 1 – Energy spectrum of α particles from a ^{228}Ra source; lines
marked with an arrow are due to decays of nuclei inside the
crystal.

down a crystal of 0.7 g of Ge which has shown good energy resolution
(<1%) on measuring alpha particles (Fig.1); it has also been
demonstrated that these devices are capable of detecting essentially
non ionizing particles. Next step will be the construction of a 10 g
germanium detector.
    The above quoted results have been reached in a very short time
and are encouraging; the time when this new technique will be applied
to a "real" experiment is not far in the future.

4. NEUTRINO FROM COLLAPSING STARS.

This subject will be deeply discussed at this same Conference,
therefore I would only mention that a new detector especially designed
for the observation of neutrinos from collapses is under costruction
at Gran Sasso. The experiment, named LVD (for Large Volume Detector)
has been proposed by a large international Collaboration and it is
described in (LVD-85, AGL-86). Essentially it is a larger and improved
version of the 90 tons detector installed at Mt. Blanc, see f.i.
(BAD-84); it consists of a large number of liquid scintillators
counters, for a total of 1800 tons, surrounded by a tracking system of
limited streamer tubes. The efficiency to detect a collapse in our
Galaxy is 100% with a signal to noise ratio of hundreds; a few tens of
events are expected in case of a collapse in the Magellanic Cloud and

therefore, also in that case, the efficiency to detect the collapse is practically 1.

It has to be mentioned that most of the newly proposed solar neutrino detectors are capable of observing neutrinos from collapses, as it has been demonstrated by Kamiokande II (HIR-87) and IMB (BIO-87).

5. SOLAR NEUTRINOS

The present situation is well known; two experiments are running, the above mentioned KAMIOKANDE II and the well known clorine experiment by R. Davis and Coll. Results from these experiments are in contrast with expectation deduced from the so called Standard Solar Model (SSM); for a description of SSM one can see (BAC-82,CAS-85). It has to be mentioned that recent, but unpublished, results by Davis are much less in contradiction with the model. It is clear that the duty of physicist in next years will be the clarification of this problem, which means the measure of neutrino flux from Sun (Fig.2) as a function of energy.

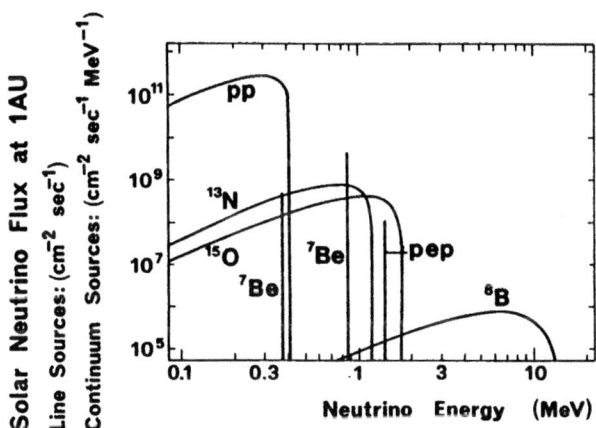

Fig. 2 - Solar neutrino flux.

In that way it will be possible to understand if the observed deficit of ν's arises from deficencies in the SSM or from a new neutrino physics, i.e. neutrino oscillation in vacuum or in matter. Measurements lasting many years could also reveal possible correlation of ν flux to the solar activity (DAV-87); geochemical measurements, like that on molybdenum presently in progress (COW-83, HAX-86) could also give information on Sun evolution in the last million of years.

Davis experiment will not be described, being so well known, while a few data on KAMIOKANDE II and on future experiments will be reported.

6. KAMIOKANDE II (KAM-87)

The Kamiokande II detector, a 3000 tons water Cerenkov detector, was
originally designed to search for proton decay; an improvement program
started in late '84 to make it capable of detecting also low energy
events induced by ^{8}B solar ν's. A 4π anticounter, made by 1.5 m of
water, viewed by 123 pm's, surrounds the inner volume, acting also as
a shield against low energy γ rays and neutrons; a more efficient
water purification system to remove radioactive impurities has also
been installed; also the electronics have been improved. Nevertheless,
background remains three order of magnitude larger than the expected
signal.

Solar neutrinos are detected through the purely leptonic
reaction:

$$\nu_e \; e^- \rightarrow \nu_e \; e^-$$

which can be induced also by ν_u and ν_τ (and $^{\nu}$); however cross section
is much larger than that of ν_u or ν_τ.

A detailed investigation on background allowed application of
many selection criteria which lowered the background to an acceptable
level. Taking advantage of the fact that the scattered electrons
roughly maintain the direction of the incoming neutrinos, the
Collaboration sat a limit of $3.2 \; 10^6$ ν's cm^{-1} s^{-1} to the ^{8}B ν flux,
the energy threshold on the scattered electrons being 9.5 MeV.

Disagreement between data and SSM is confirmed also by this
direct experiment, although limited to the high energy part of ν
spectrum.

New results are expected in a short time, based on a larger
statistics.

7. GALLIUM

The two above mentioned experiments are sensitive to the ^{8}B ν's, which
are produced in a marginal reaction chain, very dependent on the Sun
temperature. On the contrary, the flux of ν's emitted in the basic
fusion reaction of two protons is essentially independent on details
of a solar model, being strictly related to the global energy
production.

Therefore a measure of the so called p-p ν's in contrast with
expectations would have profound consequences on our knowledge of Sun
or of neutrinos.

Detection of p-p ν's is extremely difficult due to their low
energy (<420 kev) and, at present, only radiochemical methods are
suitable for such a measurement.

Among possible target nuclei, gallium, whose use was suggested by
V. Kuzmin (KUZ-66), is a suitable nucleus.

Gallium is a metal with a low melting point (about 39 C); it has
two isotopes:
^{69}Ga (i.a. 60%) and ^{71}Ga (i.a. 40%) which is the interesting one. The
reaction used to detect solar ν's is:

$$^{71}\text{Ga} \; (\nu_e, \; e^-)^{71}\text{Ge}$$

it has an energy threshold of 235 keV (Fig.3); the expected rate, according to SSM is of ≈120 SNU's.
^{71}Ge decays back to ^{71}Ga by electronic capture with a half-life of 11.4 days.

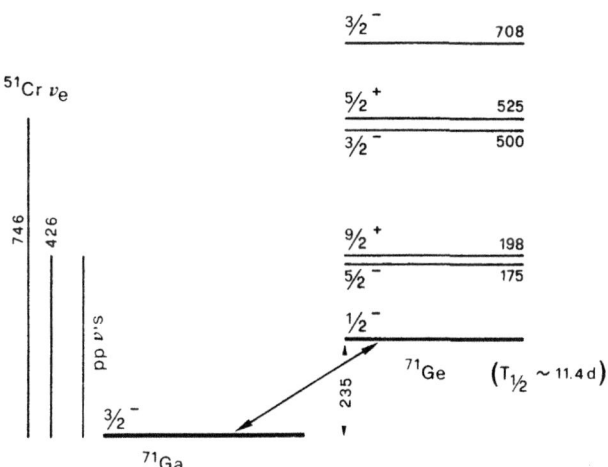

Fig. 3 – Excited levels of ^{71}Ge.

Two experiments are in the assembling stage, one at Gran Sasso and a second one at Baksan.
The Gallex European Collaboration (KIR-86), recently joined by a group from BNL, is operating at Gran Sasso.
About 30 tons of Gallium in form of $GaCl_3$ in acqueous solution of HCl will be stored in a unique tank (Fig.4).
The expected production rate is ≈1 ^{71}Ge atom/day.
Germanium ($GeCl_4$) will be routinely extracted by a nitrogen purge with a small amount of Ge carrier; through chemical reactions, $GeCl_4$ will be transformed in GeH_4 which finally will be used as a filling gas in miniaturized proportional counters.
Feasibility of the experiment has been proved in a pilot experiment carried out at BNL some years ago (DAV-81).
Another important feature of both experiments is the possibility of calibrating the entire procedure by 1 MCi ^{51}Cr source. Test on a source of natural cromium has been carried out at Grenoble; also the use of enriched cromium has been investigated and found very attractive from many points of view: reactor time, source dimensions, purity of the source etc; some tests on purification have been done at Oak - Ridge with very encouraging results.
A similar experiment, based on the use of Gallium in metallic form, is in installation at Baksan. Logic of the two experiments is very similar; however they differ in the extraction method (from a solution or from a melted metal) which is one of the delicate points on the entire experimental procedure.
First results are expected for 1990 from both the experiments;

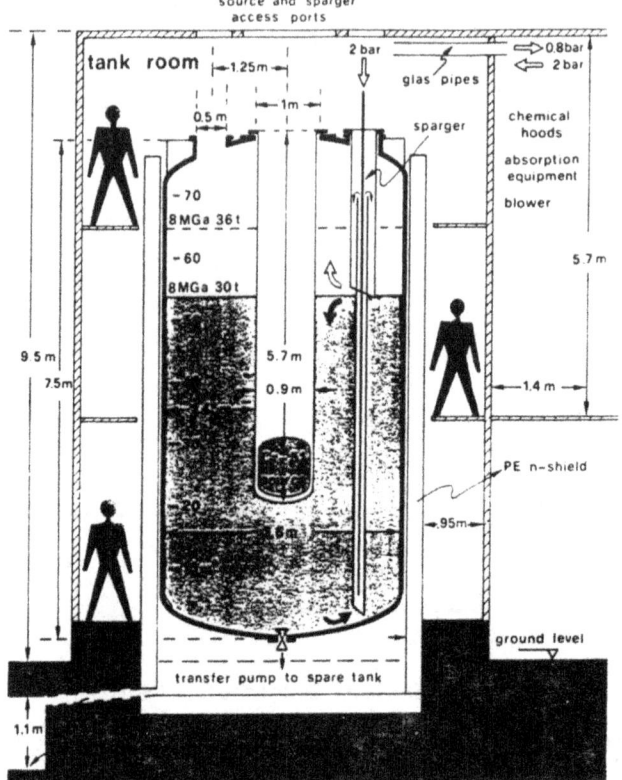

Fig. 4 - Schematic view of Gallex set-up at Gran Sasso.

however an accurate (≈10%) measurements will require at least two
years and the calibration by a Cr (or other suitable element) source.
    The three experiments which will be shortly discussed in the
following completely differ from the Gallium experiments because:
- they are sensitive only to high energy (^{8}B) ν's
- events are detected in "real" time and in two cases the direction of
  the incoming ν's can be reconstructed;
- they can observe at the same time reactions induced by any flavor
  ν's and reaction induced only by electron ν's; in that way it is
  possible, at least in principle, to measure both the total ν flux
  as well the ν_e flux and therefore to check the hypothesys of ν

oscillations independently from solar model prediction.

8. ICARUS (ICA-85)

Icarus is a project of a large liquid argon imaging chamber; by such a
detector, once built, one could afford many physical problems: proton
decay, cosmic ray physics, magnetic monopoles search, etc, flux
measurement and solar neutrino.
    Solar ν's can interact in one of the following ways: purely
leptonic interactions

$$\nu_x \ e^- \rightarrow \nu_x e^- \qquad\qquad 1)$$

which can be induced by any flavor neutrino, and charge current interactions

$$^{40}Ar \ (\nu_e, \ e^-) ^{40}\!K^* \ (4,38 \ MeV) \qquad\qquad 2)$$
$$\Big|_{\rightarrow ^{40}K + \gamma}$$

which can be induced by electrons neutrinos only.

According to the SSM, 1500 events/(Kton year) of type 1 and 1330 events/(Kton year) of type 2) are expected.

This project is very ambitious and, as a first step, the Collaboration is planning to construct a so called "baby ICARUS" of about 300 tons of liquid Ar, especially designed to detect low energy events like those due to solar ν's (ICA-88).

Tests on argon purification, electrodes structure, read-out electronic have been carried out in past years and are continuing at CERN with encouraging results (BAC-88). A 2 tons prototype is expected to be installed at Gran Sasso in 1989.

9. SUDBURY NEUTRINO OBSERVATORY (SNO)

In Canada, is under discussion the construction of a 1 Kton heavy water Cerenkov detector (EWA-87). The detector, schematically shown in Fig. 5, consists of ≈1000 tons of heavy water contained in a acrylic vessel, surrounded by lights water (acting as a shield) and viewed by a large number of photomultipliers covering at least 40% of the surface. This detector will be located very deep underground, at 2070 m of depth, in a nichel mine (INCO Limited' Chreigton mine, near Sudbury).

Background sources have been investigated and ways to lower them at an acceptable level have been proposed in a very detailed way.

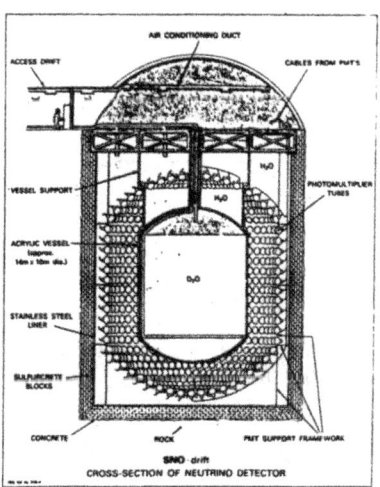

Fig. 5 - Schematic view of the SNO detector.

Obviously, such a detector can be built only in Canada, where large amounts of the expensive heavy water are available, being produced for power reactors.

High energy neutrinos from ^{8}B and He-p can be detected trough one of the following reactions:

$$\nu \, d \to e^- \, p \, p \tag{3}$$

the electron in the final state practically carry out al the neutrino energy; this reaction can be induced by electron neutrinos only; the expected yield is of the order of a few thousand of events/year (for a flux of $6 \cdot 10^6$ $\nu's/cm^2s$)

$$\nu_x e^- \to \nu_x e^- \tag{4}$$

This reaction, as discussed above, can be initiated by any flavor $\nu's$; also in this case, $\simeq 1000$ of events/year are expected, in the standard Model hypothesys

$$\nu \, d \to np \, \nu \tag{5}$$

Such a reaction can be detected by the observation of the subsequent neutron capture in d or in some nuclues (e.g. Na) diluted in heavy water. The expected number of events ($\simeq 2800$ evt's/year if SSM holds) does not depend on neutrino flavor; this reaction could provide a measure of the total neutrino flux.

Charged current (3) and purely leptonic current (4) can be "easily" observed in such a detector, while a reliable observation of neutral current events require a deep knowledge and control background, as extensively discussed in the proposal and attached documents.

10. BOREX

Recently, an Italian-American Collaboration proposed to investigate the feasibility of an experiment on solar neutrinos using ^{11}B as target nucleus. The use of boron was proposed by Raghavan et al. (RAG-86); a letter of intent to perform the "Boron Solar Neutrino Experiment" has been submitted by Borex Collaboration (BOR-88, RAG-88) to the International Scientific Committee of Gran Sasso in 1988. The interesting reactions are:

$$^{11}B \, (\nu_e, \, e^-)^{11}C$$
$$^{11}B \, (\nu_x, \, \nu_x) \, ^{11}B^*$$

It is possible to use liquid scintillators like TMB [B(OCH$_3$)$_3$] or TMBX [B$_3$O$_3$(OCH$_3$)$_3$] which contains up to 20% of Boron. The expected rates are:

128   events of nuclear excitation on 4.5 and 5.0 MeV excited levels of ^{11}B

1500   charged current events (E>3.5 MeV)

these rates are for 200 tons of nat. B and per year.

Feasibility tests are starting now; the answer to the many technical problems (transparency and stability of TMB and TMBX, purification from radioactive contaminants, etc) is expected in about one year from now.

Finally, let me remark that, all the above mentioned experiments are technically extremely difficult; however the importance of the subject justifies, at least in my opinion, such a rich experimental program which, if successful, will led to deep improvements in our knowledge of the Sun and of neutrino physics.

11. OTHER SUBJECTS

Many other subjects are usually investigated by means of underground detectors; let me only mention a few of them:

- muon physics to search for point source of high energy cosmic rays
- multimuons physics to investigate on the chemical composition of high energy primary cosmic rays
- "atmospheric neutrinos" which have been carefully investigated being the most important source of background in proton decay experiments
- high energy ν's, which can be emitted by point source; however present or planned detectors are marginally efficient because their dimensions are too small
- magnetic monopoles, which are extensively searched also underground (MAC-84, CAL-88/1).

Finally let me mention the problem of gravitational waves; it has been recognized that very sensitive resonating antennas, are disturbed by electromagnetic interactions of high energy muons (AMA-86); therefore next generation of resonating antennas to be installed deep underground.

12. THE GRAN SASSO UNDERGROUND LABORATORY

"Laboratori Nazionali del Gran Sasso" are the national Italian facility for underground physics. They are located aside the Gran Sasso tunnel, on the highway connecting Rome to L'Aquila and the Adriatic Sea (Fig.6).

They are managed by I.N.F.N., the Italian agency for basic nuclear and subnuclear physics research. The construction started in 1982, when the italian Parliament approved and partially funded the construction of the Laboratory, proposed by A. Zichichi, at that time President of I.N.F.N. (ZIC-83).

Two major italian companies, COGEFAR S.p.a. and SAEM were encharged respectively of civil works and general plants; ANAS, the national agency for road, was responsible of the entire construction.

The Gran Sasso Massif is crossed by a double tunnel more than 10 Km long; one of the two ways is opened to traffic, while the second one is not yet in operation because the highway is not yet completed. On the north side of this tunnel, about at 6.4 Km from the Assergi entrance where the rock overburden reach its maximum value (>1400 m of rock), the Laboratory has been excavated.

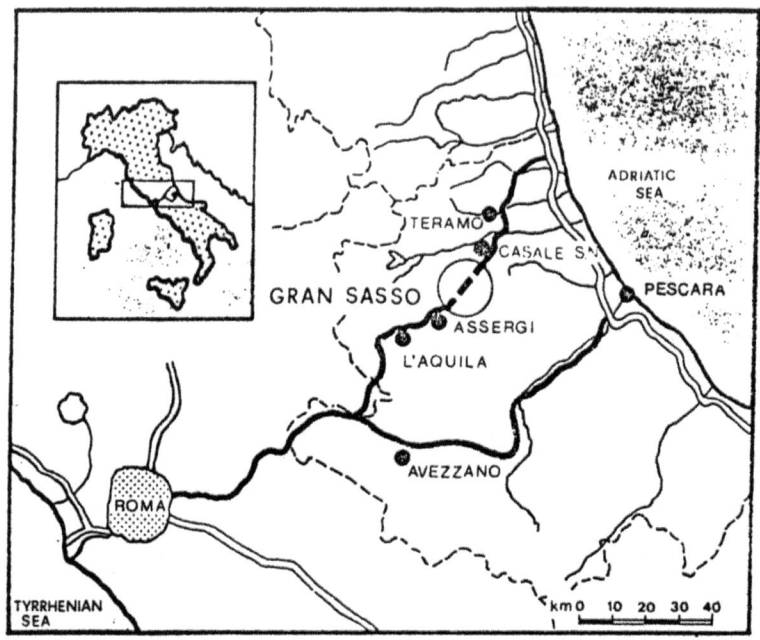

Fig. 6 - Map of the Gran Sasso area.

A general layout of the Laboratory is shown in Fig.7; there are three main halls, conventionally indicated as hall A, B and C, a system of three small galleries in the north side of the Laboratory, dedicated to geophysical researches, and a system of access tunnels, safety galleries and service areas.
Dimensions of the three main halls are reported in table II.

TABLE II

	A N wing	A S wing	B	C
length (m)	42	42	127	100
width(m) (floor level)	15.2	15.2	15.2	20.6
maximun height(m)	17.3	17.3	17.3	20.9

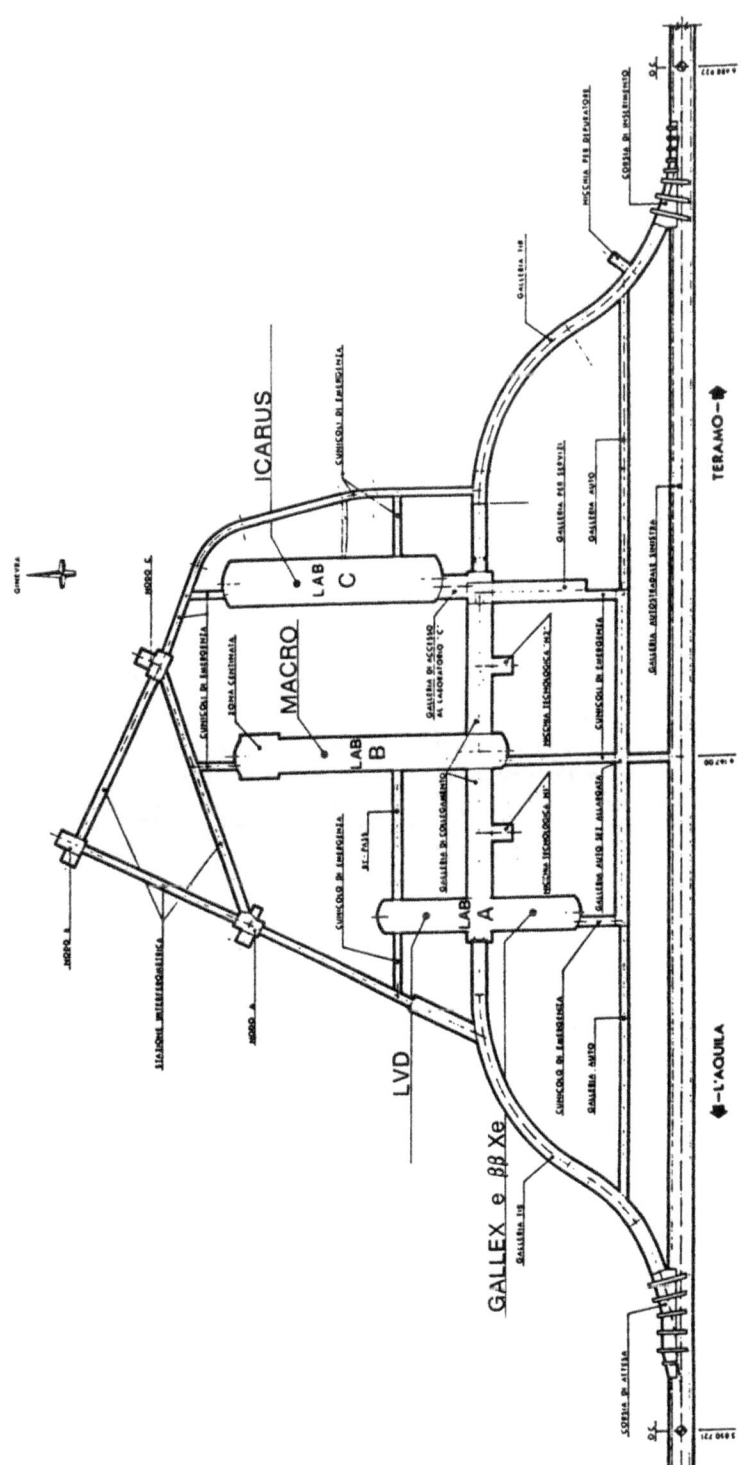

Fig. 7 - General layout of the underground laboratory.

The Gran Sasso massif is very rich of water; therefore the experimental halls are completely water proof in order to improve the environmental conditions of the laboratory and to avoid any possible pollution of water which is collected and sent to a pubblic water supply system.

Natural temperature and humidity are 6-7 C and 98% humidity; therefore walls are covered by thermal insulating pannels and a large number of fan-coils have been installed for conditioning.

A ventilation system will provide, in normal conditions, fresh air at a rate of about 40000 mc/h; an exhaust system allows to sent the "used" air not into the highway tunnel, but directly to the exterior; in case of need, a flux of 80000 mc/h of air taken from the highway tunnel can be concentrated on a single hall.

Fire and pressure (5 Kg/cm^2) resistent doors can isolate each hall from the others and the entire laboratory from the highway.

Cranes, electric power and water supplies, telephones etc. will be available in every hall; at present these plants are in operation in Hall A and B, which are essentially completed.

Data trasmission is assured by local area networks inside the underground laboratory, a 100 monomodal optical fibre cable and 100 telephone pairs, which link the underground laboratory with a computer hosted in an external building at the Assergi entrance of the tunnel.

In the past years, many preliminary measurements of the muon flux and of radioactivity level have been carried out by many groups, both along the tunnel and in the laboratory.

The muon flux is about 1 $\nu/(m^2h)$ (CAR-86).

The neutron flux is of the order of $5 \cdot 10^{-6}$ $\nu's/(cm^2s)$ for thermal neutrons and $3 \cdot 10^{-6}$ n's/(cm^{2}s) (BEL-85) for "fast" neutrons respectively; however the energy spectrum of neutrons seems definitively less energetic than that of an artifical source like Am-Be source. The most recent measurement are by the Rome II group (BEL-88) results can be summarized as follows:

Energy range	Flux (n's cm$^{-2} \cdot$s^{-1})
0 - 50 10^{-3}eV	1.01 ± 0.05 10^{-6}
50 10^{-3} - 10^3eV	1.8 ± 0.2 10^{-6}
1.-2.5 10^3keV	0.5 ± 0.1 10^{-6}
>2.5 MeV	0.3 ± 0.14 10^{-6}

Radon concentration has been measured and found of the order of 0.1 Bq/l (Measured by V. Facchini – Istituto di Fisica Generale Applicata – Universita' degli studi Milano). This measurement will be repeated when ventilation will be in operation. Every material introduced in the laboratory is measured in order to control and maintain radioactivity at present level.

At the Assergi entrance, a building host administration, staff and experimental groups offices, computers and other essential services. The computer is linked to the internal laboratory, to Campo Imperatore (see below) and to INFNET, the INFN national networks and to the pubblic network (ITAPAC); therefore it is possible to reach practically all the laboratories in Italy as well and in the world.

Three prefabricated buildings have been constructed, to satisfy the urgent needs of laboratory staff and external groups; they will be substituted in a short (2 years) time by more adequate buildings.

Finally, at Campo Imperatore - at 2000 m a.s.l. - an extensive air shower detector is being installed; it consists of 30 sheds housing plastic scintillator detectors, and of a central prefabricated building in which a hadron detector will be assembled, probably during Summer '89. A few detectors were in operation during past Winter with positive results. (NAV-88)

I would not enter in the detail on the status of different experiments, some of them will be discussed at this same Conference.

In a few words, the situation can be summarized as follow.

Double Beta decay - The xenon multicell proportional chamber is in operation by many months in a by-pass; if unnecessary the detector will not be moved from there up to the end of the experiment, leaving free room in Hall A for a bolometric device.

The Gallium experiment will be installed in Hall A (S wing); a building has been constructed to house the two main Gallium solution tanks, the Germanium extraction and processing plant, the air washing system etc.; a second building has been constructed and it will be shared between the Germanium counting system (partly in operation at present to test the quality of different kind of germanium detectors) and double beta decay criogenic apparatus.

Assembling of LVD will start in Hall A (north wing) in September.

In Hall B, the first (of six) module of MACRO has been assembled and it is under test; two other modules will be ready for the end of this year and the entire detector before the end of next year.

Installation of the interferometer and other devices will start in September.

The interest arised by the Laboratory in the scientific community and the need of a more complete decoupling of the Laboratory from the highway, dictate an important program of new works. At present the Italian Parliament is considering the allocation of more 100 billion liras to improve the general plants of the underground laboratory, to excavate an independent access to the underground laboratory through a 7 Km long tunnel, to excavate two new halls, the first one to be dedicated to experiments which make use of criogenic thecniques and the second one for experiments which require especially low radioactivity environment, new buildings outside the tunnel (assembling halls, mechanical workshop, library, computer center, canteen, etc.).

The detailed construction program as well new experiments will be decided after the Parliament discussion and - hopefully - approval of our requests.

ACKNOWLEDGEMENTS

It's a pleasure to thank my colleagues, especially Prof. E. Fiorini for interesting discussions. I would also thank Mrs. L. Brogiato, Miss. F. Masciulli and G. Sala for their help in the preparation of this paper.

250

REFERENCES

(AGL–86)   M. Aglietta et al.
             Nuovo Cimento C9,185 (1986).

(ALE–88)   A. Alessandrello et al
             Nucl. Phys. A478, 453C (1988).

(AMA–86)   E. Amaldi and G. Pizzella
             Nuovo Cimento 9C, 612 (1986).

(APR–87)   E. Aprile and V.H. – M. Ku, J. Parle Columbia
             University Preprint n. 342 (1987)
             and private communication.

(AVI–87)   F. Avignone et al.
             Proceedings of an International Workshop on Neutrino
             Physics – Ed. H.V. Klapdor and B. Povh – Heidelberg
             Oct. 20/22, 1987
             Springe – Verlag, pag. 191.

(BAC–82)   J.N. Bachall
             Review of Mod. Phys. 54, 767 (1982).

(BAD–84)   G. Badino et al.
             Nuovo Cimento C7,573, (1984).

(BAR–87)   I.R. Barabanov
             Proc.of the Second Int. Symposium on Underground
             Physics '87 – Baksan Valley (USSR) Aug. 17–19, 1987
             ed. "NAUKA", pag.279.

(BAR–87)   I.R. Barabanov et al.
             Proc. of the Second Int. Symposium on Underground
             Physics '87 – Baksan Valley (USSR) Aug. 17–19, 1987
             ed. "NAUKA" Moscow 1988, pag.279.

(BIC–88)   ICARUS Coll.
             'ICARUS I : an Optimized, Real Time Detector of Solar
             Neutrino's'
             Proposal – 1988.

(BOE–88)   F. Boehm and M.Z. Iqbal
             A Xenon Time Projecton Chamber for Double Beta Decay
             Festival – Festschrift for V. Telegdi (1988).

(BOR–86)   BOREX Coll. – AT&T Bell, Argoune, Drexel, Hawai, MIT,
             Milan, Pavia
             'Letter of intent to performe the BORON SOLAR NEUTRINO
             EXPERIMENT.'

(BUC–88)   E. Backley et al.
             'A Study of Ionization Electrons Drifting over Large
             Distances in Liquid Argon' – preprint.

(CAL-88/1) D.O. Caldwell
           Nucl. Instr. and Methods A264 (1988).

(CAL-88)   M. Calicchio et al.- MACRO Coll.
           Nucl. Instr. and Methods A264, 18 (1988).

(CAR-86)   R. Cardarelli et al.
           INFN/AE-86/11-1986.

(CAS-85)   M. Casse
           in Neutrinos and Present Day Universe
           ed. by T. Montmerle and M. Spiro
           Commissariat a l'energie atomique p.49 (1985).

(COW-83)   G.A. Cowan and W.C. Haxton
           Science 216, pag.51 (1983).

(DEG-87)   B. Degrange
           Underground Experiments
           Proc. of an International Europhysics Conference on
           High Energy Physics - Uppsala (Sweden), 1987.

(DAV-87)   R. Davis
           Proc. of the Second International Symposium on
           Underground Physics 87 - Balsan Valley (USSR) 17-19
           Aug. 1987
           Ed. "Nauka" Moscow 1988, pag.1
           also for previous reference.

(DAV-81)   R. Davis jr. et al.
           Proposal for a Fundamental Test of the Theory of
           Nuclear Fusion in the Sun with Gallium Solar
           Neutrino Detector
           BNL (1981).

(ELL-87)   S. R. Elliot et al.
           Phys. Rev. Lett. 59, 1649 (1987)
           Phys. Rev. Lett. 59, 2020 (1987).

(EWA-87)   G.T. Ewan et al.
           Sudbury Neutrino Observatory Proposal
           Report SNO - 87-12
           Queen's University, Kingston, Canada, 1987.

(FIO-88)   E. Fiorini
           'In Present Trents, Concepts and Instruments of
           Particle Physics'
           Ed. G. Baroni, L. Maiani and G. Salvini (Rome).

(FIO-88/1) E. Fiorini
           'Double Beta Decay: yesterday, today, tomorrow'
           University of Milan Preprint.

(FIS-87)   E. Fisher et al.
           Phys Lett. <u>1928</u>, 460 (1987).

(HAX-86)   W.C. Haxton et al.
           Proc. on the VIth Moriond Workshop on Massive
           Neutrinos in Particle Physics and Astrophysics
           Tignes, France, 1986 also for previous references.

(HAX-86)   W.C. Haxton et al.
           Proc. of '86 Massive Neutrinos in Astrophysics and
           Particle Physics - Ed. O. Fakler and J. Tran Thanh
           Van
           Tigne (Savoie) Jan. 25 - Feb. 1 1986, p.143.

(ICA-85)   CERN, Harward, Milan, Padoa, Rome, Turin,
           Wisconsin Coll.
           'Searching for New Underground Phenomena with High
           Resolution Visual Techinques and Magnetic Analysis.'
           A proposal for the Gran Sasso Laboratory
           INFN/AE-85/7 (1985).

(ICA-88)   ICARUS Collaboration
           'ICARUS I - An Optimized, real Time Detector of Solar
           Neutrino's'
           Proposal 1988.

(KAM-87)   Kamiokande II coll.
           'Search for ^{8}B solar neutrinos at Kamiokande II'
           preprint.

(KIR-86)   T. Kirsten
           Proc. of the 12th Int. Conf. on Neutrino Physics and
           Astrophysics
           Sendai June 3-8, 1986
           Ed. T. Kitagaki & H. Yuta.
           World Scientific, pag. 317.

(KLA-87)   H.V.Klapdor
           Proc. of the Second Int. Symposium on Underground
           Physics '87 - Baksan Valley (USSR) Aug. 17-19, 1987
           ed. "NAUKA", pag.266.

(KOT-87)   T. Kotani
           Proc. of the Second Int. Symposium on Underground
           Physics '87 - Baksan Valley (USSR) Aug. 17-19, 1987
           ed. "NAUKA", pag.260.

(KUZ-66)   V.A. Kuzmin
           Soviet Physics Jetp 22 (1966), 1051.

(KUZ-87)   V.V. Kuzminov et al.
           Proc. of the Second Int. Symposium on Underground
           Physics '87 - Baksan Valley (USSR) Aug. 17-19, 1987
           ed. "NAUKA", pag.275.

(KUZ-87)   V.V. Kuzminov et al.
            Proc. of the Second Int. Symposium on Underground
            Physics '87 - Baksan Valley (USSR) Aug. 17-19, 1987
            ed. "NAUKA", pag.

(LVD-85)   LVD Collaboration Bologna, LNF, Palermo, Turin,
            CERN, CENS, MIT, INR Moscow, ICRR Tokio, Peckin
            Proposal for a Large Volume Detector (LVD) for the
            Gran Sasso Laboratory (1985)
            see also
            C. Alberini et al.
            Nuovo Cimento 9C, (1986), 237.

(MAC-84)   MACRO Coll.
            'A Large Area Detector Dedicated to Monopole Search,
            Astrophysics and Cosmic Ray Physics at Gran Sasso
            Laboratory'
            Proposal - 1984.

(MOE-87)   S.R. Elliot, A.A. Hahn and M.K. Moe
            Neutrino Physics - Proc. of an International Workshop held
            in Heidelberg, October 20/22, 1987;
            ed. H.V. Klapdor and B. Povh -
            Springer Verlag, p. 213.

(MOS-88)   S.H. Moscley et al.
            IEEE Transaction on Nuclear Science 35, 59 (1988).

(MOR-87)   A. Morales
            Proc. of the Second Int. Symposium on Underground
            Physics '87 - Baksan Valley (USSR) Aug. 17-19, 1987
            ed. "NAUKA", pag.281.

(NAV-88)   G. Navarra
            Bollettino del 74° Congresso della Societa' di Fisica
            pag. 145 Oct. 6-11, 1988.

(RAG-88)   R.J. Raghavan et al.
            Design Concepts for BOREX
            AT&T Bell Laboratories - 1988.

(SAD-88)   B. Sadoulet
            IEE Transaction on Nuclear Science 35, 47 (1988).

(VAS-87)   A.A. Vasenko et al.
            Proc.of the Second Int. Symposium on UNDERGROUND
            Physics '87 - Baksan Valley (USSR) Aug. 17-19, 1987
            ed. "NAUKA" Moscow 1988, pag.288.

(WOL-86)   K. Wolfaberg et al.
            'Status of Molybdenum' - Technetium Solar Neutrino
            Experiment
            Isotope and Nuclear Chemistry Division Annual Report
            1986, p. 139.

(ZIC-83)   A.Zichichi
Proc. of the Int. Workshop ICOMAN 83
Prascati (Italy), 1983;
The Gran Sasso Project
INFN/AE-82/1 (1982).

DISCUSSION.

Hargrove: I have a comment. The measurement of the neutral current rate using elastic scattering is very difficult because the cross section is about 1/6 neutral current and about 5/6 charged current. To use the charged current measurements in the neutrino-Ar charged current reaction to subtract the charged current part of the elastic scattering cross section is very difficult because of the uncertainty in the charged current-Ar cross section. The other point is that the threshold in the Ar reaction is 5 MeV from the energy difference between the initial and final states. One has also to consider the background in the detector which will certainly increase the 5 MeV threshold by several MeV.

PROBING THE UNIVERSE WITH RICH CLUSTERS OF GALAXIES AS GIANT GRAVITATIONAL TELESCOPES

B.P. FORT
Visiting Astronomer, ESO

Observatoire de Toulouse
14, Avenue Édouard Belin
31400 Toulouse France

ABSTRACT. — In this short communication we report on the discovery of a new class of astronomical objects whose nature was first established with spectroscopic observations at the European Observatory: the luminous Einstein arcs (or rings) in rich clusters of galaxies. We also present more recent observations with the Canada France Hawaï telescope which prove that the cluster A370 acts on the background galaxies as a giant gravitational telescope. This observational advance will probably open a new stage in the study of gravitational lenses and their implications in cosmology.

1. Luminous arcs in rich clusters of galaxies

The discovery of a giant luminous arc in the clusters of galaxies A370 (Soucail et al. (1987, 1988 a), and Cl 2244-02 (Lynds and Petrosian, 1986) stimulated numerous hypotheses about their nature and origin (fig. 1). But, as first pointed out by Paczinski (1987), their location in the core of very massive clusters, their highly circular geometry centered on the cluster center, their extreme blue color and low surface brightness were strong pieces of evidence that we were probably observing the gravitational images of high redshift objects produced by the bending of the light rays passing through the massive cluster: the phenomenon called Einstein rings and predicted a long time ago (Zwicky 1937).

In 1987, a very intense campaign of observations took place in various observatories to get spectra of these arcs (Soucail et al, 1988b; Lynds and Petrosian, 1988; Miller and Goodrich, 1988). The observations where really challenging because the surface brightness of these arcs was hardly more luminous than 10% of the sky background. The most convincing spectrum was finally obtained for the luminous blue arc located in A370 by Soucail and collaborators (1988b) with the faint object spectrograph EFOSC on the 3.6 meter telescope of the European Southern Observatory. The spectrum showed various spectral features all along the arc which were similar to a spiral galaxy at a redshift $z = 0.724$, about twice the distance of the cluster $(Z = 0.374)$.

These ESO observations confirmed Paczinski's guess: the luminous giant arc in A370 was the first Einstein arc ever observed in the universe.

M. Caffo et al. (eds.), Astronomy, Cosmology and Fundamental Physics, 255–259.

2. Gravitational lensing by clusters of galaxies

Since the discovery of the first double QSO 0957+561 by Walsh et al (1979) a lot of theoretical work has been devoted to gravitational lensing by galaxies .

We expected to estimate some cosmological parameters and to probe the dark matter in the universe using the lens modelization. But all these attempts remained rather disappointing for the theoreticians, mainly because the number of examples was still small despite intensive searches in the sky, and because the deflecting galaxies were almost unobservable in all cases, preventing any good description of the lens (Schneider and Weiss 1986). But, using the angular separation and the number of images, theoreticians pointed out the importance of dark matter and the added effect of some clusters associated with the deflecting galaxy.

Others showed the possible importance of light amplification for QSO's located in the direction of rich foreground clusters of galaxies (Narayan et al 1984, Nottale and Hammer 1984).

In fact, our discovery of the luminous arcs comes exactly in time to prompt new theoretical studies of "extended objects" gravitationally lensed by rich clusters of galaxies. Numerous models were proposed in 1987/88 which predicted luminous arcs similar to those observed in A370 and Cl 2244-02 (P. Schneider and Weiss 1986, Soucail et al 87 b, Kovner 1987, Narashima and Chitre 1988, Grossman and Narayan 1988 etc.).

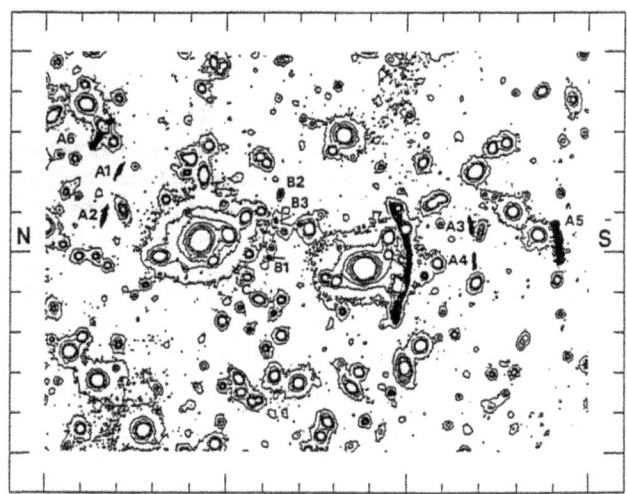

Fig. 1: Isophotal map of the deep CCD composite images with the A370 blue arc systems (see B. Fort et al, 1988 for more details).

One particularly interesting prediction for observers comes from the simulations of Grossmann and Narayan (1988). Using very realistic parameters to mimic clusters similar to A370 they show that with a realistic surface density of background galaxies (at redshift $z = 1$) the probablity of having many arcs-like images is very high. They produced simulated gravitational images that could easily be recognizable at the telescope. (Note however that a few distorted images can also have a radial coma shape instead of arcs !).

During the same period the attention of the astronomical community also focused on the possible existence of a huge population of highly redshifted galaxies. For example, very deep CCD photometry on empty fields at the south galactic pole in fact appeared quite crowded with faint blue extended galaxies having a B magnitude larger than 26 (Tyson 88).

This new population has magnitudes, surface brightness, angular sizes and extremely blue color indexes which strongly suggest a huge population of faint galaxies at large red shift $z > 0.7$ (see also Cowie, 1988).

It is clear that the simulations of Grossmann and Narayan (1988) and the possible existence of this "Tyson" population of background galaxies give us the idea of observing their possible images produced by rich cluster lenses.

3. A370 as a gravitational telescope

Indeed A370 was our first candidate because it has already acted as a gravitational Telescope for the giant luminous arc. We selected several good images taken at the CFH Telescope prime focus: two 30 min B exposures and one 30 min R exposure, all of them with an excellent seeing of 0.8 arcsec. (FWHM). Adding them, we noticed a few arc like systems very similar to what was predicted by the Grossmann and Narayan simulations (fig. 1). After a careful photometric study we derived their magnitude, color and surface brightness which were fully compatible with the "Tyson" population of background galaxies (Fort et al 1988).

This result is additional prove that A370 really acts as a gravitational lens and provides good evidence that the Tyson populations of galaxies are located at larger distance than the cluster. However, with this observation we cannot completely exclude that the new gravitational images observed in the central core of A370 do not possibly come from a chance alignment of a group of galaxies at the redshift of the giant arc $(z = 0.724$).

4. Other observational evidence for such cluster lenses

Theories tell us that the central core of a cluster of galaxies can act as a gravitational lens if the central surface density is greater than a critical value Σ_c (approximately 0.4 h.g. cm^{-2} at $z_l = 0.4$; Turner et al, 1984; Subramanian et Cowling, 1986 etc.). Calculating Σ_c with a luminosity profile and the virial mass of a cluster, it is possible to show that we can only observed arcs for rich cluster of galaxies having a small core radius or a large concentration parameter and a high velocity dispersion, say $\sigma_v > 1500$ km s^{-1}. As the velocity dispersion is strongly correlated with the X ray luminosity, this may favor clusters with $L_X > 10^{45}$erg $\times$ cm^{-2}s^{-1}.

It seems that such criteria work because in the last months several arc systems were reported to us in Cl 0024+1654 (Koo, private communication), 3C295 (Tyson, private communication), etc.

However it is important to note that the velocity dispersion of a cluster is in general rather close or below 1000 km/s. Most of the cluster core are subcritical and unable to work as lenses by themselves. More recent observations show that we can only observe arcs around a few potential singularities associated with the central cD or other sub-groups which have not yet merged in the central core. Such examples are observed in A 963 (Lavery and Henry, 1988) or in A 2218 which both present arc-like systems centered around giant galaxies (Pello et al, 1987).

Whatever the deceleration parameter q_0, the optimum distance for lensing clusters is $z = 0.7$ (Nottale et Hammer, 1984). Because of the dimming factor $(1 + z)^4$, at such a distance the galaxies members of the cluster lens become difficult to observe without very deep photometry. The gravitational images of galaxies at higher redshift will be still more difficult to observe.

It is thus possible that, without VLT's, most of the arc systems will be searched and found in rich cluster of galaxies at relatively moderate redshifts (z between 0.2 and 0.4). So far, at the date we write this paper, we know about twelve arc-like systems. If we extrapolate such a conservative view about the present day possibility to detect new cluster lenses, we estimate that, with suitable observing time on 4 meter class telescopes, we will be able to detect more than 100 new cases within a few years.

5. Implication for observational cosmology

It turns out that the observations of arcs are more promising than the "classical" QSO lenses to determine cosmological parameters and to probe the dark matter and high redshifted galaxies in the universe. The arguments in favour of that statement can be summarized in the following ways (Turner 1988):

a) The presence of arcs in rich clusters is likely to be a common phenomenon because the surface density of background galaxies is higher than that of QSO's. Indeed the surface brightness which is conserved will make detection a challenging observation both for the photometry and the spectroscopy of the few brighter arcs.

b) The angular size of the phenomena and the observational data we can gather on the cluster (galaxies photometry and redshift, profil of X-ray luminosity, etc.) allow a good description of the lens.

c) the amplification of light for longer arcs may be large because of the merging of several images when the source falls on a critical line (or better: a caustic cusp, Grossman and Narayan 1988). For example this amplification is close to 10 for the main arc in A370 (Soucail et al, 1988a, b) and is probably higher for the Cl 2244-02 arc. This offers a unique possibility of making more accurate spectro-photometry of very high redshifted (maybe primeval) objects.

In conclusion, we can reasonably presume that observations of arcs will have important implications in observational cosmology.

6. References

Blandford R., Kochanek C., Kovner I. and Narayan R. (1988), 'Gravitational lens optics.' Caltech preprint 177.

Blandford R., Lawrance C., Kovner I. and Narayan R. (1988), Caltech preprint 176. 'An iterative reconstruction technique.'

Cowie L. (1988), protogalaxies, in 'The post Recombination Universe,' Kayser et Lasenby eds., in press.

Fort B., Prieur J.L., Mathez G., Mellier Y. and Soucail, G. (1988), Astron. Astrophys. **200**, L 17.

Grossman S. and Narayan R. (1988), ApJ. **324**, L37.

Kayser R. and Schramm T. (1988), Astron. Astrophys. **191**, 39.

Kochanek C. and Soucail G. (1988), in preparation.

Kovner I. (1987), ApJ. **321**, 686.

Kovner I. (1988), Weizmann Institute preprint.

Lavery R. and Henry P. (1988), ApJ **329**, L21.

Lynds R. and Petrosian V., (1986), BAAS, **18**, 1014.

Lynds R. and Petrosian V., (1988), ApJ in press.

Miller J. and Goodrich R. (1988), Nature, **331**, 685.

Narashima D. and Chitre S. (1988) Tata Institute preprint.

Narayan R., Blandford R. and Nityananda R. (1984), Nature **310**, 112.

Nemiroff J. and Dekel A. (1988), Racah Institute preprint.

Nottale L. (1988), Ann. Phys. Fr. **4**, 13.

Nottale L. and Hammer F. (1984), A.A., **141**, 144.

Paczynski B. (1987), Nature, **325**, 572.

Pello-Descayre R., Soucail G., Sanahuja B., Mathez G. and Ojero E. (1988), Astron. Astrophys. **190**, 411.

Schneider P. and Weiss A. (1986), Astron. Astrophys. **1964**, 237.

Soucail G., Fort B., Mellier Y. and Picat J.P. (1987), Astron. Astrophys. **172**, L14.

Soucail G., Mellier Y., Fort B., Hammer F. and Mathez G. (1988) Astron. Astrophys. **184**, L7.

Soucail G., Mellier Y., Fort B., Mathez G. and Cailloux M., (1988), Astron. Astrophys. **191**, L19.

Subramanian K. and Cowling S.A. (1986), M.N.R.A.S., **219**, 333.

Turner E., Ostriker J.P. and Gott J.R. (1984), ApJ., **284**,1.

Turner E. (1988), Princeton preprint 'Dark matter in gravitational Lenses' (Moriond).

Tyson A. (1988), A.J. **96**,1.

Walsh D. Carswell R.F. and Weymann R.J. (1979), Nature **279**, 381.

Zwicky, F. (1937), Phys. Rev. **51**, 679

7. Discussion

V. TRIMBLE: What did you mean when you spoke of the rediscovery of these arcs?

B. FORT: I mean that many observers have seen some of these arcs for years; but they didn't really think it was more than chance aligment of many different objects. See a good historical review in the Lynds and Petrosian (1988), ApJ. paper in press.

B. PARTRIDGE: What is the radius of the upper arc you showed in Abell Cluster 2218?

B. FORT: This fragmented upper arc has a radius of about 10 arcsec.

DARK MATTER IN ASTRONOMY

D. LYNDEN-BELL
Institute of Astronomy
Madingley Road
Cambridge, CB3 0HA.

ABSTRACT. A sceptical eye is cast on cosmic nucleogenesis but the best data withstands scepticism. Kuijken & Gilmore's investigation shows that there is no disk dark matter in the solar neighbourhood.

Evidence for dark matter in the outer parts of galaxies is found both in extended rotation curves and in the relative motions of binary galaxies. The best evidence for dark matter remains the dynamics of the great clusters of galaxies, i.e. where galaxies are most common.

The lack of *good* evidence for dark matter in dwarf spheroidal galaxies means that any deductions from them are most doubtful.

All methods of assessing Ω from mass to light ratios give zero dark matter content to voids. Since much of the universe is empty, Ω could be much greater than the optical assessment of $0.1 < \Omega < 0.3$. The IRAS high Ω is explained as a biassing of IRAS galaxies away from high density regions.

1. Introduction

As Bologna was one of the first to emerge from the dark ages, this would be a most fitting place for dark matter to come to light! The problem of determining the density of all matter in the universe arose in the earliest application of General Relativity to cosmology. Although some assessment of the visible matter content was already possible, there were large regions where nothing was seen. The assessment of the matter content of these regions is almost as difficult today as it was in the 1920s. One of the delights of the third ESO/CERN meeting was the first data relevant to the matter content of voids was presented by Margaret Geller. The velocities on the near and far sides of a prominent void appear consistent with unperturbed Hubble flow. If this result holds up as the data improves then either Ω is low or the voids contain almost the average density of the universe. To avoid discussing all evidence for dark matter superficially, I shall concentrate on those parts in which there has been significant recent evidence. These are the Galactic disk, the discussion of Galaxy pairs, the big clusters and IRAS and optical dipoles. Before presenting that data it is only fitting to list astronomers whose initial perspicacity raised the problems and developed the methods of analysis that have led to our current knowledge.

M. Caffo et al. (eds.), Astronomy, Cosmology and Fundamental Physics, 261–277.

1922 J.C. Kapteyn Ap. J. **55**, 302	Presented a model of the Galaxy and pointed out that more refined applications of the method would allow the detection of both the visible and invisible matter in the Galaxy.
1932 J.H. Oort B.A.N. **6**, 349	Gave the "Oort limit" to the total amount of visible and invisible matter in the Galactic disk.
1936 Sinclair Smith Ap. J. **83**, 23	Gave a detailed discussion of the velocities of Virgo galaxies and deduced that there must be large quantities of mass that does not shine.
1933,37 Zwicky Helv. Phys. Acta **6**,110 Ap. J. **86**, 217	Gave the first discussions of the problems generated by high velocities in the clusters of galaxies and later and later gave a more precise discussion based on the velocities and the light from the Coma cluster of Galaxies.
1959 F.D. Kahn & Woltjer Ap. J. **130**, 705	Pointed out the problem of binding the Local Group of Galaxies and of the fact that the Galaxy and Andromeda approach one another. Considerable extra mass is necessary to reverse the initial Big Bang expansions.
1973 J.P. Ostriker & P.J.E. Peebles Ap. J. **186**, 467 1973-4 J. Einasto Proc. First European Astron. Meeting **2**, 291.	Gave evidence that galaxies are surrounded by Massive Haloes or Coronae that extend beyond the visible disks of galaxies and may increase their total masses by an order of magnitude.
1979 S.D. Tremaine & J.E. Gunn Phys. Rev. Let. **42**, 407	Gave the phase-space argument giving a lower limit to the mass of any elementary particle that could form massive haloes of Galaxies.
1983 S.M. Faber & D.N.C. Lin Ap.J. **266**, L17.	Showed that neutrinos cannot make heavy haloes for dwarf spheroidal galaxies.
1983 M. Aaronson Ap. J. **266**, L11	Gave the first measurements of velocity dispersions in dwarf spheroidals that may favour dark matter there.

Many astronomers have been concerned with measuring masses on a still longer scale by measuring deviations from a smooth Hubble flow cause by prominent clusters or by other large scale inhomogeneities. Here Ω is determined from Peeble's formula

$$v_{pec} = \frac{1}{3}\Omega^{0.6} \cdot H_0 \cdot R\left(\frac{\delta\rho}{\rho}\right).$$

There has been a secular decrease in the values deduced for the velocity induced by Virgo at the Local Group. The highest values are about 500 km/sec while the lowest are about 100. While I personally believe in values in the range 150-250 I am not yet convinced that all the complications of either the theory or the observations have been assessed. Davis & Peebles review the investigations up to 1983.

2. A Sceptical Eye on Cosmic Nucleosynthesis

The strongest evidence on the nature of dark matter comes from cosmological nucleosynthesis, so in preparing this lecture I looked at that data with the eye of an iconoclast. Deciding that $\Omega = 1$ in baryons from philosophical grounds I tried to dismiss any data that disagreed. The critical abundances are those of Li^7, Deuterium and Helium. In spite of the beautiful case that the Li^7 abundance in the high velocity stars is cosmogenic (Spite & Spite, 1982). I still found it possible that those abundances had been depleted and a recent paper by A. Boesgaard returns to the view that the population I star Lithium abundance is the more relevant one for cosmogenesis. (For a rebuttal of this view see Reeves's article in this symposium).

Turning to the deuterium abundance I find no evidence that Deuterium is universal. It has only been observed in material of normal abundances near the Sun. There is not even a determination from the Small Magellanic Cloud (where metal abundances are low). The main argument seems to be that no one has thought of a good non-cosmogenic source for deuterium. I do not have such a high opinion of mankind nor even of my nucleogenic colleagues that I admit this as a good argument. We all like our work to be fundamental and important, so few have a passionate interest in showing that deuterium is made in a more mundane fashion. This leaves only the He^4 abundance in the way of $\Omega_b = 1$ but here the modern data looks good, once that, corrupted by the adjacent sodium line or by inadequate statistical accuracy, are removed.

The helium abundance is certainly observed in low metal abundance extragalactic systems which are very widespread. It would be nice to know that metal-deficient material had precisely the same low value long ago. This might yet be found out from a high redshift system or from material blown off an old low abundance, high velocity star, but no good case is yet known. Even without that final confirmation the case for universal cosmogenic He^4 abundance is very strong. From low metal abundance extragalactic systems Pagel gives the value $Y = 23.5\% \pm .5\%$ for the primaeval value but doubling the error to allow for unknown systematic errors, my iconoclastic assessment is $Y = 24 \pm 1$.

Excepting some fine tuning of inhomogeneous theories, these new values make $\Omega_b = 1$ impossible since that requires values above 29%. Pagel stresses that the standard Big Bang model has been predictive more than once. From the observed helium abundance the deuterium abundance was predicted and found compatible. Li^7 helped to confirm that value and when the best values for the He^4 abundance were dangerously low for the standard model, the remeasurement of the neutron's lifetime indicated that it was only 10.17 minutes, lower than the previous lower limit of 10.4. The new value gives $Y = 23.6$ in amazing agreement with the observed helium (provided there are 3 neutrino species). Given that helium is cosmogenic, there is no case for giving up the observed deuterium and lithium abundances predicted by the standard model, so $.05 < \Omega_b < 0.2$ are strong limits on Ω_b.

3. The Galactic Disk and Dark Matter

Taking any easily recognisable type of star, the equation of stellar hydrostatics applied perpendicular to the galactic plane is

$$\rho_* \, \partial \psi / \partial z = \partial p_{zz} / \partial z = \partial / \partial z \, (\rho_* \sigma_{zz}^2) \tag{1}$$

where ρ_* is the partial density in that type of star and $p_{zz} = \rho_* \sigma_{zz}^2$ is the partial pressure in the stellar motions of those stars. σ_{zz} is the velocity dispersion of these stars in the z directions.

Equation (1) may be recast to give the z component of the gravitational field of all matter $\underline{g}$ which in astronomy is called K_z. Thus

$$K_z = \frac{\partial \psi}{\partial z} = \sigma_{zz}^2 \, \frac{\partial}{\partial z} \, [ln(\rho_* \sigma_{zz}^2)] \tag{2}$$

To apply this formula one must
1) Choose suitable stars that can be recognised at considerable height above the Galactic Plane.
2) Measure $\rho_*(z)$
3) Measure $\sigma_{zz}^2(z)$.
Notice that we only need the logarithmic gradient of ρ_* so the normalisation is not a problem but we nevertheless need good distances to know where we have determined K_z. Also, a spread in relative distance determination due to poor accuracy muddles stars at different distances and corrupts both the $\sigma_{zz}^2(z)$ values and the logarithmic gradient. Past investigations have been made with K giants, Ao stars, F_5 and F_8 stars, but I wish to describe to you the very recent new determination of Kuijken & Gilmore using K dwarfs. Difficulties with the past investigations can be summarised as follows.

K giants are nice and bright but there is a great spread of luminosity among giants of this type so more subtle criteria are needed to single out stars of known luminosity. It is not easy to pick the large numbers needed for good star density determinations with the luminosity precision necessary for accurate distances.

For the Ao stars the luminosity spread is smaller and the stars are bright enough to be easily seen but they are so young that there is considerable doubt that they are in equilibrium. They also have small velocity dispersion so cannot be used to get K_z at great heights above the Galactic Plane.

The F_5 and F_8 stars are in principle good candidates but on the current data they do not tell quite the same story. The best one can do with them is take an average as John Bahcall has done.

The great advantage of K dwarfs is that they have a narrow range of luminosity well separated from the giants. They are easily recognised, well-mixed and they extend to considerable heights above the Plane. The only disadvantage is that they are faint but the coming of more sensitive detectors, bigger telescopes and particularly, multiple fibre spectroscopy at the Anglo Australian Telescope, have made possible large programmes on them.

I now give a simplified account of Kuijken & Gilmore's results and analysis. Those interested should read their paper.

Figure 1a shows Kuijken & Gilmore's results for σ_{zz} as a function of height above the galactic plane while 1b shows the log of the partial pressure of the K dwarfs determined by them.

Analysing this data they deduce that the total K_z of disk matter is given by

$$| K_z | = 3.9 \ 10^{-9} cm \ s^{-2} \tag{3}$$

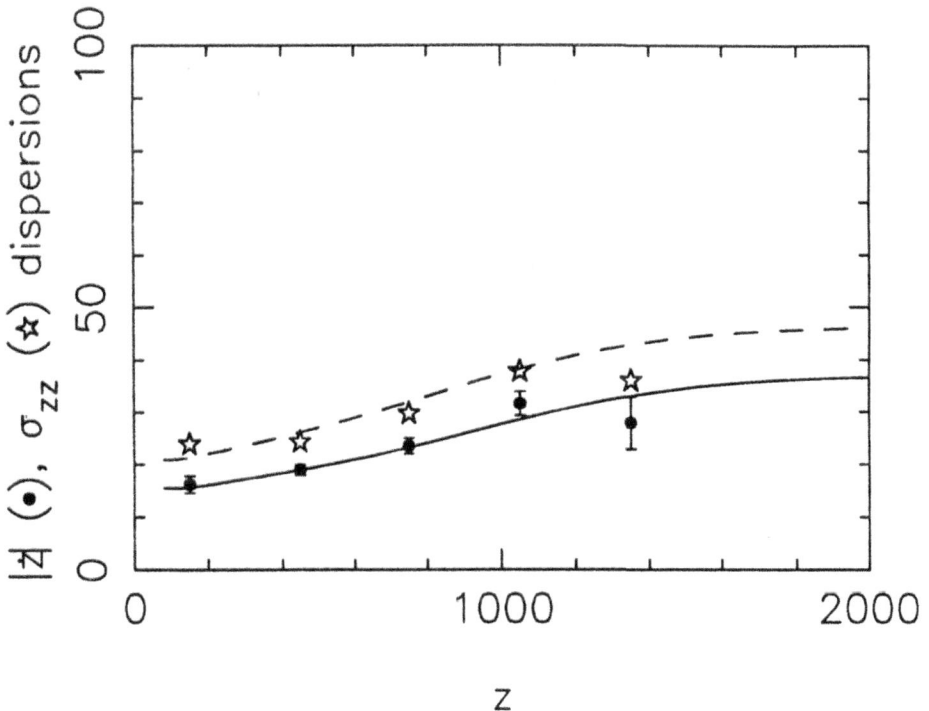

Figure 1a

This leads to a total projected surface density Σ in the Galaxy's disk near here as follows. Poisson's equation reads

$$\nabla^2 \psi = \frac{1}{R} \frac{\partial}{\partial R} \left(R \frac{\partial \psi}{\partial R} \right) + \frac{\partial^2 \psi}{\partial z^2} = -4\pi G\rho \tag{4}$$

The first term is $R^{-1} \partial/\partial R \ (V_c^2)$ but the circular velocity V_c is almost constant near here. Thus integrating through the galactic disk

$$| -K_z |_-^+ = | -\partial \psi/\partial z |_-^+ = 4\pi G\Sigma \tag{5}$$

266

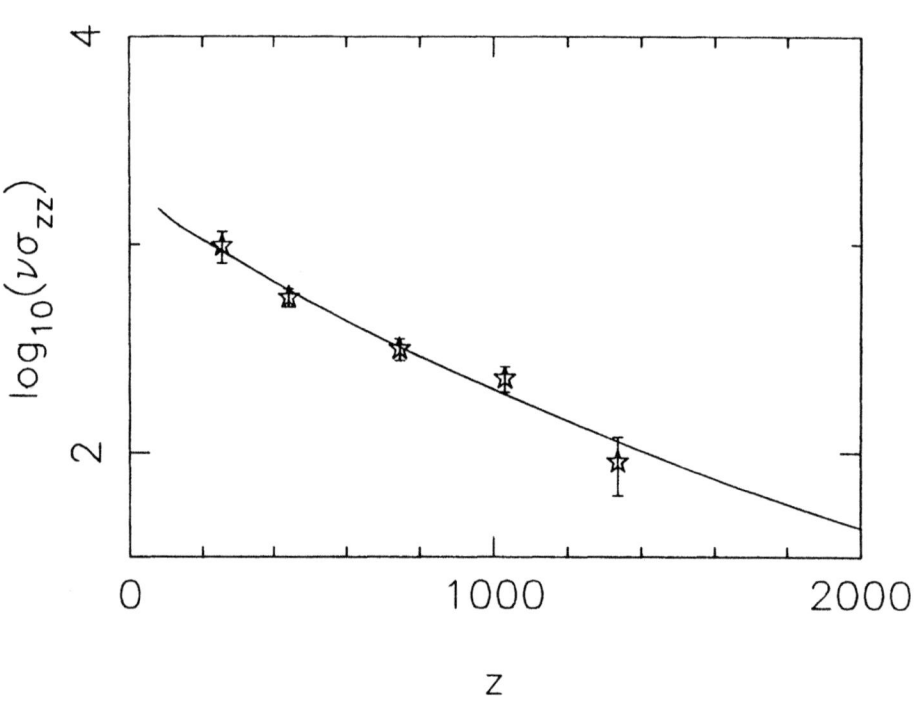

Figure 1b

Using the antisymmetry of K_z about the galactic plane we find

$$\Sigma = |\, K_z \,|/(2\pi G) \qquad (6)$$

so from (3) they deduce

$$\Sigma = 45 \pm 7 M_\odot/(pc)^2 \simeq 0.01 gm\ cm^{-2} \qquad (7)$$

This value is similar to the surface density of this piece of paper. The expected surface density found by integrating up the known constituents in the gravitational potential they deduce comes to 47 $M_\odot/pc^2$. *Thus there is no case for a disk-like distribution of dark matter in the Galaxy.*

Their result has interesting implications for the applicability of Newtonian gravity on the scale of a kpc and at accelerations as low as $10^{-9} cm/s^2$.

It also gives a mass to light ratio of the Local disk material of $M/L_B = 2.5$.

4. Rotation Curves and Binary Galaxies

The extrapolation inwards of Σ_0 deduced above with a suitable exponential scale length gives too little mass to account for the local circular velocity, even after a spheroidal component of equal mass has been added. We deduce that either
(i) The extrapolation seriously underestimates the actual total mass of the Galactic disk, or
(ii) A new spherical dark matter halo contributes half the total mass within the Sun's orbit.
    This latter interpretation which has been used in the detailed deduction of the disk K_z by Kuijken & Gilmore is in keeping with the classical results by Sancisi & van Albada which show that galaxies with extended HI distributions have velocity curves that stay high well beyond the visible disk. Velocity curves deduced from constant M/L ratio fail to reproduce this by a considerable factor.
    The careful recent study of binary galaxies by L. Schweizer extends this result somewhat further. She works with galaxies whose rotation curves have already been determined optically so the masses within the visible parts can be deduced. With galaxy velocities that are more accurate than previous studies she finds that the galaxy pairs give orbital masses four times greater than the sum of the visible parts. She also finds evidence that the dark haloes so found grow almost linearly in mass out to about 60-100 kpc in radius but do not extend to greater distances. The statistics of binaries with greater separations show a constant M/L replacing the almost linear increase of M/L with r at smaller radii. L. Schweizer deduces
For Sc pairs $< M/L_v > = 42 \pm 10$ h $(M/L)_\odot$
For E/S pairs $< M/L_v > = 78 \pm 18$ h $(M/L)_\odot$.

5. Clusters of Galaxies

The dark matter problem found by Sinclair Smith & Zwicky has only become more pressing as better data has accumulated. Dressler found $M/L_v - 547 \pm 186$ h $(M/L_v)_\odot$ from 7 rich clusters.
    Mathew Colless in a recent Cambridge Thesis used spectra taken 50 at a time with the fibres on the Anglo Australian Telescope to measure M/L ratios in a further 11 rich clusters with various flattenings he finds

$$< M/L_v > = 380 \pm 100h \ (M/L_v)_\odot$$

with no correlation with luminosity or velocity dispersion of the cluster. These clusters obey an $L \propto \sigma^2$ relationship quite well.

6. Ω from Larger Scales

Several authors have deduced $\Omega \simeq 0.1 - 0.2$ from our infall velocity into the Virgo Cluster but the value deduced is almost quadratically dependent on the infall velocity deduced on which there is not yet agreement. Perhaps the greatest interest is attached to the calculations of the IRAS dipole and the associated claim that $\Omega = 1$. The idea behind calculating dipoles is simple and interesting. A galaxy gives light whose intensity falls off like r^{-2} with distance. Its gravitational field falls in the same way. A larger galaxy will have more stars and more mass so the proportionality between light intensity and gravity field will be preserved. Adding many such contributions vectorially we see that the net gravity field on the Local Group is proportional to the net flux of light generated by external galaxies past the Local Group. Since it is the gravity field on the Local Group that we believe has generated its motion relative to the cosmic micro-wave background, we deduce that the next light flux or 'light dipole' should come from the direction of the Local Group's motion. Do these directions agree? The Sun's motion with respect to the microwave background is determined directly from its anisotropy. Subtracting the Sun's motion relative to the barycentre of the Local Group yields its motion.

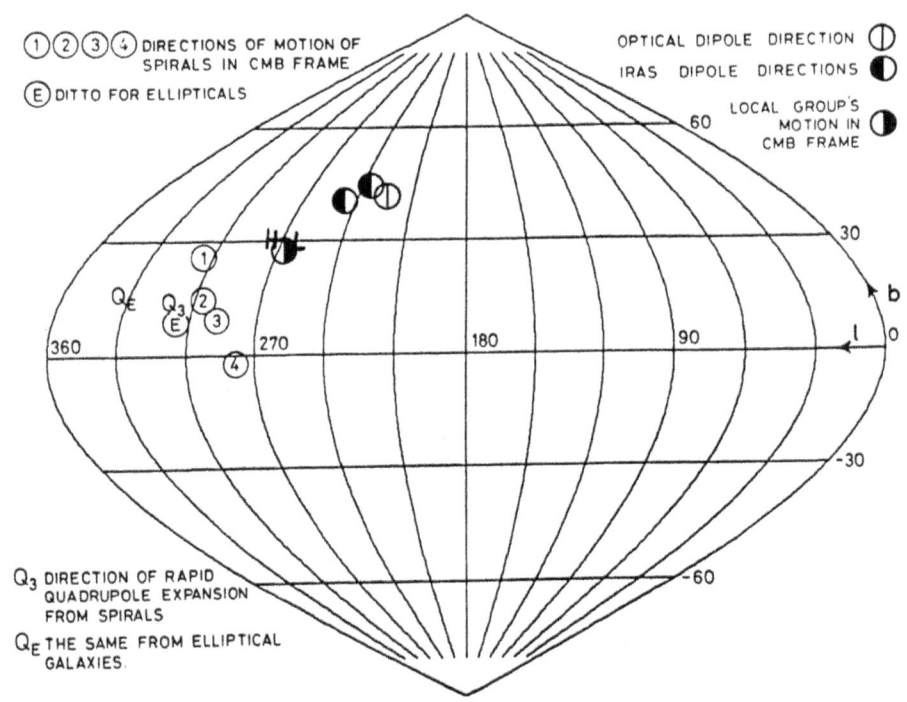

Figure 2. Interesting directions in galactic coordinates. The revised IRAS dipole direction H and the revised optical dipole direction L are very close to the Local Group's motion.

Unfortunately the luminosity function of galaxies in the 60μ IRAS band is very broad so the flux in that band past the Local Group has serious shot noise due to a few nearby galaxies that are very bright in that band because they have active star formation. Thus the direct proportionality between flux and gravity field has to be abandoned and the number-weighted dipoles calculated within each flux bin. These may then be flux weighted and added as was done by Yahil, Walker & Rowan-Robinson or merely added to get the number-weighted dipole as was done by Meiksin & Davies. The directions of the IRAS dipoles determined are shown in Figure 2. The agreement with the microwave dipole direction is plausible but not good. Unfortunately these samples of IRAS sources did not rely on IRAS observations alone but also used optional identification with bright galaxy catalogues as a subsidiary criterion. This has led to some bias to those parts of the sky when the galaxy catalogues go deepest. Harmon has shown that dropping the IRAS sources that are *only* included because of identification with optional galaxies swings the IRAS dipoles through a substantial angle in the sky. Harmon, Lahav & Meurs have determined the IRAS dipole from a new sample of IRAS sources selected from IRAS data alone to have colours and fluxes to be identified galaxies. Their dipole direction is within 8° of the cosmic microwave background - a much more satisfactory outcome.

Lahav has determined the optical dipole from calibrating the optical catalogues (see Figures 3 & 4). His first attempts to do this using a preliminary photoelectric calibration of the ESO and UGC catalogue diameters by Fonqué & Paturel, led to poor agreement. However, Lahav has since calibrated the ESO & UGC diameters of galaxies himself from the redshift surveys. The resulting dipoles are given in Lynden-Bell & Lahav and in Lahav, Rowan-Robinson & Lynden-Bell. They show good agreement $\sim 8°$ with the direction of the microwave background dipole. Thus there is now good directional agreement between the latest IRAS dipole, the optical dipole and the cosmic microwave dipole. This gives some evidence that the peculiar velocity of the Local Group is gravitationally generated and that the inverse square law of gravity is the dominant force even on this very large scale.

We performed a number of experiments to discover where the optical dipole came from. Firstly we found that more than half came from galaxies of diameter ≥ 4 minutes of arc. These are nearby galaxies typically at 2000 km/sec or less. The simplified picture in which most of the Local Group's velocity is generated by a Great Attractor at ~ 4000 km/s would predict that half the dipole came from galaxies with diameter ≥ 2 arc minutes. This suggests that the dipole comes from much nearer than the Great Attractor model predicts. Nevertheless that model gives a good description of the motion of the elliptical galaxies in our survey not just here at the Local Group but over quite large distances. A possible resolution of this discrepancy is that half of the Great Attractor may lie hidden in the Galactic plane. It then contributes more to the gravity field and induces the observed velocities in the elliptical galaxy sample but has been half omitted from the light dipole because it is too obscured. It could well be that half the velocity of the Local Group was generated by such a distant agglomeration. Some support for this view comes from Figures 5 & 6. Figure 5 shows the numbers of galaxies within 20° of the Supergalactic plane plotted from Huchra's incomplete radial velocity catalogue. Radial velocity is plotted as radius and supergalactic longitude as azimuth. In Figure 6 the incompleteness has been allowed for by comparing the numbers in the Huchra catalogue at each direction at each apparent galaxy diameter with the numbers in the diameter limited UGC and ESO catalogues. The map has been interpolated over the uncatalogued strip and the galactic plane. The large agglomeration in Centaurus Supergalactic longitude 160 appears to extend over the Galactic Plane to join Pavo Indus. Fairall has also deduced this from his radial velocity survey in the region.

Another interesting experiment is to look at the excess light coming from a cone of 90° whole angle directed at the dipole. This accounts for about 2/3 of the dipole, the other 1/3 coming from the decrement in the backward pointing cone. Interestingly the forward-pointing cone gets its greatest excess flux not in the direction of the dipole $\ell=261$, b=29 but rather when it is pointed towards $\ell=290$, b=28 which is slewed around towards Centaurus.

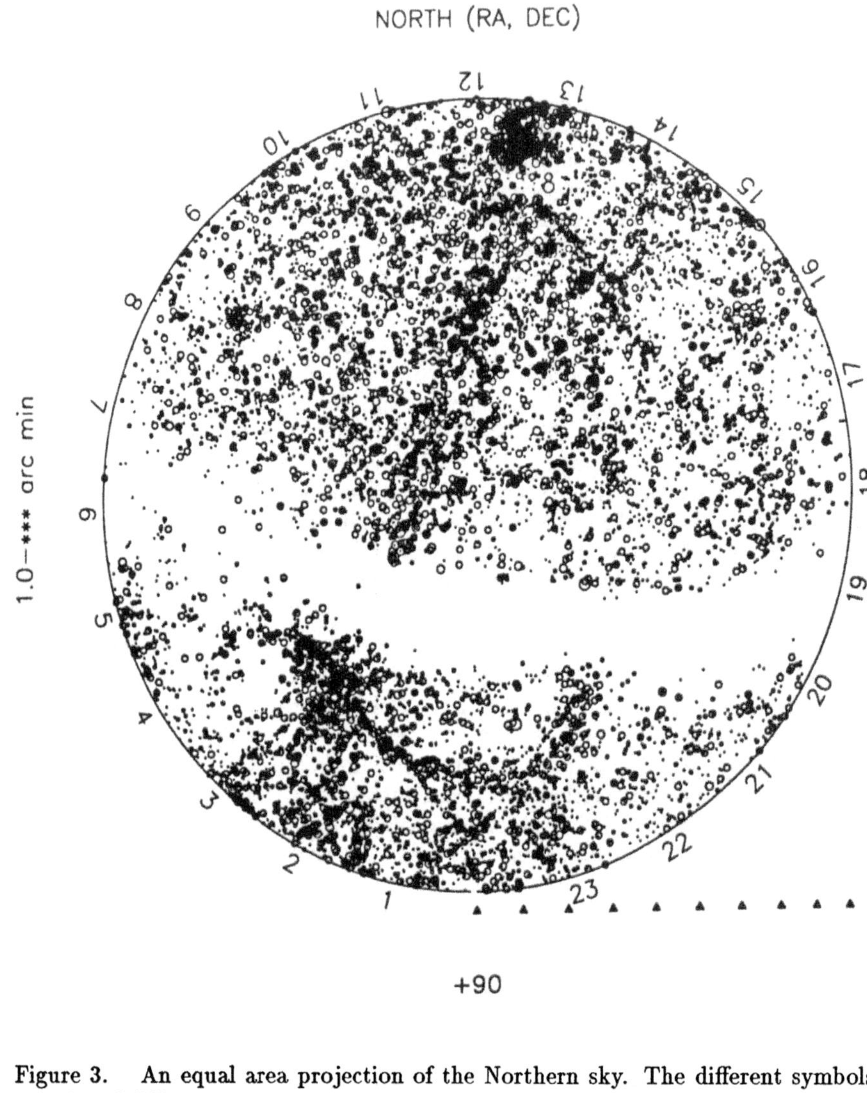

270

NORTH (RA, DEC)

1.0−∗∗∗ arc min

+90

Figure 3.   An equal area projection of the Northern sky. The different symbols indicate galaxies of different apparent diameters in Nilson's Uppsala General Catalogue. The white band is caused by obscuration in our Galaxy. The Virgo cluster is clearly visible at the top; the supergalactic band of bright galaxies stretches from it down to 2^h. The prominent arc at the bottom is the Perseus-Pisces chain.

Likewise the backward cone gives the greatest decrement not in either of the aligned positions but pointing towards $\ell=74$, $b=35$. This lies towards a prominent void in the very local distribution of Galaxies. Indeed half this void's negative contribution is from galaxies of diameter ≥ 8 arc minutes. Thus the deviation of the Local Group's motion (relative to the cosmic background radiation) from the large scale motion of all nearby ellipticals or spirals seems to be caused not just by the Virgo Cluster but also by the Local Void. The latter

SOUTH (RA, DEC)

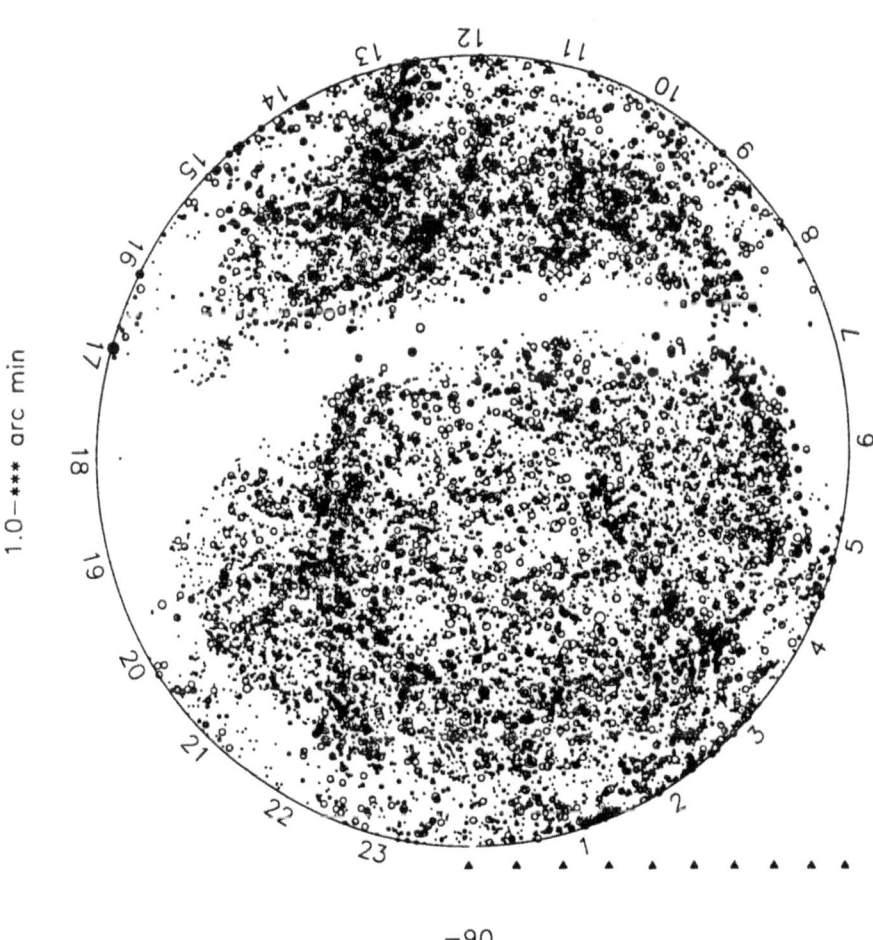

$1.0 - *** $ arc min

-90

Figure 4. A similar projection of the galaxies in the Southern sky from the ESO/Uppsala catalogue of Lauberts $\delta \leq -17\frac{1}{2}°$ and from the Morphological Catalogue of Vorontsov-Velyaminov $\delta > -17\frac{1}{2}°$. The Centaurus-Great Attractor region is prominent near the top.

probably accounts for most of the motion perpendicular to the Supergalactic Plane.

Although there are more galaxies with diameter > 1 arc minute in the optical catalogues than there are in the IRAS samples, the number-weighted percentage optical dipole is 4 times the equivalent number-weighted percentage IRAS dipole. The IRAS galaxies are significantly more uniformly distributed over the sky. When the sources are plotted on the sky and the two sky maps placed beside each other as in Figure 7, the same features can be seen but the IRAS map is a very washed out version with the features poorly emphasised while such features as the Perseus Pisces chain, Centaurus and the Pavo Indus Telescopium band of galaxies are

272

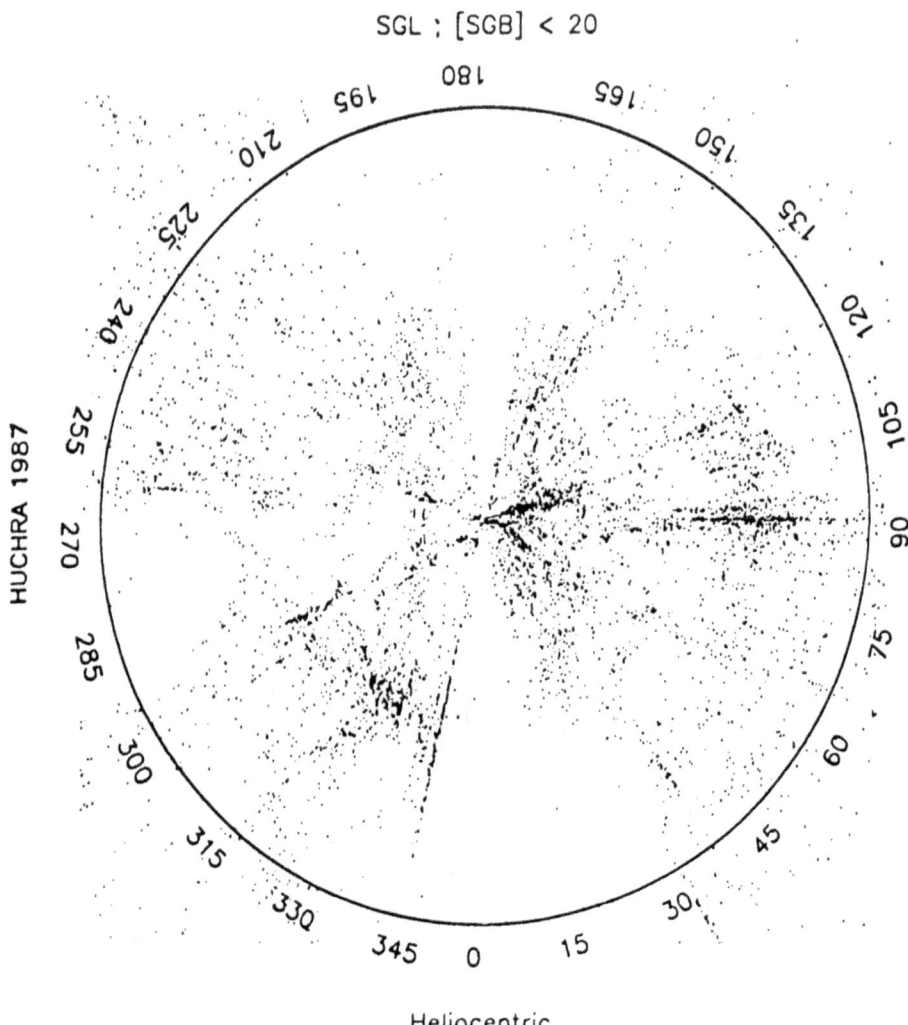

Figure 5. Galaxies in Huchra's catalogue and within 20° of the supergalactic plane are plotted with radial velocity as radius with supergalactic longitude as azimuth. The circle is at 10,000 km/s.

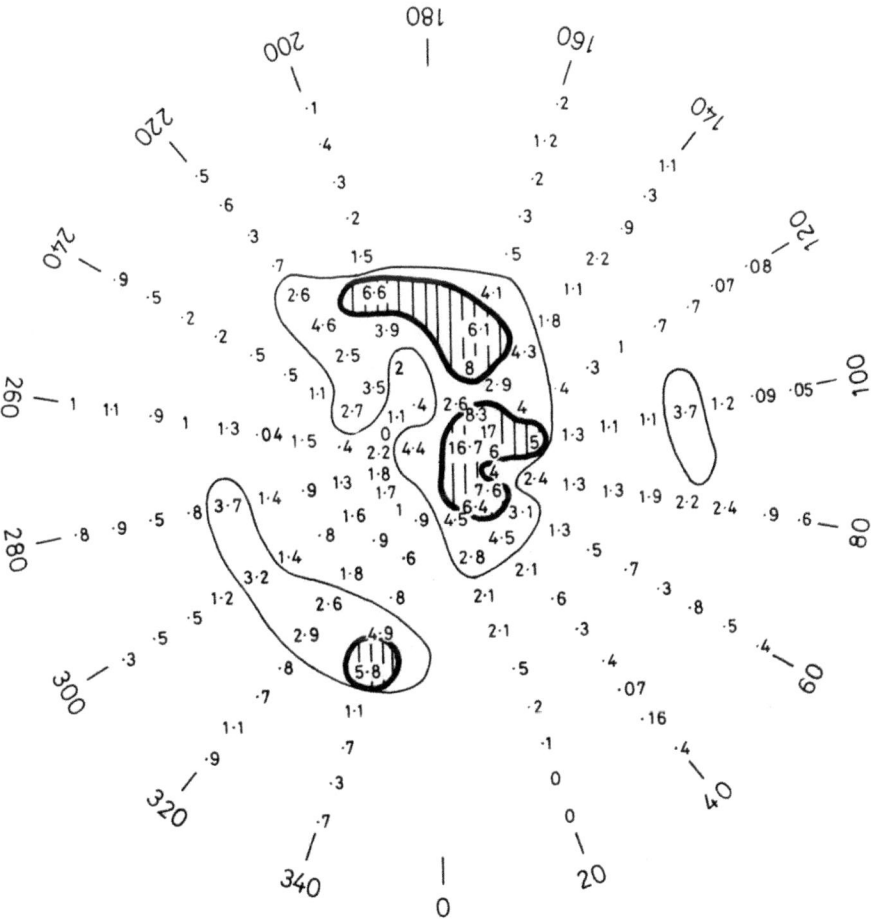

Figure 6. The density in the supergalactic plane is calculated allowing for the incompleteness of Huchra's catalogue by comparing it with the complete diameter limited UDC/ESO catalogues. The coordinates system in scale are the same as Figure 5.

prominent features of the optical sky. Relative to all galaxies the IRAS galaxies are clearly 'anti-biassed', *i.e.* more spread out. One reason for this is that elliptical & SO galaxies that are most concentrated to rich regions are almost undetectable at IRAS wavelengths due to their lack of star formation and hot dust. IT IS THIS ANTI-BIASSING THAT LEADS TO THE HIGH Ω determined from IRAS data. Of course it may be that the distribution of galaxies is biased and the anti- biassing of IRAS galaxies relative to them could be just such as to follow the dark matter distribution but there seems no reason to believe that. Rather it is likely that the broad luminosity function of IRAS galaxies has blurred the picture and by chance has done so approximately the right amount to make the apparent Ω deduced from them close to 1. In discussing the data side by side by the same oversimplified model, Lahav *et al.* get $\Omega \simeq 0.8$ from the IRAS sample and $\Omega \simeq 0.3$ from the optical data.

7. Dwarf Spheroidal Galaxies

Great interest was aroused in this subject particularly by papers by Faber & Lin and by Aaronson which used Tremaine & Gunn's phase space argument. Here the basic idea is that the velocity dispersions in these very sparse small galaxies may show that they contain dark matter. If true, the dark matter is at such a density that any particle under several hundred electron volts in mass would have a number density in phase space greater than two per cell of volume h^3, if it had less than the velocity of escape. Hence dark matter in dwarf spheroidals would significantly constrain the type of fundamental particle responsible. In particular, neutrinos could be ruled out. Unfortunately the data suggesting that there is dark matter in dwarf spheroidals has never been good and even where there has been apparent agreement over the velocity dispersions of the stars in them, there is no detailed agreement as to which stars move fast or slow (see Godwin & Lynden-Bell). It appears that radial velocity variability plus underestimates of the true errors in determining the radial velocities are still large contributors to the observed velocity dispersions. It is not yet clear that there is a dark matter problem in the dwarf spheroidals (see Kormendy), so theories should not yet be constrained by limits on particle masses derived from them. The tragedy of Mark Aaronson's untimely death will postpone the day when the observational position is clarified, but there are others following his lead in this fascinating and observationally challenging field.

8. Summary

The astronomical estimates of Ω are small, the only large one is due to anti-biasing in the IRAS data which is understood. An Ω in the range 0.1 to 0.2 is certainly compatible with the observations.

Methods that use mass-to-light ratio to assess the mass density of the universe do not fully account for the matter content of voids in the galaxy distribution. Since most of the universe is empty of galaxies, it is clear that Ω could be 1.

The strongest case for dark matter still comes from the great clusters where galaxies are densest.

The dark matter both there and around galaxies could be accounted for by superplanets

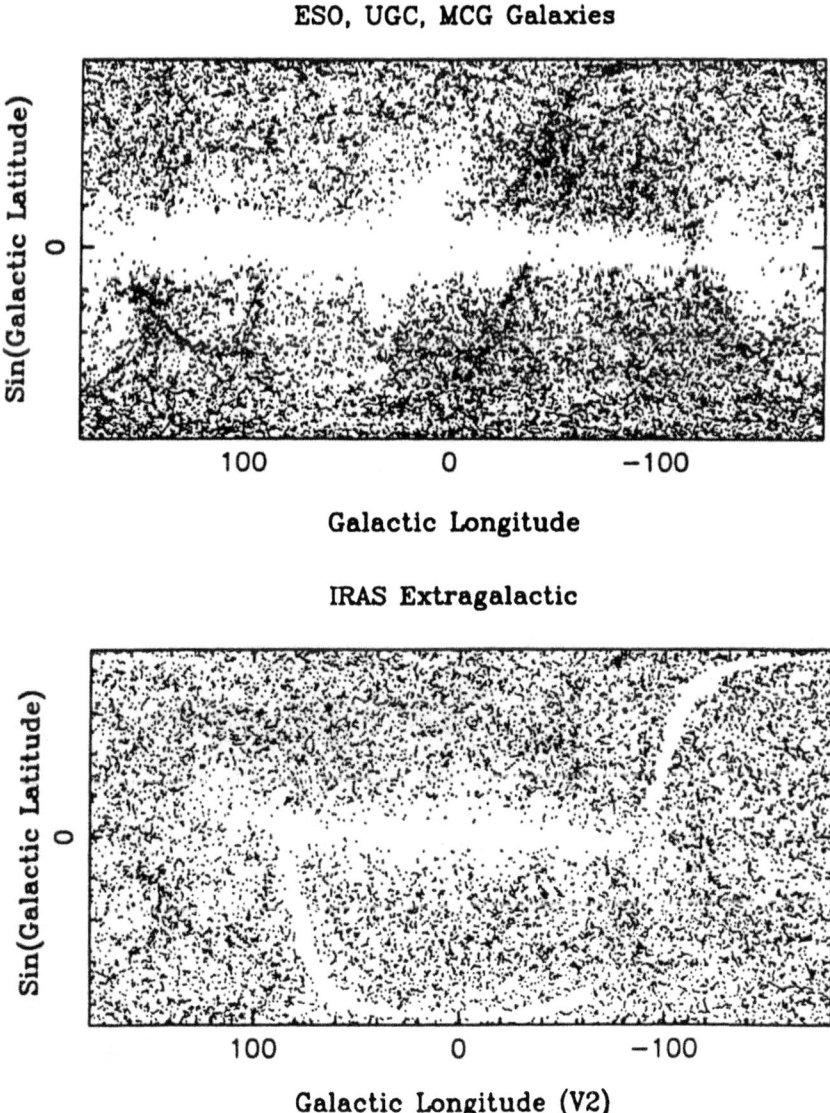

ESO, UGC, MCG Galaxies

Galactic Longitude

IRAS Extragalactic

Galactic Longitude (V2)

Figure 7.   Robin Harmon's comparison of the optical and IRAS galaxy distribution. Notice the lack of density contrast in the IRAS galaxies.

without violating the limits on Ω_b from nuclear cosmogenesis.

The M/L_v of the *known* baryons in the universe is less than 10 while the M/L_v necessary for $\Omega = 1$ is 1600 h. Hence Ω_b could be as low as 0.01. A very similar lower limit comes from the cosmic synthesis of Deuterium.

References

Einstein, A. 1917, Berlin Sitzungsberichte, p.142.

Geller, M., This symposium.

Davis, M. & Peebles, P.J.E. 1983, *Ann. Rev. Astr. Astrophys.* **21**, 109.

Spite, F. & Spite, M., 1982, *Astr. Astrophys.* **115**, 357.

Boesgaard, A.M., Budge, K.G. & Burck, E.E. 1988, *Astr. J.* **325**, 749.

Pagel, B.E.J. 1988, ESO/CERN School on Astroparticle Physics, eds. A. Rujula, D. Nanopoulos & P.A. Shaver (World Scientific Pub. Co.).

Boesgaard, A.M. & Steigman, G. 1985, *Ann. Rev. Astr. Astrophys.* **23**, 319.

Kuijken, K. & Gilmore, G. 1989, *Mon. Not. R. astr. Soc.* , submitted.

Sancisi, R.& van Albada, T.S. 1987, IAU Symposium **117**, *Dark Matter in the Universe*, eds. J. Kormendy & G.R. Knapp, p.67.

Schweizer, L.Y. 1987, *Astrophys. J. Suppl.* **64**, 427.

Dressler, A. 1978, *Astrophys. J.* **226**, 55.

Colless, M. 1987, Thesis Cambridge University, see also 1987 *Mon. Not. R. astr. Soc.* **224**, 453.

Lynden-Bell, D. & Lahav, O. 1988, Vatican Symposium, ed. V. Rubin & G. Coyne

Yahil, A., Walker, D. & Rowan-Robinson, M. 1986, *Astrophys. J.* **301**, L1.

Meiksin, A. & Davis, M. 1986, *Astr. J.* **91**, 1919.

Harmon, R.T., Lahav, O. & Meurs, E.J.A. 1987, *Mon. Not. R. astr. Soc.* **228**, 5P.

Lahav, O. 1987, *Mon. Not. R. astr. Soc.* **225**, 213.

Lahav, O., Rowan-Robinson, M. & Lynden-Bell, D. 1988, *Mon. Not. R. astr. Soc.*

Fairall, A.P. 1988, *Mon. Not. R. astr. Soc.* **230**, 69.

Godwin, P.J. & Lynden-Bell, D. 1987, *Mon. Not. R. astr. Soc.* **229**, 7P.

Kormendy, J. 1987, IAU Symposium **117**, *Dark Matter in the Universe*, eds. J. Kormendy, J. & G.R. Knapp, (Reidel).

DISCUSSION.

Goldhaber: Prof. Lynden-Bell mentioned that the baryonic fraction can be near 1 in our galaxy. How does this affect the chance of finding dark matter in the Laboratory?

Lynden-Bell: By analogy with other galaxies our galaxy probably possesses a massive halo, which probably makes a contribution at the solar radius, but direct evidence for it is rather weak and it would be baryonic e.g. superplanets. My best bet assumption would be to take a spherical halo falling like r^{-2}, making up half of the matter within the solar circle around the galaxy. Whereas the true density of dark material might be a bit greater, experimenters owe not to assume more than that.

Dressler: Conceivably half of the Great Attractor might be hidden by the Galactic Plane. If so, how much of the Local Group's 600 Km/sec motion might be attributed to such a distant concentration?

Lynden-Bell: About 300 Km/sec is easily possible if half of the Great Attractor were hidden behind the Galactic Plane, but certainly a major component of the extragalactic light flux past the Local Group comes from nearer galaxies, those with diameters $\sim 4'$.

Börner: Can you immagine a deviation from Newton's law which removes the evidence for dark matter on all scales?

Lynden-Bell: There are two types of astronomers: those who wish to find new physics by studying astronomy and those who study astronomy to find out what's there. Occams razor suggests to me that material that does not shine, superplanets for instance, is a much less dramatic hypothesis that a change in the laws of physics. If I could show that there were no dark material then I would deduce that the laws of physics need to be changed, but only then. I have no difficulty in imagining changes in physical laws, the only difficulty in believing that our changes are well motivated yet.

Trimble: Could you give a one-sentence summary of the evidence for your point one on your list of conclusions, where you didnt get to?

Lynden-Bell: If nothing hides behind the galactic plane, half of the net flux of extra-galactic light past the local group comes from galaxies of > 4 arcmin in diameter. This implies that the local group velocity arises primarily from matter ~ 2000 Km/sec, only half the distance of the Great Attractor.

Sarkar: With regard to dark matter in spirals, James Binney argued some time ago that even a 10 % systematic error in the measurements of the luminosity profile can account for much of the discrepancy between the observed rotation curve and that constructed from the luminosity profile (assuming constant M/L). Can you please comment on whether this criticism has been answered by inproved measurements of the luminosity profile?

Lynden-Bell: 10 % is a big photometric error, but quite possible given the absorbtion in galactic disks. I believe there is no objection to that as a possible explanation of discrepancies within the optically observed disks, but hard cases are the flat curves observed radio astronomically well outside the optical disk. There this type of explanation fails to keep the rotation curve so high.

DARK MATTER CANDIDATES

Michael S. Turner
NASA/Fermilab Astrophysics Center and The University of Chicago
Chicago, IL
USA

ABSTRACT. One of the simplest, yet most profound, questions we can ask about the Universe is, How much stuff is in it, and further what is that stuff composed of? Needless to say, the answer to this question has very important implications for the evolution of the Universe, determining both the ultimate fate and the course of structure formation. Remarkably, at this late date in the history of the Universe we still do not have a definitive answer to this simplest of questions—although we have some very intriguing clues. It is known with certainty that most of the material in the Universe is dark, and we have the strong suspicion that the dominant component of material in the Cosmos is not baryons, but rather is exotic relic elementary particles left over from the earliest, very hot epoch of the Universe. If true, the Dark Matter question is a most fundamental one facing both particle physics and cosmology. The leading particle dark matter candidates are: the axion, the neutralino, and a light neutrino species. All three candidates are accessible to experimental tests, and experiments are now in progress. In addition, there are several dark horse, long shot, candidates, including the superheavy magnetic monopole and soliton stars.

I. DARK MATTER IN THE UNIVERSE

The luminous matter in galaxies, as evidenced by the radiation (visible, infrared, x-ray, etc.) associated with it, contributes only a tiny fraction of closure density $(\Omega = \rho/\rho_{\rm crit};\ \rho_{\rm crit} \simeq 1.05\, h^2 \times 10^4\, {\rm eV\, cm^{-3}})$:

$$\Omega_{\rm LUM} \lesssim 0.01$$

On the other hand, there is overwhelming evidence that there is much more additional matter associated with galaxies that is not luminous. The flat rotation curves of spiral galaxies (inferred by both optical and 21 cm measurements) indicate that the typical spiral galaxy is immersed in a dark halo which contains 3–10 times the amount of matter that is associated with the luminous portion of the galaxy. This dark material whose presence is inferred by its gravitational effects alone contributes at least:

$$\Omega_{\rm HALO} \gtrsim 0.03 - 0.10$$

Since there is yet no incontravertible evidence for a rotation curve which 'turns over' the total amount of material in the halos of spiral galaxies has yet to be determined.

The mass associated with galaxies in bound systems (small groups and clusters of galaxies) can be determined by dynamical means (the virial theorem); from such measurements one infers a universal mass density of

$$\Omega_{\rm CLUSTERED} \simeq 0.1 - 0.3$$

M. Caffo et al. (eds.), Astronomy, Cosmology and Fundamental Physics, 279–286.
© 1989 by Kluwer Academic Publishers.

A note of caution; since only ~ 1 in 10 galaxies are found in clusters the mass density determined by this means may not be indicative of the true mass density.

The flat rotation curves of spiral galaxies and the dynamics of clusters provide the most convincing, I would even say irrefutable, evidence that dark component to the mass density outweighs the luminous component by at least a factor of 10.

The mass of the Virgo cluster has been determined by its influence on the dynamics of the local group, the so-called Virgo infall method, from which values of $\Omega \simeq 0.1$–0.2 have been inferred. This technique has been applied on even larger scales: the IRAS catalogue of infrared selected galaxies has been used to compute the local acceleration field and the predicted peculiar velocity (due to the inhomogeneous distribution of galaxies), and comparing this to our measured peculiar velocity values of Ω approaching unity have been obtained.[1]

Kinematical methods can also be used to determine the mean mass density of the Universe; e.g., the luminosity-red shift relation (or Hubble diagram), the angle-red shift relation, the galaxy count-red shift relation, etc. The results of the first two kinematical tests are inconclusive, largely due to concerns about galactic evolution. I should mention that there is some hope that the luminosity-red shift relation will be revived by the use of infrared observations where the evolutionary effects may be far less important. Recently, Loh and Spillar[2] have attempted to use the third test to infer the universal mass density and obtained a formal value of $\Omega = 0.9^{+0.7}_{-0.5}$. While many questions have been raised about their photometric (as opposed to spectroscopic) method of obtaining red shifts and their assumptions about the galaxy luminosity function, this technique has great cosmological leverage and at the very least is sensitive to galaxy evolution in a different way than the other two kinematic methods.

The luminous matter in the Universe must of course be baryons! However, not all baryons are necessarily luminous. Our best knowledge of the baryonic mass density derives from primordial nucleosynthesis. In the standard model of big bang nucleosynthesis (BBN) concordance of the observed abundances of D, ^{3}He, ^{4}He, and ^{7}Li require the baryon-to-photon ratio to lie in the narrow interval $\eta = 3 - 7 \times 10^{-10}$, or equivalently,[3]

$$0.011 \lesssim 0.011 h^{-2} \lesssim \Omega_B \lesssim 0.025 h^{-2} \lesssim 0.15$$

where h is the present value of the Hubble constant in units of 100 km sec^{-1} Mpc^{-1} and $0.4 \lesssim h \lesssim 1$. Since luminous matter contributes at most 1% of critical density there is already evidence that some of the baryons in the Universe are dark—likely in the form of jupiters, white dwarfs, neutron stars, and black holes.

What is one to conclude from this? First, it is a certainty that the dominant component of matter in the Universe is dark—by at least a factor of 10. Second, if the universal density is greater than 15% of critical, then there must be non-baryonic dark matter. I should caution that at present there is no irrefutable case for $\Omega \gtrsim 0.15$, and so it is still possible that baryons are the whole story!

From this point forward I will assume that Ω is indeed equal to unity. I believe that the theoretical reasons for believing such are very compelling. Briefly, those are reasons are: (1) Structure formation—Structure formation in the Universe begins when the Universe becomes matter-dominated and ceases when the Universe begins its coasting phase (i.e., when the curvature term dominates the matter density). The red shift at which the Universe becomes matter dominated is proportional to Ω; the red shift at which the coasting phase commences is proportional to Ω^{-1}. In a low-Ω Universe the growth of density inhomogeneities gets squeezed at both ends, thereby requiring larger initial inhomogeneities which in turn lead to larger temperature fluctuations in the microwave background. Conventional scenarios of structure formation are all but ruled out by the smoothness of the microwave background for $\Omega \lesssim 0.2$. (2) Naturalness/good taste—Ω does not remain constant as the Universe evolves, unless it was precisely unity initially. Rather, as time goes on it deviates more and more from unity. That Ω today is still of order unity implies that the value of Ω at the Planck epoch must have been unity to within a part in 10^{60}. (3) Inflation—The inflationary paradigm is a very attractive early Universe scenario based upon plausible (albeit speculative) microphysics. It provides a means of understanding a number of cosmological puzzles including the present value of Ω being of order unity. In the inflationary scenario Ω is reset to a value very, very close to unity during inflation, so close that an inescapable prediction of inflation is that Ω today should be unity (more precisely, that the curvature of the Universe should be negligible).

With the assumption that Ω is 1 today it follows that the dominant form of matter in the Universe must be non-baryonic. Furthermore, if Ω is unity (as theoretical prejudice would have), then there is strong indication for a component of the matter density which is not associated with bright galaxies, is less clustered, and contributes about 0.8 of critical, a fact which should be kept in mind. To summarize then, theory and observation indicate: $\Omega = 1$, $\Omega_B \sim 0.1$, $\Omega_X \sim 0.9$, $\Omega_{\text{CLUSTERED}} \simeq 0.1-0.3$, and a local density of dark matter (in our halo) $\simeq 0.3 \, \text{GeV} \, \text{cm}^{-3}$.[4]

Before proceeding to relic WIMP dark matter, I should comment on the possibility that $\Omega = \Omega_B = 1$. It is a great leap to assume that $\Omega = 1$; it is an even greater leap to invoke a new form of matter to explain it. Just how secure is the BBN constraint, $\Omega_B \lesssim 0.15$? Recently, two non-standard scenarios have been suggested to circumvent this important constraint: (1) A second, late period of nucleosynthesis which 'resets' the light element abundances and is triggered by the hadronic decay products of a particle which decays when the Universe is $\sim 10^5 - 10^6$ sec old;[5] (2) Large inhomogeneities in the local baryon-to-photon ratio at the epoch of BBN arising due to the effects of a strongly first order quark/hadron phase transition which modify the standard picture. Both scenarios invoke new parameters to adjust (although ultimately fixed by experiment and not cosmology), and yet neither is able to reproduce the success of the standard scenario of primordial nucleosynthesis. In the first scenario ^{7}Li is underproduced, while ^{6}Li is overproduced (although observations have not yet definitively ruled this scenario out). The second scenario relies upon the quark/hadron transition being strongly first order (which seems unlikely) with a very low transition temperature ($T_C \lesssim 125$ MeV). Moreover, at present, the most detailed simulations of nucleosynthesis with such inhomogeneities indicate that *none* of the 4 light element abundances are concordant, with ^{7}Li being overproduced by almost two orders-of-magnitude.[7] It is certainly important to keep an open mind to the possibility that all of the dark matter is baryonic; however, at present the case against $\Omega_B \sim 1$ seems quite compelling. Further, if Ω_B were one would have to work hard to explain where all those dark baryons (99%!) are—which is not an easy task.

II. RELIC WIMPS AS THE DARK MATTER

There is ample evidence which indicates that the early Universe was very hot and throughout most of its early history in thermal equilibrium. Therefore, at early times all kinds of exotic particles (which theorists are certain exist and experimentalists struggle to find) should have been present in great abundance (comparable to the photons). Moreover, there is strong reason to believe that during its early history the Universe went through several, if not many, symmetry breaking phase transitions, during which topological relics can be produced (monopoles, cosmic string, soliton stars, etc.). Many of these exotic particles and objects (if they exist) are very likely to still be with us in interesting numbers. And, for some of these relics, their abundance provides closure density for plausible values of the parameters of the theory, so that there are numerous attractive particle physics candidates for the dark matter in the Universe! The candidates can be organized into 4 categories: Thermal Relics (hot and cold); Asymmetric Relics; and Non-Thermal Relics.

• **Thermal Relics**—At very high temperatures $(T \gg m_X)$, the equilibrium abundance of a species X is comparable to that of the photons, while at low temperatures $(T \ll m_X)$ the equilibrium abundance is exponentially small, $n_{XEQ}/n_\gamma \simeq (m_X/T)^{3/2} \exp(-m_X/T)$. If equilibrium were the whole story thermal relics would be very uninteresting indeed. However, equilibrium can only maintained so long as the interactions which control the abundance of the species (decays, annihilations and their inverse processes) are occurring rapidly on the expansion time scale $(\Gamma \gtrsim H)$. Consider a weakly-interacting massive particle species (or *WIMP*, for short) which is stable. Eventually its annihilation rate falls below the expansion rate $(\Gamma \lesssim H)$; annihilations *freeze out*, and the particle's relic abundance freezes in, at about the equilibrium value for the freeze out epoch.

If the speicies is relativistic at the time of freeze out, its relic abundance relative to photons will be of order unity. Such a relic is referred to as a *hot*, thermal relic; the interactions of ordinary neutrinos freeze out at a temperature $\sim$ few MeV, so a light neutrino species $(m_\nu \lesssim \text{MeV})$ is an example of a hot, thermal relic. Note, that for a hot relic the contribution to present energy density scales as the mass; for a neutrino species, $\Omega_\nu \simeq (m_\nu/91h^2\,\text{eV})$.

If the species is non-relativistic at the time of freeze out, its relic abundance will be exponentially less than that of the photons. Such a relic is referred to as a *cold*, thermal relic. Examples include a heavy, stable neutrino species and the lightest superpartner (usually the neutralino[8]) in supersymmetric extensions of the standard model. Interestingly enough, the relic abundance of a cold relic is inversely proportional to the annihilation cross section of the species

$$\Omega_X \sim 10^{-36}\,\text{cm}^2/ < \sigma|v| >_{\text{ann}}$$

This means that the more weakly-interacting a species is, the greater its relic abundance, and that a cold relic which contributes closure density must have an annihilation cross section (and interaction cross sections with ordinary matter) which is roughly $10^{-36}\,\text{cm}^2$, characterisic of weak interactions. Note too, that for a thermal relic, the numbers of relic particles and antiparticles should be equal.

• **Asymmetric Relics**—Above it was tacitly assumed that the abundance of particle and antiparticle species were identically equal, so that the annihilation rate (and hence cross section) determines freeze out and the relic abundance. If an asymmetry exists between particle and antiparticle species, say more particles than antiparticles, then the relic abundance can actually be determined by the size of the asymmetry, in which case the relic population consists only of particles. [The criterion for this to occur is that the asymmetry be greater than the relic abundance that would result from the freeze out of annihilations.] A familiar example is baryons; were baryons and antibaryons initially present in equal numbers, their relic abundance would be: $n_b/n_\gamma = n_{\bar{b}}/n_\gamma \simeq 10^{-18}$, some 8 or so orders-of-magnitude smaller than the observed $n_b/n_\gamma \simeq 4-7 \times 10^{-10}$. [It is amusing to note that even for an asymmetric relic annihilations become impotent before all the antiparticles are exhausted; for antiprotons the predicted relic abundance is: $n_{\bar{b}}/n_\gamma \sim 10^{18} \exp(-9 \times 10^5)$.] The general framework which explains the baryon asymmetry, baryogenesis, suggests that there might be similiar asymmetries associated with other species which carry approximately conserved quantum numbers.

• **Non-thermal Relics**—There are a handful of very interesting potential relics whose interactions are so feeble that they should never have been thermal equilibrium at early times. Nevertheless, such relics may have been produced by other, very interesting processes. Included in this list are superheavy magnetic monopoles, axions, and soliton stars. Monopoles are topological solitons associated with the symmetry breakdown of a semi-simple group to one which contains a $U(1)$ factor, e.g., $SU(5) \rightarrow SU(3)_C \otimes SU(2)_L \otimes U(1)_Y$, and are produced in the SSB phase transition as topological defects, owing to the existence of particle horizons in the standard cosmology. And, as is well-known, in the standard cosmology the relic abundance of monopoles produced by this mechanism is catastrophically large—so large that the Universe would reach the temperature of 3 K at the youthful age of 30,000 yrs. One of the attractive features of inflation is that monopoles are naturally diluted to a safe (and most likely uninteresting) relic abundance.

A second example of a non-thermal relic is the axion. Axions are produced cosmologically by the initial misalignment of the axion field; when the axion develops a mass due to instanton effects $(T \sim \Lambda_{\rm QCD})$ the axion field then begins to oscillate due to this initial misalignment.[9] These oscillations correspond to a condensate of zero momentum axions, whose relic density is $\Omega_a \sim (m_a/10^{-5}\,{\rm eV})^{-1.2}$. From the peculiar scaling of the mass density and the fact that $m_a \sim 10^{-5}$ eV corresponds to closure density it is clear that axions are produced in highly non-thermal numbers. [Were axions present in thermal numbers, $\Omega_a \sim (m_a/100\,{\rm eV})$, some 7 orders-of-magnitude smaller than the coherent abundance. For axion masses $\gtrsim 0.01$ eV there is also a thermal population of axions, whose abundance is greater than that of the coherently produced population.[10] It has also been pointed out that axions produced by another coherent process, the decay of axionic strings, may contribute significantly to the relic abundance of axions.[11]]

Soliton stars are a generic class of non-topological solitons whose stability owes to dynamics rather than topology. The simplest example is a region of false vacuum which is stabilized against collapse by the presence of particles which carry a conserved charge and whose mass in a false vacuum region is much less than it is in the true vacuum. Whether or not a plausible mechanism exists to produce such objects in interesting numbers remains to be seen.[12]

• **Truly Exotic Relics**—There are even more exotic possibilities for the dominant form of matter in the Universe. For example, if the relic WIMP is unstable and decays on a cosmological time scale (say $\tau \sim 10^9$ yrs) into light particles which are still relativistic today, then the bulk of the mass density in the Universe would be in the form of relativistic debris. Such a scenario would neatly explain why most of the material in the Universe appears to be unclustered; however, it is not without its difficulties: a youthful Universe—in a Universe dominated by relativistic particles the age is $\sim 1/2H_0^{-1}$; and the formation of structure—since density inhomogeneities do not grow during a radiation-dominated phase, all the structure must be produced before the WIMP's decay. Another exotic possiblity is that most of the mass density in the Universe exists in the form of vacuum energy (i.e., a relic cosmological term). This scenario too accounts for most of the material in the Universe being unclustered. Of course, the crucial question here is why the cosmological term would have such a small, but non-zero, value compared to its natural scale, m_{pl}^4.

• **A New Cosmic Ratio**—Cosmology is a science in which the data are few and far between. Every piece of information we have is important, and especially dimensionless ratios (like the baryon-to-photon ratio, the relative abundances of the light elements, etc.). If the mass density of the Universe is indeed comprised of 1 part baryons and 9 parts exotica, then a new cosmic ratio exists, $r = \Omega_{baryon}/\Omega_{exotic} \sim$ 0.1. Why should this ratio have the value close to unity, rather than 10^{-20} or 10^{20}? While it is not possible to answer this question in a definitive way at present, it is intriguing to speculate as to why.[13] According to baryogenesis, the present baryon density traces to the dynamical evolution of a baryon asymmetry, and we have just discussed how the density of various relic WIMP's might arise. In the case of light neutrinos as the dark matter, the near equality of neutrino and baryon densities involves the smallness of neutrino masses relative to other fermion masses. The most attractive scenario for neutrino masses is the see-saw mechanism, where in $m_\nu \sim m_{fermion}^2/M$ and M is some superhigh energy scale (associated with unification). Here then, the near equality of neutrino and baryon densities traces to the very large value of the unification scale M. For a heavy neutrino or neutralino, the near equality traces to the large discrepancy between the weak scale (which sets the annihilation cross section) and the Planck scale. [This same discrepancy in scale 'explains' why stars shine.] For the axion, the near equality traces to proximity of the PQ symmetry breaking scale and the Planck scale. Finally, for an asymmetric relic, say of mass comparable to the baryon, the near equality of densities would trace to similar asymmetries in baryons and exotics.

There are almost too many candidate WIMP's for the dark matter in the Universe to list, and many are very well-movitated. Let me use my own personal prejudice to pare the list down to 3 most promising candidates. They are (not in any special order): *the axion*—the axion is perhaps the most compelling and simplest extension of the standard model, and the relic axions which would be the dark matter are accessible to experimental detection; *the neutralino*—supersymmetry is a very well-motivated extension of the standard model (which will be tested with the next generation of accelerators, if not the present), and in the simplest versions of supersymmetry the lightest superpartner is stable; relic neutralinos too are detectable; *a light neutrino*—the neutrino is known to exist! and in three flavors!; the electron neutrino mass is accessible to laboratory experiment (and at present the ITEP group still report a positive result); a supernova in our galaxy would likely make a determination of the μ and τ neutrino masses possible if one should be in the cosmologically interesting range. Finally, one should not forget about longshots! My favor longshot this year is the superheavy magnetic monopole. The *MACRO* experiment in the Gran Sasso Laboratory will start operating in the next year or so, and will ultimately reach a sensitivity level of 10^{-16} cm^{-2} sr^{-1} s^{-1}—a full order of magnitude below the Parker limit.

III. DARK MATTER DETECTION

The past decade has witnessed a renaissance in cosmology, triggered in large measure by the infusion of new ideas about the earliest history of the Universe, ideas which are based upon very attractive and well-founded speculations about fundamental physics at energies well beyond the weak scale. If progress in cosmology is to continue we must have experimental and observational data to test these ideas and to help theorists to narrow their future speculations. Cosmological data are hard to come by, more often than not requiring heroic undertakings. There are a myriad of interesting and attractive ideas to be tested, and as theorists we must sort the wheat from the chafe for our experimental and observational colleagues.

The relic WIMP hypothesis is an idea most worthy of testing, and fortunately is one which is amenable to testing. Already numerous experiments/observations are being carried out or planned. They include a variety of laboratory experiments (accelerator searches for supersymmetric particles, ν mass and oscillation experiments, $\beta\beta0\nu$ experiments), searches for the relic WIMP's themselves (axion searches, monopole searches, cryogenic searches for cold, thermal relics), indirect searches for WIMP annihilation products ($\nu\bar{\nu}$'s from WIMP annihilations in the earth and sun, $\bar{p}$'s, γ's, e^{+}'s from WIMP annihilations in the halo), and cosmological observations (kinematic tests for Ω, structure formation). The structure formation test is a most interesting one; given the amount and composition of matter in the Universe, as well as the nature of the primeval inhomogeneites, the structure formation problem is a well-defined initial data problem. At present the agreement between the numerical simulations of cold dark matter (with inflation-produced, adiabatic primeval fluctuations) is, save for two observations (cluster-cluster correlations and the peculiar velocity field), remarkable. And I believe there is still time for those two observations to come around!

All of these experiments are of the utmost importance for both cosmology and particle physics—the first evidence for new physics beyond the standard model may well come from the discovery of relic WIMP's. Finally, we should not forget cosmological particle relics with abundances less than critical (remember that $\Omega_{3K} \sim 10^{-5}$); if low energy SUSY is correct it is difficult to escape having a relic superpartner with $\Omega \lesssim 0.01$ (see Griest[8]), perhaps too low to close the Universe, but sufficient to detect. Likewise, relic thermal axions of mass $\sim$ few eV would not close the Universe, but could be detected by their cosmological decays.[10]

ACKNOWLEDGMENTS, THANKS, and APOLOGIES

This work was supported in part by the DoE through contract DE AC02-80ER 10587. My sincere thanks to the organizers of this Third CERN/ESO Symposium and to the city and University of Bologna for their hospitality. Finally, my apologies for the brevity of this report and the paucity of references; for more complete reviews of the topic of Dark Matter in the Universe, see Refs. 4.

REFERENCES

1. M.A. Strauss and M. Davis, UC Berkeley preprint (1988).
2. E. Loh and E. Spillar, *Ap. J.* **307**, L1 (1986).
3. J. Yang, et al, *Ap. J.* **281**, 493 (1984).
4. For more complete reviews of dark matter see, e.g., V. Trimble, *Ann. Rev. Astron. Astrophys.* **25**, 425 (1987); J. Primack, B. Sadoulet, and D. Seckel, *Ann. Rev. Nucl. Part. Sci.* **39** (1989); M.S. Turner, in *Dark Matter in the Universe*, eds. J. Kormendy and G. Knapp (Reidel, Dordrecht, 1987).
5. S. Dimopoulos, et al, *Phys. Rev. Lett.* **60**, 7 (1988); *Ap. J.* **330**, 545 (1988).
6. J. Applegate, et al, *Phys. Rev.* **35**, 1151 (1987); C. Alcock, et al, *Ap. J.* **320**, 439 (1987).

7. H. Kurki-Suonio, et al, *Phys. Rev.* **D38**, 1091 (1988).
8. J. Ellis, et al, *Nucl. Phys.* **B238**, 453 (1984); K. Griest, *Phys. Rev.* **D38**, 2357 (1988).
9. J. Preskill, F. Wilczek, and M. Wise, *Phys. Lett.* **120B**, 127 (1983); L. Abbott and P. Sikivie, *ibid*, 133; M. Dine and W. Fischler, *ibid*, 137.
10. M.S. Turner, *Phys. Rev. Lett.* **59**, 2489 (1987).
11. R.L. Davis, *Phys. Lett.* **180B**, 225 (1986).
12. J. Frieman, et al, *Phys. Rev. Lett.* **60**, 2101 (1988).
13. M.S. Turner and B.J. Carr, *Mod. Phys. Lett.* **A2**, 1 (1987).

DISCUSSION

Vauclair: This is a comment about the lithium primordial abundance. When we say that the lithium abundance observed in population II stars is the primordial one, we assume that the lithium abundance has not changed in these stars during 15 billion years. In reality there are two processes which may deplete lithium in these stars. If there is no turbulence at all, lithium is depleted by gravitational settling. In case of large turbulence, lithium is destroyed by nuclear reactions. It is possible that the lithium abundance remains constant, but this needs a fine tuning of turbulence so that gravitational settling may be prevented and the nuclear destruction timescale be still longer than the stellar age. This is not excluded, but it is also not excluded that some destruction occurred in these stars, in which case the lithium primordial abundance would be between 10^{-10} and 10^{-9}. I have done some computations using Zahn's theory of turbulent mixing induced by stellar rotation, and I find that the destruction process may lead to a "plateau shape" of the lithium abundance as observed, due to the rapid variation of the turbulent diffusion coefficient with radius inside the stars.

Sarkar: You quoted the nucleosynthesis limit on neutrino families $N_\nu < 4$. This assumes that the neutron half-life exceeds 10.4 min. and that the nucleon density is high enough that the ^{7}Li abundance is $\simeq 10^{-10}$. But recent experiments suggest that 10.4 min. is more likely an upper limit, not a lower bound to the neutron half-life. Also it has been argued that the primordial ^{7}Li abundance may be the Pop I value of $\simeq 10^{-9}$, which would allow a lower nucleon density. Therefore the limit $N_\nu < 4$ should be perhaps relaxed. This is very important because this cosmological limit on light neutrino types cannot be tested in the laboratory, since it includes particles which do not necessarily couple to the Z^0. Hence it should be critically examined on its own terms, given that it is a powerful constraint on new physics beyond the standard model.

Turner: The dependence of the BBN limit to N_ν upon $\tau_{1/2}$ is slight: dropping $\tau_{1/2}$ to 10.2 min increases the bound to N_ν by 0.2. While I do not agree that the Pop I ^{7}Li abundance reflects the primordial ^{7}Li abundance, the lower limit to η required to derive the limit to N_ν follows from the abundances of D and ^{3}He (and not ^{7}Li). I believe that the BBN limit to N_ν stands firm at $N_\nu \leq 4$ (note the '$\leq$' rather than '$<$'), the limit established in 1984 (for further discussion, see, G. Steigman, et al, *Phys. Lett.* **176B**, 33 (1986)).

LIMITS ON WIMPS FROM GLOBULAR CLUSTER STARS

Robert T. Rood
University of Virginia
and
Alvio Renzini
Università di Bologna

ABSTRACT. We describe how the Weakly Interacting Massive particles (WIMPs) which have been suggested as a solution to the solar neutrino problem might affect the evolution of globular cluster stars. We show that even at this preliminary stage of our investigation the possible range of WIMP parameters is severely limited.

1. INTRODUCTION

If the universe is closed, 80–90% of its mass may consist of nonbaryonic WIMPs. With the correct properties WIMPs can be captured by and then orbit inside stars. This can significantly alter the internal structure of some stars. In particular, Faulkner, Gilliand, Press, and Spergel in various combinations (see references in Spergel and Faulkner 1988) have suggested that WIMPs could effectively transport energy in the core of the sun. This could lead to an isothermal core, and under some conditions bring the predicted solar neutrino flux into line with the observed flux. This "solution" to the long standing solar neutrino problem stands alone among the other "nonstandard" solar model solutions in that it solves another potential problem in solar structure—the frequency splitting of low degree p-modes. Because of this it merits serious considerarion.

Even though the WIMP solution (unlike most of the other solutions) does not do obvious violence to the rest of stellar interior theory, one should expect WIMPs to affect the evolution of other stars like the sun. The best laboratories for investigating such effects are globular cluster stars: The physics involved is virtually identical to that in the sun, and these stars are found in large homogeneous samples which allow a very detailed comparison to interior theory. This comparison yields impressive agreement between observation and the standard theory. We discuss briefly here how one might search for WIMPs in globular cluster stars and what constraints might be placed on WIMP parameters.

2. OVERVIEW

For WIMPs to solve the solar neutrino problem they must carry enough energy to make the central core where the ^8B neutrinos are produced isothermal. This leads to a relatively narrow range of physical parameters. If the mass of the WIMP (m_{WIMP}) is too small, they evaporate from the sun. If it is too large, they collect

M. Caffo et al. (eds.), Astronomy, Cosmology and Fundamental Physics, 287–292.

so close to the center of the sun that the isothermal core is too small. The ability of WIMPs to carry energy through a star depends on the WIMP-nucleon cross section and the number of WIMPs present in the star. The number of WIMPs is set by the capture rate from the unseen massive halo of the Galaxy. The transport efficiency is highest if the WIMPs scatter roughly once per orbit. If the cross section for WIMP nucleon interaction is too small then many WIMPs are required to carry the necessary amount of energy. If it is too large, WIMP energy transport becomes similar to radiative diffusion. The WIMP-nucleon cross section enters the evaporation rate and capture rate as well as the efficiency of energy transport (see Spergel and Faulkner 1988, hereafter SF).

Depending on the type of WIMPs one imagines might be present in stars, the interaction with hydrogen ($\sigma_{H,WIMP}$) and helium ($\sigma_{He,WIMP}$) might be quite different. Candidate WIMPs have been suggested with $\sigma_{He,WIMP}/\sigma_{H,WIMP}$ ranging from 0–16. This unfortunately introduces yet another dimension to an already rich parameter space.

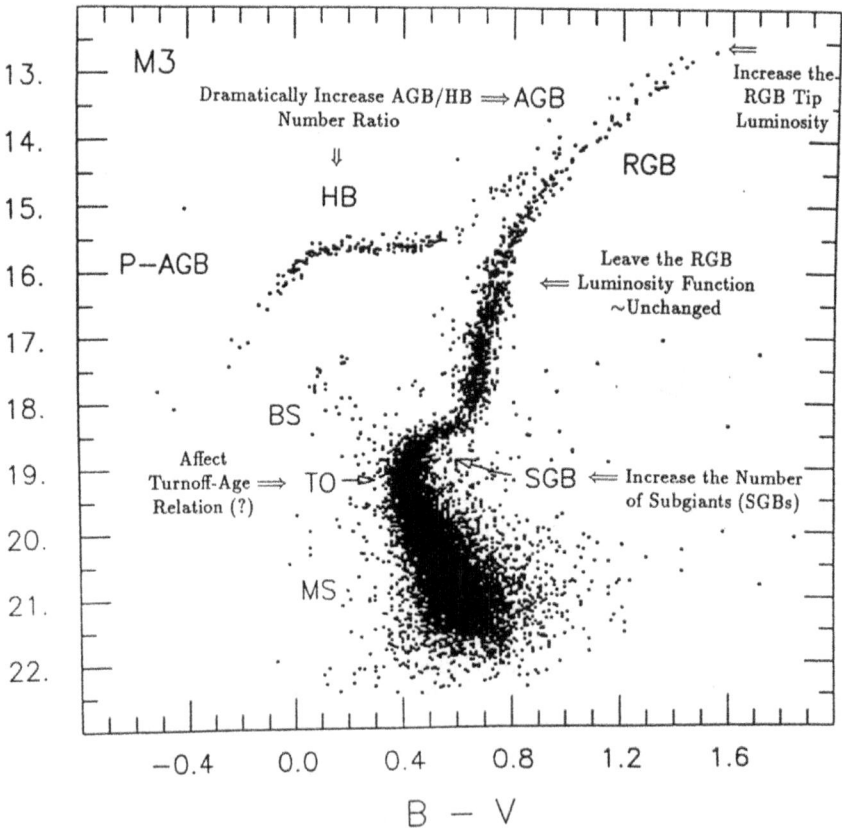

Fig. 1. A schematic indication of what WIMPs might do to the H-R diagram of a globular cluster. We acknowledge the courtesy of Buonanno *et al.* (1988) for their H-R diagram of M3.

We can identify a number of places where WIMPs might affect globular cluster stars (see Fig. 1).

• *The main sequence.* Cores would be isothermal and neutrino fluxes reduced. This is safely unobservable.

• *The subgiant branch.* Faulkner and Swenson (1988) note that WIMPy stars leave the main sequence with substantial hydrogen remaining at their centers. The hydrogen remaining in the center at turnoff will eventually burn presumably on the subgiant or lower giant branch extending the lifetime and thus the number of stars in these phases. The total amount of hydrogen burned during the subgiant branch (SGB) phase is $\sim 0.05\,M_\odot$, compared to roughly $0.03\,M_\odot$ remaining at turnoff in the isothermal cores of Faulkner and Swenson models. Thus the number of subgiants could be increased by $\sim 60\%$ over standard, non-WIMPy models. There is however no direct evidence for such an excess of SGB stars in well studied globular clusters (cf. Renzini and Fusi Pecci 1988 for a comparison of predicted and observed $N_{\mathrm{SGB}}/N_{\mathrm{HB}}$ ratios), and we conclude that existing globular cluster data certainly do not support the presence of extended isothermal cores in old main sequence stars. Note also that this is the only test for WIMPs in globular cluster stars which applies if $\sigma_{\mathrm{He,WIMP}} = 0$.

• *Near the red giant tip.* Superficially one would not expect WIMP energy tranport to be important in giants where energy transport via electron conduction is already very efficient. However, preliminary estimates of the WIMP *opacity* (in analogy to the conductive opacity) indicate that it can be considerably smaller than the conductive opacty. This coupled with the nonlocal nature of WIMP energy transport may lead to significant effects. The agreement between the standard model and helium core flash luminosity is impressive (Frogel, Cohen, and Persson 1983), and therefore additional study of WIMP effects near flash could lead to significant constraints on WIMP properties.

• *The horizontal branch.* Renzini (1987) has suggested that WIMPs would carry energy so efficiently that convection in the cores of horizontal branch (HB) stars would be suppressed. This would sharply decrease the amount of helium available during the HB phase and thus greatly decrease the HB lifetime, at variance with observed ratios of HB to asymptotic giant branch stars. Recently, SF have suggested that Renzini's arguments ignore important factors. We consider this point in the next section.

3. WIMPS IN HORIZONTAL BRANCH STARS

SF have argued that WIMPs either evaporate from HB stars or cannot transport the requisite amount of energy. From their presentation one might be led to the conclusion that WIMPs from a vast area of parameter space had no bearing on HB evolution while still solving the solar neutrino problem. We feel that their conclusions are overstated.

Our results are summarized in Fig. 2 where we show areas in the m_{WIMP} $-\sigma_{\mathrm{He,WIMP}}$ plane which are forbidden for various reasons. Our results are derived from precisely the same expressions for WIMP capture, escape, and heat transport as given in SF [their eq.s (1), (2), and (7), respectively] but applied to detailed stellar models rather than polytropes. The polytropic models used by SF are reasonably accurate, but because the model quantities sometimes appear in exponents surprizing large differences can result. We have adopted $\sigma_{\mathrm{He,WIMP}}/\sigma_{\mathrm{H,WIMP}} = 16$.

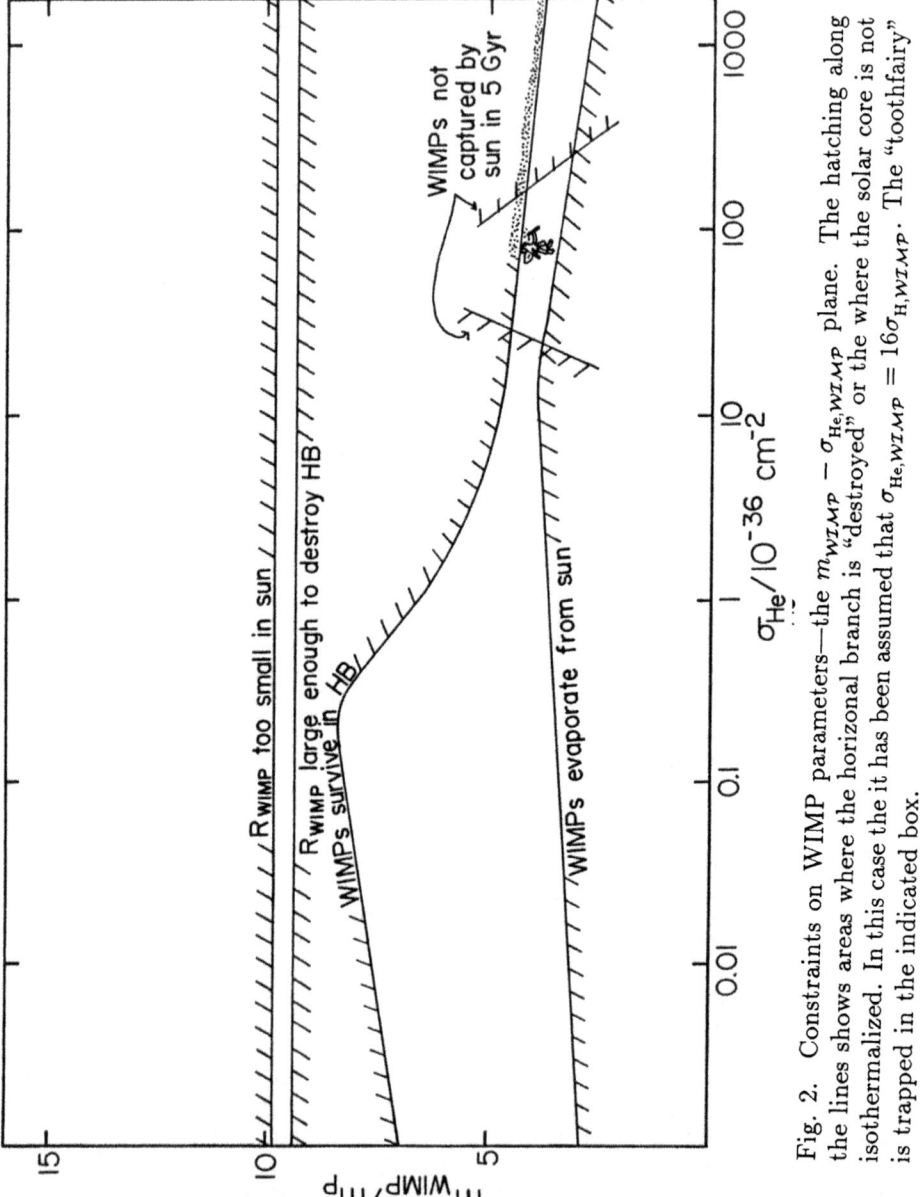

Fig. 2. Constraints on WIMP parameters—the $m_{WIMP} - \sigma_{He,WIMP}$ plane. The hatching along the lines shows areas where the horizonal branch is "destroyed" or the where the solar core is not isothermalized. In this case the it has been assumed that $\sigma_{He,WIMP} = 16\sigma_{H,WIMP}$. The "toothfairy" is trapped in the indicated box.

In Fig. 2 the two horizontal lines near $m_{WIMP} = 10\,m_p$ show the restrictions placed by the size of the sphere occupied by the WIMPs. Above the upper line the solar WIMPsphere is too small to get rid of the ^{8}B neutrinos. Below the lower line the WIMPsphere in HB stars encompasses the entire energy producing region. The essential coincidence of these two lines was what motivated Renzini's paper. The lowest line extending across the plot shows the mass below which WIMPs evaporate from the sun. Above that by a factor of 2 in m_{WIMP} on the left and 25% on the right is the HB WIMP evaporation line. Above this line WIMPs remain in HB stars for 10^8 yr and thus destroy the HB. Along (and above) the hatched part of this line WIMPs can carry more than half the core luminosity. Here we have taken the total number of WIMPs to be three times (the ratio of HB precursor life to solar life) the number required to carry the solar luminosity in a ZAMS solar model. This is not the same assumption made by SF, but we feel the sun should play by the same rules as the HB. At the shift to dots the WIMP energy tranport might fall below that necessary to suppress convection. For WIMPs to "survive" above this line there would have to be a fine tuning in the number of WIMPs—just enough to solve the solar neutrino problem, but not to many to suppress convection in the HB. We finally tie these stellar WIMPs to cosmological WIMPs by demanding that the sun capture from a halo with equivalent density of $0.01\,m_p$ cm^{-3} . (We think this is the density assumed by SF.) This leads to the two more vertical lines. In the end we find this *toothfairy* trapped in quite a small box (unless she yet escapes into another dimension or vaporizes...). However, if WIMPs are eventually found with just these properties we'll be glad to take credit for the prediction. At the moment it strikes us as bullet dodging.

4. CONCLUSIONS

Despite claims to the contrary globular cluster stars place strong limits on the physical parameters of astrophysically interesing WIMPs. More careful consideration may lead to even more constraints or more stringent ones. For example, we would have grossly underestimated the number of WIMPs if globular clusters were really to contain a large amount of indigenous WIMP dark matter (Rees 1988).

REFERENCES

Buonanno, R., Buzzoni, A., Corsi, C.E., Fusi Pecci, F., Sandage, A.R. 1988, *Globular Cluster Systems in Galaxies*, ed. J.E. Grindlay and A.G.D. Philip (Dordrecht: Kluwer), p. 621.
Faulkner, J., and Swenson, F. J. 1988, *Ap. J. Letters*, **329**, L47.
Frogel, J.A., Cohen, J.G., and Persson, S.E. 1983, *Ap. J.*, **275**, 773.
Rees, M. 1988, *Globular Cluster Systems in Galaxies*, ed. J.E. Grindlay and A.G.D. Philip (Dordrecht: Kluwer), p. 453.
Renzini, A. 1987, *Astr. Ap.*, **171**, 121.
Renzini, A., and Fusi Pecci, F. 1988, *Ann. Rev. Astr. Ap.*, **26**, 000.
Spergel, D.N., and Faulkner, J. 1988, *Ap. J. Letters*, **331**, L21. (SF)

DISCUSSION

SCIAMA: Would WIMPs in globular cluster stars appreciably reduce their ages?

ROOD: Faulkner has argued that ages might be reduced somewhat (a few 10%). This is quite plausible purely from what he finds on the main sequence. The cluster age question is quite complicated and a self consistent picture of WIMPs in cluster stars has yet to be produced. Often things tend to cancel out. I would be very surprized if even WIMPy cluster ages were consistent with $H_{\circ} = 100$, whatever Ω is.

UNKNOWN: How restrictive is your assumption that $\sigma_{He,WIMP}/\sigma_{H,WIMP} = 16$?

ROOD: If $\sigma_{He,WIMP}/\sigma_{H,WIMP} = 1$, then the permited box is a thinner in height and displaced by about a factor of 10 toward lower $\sigma_{He,WIMP}$. If $\sigma_{He,WIMP} = 0$, then they probably don't affect the HB. In that case the only possible test is using the subgiants.

IMPROVED LIMITS FROM THE GALACTIC AXION SEARCH

B. E. Moskowitz, S. De Panfilis, A. C. Melissinos,
J. T. Rogers, Y. K. Semertzidis and W. U. Wuensch
Department of Physics and Astronomy
University of Rochester, Rochester, NY 14627 U.S.A.

H. J. Halama and A. G. Prodell
Brookhaven National Laboratory, Upton, NY 11973 U.S.A.

W. B. Fowler and F. A. Nezrick
Fermi National Accelerator Laboratory, Batavia, IL 60510 U.S.A.

ABSTRACT
Cold, light axions are a leading candidate for the composition of dark matter in the galactic halo. An experiment designed to detect the microwave conversion signal from galactic axions has searched the frequency range 1.1 - 2.5 GHz. The updated limits on axion density and coupling are presented, and improvements for a possible second-generation experiment are discussed.

1. INTRODUCTION

Axions were first predicted [1] in 1978 by Weinberg and by Wilczek as the Goldstone boson of the Peccei-Quinn mechanism [2] used to explain CP conservation in strong interactions. There is one free parameter in the theory, namely m_a , the mass of the axion, which is inversely proportional to the Peccei-Quinn symmetry-breaking scale. Upper limits on m_a come from accelerator and reactor searches [3], the lifetimes of red giant stars [4] and the observed SN1987a neutrino signal [5]; a lower limit on m_a derives from the requirement that the universe not be over-closed [6]. The mass region that is still allowed may be summarized as:

$$10^{-5} \, eV \leq m_a \leq 10^{-3} \, eV \qquad (1)$$

with the lower value of 10^{-5} eV preferred for closing the universe.

Axions are a leading candidate for the galactic dark matter since they should have cooled after spontaneous symmetry breaking to form a cold, Bose gas which condensed into galactic halos. Using a simple isothermal model by Turner [7], the calculated local density is

$$\rho_a = 5 \times 10^{-25} \, g/cm^3 \qquad (2)$$

and the r.m.s. velocity of the axions is of order of the galactic virial velocity $\approx$ 10^{-3} c. The resulting narrow line in the energy spectrum is characterized by:

293

M. Caffo et al. (eds.), Astronomy, Cosmology and Fundamental Physics, 293–296.
© *1989 by Kluwer Academic Publishers.*

$$Q_a = \frac{E_a}{\Delta E_a} = \frac{m_a c^2}{\frac{1}{2} m_a v^2} \approx 10^6 \tag{3}$$

2. DETECTION

The axion couples to two photons via a triangle diagram of quarks and/or leptons; however the coupling is too weak for the free decay of axions to be observed. Using an idea of Sikivie [8], one may instead detect the conversion of the axion into a single photon in the presence of a magnetic field. For an axion mass of 10^{-5} eV, the frequency of the outgoing photon will be 2 GHz, which is in the microwave region. The interaction Lagrangian for this process is given by:

$$L_{int} = -g_{a\gamma\gamma} \ E_\gamma \cdot B_0 \ \phi_a \tag{4}$$

where $g_{a\gamma\gamma}$ is the coupling constant, E_γ is the electric field of the photon, B_0 is the applied external magnetic field, and ϕ_a is the pseudoscalar axion field. To enhance the reaction, a resonant electromagnetic cavity may be used to trap the photon. The mode which maximizes $E_\gamma \cdot B_0$, and therefore the conversion rate, is the TM_{010} mode of a right, cylindrical cavity. Using the DFS model [9], the power expected in this mode [10] for a photon frequency f_γ , a volume V, and a loaded quality factor Q_L is given by:

$$P_S = (4 \times 10^{-24} \text{W}) \left[\frac{\rho_a}{5 \times 10^{-25} \text{ g/cm}^3} \right] \left[\frac{f_\gamma}{1 \text{ GHz}} \right] \left[\frac{B_0}{6 \text{ T}} \right]^2 \left[\frac{V}{10^4 \text{ cm}^3} \right] \left[\frac{Q_L}{10^5} \right] \tag{5}$$

Since the mass of the axion, and therefore the frequency of the outgoing photon, is unknown, the cavity needs to be tuned in frequency during the search.

3. APPARATUS

The detector built by the Rochester-Brookhaven-Fermilab group is shown in Fig. 1. The magnet is a superconducting Nb-Ti solenoid with a 20 cm bore and a peak field of 6.6 T; an insert is available for boosting the peak field to 8.0 T with a bore of 15 cm. The cavity is an electro-polished, oxygen-free copper cylinder of length 40 cm. Five cavities of different diameters have been used to date to cover the frequency range 1.1 - 2.5 GHz. Tuning is accomplished by raising a single-crystal sapphire rod from the cavity via a motor drive. The signal is read through an induction pickup loop connected to a cryogenic preamplifier. The preamp is followed by a two-stage mixer and a 64-channel multi-channel analyzer which is scanned and read by a computer. The channel width is 400 Hz. A more detailed description of the apparatus may be found elsewhere [11].

4. UPDATED RESULTS

No axion signal has yet been observed to a confidence level of 4 standard deviations. The experimental limits obtained on the quantity $((g_{a\delta\delta} / m_a)^2 \rho_a)$ are summarized in Fig. 2, where the density in equation (2) has been assumed. These limits extend the mass range of previously reported results [11]. Also shown for comparison are the theoretical mass-coupling relations of the DFS model [9] and of Kim's hadronic model [12]. The plan of the Rochester-Brookhaven-Fermilab group is to complete the survey up to 6 GHz at equivalent sensitivity.

5. SENSITIVITY IMPROVEMENTS

Although the present experimental result is more than one order of magnitude away in sensitivity to $g_{a\delta\delta}$ from the popular models, improvements in the apparatus are certainly possible. The limit in sensitivity for a given signal-to-noise ratio [S/N] is given in MKS units by:

$$\left[\left(\frac{g_{a\delta\delta}}{m_a} \right)^2 \rho_a \right] = \frac{1}{2\pi \epsilon_0 c^2} \left[\frac{S}{N} \right] \frac{k_B T_N}{0.7 \cdot f_\delta B_0^2 V} \left(\frac{df/dt}{Q_a Q_L} \right)^{1/2} \tag{6}$$

where k_B = Boltzmann's constant, T_N = total system noise temperature, df/dt = the sweeping rate in Hz/sec; the above assumes that $Q_L \leq Q_a$.

Clearly large gains can be made by raising the magnetic field B_0 while maintaining a large bore volume V, and by lowering the noise temperature T_N. Improvement of Q_L above the achieved value of 10^5 helps only slightly. Finally the sweep rate df/dt can in principle be made arbitrarily small, but in practice is limited by the 1 - 2 year lifetime of the search. A possible second-generation experiment that has B_0 = 15 T, V = 20 liters, T_N = 4 K (2 K HEMT amplifiers + 2 K pumped cryostat) and $Q_L = 10^6$ would improve the present limit on $g_{a\delta\delta}$ by a factor of ten. Such an experiment would test the upper range of interesting astrophysical models.

REFERENCES

[1] S. Weinberg, Phys. Rev. Lett. **40**, 223 (1978); F. Wilczek, Phys. Rev. Lett. **40**, 279 (1978).
[2] R. Peccei and H. Quinn, Phys. Rev. Lett. **38**, 1440 (1977).
[3] L. Vuilleumier, et al., Phys. Lett. **101B**, 341 (1981); A. Zehnder, Phys. Lett. **104B**, 494 (1981); C. Edwards, et al., Phys. Rev. Lett. **48**, 903 (1982).
[4] G. Raffelt and D. Dearborn, Phys. Rev. **D36**, 2211 (1987).
[5] G. Raffelt and D. Seckel, Phys. Rev. Lett. **60**, 1793 (1988); M. S. Turner, Phys. Rev. Lett. **60**, 1797 (1988).
[6] J. Preskill, et al., Phys. Lett. **120B**, 127 (1983); L. Abbott and P. Sikivie, Phys.

Lett. **120B**, 133 (1983); M. Dine and W. Fischler, Phys. Lett. **120B**, 137 (1983).

[7]   M. S. Turner, Phys. Rev. **D33**, 889 (1986).

[8]   P. Sikivie, Phys. Rev. Lett. **51**, 1415 (1983) and **52**, 695 (1984).

[9]   M. Dine, *et al.*, Phys. Lett. **104B**, 199 (1981).

[10]   L. Krauss, *et al.*, Phys. Rev. Lett. **55**, 1797 (1985).

[11]   S. De Panfilis, *et al.*, Phys. Rev. Lett. **59**, 839 (1987); B. E. Moskowitz, Nucl. Instr. and Meth. **A264**, 98 (1988).

[12]   J. E. Kim, Phys. Rep. **150**, 1 (1987).

FIGURES

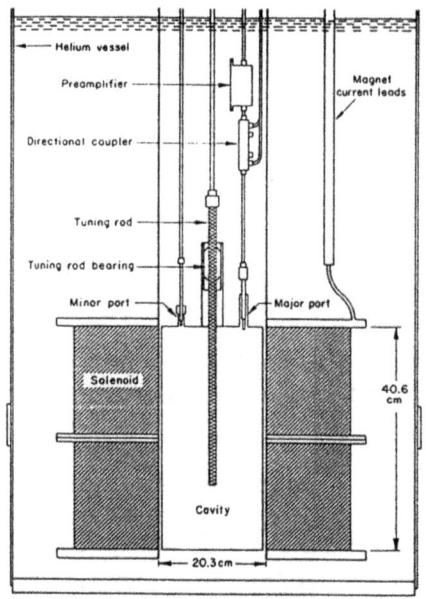
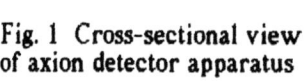

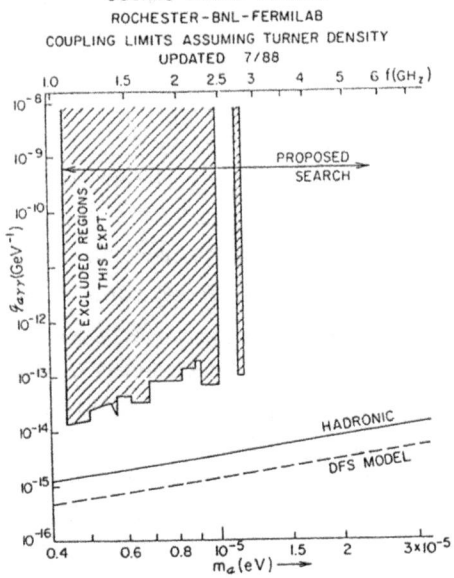

Fig. 1 Cross-sectional view of axion detector apparatus

Fig. 2 Experimentally excluded regions in mass-coupling phase space, assuming eq. (2); also shown are theoretical relations from references [9] and [12].

LOW TEMPERATURE DETECTORS FOR NEUTRINOS AND DARK MATTER

L. GONZALEZ-MESTRES and D. PERRET-GALLIX

L.A.P.P.
B.P. 909, 74019 Annecy-le-Vieux Cedex, France

Abstract: We briefly review the physics motivation and the present status of cryogenic detectors, with special emphasis on applications at the cross-disciplinary frontier between particle physics and astrophysics.

1. INTRODUCTION

Cryogenic detectors have been extensively used in astrophysics to detect low energy photons, a well-known application being the study of possible anisotropies of microwave background radiation [1]. Their appearance in particle physics has been motivated by several proposals to study low energy neutrinos and, possibly, detect dark matter candidates. The search for magnetic monopoles with induction loops is by now a running program [2], and a beautiful search for cosmic axions is being presented at this Symposium by B. Moskowitz. We review here several developments that are being carried on, focusing mainly on applications to neutrino mass measurements, low energy solar neutrino detection, the study of rare decays such as neutrinoless double β, and the search for galactic WIMP (weakly interacting massive particles, that may provide the Universe missing mass). The use of low temperature devices is expected to bring higher sensitivity and energy resolution, for the following reasons:

1) The lower energy of elementary excitations. A single phonon has energy $E < \hbar\omega(\text{Debye}) < 0.1$ eV, and will further degrade to thermal phonons in bolometers or to quasi-particles in a superconducting system. Quasiparticles in superconductors have excitation energies in the range 10^{-6}-10^{-3} eV, while the semiconductor energy gap is of the order of 1 eV and visible photons produced in scintillators have an energy of around 3 eV. Other ways to obtain a spectrum of very low energy excitations at low temperature can be found, a recent example being the splitting of a Kramers doublet by a magnetic field [3], which appears to lead to $\sim 10^{-6}$ eV excitation energy in well suited spin systems.

2) The fast decrease of specific heats:

$$c \ (\text{lattice}) \simeq a \ T^3 \qquad \{1a\}$$

$$c \ (\text{superconducting electrons}) \simeq b \ \exp(-\Delta/kT) \qquad \{1b\}$$

where Δ is the superconducting energy gap. This makes possible thermal detection, where a small increase in temperature can be detected with the help of special thermometers. An appealing feature of thermal detection is that the energy resolution is free from the energy-dependent statistical component, due to the very large number of elementary excitations involved (thermal phonons) and is only limited by the read-out noise and thermal fluctuations

3) The lower thermal noise may be crucial for the detection of very small signals when using cold electronics. This may often determine the sensitivity threshold.

297

M. Caffo et al. (eds.), Astronomy, Cosmology and Fundamental Physics, 297–302.
© 1989 by Kluwer Academic Publishers.

In addition, some low temperature effects provide specific signals (e.g. change in magnetization) or amplification effects (e.g. metastable phase transition in superconductors, latent heat release or quasi-particle multiplication). The wide variety of superconductive materials makes low temperature techniques attractive when active targets are needed.

Low temperature detectors are still at the stage of feasibility studies, but have already provided some interesting results. We present, here, a brief introduction to these new techniques and report on the present status of the art.

2. CRYSTAL CALORIMETERS (BOLOMETERS)

The specific heat of an insulating crystal at low temperature is dominated by lattice vibrations (phonons). An energy deposition E converted into heat will lead to an increase in temperature $\Delta T \simeq E/C$ (C = heat capacity), that can be detected with a resistive thermometer (thermistor) implanted on the crystal. If a careful design allows to minimize Johnson noise from the thermistor and electronic read-out, energy resolution will be given by thermal fluctuations of the phonon system:

$$\Delta E_{rms} \simeq \zeta \ (C/k)^{1/2} \ kT \propto T^{5/2} M^{1/2} \qquad \{2\}$$

where k is the Boltzmann constant and M the mass of the crystal. The heat capacity of the thermistor has been neglected, which may not be correct for small bolometers. C/k is actually the number of phonon modes and the coefficient ζ depends on the details of detector architecture, but is often estimated to be in the range 1.5 - 2 [4]. In such an optimal scenario, it is expected that a sizeable increase in detector mass can be compensated by a moderate decrease in temperature, according to the $T^{5/2} M^{1/2}$ law.

Work with small bolometers has been pursued for years, having particularly in mind the measurement of the electron neutrino mass from the ^{3}H Kurie plot, which can be made with detectors smaller than 1 mm^3 . If a better experiment than the existing ones is to be performed, energy resolution of 10 eV or less on 18.6 keV electrons (tritium β decay) should be achieved. The performances already reached by such small bolometers are rather encouraging: a diamond bolometer (0.25 mm^3) at T = 1.3 K reached FWHM energy resolution of 36 keV on 5.5 MeV α particles [5], and the choice of temperature (rather high for bolometric techniques) was only due to limited equipment. At lower temperatures T $\simeq$ 100 mK, much better results have been achieved with a Si crystal (500$\times$ 500$\times$ 25 μm^3) reaching 17 eV FWHM energy resolution on 6 keV γ's [6].

More recently, the study of large bolometers has been undertaken with some success. Using a 0.7 g germanium bolometer at T $\simeq$ 44 mK, the Milano group [7] has obtained 1% energy resolution on α particles from a ^{228}Ra source in radioactive equilibrium with its daughters. Furthermore, a previous high flux irradiation allowed to implant daughter nuclei in the crystal, which produced satellite peaks shifted upwards by 100 keV, value compatible with the nucleus recoil energy. As the implanted nuclei decayed, the satellite peaks disappeared and only the single peaks from external α's remained. These authors conclude that the bolometer was sensitive to nucleus recoil energies of 100 keV, as expected from the 50 keV energy resolution.

A new promising idea is the so-called "magnetic bolometer" [3]. There, half of the deposited heat is converted into very low energy spin excitations and a small change in magnetization can be detected by a SQUID read-out. Although further work on the subject is required, a 30 keV energy resolution on 5.5 MeV α's is claimed by the authors on a 7.35 g sapphire absorber with a 135 mg YAG:Eb^{3+} magnetic bolometer implanted on the sapphire.

In some applications, time resolution may be important for event identification and background rejection. Large bolometers are not by themselves fast detectors, due mainly to the slow propagation of phonons and heat. As an example, the bolometer from the Milano group gives rise times of the order of 200 μs and the one from [3] exhibits a 200 ms rise time. Perhaps bolometry should in some cases be combined with other detection principles (light propagation?) in order to introduce a fast component in the detector response.

To date, the main physics motivation for the development of large bolometers (100 g — 1 Kg) lies in neutrinoless double β decays [8], where energy resolution is a crucial ingredient for background rejection, and dark matter searches through nuclear coherent scattering [9], where sensitivity to energy deposition below 1 keV appears to be required. More difficult, because of background rates, appears to be a solar neutrino experiment based on $\nu-e^-$ scattering using several tons of bolometric detector [10]. The theoretically expected performance from bolometers can be as good as $\Delta E \approx 100$ eV on a 1 Kg absorber cooled at $T = 10$ mK, but several problems must be solved before actually demonstrating such a sensitivity.

3. SUPERCONDUCTING TUNNELING JUNCTIONS (STJ)

Superconductors provide the unique possibility of producing diodes with about 10^{-3} eV current carrier excitation energy. Then, a statistical $N^{1/2}$ law (Poisson distribution) for energy resolution leads again to exceptional performances for the detection of low energy particles. In a STJ with a small bias voltage, quasiparticles and holes tunnel across a thin insulating layer separating two superconducting samples, and the current can be read with conventional low noise pre-amplifiers. Usually, STJ are made of two metallic films separated by the insulating layer, and are not expected to be massive detectors. However, new ideas have recently emerged (e.g. quasiparticle trapping, which also provides multiplication [11]) to incorporate bulk superconducting specimens.

The bias voltage creates a thermal current $I_{th} \propto \exp(-\Delta/kT)$ that can be lowered by working at low reduced temperature ($t = T/T_c$). In order to prevent Cooper pair tunneling (DC-Josephson current), a magnetic field parallel to the oxide barrier is applied. An incoming particle will excite mainly electrons of energy much larger than Δ, but these electrons will later relax emitting phonons. At $t \ll 1$, phonons mainly excite quasiparticles, which can then tunnel across the junction or recombine.

The expected energy resolution in a STJ is:

$$\Delta E_{rms} \simeq (f E \varepsilon)^{1/2} \qquad \{3\}$$

where f is the Fano anti-correlation factor ($0.1 < f < 1$) and ε the effective quasiparticle excitation energy ($\varepsilon > \Delta$). Potentially, a Sn-SnO-Sn detector with $f \simeq 1$ and $\varepsilon \simeq \Delta \simeq 0.6$ meV, should reach 0.1% energy resolution on 6 keV γ's. Experimental results have not yet obtained this limit, but the SIN group has achieved [12] 48 eV FWHM energy resolution on the 5.89 keV ^{55}Mn K_α peak, whereas the Garching group [13] reports 88 eV resolution, determined from the energy difference between the K_β (6.49 keV) and K_α peaks. A typical signal rise time from existing STJ is of the order of 15 μs.

Apart from the detection of low energy γ rays, a possible use of small STJ would be neutrino mass measurements [14], but if larger devices can be made, they could be used [11] to detect low energy solar neutrinos through the ^{115}In Raghavan's reaction [15]. Furthermore, STJ provide an interesting read-out for crystal phonon detectors, where ballistic phonons would be converted into quasiparticles by a STJ read-out installed on the surface of the crystal. Since ballistic phonons propagate along the main crystallographic axis, a STJ read-out would extract information on the position of the event inside the crystal [10, 16].

The development of STJ also brings interesting information concerning the practical possibility to detect low temperature excitations. The observed energy resolution of 40 eV for 6 keV γ's (obtained from the 48 eV after subtracting 24 eV uncorrelated baseline noise) corresponds to an effective energy excitation of 50 meV ($\approx 80 \Delta$) per detected carrier, if the $N^{1/2}$ law is applied. The 10% energy resolution of a NaI(Tl) scintillator for a 100 keV γ indicates in turn about 0.16

keV ($\approx$ 55 $E_{\gamma \text{ visible}}$) per photoelectron. Furthermore, if STJ are used as phonon detectors, high frequency phonons in the range 0.01-0.1 eV are above the effective quasi-particle excitation energy.

4. SUPERHEATED SUPERCONDUCTING GRANULES (SSG)

A type I superconductor with low enough κ (the Ginzburg-Landau parameter) can exhibit metastable states, due to the positive normal-superconducting interface energy. In particular, a superconducting sample may remain in this state for values of the external magnetic field larger than the critical field H_c. This metastable phase is called superheating, and has been obtained for pure metal microspheres of 1-100 μm diameter. It was proposed long ago [17] to use SSG as a particle detector in the form of a suspension of small microspheres into some dielectric material, with a read-out of current loops oriented in the plane normal to the applied magnetic field H_0. Then, the energy E released by an incident particle would be converted into heat and originate a fast, irreversible transition of one or several granules. This phase transition could be detected through the disappearance of the Meissner effect, which as the Meissner magnetic dipole decreases will induce a current in the read-out loops. Although the basic principle of grain flipping under irradiation was already demonstrated in the original paper, many problems remained unsolved. In particular, the signal produced by single grain flips was not detectable in real time due to the small size ($\simeq 2\mu$m diameter) of the grains used. Furthermore, the fact that low energy particles have a short path in matter led to the flip of only a few grains and made impossible to obtain a proportional response. More recently, progress has been made in the SSG real time read-out and granules of sizes 10-400 μm have been shown to be sensitive to all kinds of low energy sources, down to 6 keV γ's [18]. The observed sensitivity can be understood on theoretical grounds and, when extrapolated to very small grains, gives rather encouraging figures: 1 μm diameter In grains at T = 200 mK would be sensitive to about 300 eV with 80% efficiency, whereas Ga grains of the same size cooled down to 100 mK would achieve a similar performance for 4 eV energy deposition. Unfortunately, for realistic detection purposes, the above result is not by itself sufficient. Two examples:

1) SSG have been proposed [19] as an indium detector due to their potentiality in segmentation (crucial for background rejection). A X-Y current loop read-out would allow to segment a 4 ton indium detector into 10^7 elementary cells, with only 10^5 electronic channels. However, such an instrumentation would require 5mm $\times$ 1m current loops, which makes extremely difficult the detection of 116 keV secondaries.

2) Dark matter detection through nucleus recoil energy encounters an even more severe difficulty, since only single grain flips are expected. We therefore have only a threshold detector, without any energy resolution.

To cure both diseases, we have recently proposed a new operating principle, based on the concept of "amplification by thermal micro-avalanche" [20]. Metastability implies the possibility of a positive latent heat released by the superconducting to normal phase transition. Theoretical calculations for an isolated single sphere give: q^ℓ (latent heat) > 0 for $T < 1/3 \, T_c$ at H_0 (applied magnetic field) $= H_c$ (the sphere is in a metastable state due to the demagnetization coefficient). Then, the flip of a single granule can release latent heat which, together with the deposited energy, will be dispersed in the detector. If heat exchanges between granules through the dielectric material are efficient enough, new flips will be produced which in turn will release more latent heat. If such a scenario can be achieved with sufficiently small grains (1 μm in diameter), a signal $\Delta\Phi$ proportional to the deposited energy is expected even for a nucleus recoil, which restores energy resolution. Furthermore, the appearance of extra flips will lead to an important amplification effect (one or two orders of magnitude), which may solve the basic problems for a [115]In experiment. Time resolution is expected to be in the range 10-100 ns using small granules.

Other uses would then be possible: double beta decays [21], X-ray imaging [22], dark matter searches through inelastic scattering with a [119]Sn target [20]. Furthermore, the dielectric material may then be used as an active target (hydrogen for dark matter searches [20])... However, if experimental evidence for global avalanches already exists [23], precise measurements of heat exchanges in the detector are required to better evaluate the real performance of the micro-avalanche effect. Another crucial problem is large scale production of very small grains. Production of 25 μm average size tin granules has already been demonstrated [24] at a rate of 5Kg/hour using a 40 KHz ultrasonic atomizer. A new development is underway in order to adapt the existing procedure to higher ultrasonic frequencies (up to $\approx$ 5 MHz).

An independent application, using large granules, would be the detection of magnetic monopoles [25], where the flux tube injected by the monopole would destroy the superconductivity of one or several granules. The main advantage of the SSG technique would then be a comparatively large signal (several orders of magnitude larger than in induction experiments), a much better background rejection due to the large size of the grains used, and a measurement of the monopole speed and direction.

5. OTHER DETECTORS

Energy deposited in superfluid ^{4}He at low temperature (100 mK) would produce rotons (Δ/k = 8.65 K). A 200 keV electron from neutrino scattering is expected to produce $\approx$ 10^8 elementary excitations, which will propagate ballistically in all directions. Some will hit the surface of the liquid and evaporate a sizeable number of helium atoms, that may be detected by bolometric techniques [26]. No experimental result exists yet on this technique, but a development is being carried on at Brown University. Even more ambitious is a proposal from the Lancaster group [27], where superfluid ^{3}He (cooled below 1mK) would produce $\approx$ 10^7 quasiparticles per deposited eV. Unfortunately, such quasiparticles are neutral and no known technique exists to detect them in a reasonably small number, but obviously further work is required.

Several new ideas on the possible use of devices operating below 1 mK have been put forward by T.O. Niinikoski [28], who was able to obtain interesting bounds on dark matter from measured heat leaks in Cu adiabatic nuclear demagnetization refrigerators. This is a beautiful illustration of the room that is still left for innovation in the field of low temperature detectors.

6. CONCLUSION

Low temperature detectors are a new growing field in astrophysics and particle physics instrumentation, and tend to cover the cross-disciplinary domain between both sciences. The expected cost of such detectors looks moderate, as compared with the main experimental activities currently in progress. However, the feasibility of cryogenic astro-particle experiments remains in most cases to be demonstrated. The progress made in the last years is particularly encouraging, and further effort is required to carry on the basic studies. The results of the next two years developments are likely to be crucial to decide on the future of this original and exciting research activity.

Further information on cryogenic detectors can be found in the Proceedings of two recent European Workshops [29], [30].

302

References

[1]   See, for instance, P. de Bernardis in ref. [30].
[2]   See, for instance, J. Incandela, same reference.
[3]   M. Buhler and E. Umlauf, Europhys. Lett. 5, 297 (1988).
[4]   S.H. Moseley, J.C. Mather and D. Mc Cammon, J. Appl. Phys. 56, 1257 (1984).
[5]   N.Coron et al. , Nature 314, 681 (1984).
[6]   D. Mc Cammon et al., Proc. 18th Conference on Low Temp. Phys., Kyoto 1987; Jap. J. Appl. Phys. 26 Suppl., 26.
[7]   A. Alessandrello, D.V. Camin, E. Fiorini and E. Giuliani, Phys. Lett. 202, 611 (1988).
[8]   E. Fiorini, in ref [29].
[9]   M.W. Goodman and E. Witten, Phys. Rev. D31, 3059 (1985).
[10]   See, for instance, B.Cabrera in "Superconductive Particle Detectors",, Ed. A. Barone, World Scientific Pub. 1987.
[11]   See, for instance, N. Booth in ref. [29].
[12]   See, for instance, D. Twerenbold and W. Rothmund and A. Zehnder in "Superconductive Particle Detectors".
[13]   Th. Peterreins, F. Probst, F. von Feilitzsch, and H. Kraus in "Superconductive particle detectors".
[14]   F. Cardone, ref. [30].
[15]   R.S. Raghavan, Phys. Rev. Lett. 37, 259 (1976).
[16]   Th. Peterreins, F. Probst, F. von Feilitzsch and H. Kraus in [30].
[17]   H. Bernas et al. , Phys. Lett. 37, 359 (1967).
[18]   For a review of recent results and possible uses, see:
        L. Gonzalez-Mestres and D. Perret-Gallix, in "Superconductive Particle Detectors" and K. Pretzl, same reference.
[19]   Rapport de la Jeune Equipe "Neutrino-Indium" du CNRS (ENS Paris, LPC College de France, Ecole Polytechnique, DPhPE Saclay, IP Strasbourg, LAPP Annecy), January 1987.
[20]   L. Gonzalez-Mestres and D. Perret-Gallix, Proceedings of the Moriond Meetings on "Neutrinos and Exotic Phenomena" (January 1988) and "Dark Matter" (March 1988), Ed. Frontieres.
[21]   P. Andreo, J. Garcia-Esteve and A.F. Pacheco, in ref [29].
[22]   C. Valette, G. Waysand, Orsay report (1976).
[23]   F. von Feilitzsch et al. in [29].
[24]   EXTRAMET, Zone Industrielle, Annemasse (Haute-Savoie), France.
[25]   L. Gonzalez-Mestres and D. Perret-Gallix, Proceedings of Underground Physics 85, Ed. Il Nuovo Cimento (1986).
[26]   R. Lanou, H. J. Maris and G. Seidel, ref. [29].
[27]   G. Pickett, ref. [30].
[28]   T.O. Niinikoski, in preparation and Proceedings of the "Rencontre sur la Masse Cachée dans l'Univers et la Matière Noire", Annecy July 8-10, 1987. Ed. L. Gonzalez-Mestres and D. Perret-Gallix, (Annales de Physique, France).
[29]   "Low Temperature Detectors for Neutrinos and Dark Matter", Ed. K. Pretzl, N. Schmitz and L. Stodolsky, Springer-Verlag 1987.
[30]   "Low Temperature Detectors for Neutrinos and Dark Matter-II", Ed. L. Gonzalez-Mestres and D. Perret-Gallix, Frontieres 1988 (in preparation).

IS THERE DARK MATTER IN ELLIPTICAL GALAXIES?

G. Bertin, R.P. Saglia, and M. Stiavelli
Scuola Normale Superiore
56100 Pisa
Italy

ABSTRACT. The presence and distribution of dark matter in elliptical galaxies is expected to be constrained not only by the properties of the overall force field (as could be traced by an HI rotation curve), but also by the density and velocity dispersion profile of the luminous stellar component, since stellar orbits are determined by the *total self- consistent* potential. In order to study these dynamical constraints, we have constructed self-consistent, two-component, spherically symmetric models of elliptical galaxies, under the assumption that the distribution functions for the dark and luminous components are of the same general analytic form.

Previously we have studied a simple, one-parameter, sequence of equilibrium models of elliptical galaxies (f_∞) that incorporates the essential features of dissipationless collapse. Under the assumption of a constant mass-to-light ratio, these models have been shown to accurately fit both the photometry and the velocity dispersion of eight bright ellipticals, performing even better than the empirical de Vaucouleurs $R^{1/4}$ law (Bertin et al. 1988). This result shows that dark matter is *not* required in order to explain the observed luminosity of elliptical galaxies nor the present stellar kinematical data.

However, currently available (stellar) velocity dispersions do not allow us to make precise statements on the amount of dark matter in ellipticals since they are subject to sizable uncertainties and extend usually out to the half luminosity radius R_e only. In addition, in a few cases cold gas is detected and the derived rotation curves extend well beyond R_e (see, e.g., NGC 5666, Lake et al. 1987); these give strong indications that dark halos are indeed present. In other cases, evidence for dark matter derives from extended X-ray emission, which seems to require the existence of a deep potential well (e.g., see Fabian and Thomas 1987).

In order to set specific dynamical constraints on the existence and extent of dark matter in ellipticals, we have constructed self-consistent, two-component, spherically symmetric models by assuming that the relevant distribution functions are both of the f_∞ form. Such an assumption is not unreasonable since the dark halo was probably formed via dissipationless collapse as well as the luminous stellar system. Note that we are studying *fully* self-consistent two-component models, in contrast to the more common procedure of computing a self-consistent solution for one component while keeping the potential of the second component fixed. The

M. Caffo et al. (eds.), Astronomy, Cosmology and Fundamental Physics, 303–307.

distribution functions for the stars, f_L, and for the dark matter, f_D, are:

$$f_L = A_L(-E)^{3/2} exp[-a_L E - c_L J^2/2] \qquad (1a)$$

$$f_D = A_D(-E)^{3/2} exp[-a_D E - c_D J^2/2] \qquad (1b)$$

where E and J^2 are respectively the energy and the angular momentum square per unit mass. Different choices for the form of f_L and f_D are also under investigation. Self-consistent solutions are obtained by solving the Poisson equation:

$$\nabla^2 \Phi = 4\pi G(\int f_L d^3v + \int f_D d^3v). \qquad (2)$$

Each solution is characterized by four dimensionless parameters: a_D/a_L, A_D/A_L, $\Psi = -a_L \Phi(0)$, and c_L/c_D. In turn, these models can be described in terms of the total mass ratio $\overline{\mu} = M_D/M_L$, of the half-mass radius ratio $\overline{\rho} = R_{MD}/R_{ML}$, of the rotation flatness parameter $\overline{V} = V(r = 3R_{ML})/V_{max}$, and of the dispersion flatness parameter $\overline{S} = \sigma(r = 3R_{ML})/\sigma(0)$. Here V is the rotation curve defined by $V^2 = rd\Phi/dr$ and $\sigma = \sqrt{\sigma_{rr}^2 + \sigma_{\theta\theta}^2 + \sigma_{\phi\phi}^2}$ is the total unprojected velocity dispersion of the luminous component. The correspondence between $\overline{\mu}$, $\overline{\rho}$, $\overline{V}$, $\overline{S}$ and a_D/a_L, A_D/A_L, Ψ, c_L/c_D is non-trivial. Understanding such correspondence is the first goal of an extensive survey of two component models that we have undertaken. Previous experience with one component models and some analytic estimates of various quantities are a guide in the numerical investigation. So far we have collected data on about 600 models with $0.05 \leq a_D/a_L \leq 1$, $0.1 \leq A_D/A_L < 50$, $6 \leq \Psi \leq 18$ and a wide range for the self-consistent values of c_L/c_D. Basically, we expect that photometric and kinematical data in specific objects will be able to isolate well defined regions of the parameter space. Further constraints should also arise from stability arguments, but these are very hard to assess at this stage.

When the dark component is sufficiently diffuse, models are obtained characterized by a projected luminosity distribution dependent on the depth of the central gravitational potential well Ψ but fairly independent of the amount of dark matter (see Fig. 1). The velocity dispersion and especially the rotation curves, instead, depend appreciably on the amount of dark matter present and show a much slower decrease with radius than observed in one-component models. This is illustrated in Fig. 2, where we compare velocity dispersions and rotation curves of two models characterized by the same values of the dimensionless gravitational potential Ψ but by a different amount of dark matter ($\overline{\mu} = 0$ and $\overline{\mu} \simeq 4.$, respectively). In Fig. 3 we show, for the model with $\overline{\mu} \simeq 4.$, the value of the integrated mass to luminous mass ratio, as a function of radius. This quantity is independent of the mass-to-light ratio referred to the luminous component alone. Note that inside $R \simeq 2.5 R_e$ the amount of dark mass approximately equals $2 M_L$. This is roughly equivalent to what is expected in spiral galaxies. For comparison with the well known "maximum disk analysis" in spiral galaxies (van Albada and Sancisi 1986), in Fig. 4 we illustrate for the same model the decomposition of the rotation curve into the two separate contributions of dark and luminous matter; from this we can better appreciate how the two components may conspire to produce a fairly flat rotation curve. In spiral galaxies, the rising part of the rotation curve is a key factor in the determination of the maximum disk solution. In contrast, the inner part of $V(r)$ is generally not observed in elliptical galaxies even when gas data are available.

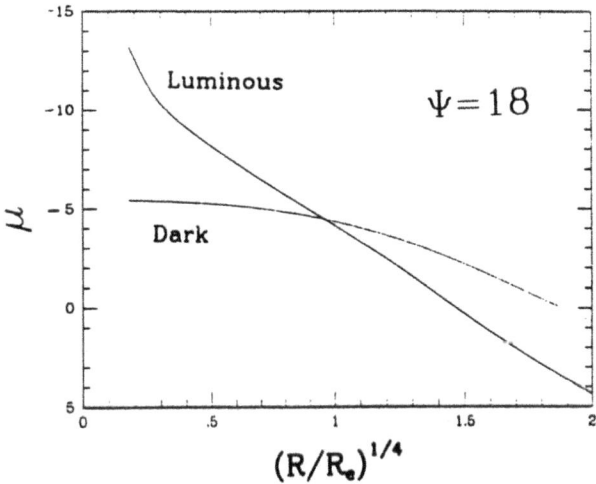

Figure 1: Projected density distribution of the luminous component (solid line) of the model with $\Psi = 18$ and $M_D \simeq 4M_L$. The departures from the $R^{1/4}$ law are smaller than 0.2 mag over a range of 10 mag. We also show the dark projected density distribution (thin line) for comparison. This model , characterized by $\bar{\rho} = 0.15$, $\bar{V} = 0.79$, $\bar{S} = 0.55$, will be further illustrated in the following Figures.

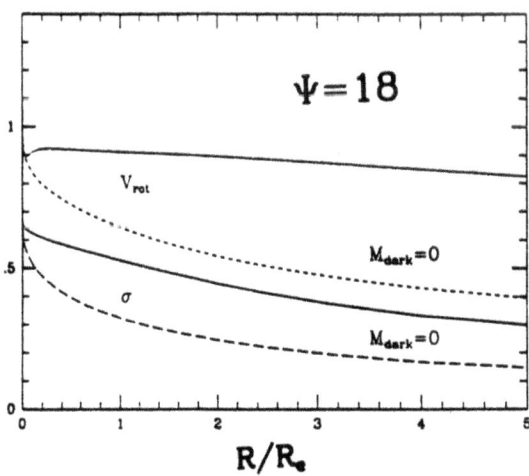

Figure 2: Projected velocity dispersions (long dashed and solid lines) and the rotation curves (short dashed and thin lines) for the models with $\Psi = 18$ and $\bar{\mu} = 0$ and $\bar{\mu} \simeq 4$. The radius is normalized to the half projected luminosity radius R_e. Velocities are given in arbitrary units. Note that the inner peak in V need not be easily recognized by observations.

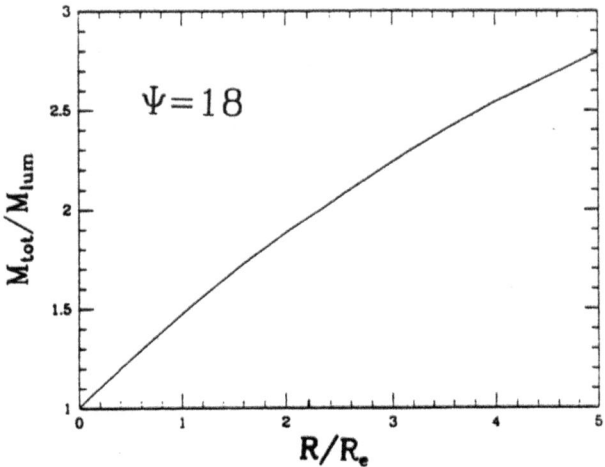

Figure 3: Projected total mass to projected luminous mass ratio vs R/R_e. This mass ratio is proportional to M/L once the mass-to-light ratio of the luminous component is specified.

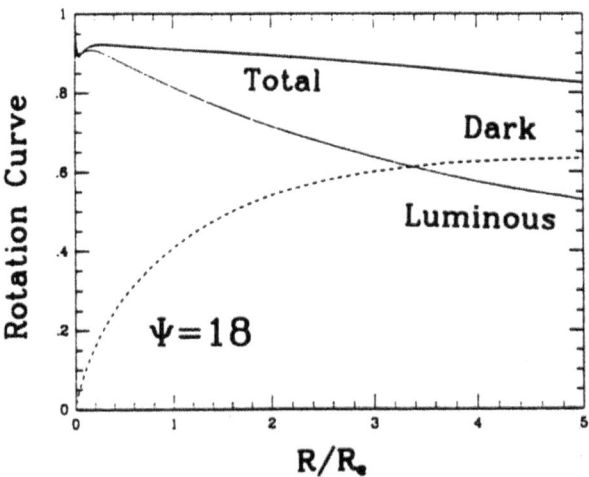

Figure 4: The rotation curve $V = \sqrt{V_D^2 + V_L^2}$ (solid line) is separated into the contributions V_L of the luminous component (thin line) and V_D of the dark compo-nent (short dashed line).

REFERENCES

T. S. van Albada, and R. Sancisi, 1986, *Phil. Trans. R. Soc. London A*, **320**, 447
G. Bertin, R.P. Saglia, and M. Stiavelli, 1988, *Astrophys. J.* **330**, 78
A.C. Fabian, and P.A. Thomas, 1987, in *IAU Symp.* **127**, Reidel, Dordrecht p. 155
G. Lake, R.A. Schommer, J.H. van Gorkom, 1987, *Astrophys. J.* **314**, 57

DISCUSSION.

Lynden-Bell: Is it not true that the strongest evidence for halos about ellipticals comes from the X-ray, especially the M87 data?

Bertin: Yes, I agree that X-rays in special cases give strong evidence. Still it is remarkable how well one can do in many elliptical galaxies, such as NGC3379, with constant M/L ratio. The point of view I presented is to try to move in a quantitative and conservative way from the center outward.

OPTICAL STUDIES OF THE SUPERNOVA 1987A

L. Woltjer
European Southern Observatory
Karl-Schwarzschild-Str. 2
D-8046 Garching bei München
Federal Republic of Germany

SUMMARY

Optical studies of SN 1987A have shown that the luminosity at maximum was unusually faint; that some time after maximum the luminosity started to decline exponentially in a way which was not very different from what would have been expected if 0.07 solar masses of Ni^{56} had been synthesized to subsequently decay into ^{56}Co and ^{56}Fe; and that the corresponding amount of ^{56}Co appears actually to have been detected spectroscopically. The identification of other products of nucleogenesis is in progress.

INTRODUCTION

Several hundred supernovae have been detected in galaxies. Photometric and spectroscopic studies have led to the recognition of two types. Those of type I are found in all types of galaxies, have maximum visual luminosities within a narrow range around 1×10^{10} times the luminosity of the sun $(L_\odot)$ and have spectra devoid of hydrogen lines. Type II supernovae are only found in galaxies where massive stars are still being formed, have maximum luminosities within a large spread (factor of five) around 0.2×10^{10} $L_\odot$ and have spectra with strong hydrogen lines. Type II supernovae generally have been associated with the end point of the evolution of relatively massive (> 6-8 solar masses) and therefore short-lived (few million years) stars. The nature of type I supernovae is less clear and different subgroups may have very different origins; some may be associated with the explosion of white dwarfs which have been pushed over the Chandrasekhar limit by mergers in binary systems. The frequency of occurrence of supernovae in large spiral galaxies is of the order of a few per century with type II perhaps twice as frequent as type I.

During the last thousand years five supernovae (AD 1006, 1054, 1181, 1572, 1604) have been seen in our own galaxy. All five are nearer than 4 kpc to the sun, as is CAS A the remnant of a supernova which exploded around 1700, but which has not been seen with certainty. It is interesting to note that two of the six (1181, CAS A) were very faint

309

($\lesssim 1 \times 10^8$ $L_\odot$). Since such faint objects are more easily missed than brighter ones, they could actually be quite numerous (Woltjer, 1985). In the Large Magellanic Cloud - a satellite galaxy of our own with a luminosity about one tenth of ours - some forty supernova remnants have been detected, primarily by their X-ray emission which results from the interaction between the supernova ejecta and the interstellar gas in the galaxy. From these the supernova frequency in the Large Magellanic Cloud has been estimated as one or two per thousand years (e.g. Hughes et al., 1984). The appearance of a supernova in the LMC thus is a rare event and we have been lucky to witness SN 1987A, which was so brilliant that many observations could be made, which it is still impossible to obtain for other, more distant supernovae.

SUPERNOVAE OF TYPE II

When a massive star evolves, its interior tends to become denser and hotter. As a consequence nuclear reactions become possible with increasing Coulomb barriers. The H→He transformation is followed by He→C→O...Si... iron group elements. At this point no further nuclear energy can be liberated, while the star continues to lose energy by radiation. Upon further contraction of the interior the pressure is limited owing to neutrino cooling or cooling associated with the breakup of iron peak elements by energetic gamma rays. As a consequence the pressure gradient becomes insufficient to balance the gravitational forces and the core collapses - in many cases to form a neutron star. The envelope will also lose its support and fall in. At some moment, however, it will bounce against the core and an outwards moving shock is produced which may be further energized by neutrino absorption and by nuclear burning. Explosively nuclear reactions occur in the inner parts of the envelope which terminate with the formation of ^{56}Ni. While theoretically the outward propagation of the shock is still only partly understood, it apparently causes the ejection of the envelope with speeds of several thousands of km/sec for the bulk of the material, and higher speeds for the more tenuous outer part. Behind the shock the temperature is high and as a consequence much of the thermal energy in the envelope is in the form of radiation.

The envelope cools by expansion and by the diffusive loss of radiation. The radiative thermal energy varies with the inverse of the radius in the case of a homologous adiabatic expansion, while the expansion velocity is determined by the gross energetics of the event and does not depend too much on the detailed structure of the envelope. If the envelope is very extended it takes a long time for the expansion loss to become important and much radiation can diffuse out before this. If the envelope is more compact, expansion losses become dominant and less energy is left to be radiated. While this reasoning may be a bit too simple, it makes it plausible that to obtain the higher supernova luminosities red giant envelopes have been invoked. The star that became SN 1987A, however, was blue and therefore of relatively small radius; not surprisingly it turned out to be of low luminosity.

THE LIGHT CURVE OF SN 1987A

The star catalogued by Sanduleak (1970) as -69 202 was a bluish supergiant, without important variations in its luminosity, of about 2.5×10^4 $L_\odot$. Within some hours after the core collapse (as inferred from the neutrino burst) its brightness increased more than hundredfold to peak at 1×10^8 $L_\odot$ three months later. The supernova has been observed in many different wavelengths bands (figure 1) and a crude fit to a black-body spectrum is possible. Shortward of 0.5 micron the flux is well below the black-body curve because of absorption by various atoms and ions of iron and other metals. Around 10 microns there appears to be an excess which may be due to dust heated by the initial supernova flash. Integrating under the spectral curves we may obtain the total (bolometric) luminosity as a function of time (figure 2). The initial phases depend on the detailed hydrodynamical processes taking place for which models may be constructed which represent the observations rather well (Arnett, 1988). After four months a more regular exponential decline sets in.

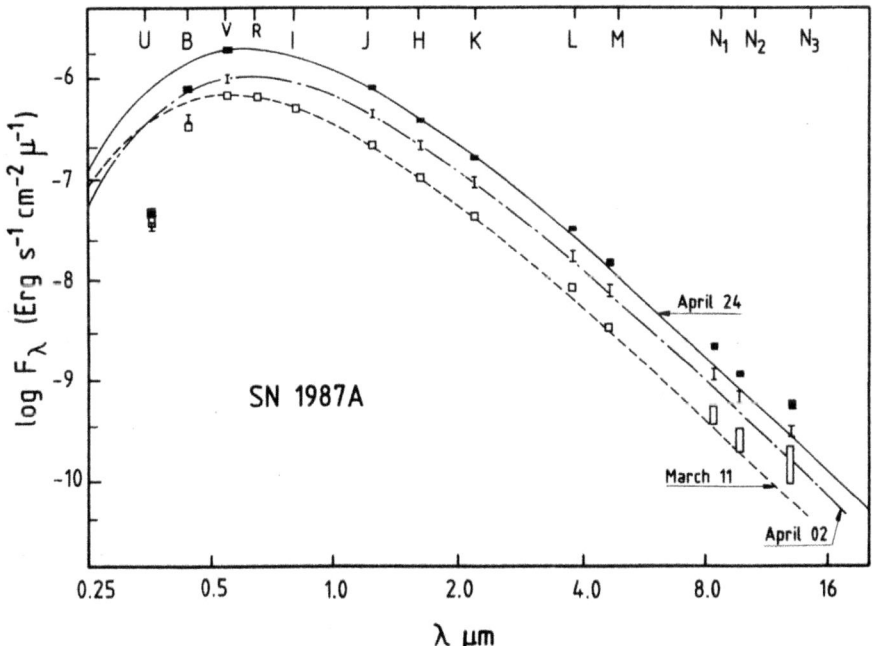

Figure 1: Fits of black-body curves to the photometric data obtained at ESO (Bouchet, 1988). Shortward of 0.5 micron metal absorption depresses the flux, while around 10 microns an excess is seen which could perhaps be due to heated dust at some distance from the supernova.

312

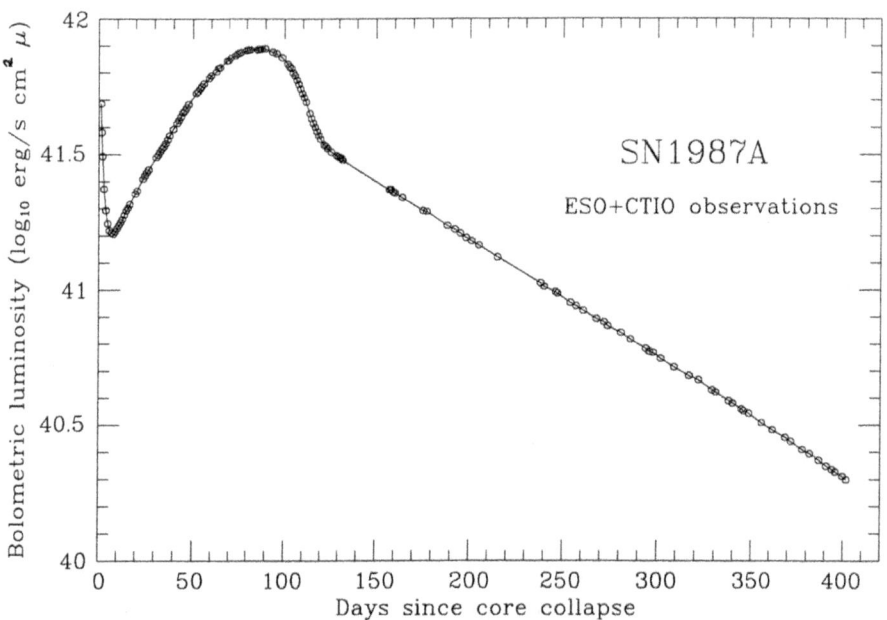

Figure 2: Evolution of the bolometric luminosity of SN 1987A from data obtained at ESO and at Tololo (Bouchet and Suntzeff, 1988). The best fit to the data between 150 and 300 days is with a half life of 70 days.

We have already noted that the envelope cools by radiative losses and by adiabatic expansion. If this were all the supernova would decline very rapidly after these four months. The fact that it does not shows that compensating heat sources must be available, most probably either radio-active decay of ^{56}Ni or pulsar based.

^{56}Ni is the plausible end result of explosive burning at the base of the envelope. It decays (by electron capture) with a half life of 6 days to ^{56}Co. The latter decays (by electron capture and positron emission) with a half life of 77 days to ^{56}Fe. The decay of ^{56}Ni being so rapid, little of it is left after a few months, and its main effect is to make the envelope temperature in the first month somewhat higher than it would have been if shock heating were the only factor. The ^{56}Co decay should be important much longer and allow the envelope to continue radiating after it has become more transparent and the initial thermal energy has vanished. If the energy of all of the ^{56}Co decay products is thermalized and converted quickly to radiation which can escape from the envelope, the luminosity in these later phases should be equal to the release rate of radio-active energy. In fact, between 200 and 300 days the half life of the luminosity output is estimated by different observers to be between 70 and 80 days, remarkably close to the ^{56}Co

half life of 77 days. While it will be of much importance to establish
the bolometric light curve with more precision than has been possible so
far and while the light curve may contain components (like the 10 micron
excess) which are unrelated to the cobalt decay, the relatively close
agreement between the two numbers is highly suggestive.

After further expansion the envelope becomes so tenuous that not
all the γ or e^+ emitted in the decay process will be absorbed and
thermalized and the luminosity could be expected to decline more
rapidly. There is, in fact, some weak evidence for a steepening of the
light curve after 300 days.

From the luminosity of the supernova during the exponential phase,
the original amount of ^{56}Ni may be estimated to have been 0.07 times the
mass of the sun. The X-ray data discussed elsewhere in this volume are
compatible with this result provided the ^{56}Ni has been mixed through a
good part of the envelope - presumably by Rayleigh-Taylor type
instabilities.

A new born pulsar could also provide an important energy input into
the envelope. While a more decisive test of the possible role of a
pulsar should come somewhat later during the evolution of the envelope,
we note that while the 77 day half life of the luminosity decline is a
specific prediction of the ^{56}Ni scenario, in the case of the pulsar it
could have been very different depending on the parameters
characterizing its rotation and magnetic field.

SPECTROSCOPIC RESULTS

An infrared emission line at 10.53 microns, attributed to Co^+
appeared in spectra of SN 1987A some 100 days after the explosion. It
reached a maximum 150 days later and subsequently declined. The most
plausible interpretation is that the cobalt was situated deeper down in
the envelope and became fully visible only after the envelope became
transparent. Because of the decay of the ^{56}Co, the line subsequently
became weaker again. According to Danziger et al. (1988) at day 280
there were 0.0044 solar masses of Co^+ and at day 400 about 0.0020. Since
Co^+ should be the dominant stage of ionization, this should also be the
total amount of cobalt. Extrapolating back to the time of the explosion
we see that this corresponds to 0.05 respectively 0.07 solar masses of
^{56}Ni, remarkably close to what had been inferred from the light curve.
Clearly the detection of cobalt in about the right amount provides
strong support for the radio-active models.

Further evidence is provided by the line of Fe^+ at 1.64 micron.
Again most of the iron should be Fe^+. Estimating the temperature from
the models, Danziger et al. infer at day 280 the presence of 0.06 solar
masses of iron. With standard elemental abundances for the Large
Magellanic Cloud there should have been no more than 0.003 solar masses
of iron before the explosion in a 5 solar mass envelope. Adding also the
cobalt, we see that about 0.061 solar masses of Ni^{56} should have been
synthesized, again in remarkable agreement with expectation.

Although the overall picture seems to be remarkably coherent some
caveats should be noted. First of all the identification of the broad
lines is not always unambiguous since an expansion velocity of 3000

km/sec corresponds to a line broadening equal to 0.02 wavelengths. In the case of the Co^+ line, S^{+++} might provide an alternative identification which is, however, less plausible because of the higher ionization conditions which are required. The line identified with Fe^+ could in principle also be due to un-ionized Ni. Apart from the problems of line identification, uncertainties in the temperatures could influence the derived abundances (in particular for Fe^+), while it also is not yet clear if the envelope is fully transparent at the relevant wavelengths.

Other abundance anomalies have been inferred. There is some evidence from a presupernova spectrum for nitrogen enhancement, which is confirmed by the appearance of strong narrow nitrogen lines in the ultraviolet some months after the explosion (Fransson et al., 1988). The fact that these lines have widths less than 30 km/sec shows that they are not formed in the supernova envelope, but in matter lost by the presupernova in a more gentle outflow and ionized by photons from the supernova. Presumably this enhanced N is the result of CNO cycling in the presupernova. More interestingly the early supernova spectra gave indications for overabundances of Ba and Sr (Williams, 1987) elements synthesized by slow neutron capture; most probably these have been synthesized in the presupernova. Other indications of nuclear processing have been found in the abundances of C and O. It will take a long time to sort out all these effects, especially since in the crowded parts of the spectra only a very detailed modeling of the envelope can give reliable abundances (Lucy, 1987).

The SN 1987A has given us for the first time a chance to study in detail the nucleogenetic processes responsible for the formation of many of the elements. Continuing observations will be needed for many years to obtain a complete picture.

Most of the results obtained for SN 1987A are due to the combined efforts of many scientists at ESO and elsewhere. I am particularly indebted to P. Bouchet and J. Danziger for providing me with the most recent data.

REFERENCES

Arnett, W.D., 1988: Proc. B. Strömgren memorial symposium, to be published.

Bouchet, P., 1988: Private communication.

Bouchet, P. and Suntzeff, N.V., 1988: Private communication.

Danziger, I.J., Bouchet, P., Gouiffes, C. and Rufener, F., 1988: Proc. Yamada Conf., to be published.

Fransson, C., Cassatella, A., Gilmozzi, R., Panagia, N., Wamsteker, W., Kirshner, R.P. and Sonneborn, G., 1988: Astrophys. J., to be published.

Hughes, J.P., Helfand, D.J. and Kahn, S.M., 1984: Astrophys. J. **281**, L25.

Lucy, L., 1987: ESO Conference Proceedings **26**, 417.

Sanduleak, N., 1970: Cerro Tololo Contr. 89.

Williams, R.E., 1987: Astrophys. J. **320**, L117.

Woltjer, L., 1985: Supernovae, their Progenitors and Remnants (G. Srinivasan and V. Radhakrishnan, eds.; Indian Ac.Sci. Bangalore), p. 172.

DISCUSSION.

Sarkar: Could you comment on the possibility that the SN1987a progenitor was a member of a binary system?

Woltjer: The possibility of an invisible binary companion has been discussed, but there is no positive evidence in its favor.

NEUTRINO ASTROPHYSICS: ITS BIRTH AND FUTURE
(as seen by KAMIOKANDE)

M. Koshiba
CERN, Geneva, Switzerland
Tokai University, Japan

ABSTRACT. The birth of observational neutrino astronomy/astrophysics is described and its future possibilities are discussed.

1. Introduction

A newly born branch of basic science eagerly awaits the active participation of young scientists and this talk is meant to encourage especially them to take part in the exploration of this NEU-astrophysics by pointing out that it was born only in 1987. I think this statement can be made despite the existence of the pioneering work of Davis[1] in the US on the solar ^{8}B neutrinos, which gave rise to the so-called Solar Neutrino Problem (SNP); i.e. the observed neutrino flux which is about 1/3 of that expected from the Standard Solar Model SSM.[2] This is because one needs the arrival direction and time of the signal in order to construct an astronomy and, in addition, the spectrum to construct an astrophysics. The Davis' experiment using the inverse beta decay, ^{37}Cl to ^{37}Ar, to detect the solar neutrinos did not satisfy these requirements. The observation of solar ^{8}B neutrinos by KAMIOKANDE-II[3] uses the elastic scattering of these neutrinos off electrons for their detection and hence it is a real-time, directional, and spectral observation. The observation by the same experiment[4], and by the IMB Collaboration[5], of the neutrino burst from the supernova SN1987a was based on the reaction $\bar{\nu}_e + p \rightarrow e^- + n$, and it is real-time and spectral, though not directional.

We have learnt a lot about our Universe through the long history of optical astrophysics, e.g. the universal expansion and/or the chemical abundance. The post-war development of radio astronomy has shown, among other things, the existence of the cosmic microwave background at 2.7 K, thereby making the Big Bang origin of our Universe quite plausible. The X-ray astronomy has given some evidence for neutron stars and possibly for black holes. The more recent infrared astronomy is beginning to tell us about the birth of stars. However, all these signals are electromagnetic waves and, as such, interact rather strongly with matter. This means that only the information on the thin surface of stellar objects and/or on diffuse gaseous objects can be conveyed by these signals.

If we were to probe deep inside such a stellar object, be it a star or a galaxy, it is clear that we would have to use a much more penetrating signal interacting only through the weak force with matter. Such a signal exists in nature and it is the neutrino.

317

M. Caffo et al. (eds.), Astronomy, Cosmology and Fundamental Physics, 317–326.

318

There is one more thing I should mention here. That is the possibility of neutrino-flavour oscillation in the case of non-zero mass neutrinos. The mass eigenstates are, in general, not identical with the flavour eigenstates; and therefore there arises, even in the absence of material media, the quantum-mechanical evolution of an initially non-existent flavour state as time goes on. This possibility was used by Pontecorvo[6] to explain the SNP; the initially produced ν_e's were converted into other flavour states, ν_μ and ν_τ, which cannot produce signals in the ^{37}Cl experiment of Davis. When the neutrinos are traversing matter there is a more efficient way of flavour changing. This was pointed out by Mikheyev and Smirnov[7] following the formalism of Wolfenstein.[8] It is quite similar in mechanism to the K_S^0 regeneration by K_L^0 in matter.

2. Solar neutrino observation by KAMIOKANDE (see Fig. 1)

The most difficult task in this observation is obviously that of background rejection. There are four main causes of background involved here. The first is the Rn dissolved in the

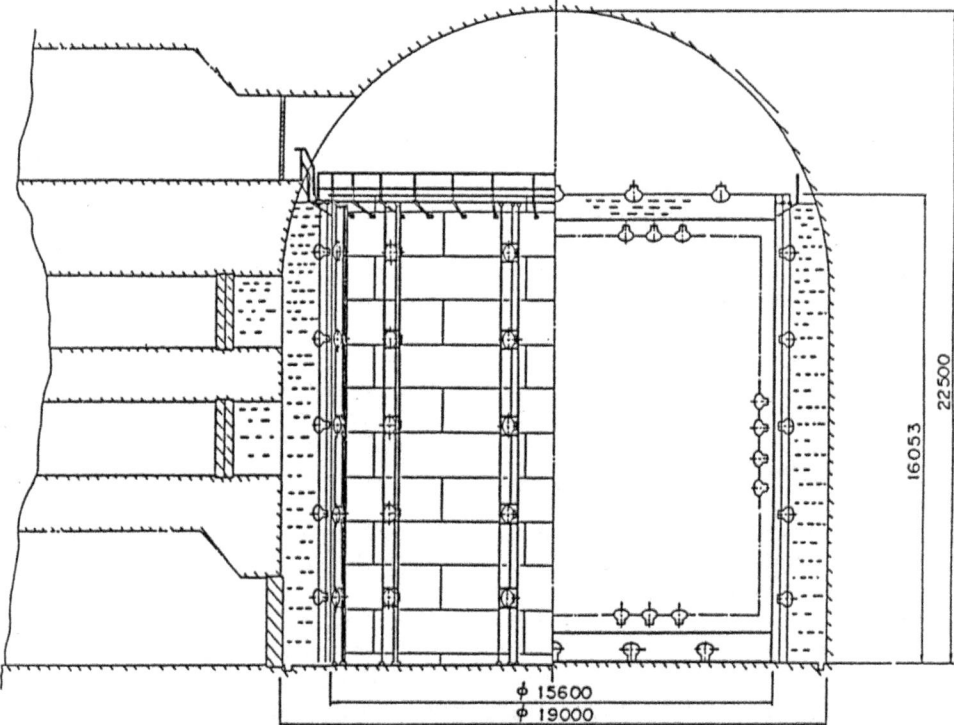

Figure 1.   Schematic drawing of the KAMIOKANDE detector. By observing the intensity as well as the arrival time of the Cherenkov light produced by fast particles one can reconstruct the event vertex, the direction of motion, and the energy. Note that 20% of the entire inner surface of the inner detector is covered by a photocathode, resulting in 3.4 photoelectrons per 1 MeV energy lost by electrons.

water. The second is the low-energy delayed β-emitter among the fragments produced by high-energy μ interactions in the water, and the third is the γ's and the neutrons from the surrounding rocks. The fourth cause is the U and Th contamination in the water. The Rn has as one of its descendants ^{214}Bi, which gives 3.26 MeV β. This low-energy β triggers the 7.5 MeV threshold with about 10^{-5} probability because of the finite-energy resolution. In the early days of operation, once in a while, 10 t of fresh water were added to compensate for the loss due to evaporation. This addition of 0.3% water resulted in the jump of trigger rate as can be seen in Fig. 2. The 3.8 d decay of the ancestor nucleus ^{222}Rn is clearly seen. This cause of background was taken care of by making airtight the inner water-circulating system and by providing an airtight reservoir tank for replenishing the water.

The second cause, ^{12}N and ^{12}B, etc., from the oxygen breakups, had to be deleted in the off-line analyses by using the time- and space-correlations with the parent μ. At this underground depth of 1000 m the cosmic-ray muon rate is 0.37 Hz and we cannot just kill the detector after every passage of muons; some of the possible radioactive fragments of oxygen, such as ^{11}Be, ^{15}C, and/or ^{16}N, have half lives in seconds. This analysis of fitting the muon-induced low-energy events with respect to the time dependence and the energy spectrum gave us one means of carrying out the absolute energy calibration. The absolute energy calibration was supplemented by the observation of μ–e decay electrons and finally by the observation of Compton electrons from the known monochromatic γ-rays resulting from the thermal neutron capture by Ni-foil. The thermal neutrons were from a ^{252}Cf fission source immersed in the water. The resulting absolute energy calibration is better than 3%.

The third cause, γ's and n's from the surrounding rocks, has been dealt with by installing Cherenkov anticounters of water, > 1.4 m thick, all around the main detector, and also deleting the peripheral region, < 3.14 m from the top and < 2.0 m from the other

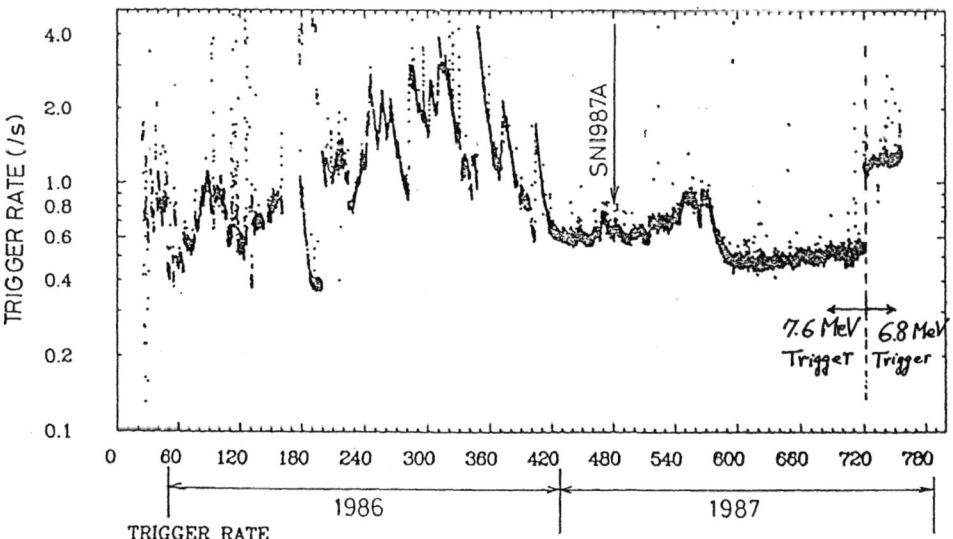

Figure 2. The improvements in trigger rate of low-energy events. See text.

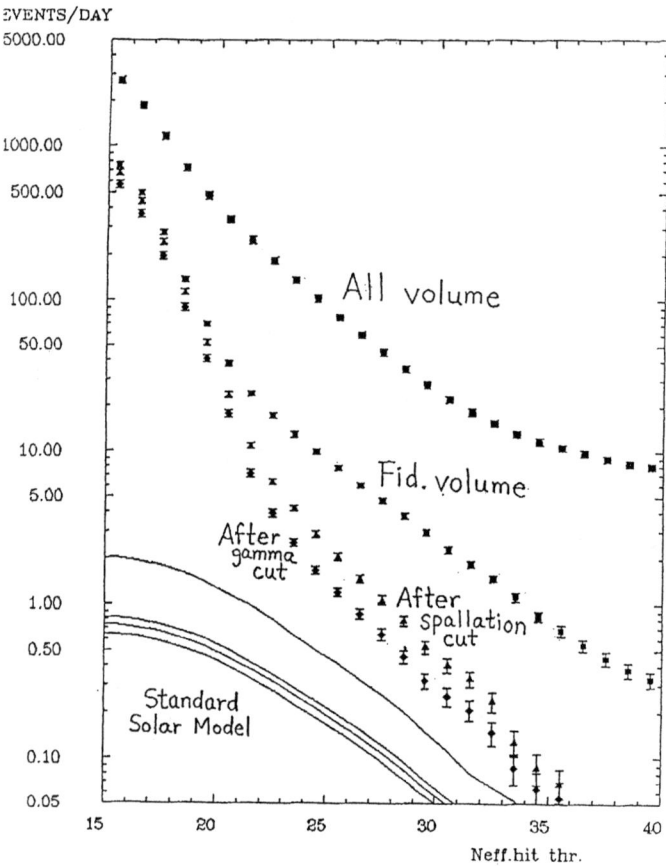

Figure 3.   The reduction of the data to obtain the solar ^{8}B neutrinos. See text.

boundaries of the inner photomultiplier tube (PMT) arrays, in the analysis of solar neutrino events; the resulting fiducial mass is 680 t. There still seem to remain some γ-events, even after the fiducial cut, and an additional cut was made for those excess events in the direction normal to and in the vicinity of the container walls.

To overcome the last cause of background, U and Th contamination in the water, we installed columns of ion-exchange resins and the contamination was reduced from the original 10^{-9} to the present 10^{-12} level and is still improving gradually.

The effect of these background cuts is schematically shown in Fig. 3. The remaining events are shown in Fig. 4 in the form of the angular distribution with respect to the sun–earth direction, which were then fitted by the maximum-likelihood method to estimate the number of solar neutrino events in the sun–earth direction region. Figure 5 shows the latest result, still preliminary, of the solar neutrino observation by KAMIOKANDE; one sees that the observed flux is away from the null flux and is definitely smaller than the SSM expectation.

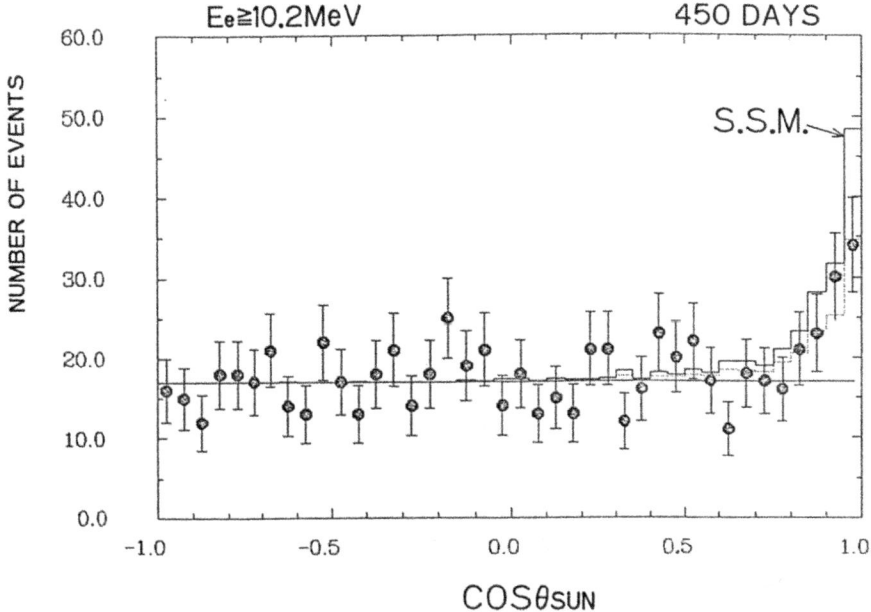

Figure 4.   The use of the directionality of the solar neutrino events.

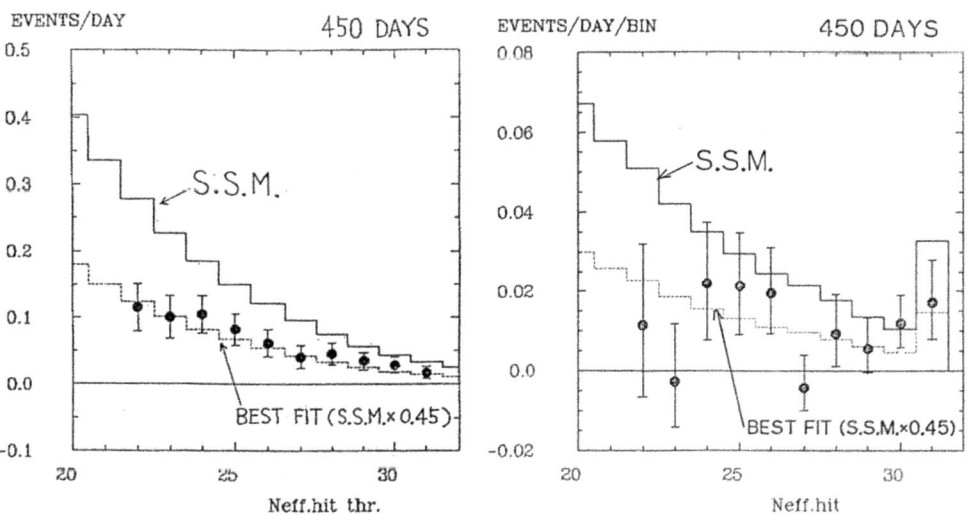

Figure 5.   The latest result, still preliminary, on the solar ^{8}B neutrinos.
a) The integral energy spectrum of the recoil electrons produced by the solar ^{8}B neutrinos.
b) The same results in the differential form.

322

3. Neutrino burst from supernova 1987a

The observation of the neutrino burst from this supernova has already been published and discussed in detail[9]. Therefore, I will just show the raw data of the low-energy events as a laser printout (see Fig. 6). Here I wish to point out that with this type of detector the

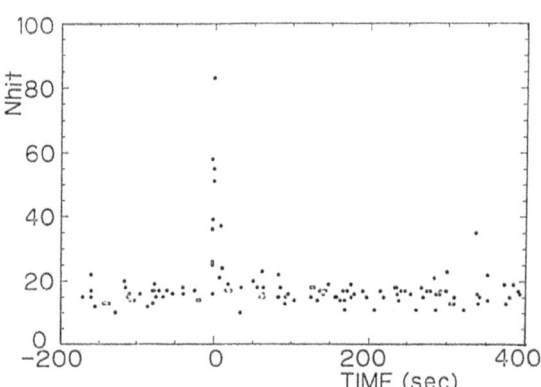

Figure 6. The neutrino burst signal from SN1987a as observed by KAMIOKANDE. The N_{hit} on the vertical axis is the number of photomultipliers which gave the light signals and to a good approximation represents the event energy.

discovery of the supernova signal is quite easy and that if the computer was on the site to analyse the data real-time it would be possible to give warning of a supernova explosion to the optical observatories in the world so that they can wait for the early phase flare-up in ultraviolet (UV). Note that with a next-generation detector such as SuperKAMIOKANDE[10] the number of the direction-indicating events, ($\nu_e + e^-$) elastic scattering, in the first 10 ms, would be about 200 for a supernova at the Galactic centre, and this is enough to point out the direction with 2° accuracy, and also that the subsequent observation of a neutrino spectrum as a function of time does not depend on the optical visibility of the object, which can be very bad in the vicinity of the Galactic centre.

4. Search for point sources of high-energy neutrinos[11]

The search for high-energy ν_μ's from the direction of SN1987a was made with negative results so far. A search was also made for point sources of high energy ν_μ's, again with negative results. The 90% CL upper limit fluxes obtained for the neutrino-produced upward-going μ's from the direction of possible sources are, in units of 10^{-13} cm^{-2} s^{-1}: 1.2 for SN1987a, 0.99 for Cyg.- X-3, 1.44 for Her.-X-1, 0.66 for Crab, 0.63 for Geminga, 0.56 for SS433, 0.70 for Gal.Cen., 0.34 for Vel.-X-1, and 0.26 for LMC-X-4. These values are still preliminary and the final results will be published shortly by the KAMIOKANDE Collaboration. It is clear, however, that detectors with a sensitive area orders of magnitude larger will be needed to detect the high-energy neutrinos from point sources.

5. Search for Heavy Relic Particles in the sun

The search for possible neutrino signals from the direction of the sun have been made in the contained events as well as in the upward-going μ events. The preliminary results so far obtained by KAMIOKANDE are the 90% CL excluded mass regions (in GeV): for Majorana ν, > 76; for Dirac ν, 10–20 and > 60; for scalar ν_e, 3–25; for scalar ν_μ, > 3; for scalar ν_τ, 4–25; and for the photino no constraint as yet. The final results will appear shortly[12].

6. Hint of a positive result on neutrino oscillation

Recently a careful analysis of the atmospheric neutrinos by KAMIOKANDE revealed some interesting results[13]. From the early days of operation, both KAMIOKANDE and IMB (a large water Cherenkov detector in the US), noticed something inexplicable at the level of 2 standard deviations. Although the overall flux of atmospheric neutrinos has a theoretical uncertainty of about 20%, the number ratio R, or at least its lower limit, of $(\nu_\mu + \bar{\nu}_\mu)$ to $(\nu_e + \bar{\nu}_e)$ can be rather well estimated. That is, disregarding the difference in the energy, the ratio R is 2 because π-decay gives $1\nu_\mu$ and 1μ, and this 1μ when it decays gives $1\nu_\mu$ and $1\nu_e$ together with 1e. The energy consideration will raise R somewhat and so does the existence of undecayed μ's. On the other hand, the K decay into an electron will increase the number of ν_e's, thus reducing R. However, the branching ratio of this decay mode is small and even if one assumes the equal production of π and K the reduction of R is less than 5%.

Among the totally contained events, the percentage of those events with a μ–e decay electron signal is essentially determined by this ratio R, although the inclusion of the inelastic events with π–μ–e decay will dilute the effect somewhat. The IMB experiment expected (34 ± 1)% and observed (26 ± 3)%, while KAMIOKANDE expected (77 ± 2)% and observed (61 ± 7)%. This seeming reduction of ν_μ events was a 2.6σ effect in IMB and 2.2σ in KAMIOKANDE. The IMB Collaboration quoted[14] as the cause of this effect the following three possibilities; i) the assumed input ν_μ/ν_e being too large, ii) the detection efficiency of μ–e decay electrons being overestimated, or iii) some as-yet-unaccounted-for physics. The possibility (i) is very unlikely, as explained above, and the possibility (ii) is also unlikely, especially in KAMIOKANDE where the detection threshold has been 7.5 MeV for electrons. The effect will be more clearly visible in the pseudo-elastic charged-current events. We thus look at the sample of single-ring contained events, 0.2 to 2.0 GeV/c, and analyse whether the secondary is an electron or a μ. The μ–e identification was made with 98% efficiency. The result is (No. of μ)$_{data}$/(No. of μ)$_{MC}$ = (0.59 ± 0.064), a 6σ effect; while (No. of e)$_{data}$/(No. of e)$_{MC}$ = (1.12 ± 0.18), which is quite normal. The comparisons are shown graphically in Fig. 7. In order to get rid of the 20% uncertainty in the overall flux of atmospheric neutrinos, we take the ratio of the two types of events and obtain (No. of μ/No. of e)$_{data}$/(No. of μ/No. of e)$_{MC}$ = (0.56 ± 0.09), a 5σ effect.

One of the promising explanations of this observed anomaly can be sought in the neutrino oscillation, in vacuum and/or in matter, and the full analysis along this line will appear shortly[15]. The consistency check with the upward-going μ data of the same experiment is being made. If this were the correct explanation, the responsible parameter range of neutrino oscillation would be Δm^2 of 10^{-2} to 10^{-3} (eV)^2 and $\sin^2 2\theta$ of 0.3 or larger, with a hint of preference for ν_μ–ν_τ oscillation over that of ν_μ–ν_e.

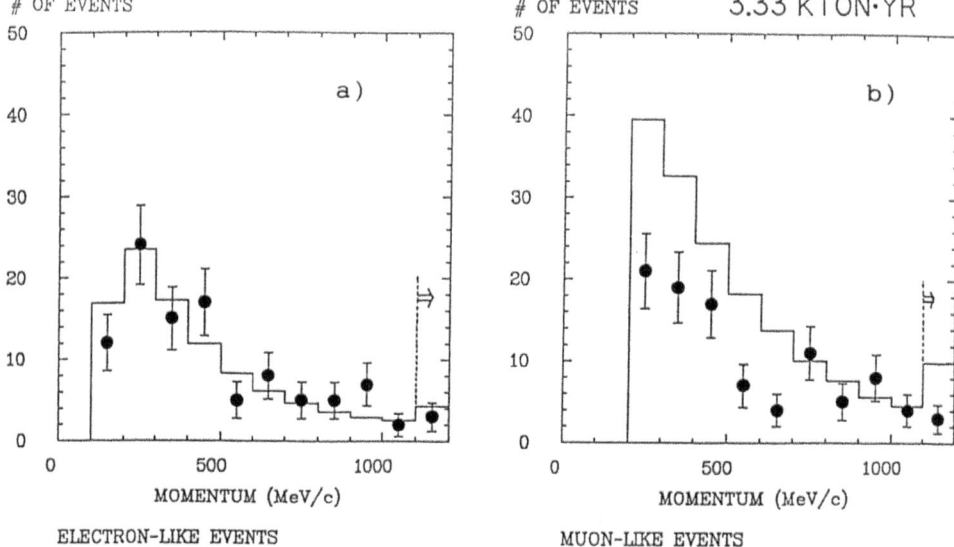

Figure 7. The momentum spectra of electron-like events (a) and of muon-like events (b) as observed by KAMIOKANDE.

7. Future outlook

The proposed SuperKAMIOKANDE is an upgrade of KAMIOKANDE in the sense that it has 25 times the fiducial mass and 2 times the photon-detection efficiency. It is described in detail in other papers and I will just mention here again that with this detector one can expect 200 (ν_e + e) elastic-scattering events as well as about 4000 ($\bar{\nu}_e$ + p) events for a supernova occurring near the centre of our Galaxy so that the directional accuracy of 2° and the spectrum change with time would be obtained, not to mention the accurate observation of solar neutrinos and the search for Dark Matter. It is to be noted also that, based on the solar neutrino result of KAMIOKANDE, the variation of the central temperature of the sun can be monitored with an accuracy of better than 1% per week by this SuperKAMIOKANDE.

Now we go on to the near-future prospect for high-energy neutrino astronomy. The proposal for the Lake Experiment on Neutrino Activity (LENA), was made along these lines. In what follows, a version of LENA conceived for installation near Gran Sasso is described in some detail (see Fig. 8). The physics aims of the experiment are:
1) Appearance experiments on ν_μ/ν_e and ν_μ/ν_τ oscillations at the parameter range down to a Δm^2 of 10^{-3} (eV)^2 and a $\sin^2 2\theta$ of 0.1. The CERN neutrino beams, either from the SPS or the PS, could be directed toward the Gran Sasso. The distance is 730 km, to be compared with the oscillation length of 1240 km for 0.5 GeV neutrinos for Δm^2 of 10^{-3} (eV)^2. The event rate for a 1 Mt detector is 60 for 10^{19} protons on the target at CERN, even for the low-energy focused neutrino beam as used for the BEBC experiment, and is orders of magnitude larger for the high-energy focused neutrino beam. The identification of electrons and muons is excellent, even at the low energy of 200 MeV. The identification

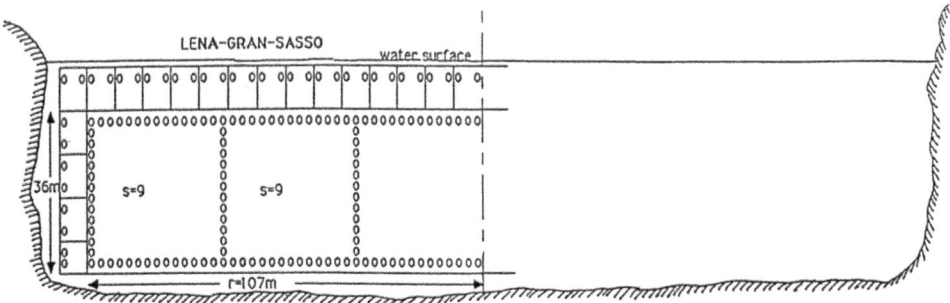

Figure 8. The schematic cross-section view of LENA–Gran-Sasso. The small ovals each represent one 50 cm $\varnothing$ photomultiplier and s is the area to be covered by a single 50 cm $\varnothing$ photomultiplier on the surface.

of τ could probably be made by measuring the invariant masses of π^0, ϱ, and A1 mesons in the $\nu_\tau + 3\pi$ decay mode of the produced τ leptons.

2) Study of atmospheric neutrinos. This type of study by KAMIOKANDE gave the first hint of neutrino oscillation in the parameter range mentioned above. The detector here can expect ~ 14,000 upward-going muons per year to be compared with 40 in KAMIOKANDE.

3) One can initiate also high-energy neutrino astronomy with a realistic sensitive area of 3.6×10^4 m^2. The angular measurement accuracy is better than 1°.

4) Search for Heavy Relic Particles in the sun by taking advantage of 1 Mt fiducial mass together with the μ–e discrimination capability.

5) Search for the point sources of high-energy γ-rays. The top layer modules, which act as the anticounters for other studies, are good energy-flow detectors and with the complete back-up by the inner detector as the muon detector can detect the high-energy γ-rays cleanly separated from the otherwise overwhelming background of cosmic-ray hadron showers. The accuracy in the angular measurement is better than 1°. The threshold energy of γ-rays will be 10^{13} eV at 2000 m a.s.l.

6) Study of the primary cosmic-ray composition at and above 10^{16} eV. The detector dimensions would be: size of the cylindrical water reservoir R = 113 m and D – 46 m; size of the inner detector R = 107 m and D = 36 m. Fiducial volume, 3 m inside the PMT array: 1.02×10^6 m^3.

The PMTs are installed over the inner surface of the inner detector, as well as over the vertical membranes of concentric cylinders of radii 34 and 69 m, with 1 PMT per 9 m^2. The inner detector is surrounded on the top and the sides by anticounter modules of 36 m^2 × 10 m height each containing 2 PMTs. The threshold detection energy is 60 MeV over the entire inner volume of 1.02×10^6 m^3. The approximate cost would be: 15,000 PMTs with electronics @ SF2500 = 37.5 MSF, plus civil engineering costs (depending very much on the chosen terrain).

Acknowledgements

The experimental data used in this paper are mostly from KAMIOKANDE-II, and my collaborators in this experiment are as of June 1988: K.S. Hirata, T. Kajita, T. Kifune,

K. Kihara, M. Nakahata, K. Nakamura, S. Ohara, Y. Oyama, N. Sato, M. Takita, Y. Totsuka and Y. Yaginuma from the Institute for Cosmic Ray Research, University of Tokyo; M. Mori, A. Suzuki, K. Takahashi, T. Tanimori and M. Yamada from the National Laboratory for High Energy Physics (KEK); T. Suda from the Department of Physics, University of Kobe; K. Miyano, H. Miyata and H. Takei from the Department of Physics, University of Niigata; K. Kaneyuki, Y. Nagashima and Y. Suzuki from the Department of Physics, University of Osaka; E.W. Beier, L.R. Feldscher, E.D. Frank, W. Frati, S.B. Kim, A.K. Mann, F.M. Newcomer, R. Van Berg and W. Zhang from the Department of Physics, University of Pennsylvania. It is a pleasure to acknowledge the generous and timely support for KAMIOKANDE given by the Ministry of Education, Culture and Science of Japan. The experiment was supported, in part, also by the Department of Energy of the USA. It was indeed an honour as well as a pleasure for me to be invited to the third ESO–CERN Symposium on Astronomy, Cosmology and Fundamental Physics on the occasion of the 900th anniversary of the University of Bologna. The author is grateful to Professors G.C. Setti and G. Giacomelli for their kind hospitality.

References

1) R. Davis Jr. et al., *Phys. Rev. Lett.* **20** (1968) 1205.
   J.K. Rowley et al., *in Solar Neutrinos and Neutrino Astronomy,* eds. M.L. Cherry, W.A. Fowler and K. Lande (AIP Conf. Proc. No. 126, New York, 1985), p. 1.
2) J.N. Bahcall and R.K. Ulrich, *Rev. Mod. Phys.* **60** (1988) 297.
3) For details of this observation, see M. Nakahata, doctoral thesis, Univ. of Tokyo preprint, UT–ICEPP–88–01, Feb. 1988.
4) K.Hirata et al., *Phys. Rev. Lett.* **58** (1987) 1490. The full paper will appear shortly in *Phys. Rev. D.*
5) R.M. Bionta et al., *Phys. Rev. Lett.* **58** (1987) 1494.
6) B. Pontecorvo, *Sov. Phys.–JETP* **26** (1968) 984;
   V. Gribov and B. Pontecorvo, *Phys. Lett.* **28B** (1969) 495.
7) S.P. Mikheyev and A.Yu. Smirnov, *Nuovo Cimento* **C9** (1986) 17.
8) L. Wolfenstein, *Phys. Rev.* **D17** (1978) 2369 and **D20** (1979) 2634.
9) For the implications of this event concerning particle physics and astrophysics, see for instance, D.N. Schramm, *Nucl. Phys.* **B3** (1988) 471; also, A. Burrows, Arizona Theoretical Astrophysics Preprint No. 88–16, University of Arizona (1988).
10) Y. Totsuka, Proc. 7th WOGU/ICOBAN'86, Toyama, Japan (World Scientific, Singapore, 1987), p. 118.
11) Y. Oyama et al., *Phys. Rev. Lett.* **56** (1986) 991 and **59** (1987) 2604.
12) N. Sato, doctoral thesis, University of Tokyo (1988).
13) K.S. Hirata et al., *Phys. Lett.* **B205** (1988) 416.
14) T.J. Haines et al., *Phys. Rev. Lett.* **57** (1986) 1986.
15) M. Takita, doctoral thesis, Univ. of Tokyo (1988). See also V. Barger and K. Whisnant, Univ. of Wisconsin preprint, MAD/PH/414 (1988), and J.G. Learned et al., Univ. of Hawaii preprint, UH–511–643–88 (1988).

Models of Type II Supernovae and Supernova 1987A

Wolfgang Hillebrandt, Peter Höflich, Hans-Thomas Janka,
and Ralph Mönchmeyer
Max-Planck-Institut für Physik und Astrophysik,
Institut für Astrophysik
Karl-Schwarzschild-Str. 1, D-8046 Garching, FRG

Abstract. Implications from the observations of SN 1987A for the standard model of type II supernova explosions are discussed. In particular, we find evidence for strong mixing prior to the explosion of SN 1987A, which was not expected from simple spherically symmetric models. Moreover, because there are indications that the progenitor may have been a rotating star, we will discuss possible effects of rotation on the outcome of numerical simulations. Finally, we present some results of Monte Carlo simulations of neutrino transport in type II supernovae. It is demonstrated that the neutrino spectra are rather non-thermal, and that the mean neutrino energies differ significantly from those obtained from simple transport schemes. Our results may be of importance for interpretations of the observed neutrino signals.

1. Introduction

The supernova in the Large Magellanic Cloud, SN 1987A, is unique with respect to the facts that the progenitor has been identified, and that we know many of its properties fairly well. From its luminosity class and spectral type it has been concluded that its main sequence mass must have been approximately 20 $M_\odot$ (Arnett, 1987, Hillebrandt et al., 1987, Woosley et al., 1987), in agreement with expectations for type II supernova progenitors. Stars with masses around 20 $M_\odot$ are believed to form He-cores of about 6 $M_\odot$ and Fe-cores close to 1.4 $M_\odot$ during subsequent phases of hydrostatic nuclear burning (Arnett, 1977, Wilson et al., 1986, Hashimoto and Nomoto, 1987). They should then collapse to neutron star densities, because the entropy at the onset of collapse is predicted to be sufficiently low such that the relativistic leptons dominate the pressure. The facts that neutrinos from SN 1987A have indeed been observed (Hirata et al., 1987, Bionta et al., 1987), and that the energy emitted in neutrinos was of the order of the binding energy of a newly born neutron star (Sato and Suzuki, 1987, De Rujula, 1987) indicate that this has actually happened. SN 1987A, therefore, establishes the first proof that the core-collapse scenario is essentially correct for type II supernovae.

There are, however, many questions and uncertainties related to those conclusions. First of all, from the evolution of the light curve and the spectra up to June 1987 one can conclude that the progenitor of SN 1987A has undergone moderate mass loss only and possessed a hydrogen-rich envelope of about 10 $M_\odot$ at the time of explosion (Höflich, 1987b, Woosley, 1987, Arnett and Fu, 1987). Therefore, it is

327

M. Caffo et al. (eds.), Astronomy, Cosmology and Fundamental Physics, 327–340.

difficult to understand, why in the early spectra one finds evidence for significant He- and Ba-enrichments (Höflich, 1987b, 1988), and why the UV-spectra indicate the presence of a low-velocity circum-stellar shell with very large nitrogen over-abundance (Casatella et al., 1987, Fransson et al., 1988). We will address these questions in Sec. 2. One possible interpretation is that mixing was caused by large scale motions induced by rotation (Weiss et al., 1988).

We have computed, therefore, core-collapse models for rotating stars in order to see, whether rotation may lead to prompt explosions. Some of our results are presented in Sec. 3. We find, however, that for the angular momentum distributions discussed in the present paper, rotation rather weakens the chance of prompt explosions, as compared to spherically symmetric models.

In Sec. 4 results of Monte Carlo simulations of neutrino transport in type II supernovae will be presented. Because during the first few seconds of its life, a newly born neutron star will still be surrounded by a rather extended atmosphere, such extensive computations seem to be necessary if one is interested in accurate neutrino spectra and average neutrino energies. Implications from our results for the interpretation of the neutrino events seen from SN 1987A will also be discussed.

Finally, a summary and conclusions follow in Sec. 5.

2. Interpretation of the Early Spectra of SN 1987A

The observed spectra give direct information on the physical and chemical conditions of the supernova photosphere at a given time. Deeper layers of the expanding envelope are observable at later times. Therefore, a detailed analysis of the observed spectra will allow a determination of the density structure and the chemical composition as a function of depth, and the mass of the H-rich envelope. Furthermore, the chemical profiles yield information about the mixing of different layers during stellar evolution or during the explosion. In order to investigate these questions we have calculated scattering dominated photospheres for type II supernovae and have applied them to SN 1987A. Because the importance of a sophisticated treatment of the radiation transport (including extension effects) and of the occupation numbers has been demonstrated (Hempe, 1981; Höflich, 1987a), detailed atomic models were used for a number of elements (H, He, C, N, O, Ne, Na, Mg, K, Ca and Ba) (see Höflich (1987a,1987b,1988) for more details of the models).

Generally speaking, our models are described by the following parameters: i) the initial density and global chemical profile, which we took from the stellar structure of a B3I star as calculated by A. Weiss (personal communication), and from a power law for deeper layers than the hydrogen rich region, ii) an expansion parameter λ, which describes the homologous expansion of the stellar envelope, iii) the photospheric expansion velocity, iv) a statistical component of the velocity field, and v) the total luminosity, which was taken directly from observations (Menzies et al., 1987; Catchpole et al. 1987).

A comparison of synthetic and observed spectra at a specific time allows us to deduce the photospheric radius R_{ph}, the corresponding effective temperature T_{eff}, the particle density N_o, the velocity field v at R_{ph}, the distance R_{HII} at which most of the hydrogen becomes neutral, the chemical composition and the mass of the envelope which has passed through the photosphere (see Table 1).

One to two weeks after the explosion, the geometrical dilution of the material and the recombination of hydrogen outside a certain distance R_{HII} result in a much slower change of R_{ph} and, consequently, in smaller changes of the spectral energy distribution as long as the continuum is determined by electron scattering,

Table 1: Distance R_{ph} at which the optical depth is 1 for true absorption at 5000 Å, effective temperature T_{eff}, particle density N_o, velocity field v in km/s and mean value of $\tilde{n}$ corresponding to a density profile $N(r) \propto r^{-n}$ of the continuum forming region $(0.1 \leq \tau_{5000} \leq 1.0)$ are given for the models (for the density profile see text). All quantities are given in cgs-units. $\dot{M}$ is the mass in solar units which has gone through the photosphere. The radius R_{HII} of the HII-region and the corresponding date are given for SN1987A. Note, that the given velocity v includes a statistical term in the order of 5 to 20 %.

No.	R_{ph}	T_{eff}	N_o	v	$\tilde{n}$	R_{HII}	Date	$\dot{M}$
I[1]	1.28E14	12400 K	3.8E12	20000.	13.	-	Feb.25	7.E-3
II[1]	2.4E14	9150 K	1.3E12	16500.	10.	3.9E14	Feb.26	1.2E-2
III[1]	4.9E14	6500 K	4.8E11	12000.	7.2	7.8E14	Mar.02	2.5E-1
IV[1]	8.5E14	5300 K	3.6E11	8000.	5.6	1.15E15	Mar.09	8.E-1
V	1.07E15	5800 K	2.7E11	4900.	4.3	1.25E15	Mar.23	2.3
VI	1.17E15	5900 K	2.2E11	3300.	2.8	1.48E15	Apr.16	5.2
VII	1.21E15	6160 K	2.0E11	1800.	2.6	1.58E15	May 14	7.1
VIII	9.40E14	6300 K	1.9E11	1100.	2.2	1.30E15	June05	9.3
IX[2]	7.8E14	4900 K	1.5E11	800.	1.9	1.17E15	Oct.02	11.7

[1] pure hydrogen (and helium) model [2] outer density structure is given by model I-VIII

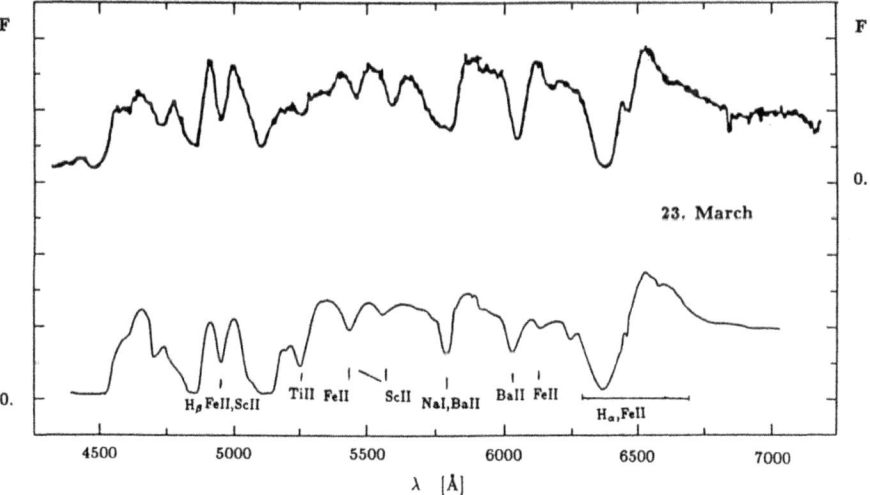

Figure 1: Spectrum as observed by Menzies et al. (1987) on May 23th in comparison with the reddened, synthetic spectrum (lower spectrum, $E_{B-V} = 0.15^m$) as calculated in model V (see Table 1). In addition, identifications of some strong features are given.

i.e. the hydrogen envelope can be observed at the continuum forming region. The comparison between calculated and observed colours in Johnson's system (1966) at these stages shows agreement in the optical and IR-wavelength range (Höflich, 1988). The IR-excess is due to free-free radiation and extension effects of the order of 10 to 20 % for the IR-continuum forming region. Note that, at least in the nearer IR, no additional dust emission is needed to explain the observed excess. Generally, the spectra can be fitted by models with a third of solar abundances for most of the heavy elements (see, e.g., Figure 1). This value is consistent with the chemical composition of the LMC (Dufour, 1984; Kudritzky et al., 1987). However, the s-process elements Sc,V,Cr,Sr ($\approx 2 \cdot$ solar), Ti (solar), Ba ($\approx 10 \cdot$ solar), and Na ($\approx 5 \cdot$ solar) are enriched. The IR-spectra also indicate an overabundance of He by about a factor of 2 in the hydrogen rich layers (Höflich, 1988). Within the uncertainties (factor 2) no chemical gradient of elements other than H and He could be confirmed by ours analysis. Helium rich layers are seen at the photospheric region in October 87, but a certain amount of H is still seen at the photosphere. This and the overabundances of processed matter in the hydrogen rich layers are strong evidence that the envelope has undergone nuclear processing ornd some mixing with deeper layers.

The decrease of R_{ph} after the maximum on May 14th is caused by the decreasing luminosity and by the changing hydrogen abundances in the continuum forming region (Y=0.27-0.70), as indicated by the adjusted initial chemical profile. The resulting decrease of the continuum flux and the changes in the hydrogen line profiles, especially the increase of the peak height of the H_α emission component, allow us to determine the mass of the hydrogen rich layers by adjusting the total mass of our initial stellar model. We find roughly 9-11 $M_\odot$ for this layer (Höflich, 1987b, 1988), indicating only moderate mass loss of the progenitor star, in agreement with recent light curve models (e.g. Woosley, 1988).

3. The Collapse of Rotating Stellar Cores

Because the explosion mechanism for stars as massive as Sk -69 202 is still not understood and because there are indications that spherically symmetric models cannot fully explain the observations, we have studied the influence of rotation on collapsing stellar cores. In particular, we have computed two models, which had an iron core of $1.36 M_\odot$ (Weaver, Woosley, and Fuller, 1982) and differed only in their amount of initial rotational energy.

We followed the change of electron and neutrino concentrations by taking into account electron captures on protons, simplifying neutrino transport by a trapping scheme. For the trapping densities we chose $3 \cdot 10^{11} g/cm^3$ during collapse and $10^{10} g/cm^3$ after the shock formation. The properties of the core matter were described by the equation of state of Wolff (Hillebrandt and Wolff, 1985).

The initial distribution of the angular momentum was chosen such that a large fraction of the Fe-Ni-core resembled a rigid rotator. The angular velocity varied only with radius according to

$$\omega(r) = \omega_0 \cdot \frac{R^2}{r^2 + R^2} , \quad with \ R = 10^8 cm , \qquad (3.1)$$

where the value of R corresponds to an interior mass of roughly $1.31 M_\odot$. Model A and Model B were defined by $\omega_0 = 4.0$ (Model A) and $\omega_0 = 8.0$ (Model B), corresponding to ratios of rotational to gravitational energy of $\beta_A = 0.005$ and $\beta_B = 0.02$, respectively.

We now briefly summarize the results of our calculations. Note that in rotating cores a nonspherical surface separates subsonically from supersonically falling regions during collapse. We will call the subsonic region, which after bounce corresponds to the unshocked central core region, the "inner core".

Model A bounced at 189 ms, at a central density of $2.58 \cdot 10^{14} g/cm^3$ and a central entropy of $0.89 \, k_b/nucl..$ The angular velocity ω reached a maximum value of $1.4 \cdot 10^4/s$ in the "bottle-neck" shaped density dip and was in the range of $(1-2)\cdot 10^3/s$ at the surface of the inner core. The ratio of rotational to gravitational energy for the total core at bounce was $\beta \approx 0.08$, corresponding to a rotational energy of approximately $8 \cdot 10^{51} erg$. The electron and neutrino concentrations at the center were $Y_e^c \approx 0.32$ and $Y_\nu^c \approx 0.08$. The mass of the inner core at bounce was $0.9 M_\odot$, and its kinetic energy was roughly $6 \cdot 10^{51} erg$. The ratio of maximum infall velocity at the pole to maximum equatorial velocity was roughly 2.2. The peak value for the infall velocity in the bottle-neck reached $10^{10} cm/s$, a value close to the local free fall velocity, indicating a small pressure gradient compared to the gravitational acceleration in the outer axis regions.

The inner core was only slightly compressed at bounce. Its asymmetric expansion generated a deformed, but closed, shock surface, which separated from the inner core. The large infall velocities at the axis caused a high dissipation rate and as a result blobs of high entropy appeared behind the shock, while it moved outwards. These blobs supported the shock movement at the axis for 3 ms due to buoyancy forces. Simultaneously a circulation flow from the equatorial plane to the symmetry axis evolved due to a vortex generating inclination of isobaric to isopycnic surfaces. A smaller vortex instead lead to the "mushroom shape" of the entropy blobs (Kelvin-Helmholtz instability). The main damping effect on the shock propagation resulted in these stages from the photodisintegration of nuclei.

2 ms after bounce the inner core began to contract again and an overall collapse motion towards the center evolved, which as a rare-faction wave weakened the shock. The described expansion and subsequent contraction of the core were only the first major amplitudes of an damped overall radial oscillation of the inner core with a mean period of 2.5 ms.

When the density in the shock front finally became smaller than the trapping density, energy losses via escaping neutrinos additionally weakened the shock propagation. Both velocity and entropy generation of the shock were drastically reduced. As soon as the neutrino losses became significant, the shock propagation stalled, and an accretion shock formed after roughly $0.31 M_\odot$ had been disintegrated.

The final configuration, shown in Fig.2, consisted of a central condensed object of $1 M_\odot$ with low entropy, surrounded by a region of shock heated material with higher entropy. This region was separated from a deformed accretion shock by a strip with a width of 40 km, where due to mixing the entropy was approximately constant at a level of $2 k_B/nucl..$ Only at the axis, a region existed, where the entropy reached values around $9.5 k_B/nucl..$ The shock stalled at 130 km in the equatorial plane and at 70 km in the polar direction. Between the inner core and the front large scale meridional circulations finally evolved. The mass inside the shock front was $1.31 M_\odot$ at the end of the computation and the final value of β was 0.09 .

While the evolution of Model A was in many respects similar to non-rotating models (see Hillebrandt and Wolff, 1985), the quite different evolution of Model B, which will be discussed next, can be understood from the fact that the adiabatic index of the equation of state used here is very close to the critical value of 4/3 and, therefore, a relatively small amount of rotation can stabilize the core (see, e.g., Tassoul, 1978).

332

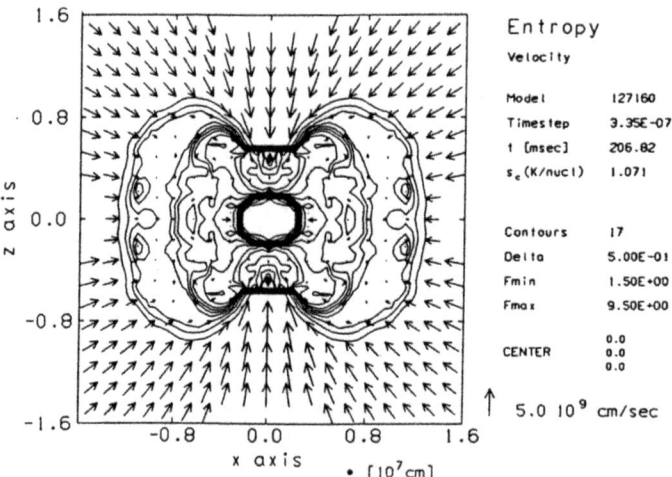

Figure 2: Entropy profiles (in units of $k_B/nucl$) and flow pattern for model A at the end of the calculation about 18 ms after bounce. A blob of high entropy survived neutrino cooling at the axis. The shock is separated from the inner core by a strip of low entropy surrounding an extended region of shock heated matter.

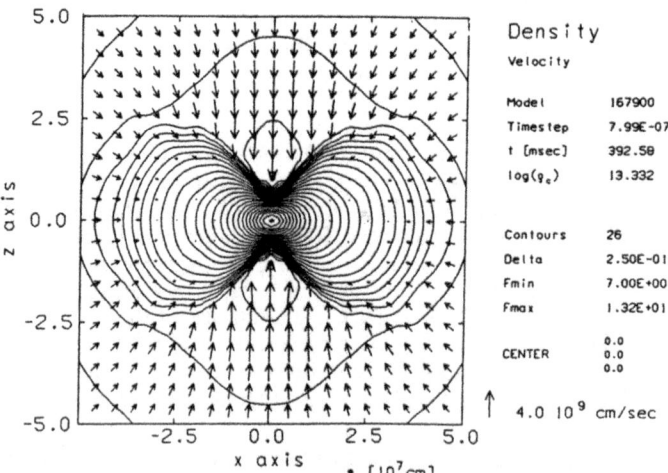

Figure 3: Density profiles and flow pattern for model B about 84 ms after bounce at the end of the calculation, showing a "butterfly" shaped configuration in approximate equilibrium and the deformed accretion shock at the axis.

Due to larger decelerating centrifugal forces the timescale of Model B was stretched significantly in comparison to Model A. Flattening of the inner core started at a central density of $10^{11} g/cm^3$ after 250 ms, just before neutrino trapping set in. At 306 ms a compact inner core of approximately $1.15 M_\odot$ had evolved. A large extended "bottle-neck channel" formed, in which the infall velocities reached approximately free fall velocities.

The model bounced after 308 ms. The inner core mass was at that time roughly $1.2 M_\odot$, its equatorial diameter approximately 190 km, its polar diameter 90 km. The central lepton fraction was again approximately 0.41 with a neutrino fraction of 0.08. The maximum infall velocities were $5.2 \cdot 10^9 cm/sec$ at the pole and $2 \cdot 10^9 cm/sec$ in the equatorial plane, respectively. The central density reached a peak value of $3.0 \cdot 10^{13} g/cm^3$ only.

A strong shock formed at the polar axis, and a diffuse region of high entropy $(S_{max} \approx 13 k_b/nucl.)$ evolved at the bottom of the "bottle-neck channel". But the dissipated energy was directly carried away by neutrinos as the densities in the polar shock region were below the trapping density. Therefore the shock could not propagate along the axis against the high ram pressure of the infalling material. It remained an accretion shock located at 60 km.

In the equatorial plane a weaker shock reached 250 km, instead. Its propagation was supported by an expansion of the inner core in the equatorial plane until the central density value had dropped to approximately $2 \cdot 10^{13} g/cm^3$. At 250 km in the equatorial plane the infall velocities were $1.2 \cdot 10^9 cm/sec$ only, because centrifugal forces and the local pressure gradient could decelerate the infalling matter adiabatically after the shock stalled.

From roughly 30 ms after core bounce on the evolution was characterized by the formation of a matter accreting equilibrium configuration with growing radius in the equatorial plane, while the bottle neck structure at the pole was adjusted to the ram pressure and continuous neutrino cooling. The bottom of the bottleneck was only slightly lifted to 66 km. The final configuration at the end of the computation had a rather deformed "butterfly" shape (see Fig.3). The total mass being in approximate equilibrium was roughly $1.36 M_\odot$. The angular velocitiy was constant on cylindrical surfaces in the final configuration. At the surface of the equilibrium core the value of the angular velocity was $200/s$ in the equatorial plane and the value of β for the total core was 0.14. But again we found no indication for a prompt explosion.

The further evolution of this equilibrium configuration is probably determined by two processes :

First note that the value of β for the $1.36 M_\odot$ core indicates that it is on the edge of secular instability to triaxial deformations.

Secondly, the inner $1.1 M_\odot$ will loose neutrinos due to diffusion, and thus cool down while getting neutronized. Due to lower average densities the typical neutrino diffusion time scale will be smaller than typical values for non-rotating models. It is not clear at all whether the resulting pressure reduction will lead to a second dynamical collapse or only to a slow quasistationary contraction. But every contraction will lead to an increase of the rotational energy thereby strengthening the tendency to triaxial deformations.

In any case, the long term evolution of rapidly spinning stellar cores deserves more attention and should be investigated more carefully. Unfortunately, the time scales of interest will be of the order of several seconds and can only be treated with implicit hydro codes.

4. Spectral Distributions of Neutrinos Emitted from Type II Supernovae

Commonly used transport methods for neutrinos in core collapse calculations are either simple (non-)equilibrium transport schemes or (multigroup) flux-limited diffusion algorithms (e.g. Myra et al., 1987, Cooperstein et al., 1986, Mayle et al., 1987, Burrows and Lattimer, 1986), which may not be sufficiently precise if detailed spectral information or accurate values of the total neutrino flux are to be determined.

Because at present it is far from being practicable to combine hydrodynamical simulations with accurate neutrino transport methods, we have used computed core collapse models as a background and superimposed a Monte Carlo transport in order to investigate the particle flux in the extended region of neutrino-matter decoupling more thoroughly (Janka, 1987, Janka and Hillebrandt, 1988). The Monte Carlo code incorporates non-equilibrium phase space blocking effects and includes the full angle-energy dependence of the reaction rates as well as of the neutrino flow in a spherically symmetric geometry. The interaction processes with charged leptons were sampled from the 'exact' reaction rates.

We used a core collapse simulation for a $20 M_\odot$ star with a $1.36 M_\odot$ iron core (Hillebrandt, 1987) as a background model. In this computation the shock failed to produce a supernova explosion and transformed into an almost standing accretion shock at a radius of about 150 kilometers. The matter below quickly settled into hydrostatic equilibrium. Two stages of evolution were investigated. The early one, about 12 milliseconds after core bounce, was characterized by high electron capture rates in the shocked matter and a lepton fraction between 0.15 and 0.5 in the semi-transparent region from densities of $10^{12} \mathrm{g cm}^{-3}$ down to $10^{10} \mathrm{g cm}^{-3}$. The later model showed significant deleptonization ($Y_e \approx 0.1...0.15$) and the lepton degeneracy had dropped to values in the vicinity of 1. Therefore electron antineutrinos considerably participate in the energy transport, whereas in the first situation electron neutrinos were dominant.

Our results emphasize the importance of an elaborate transport treatment in the decoupling region of neutrinos and matter. The strong energy dependence of neutrino-matter interactions leads to a clear correlation between angular dispersion of the flux and particle energy. This causes a narrowing of the spectra around the peak value and an enhancement of the low energy component relative to the high energy tail. This effect shows up for electron neutrinos and antineutrinos, but increases in strength for muon and tauon neutrinos (which are treated equally as far as transport is concerned due to only small differences in their reaction cross sections), because they are generated exclusively in the high density regions and disentangle from emission-absorption balance with the matter deep inside. While diffusing outward they stay near chemical equilibrium only for densities above roughly $10^{12} \mathrm{g cm}^{-3}$, and their (average) 'energy sphere' is localized at about $5 \cdot 10^{11} \mathrm{g cm}^{-3}$, which they leave with a mean energy around 22 to 24 MeV. Nevertheless, their spectrum is shifted towards lower energies by another 5-7 MeV when they traverse the semi-transparent layers down to densities of about $10^{10} \mathrm{g cm}^{-3}$: The high energy ν_μ's are hindered by multiple scatterings, while the low energy ones rapidly evolve into a freely streaming flow, which is permanently fed by particles downscattered by neutrino-lepton collisions at the 'neutrino sphere'. Table 2 lists the values of the number fluxes and average energies we computed for electron neutrinos and antineutrinos and for muon type neutrinos leaving the surface of our stellar model. Gravitational redshift effects play no role in the situations discussed here.

	Model I (t = 12 ms after core bounce)			Model II (t = 315 ms after core bounce)		
	F_ν	L_ν	$\langle\epsilon_\nu\rangle$	F_ν	L_ν	$\langle\epsilon_\nu\rangle$
	$[\nu's \cdot s^{-1}]$	$[\text{ergs}\cdot s^{-1}]$	[MeV]	$[\nu's\cdot s^{-1}]$	$[\text{ergs}\cdot s^{-1}]$	[MeV]
ν_e	$1.6\cdot 10^{58}$	$2.4\cdot 10^{53}$	9.4	$2.4\cdot 10^{57}$	$3.0\cdot 10^{52}$	7.8
$\bar\nu_e$				$1.0\cdot 10^{58}$	$2.2\cdot 10^{53}$	13.8
ν_μ				$2.5\cdot 10^{57}$	$6.5\cdot 10^{52}$	16.5

Table 2 : Values of the different number (F_ν) and energy fluxes (L_ν) and the mean energies ($\langle\epsilon_\nu\rangle$) of neutrinos leaving the surface of our models.

The spectral shapes are actually non-thermal. However, they can be approximated by Fermi-Dirac-distribution functions by introducing an effective chemical potential, derived from the relation

$$\frac{F_2(\eta_\nu)\cdot F_4(\eta_\nu)}{F_3(\eta_\nu)\cdot F_3(\eta_\nu)} = \frac{\langle\epsilon_\nu^2\rangle}{\langle\epsilon_\nu\rangle^2} \quad , \tag{4.1}$$

where $F_n(\eta_\nu)$ are the usual relativistic Fermi-integrals and $\langle\epsilon_\nu\rangle$, $\langle\epsilon_\nu^2\rangle$ are the neutrino mean energy and mean square energy, respectively. The temperature

$$T_\nu = \langle\epsilon_\nu\rangle \cdot \frac{F_2(\eta_\nu)}{F_3(\eta_\nu)} \quad , \tag{4.2}$$

will be determined by the stage of evolution and the details of the physics employed in the stellar model computations (e.g. the equation of state) and thus may experience strong changes with time. In contrast the effective degeneracy parameter η_ν typifies the spectral shape and should be a quantity reflecting the integral result of the neutrino-matter interactions more genuine. Figures 4 a-d show the Monte Carlo spectra compared to blackbody distributions and our non-blackbody fits; table 3 displays the corresponding parameters.

	$\nu_{e,I}$	$\nu_{e,II}$	$\bar\nu_{e,II}$	$\nu_{\mu,II}$
T_ν [MeV]	1.8	1.9	3.4	4.5
η_ν	5.4	3.5	3.4	2.4
$F_3(\eta_\nu)/F_2(\eta_\nu)$	5.2	4.1	4.1	3.7

Table 3 : Parameters for Fermi-Dirac-distribution fits of the Monte Carlo neutrino spectra.

In model I the electron neutrinos have to pass through matter with moderately degenerate electrons ($\eta_{e^-} \approx 3...5$) and a neutron mass fraction of $Y_n \approx 0.5...0.8$, whereas in the later model one has nondegenerate leptons and a larger concentration of neutrons. This accounts for a reduced mean neutrino energy and explains the drop of η_ν from 5.4 to 3.5. The second number also holds for electron antineutrinos, while muon neutrinos with an equilibrium chemical potential of zero and their weaker energetic coupling to the matter have $\eta_\nu \approx 2.4$.

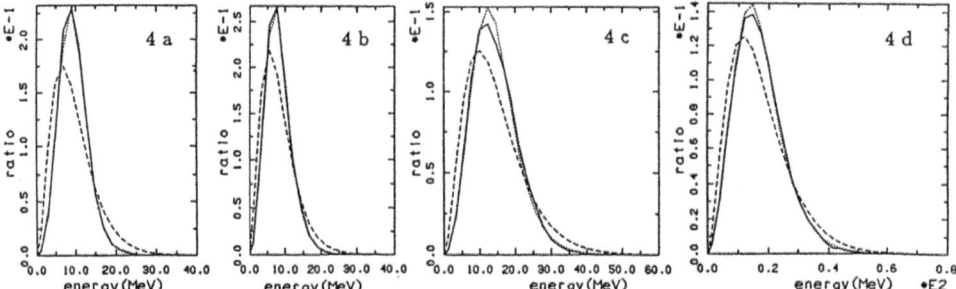

Figure 4: Monte Carlo spectra in comparison with blackbody spectra representing the mean neutrino energy correctly (dashed lines) and our Fermi-Dirac fit distributions with an effective chemical potential (parameters as in table 3) (dotted lines). Remaining discrepancies indicate the actually non-thermal nature of the spectra.

a : $\quad \nu_e$, mod. I $\quad$ (12 $\quad$ msec. after core bounce) ,
b : $\quad \nu_e$, mod. II $\quad$ (315 msec. after core bounce) ,
c : $\quad \bar{\nu}_e$, mod. II $\quad$ (315 msec. after core bounce) ,
d : $\quad \nu_\mu$, mod. II $\quad$ (315 msec. after core bounce) .

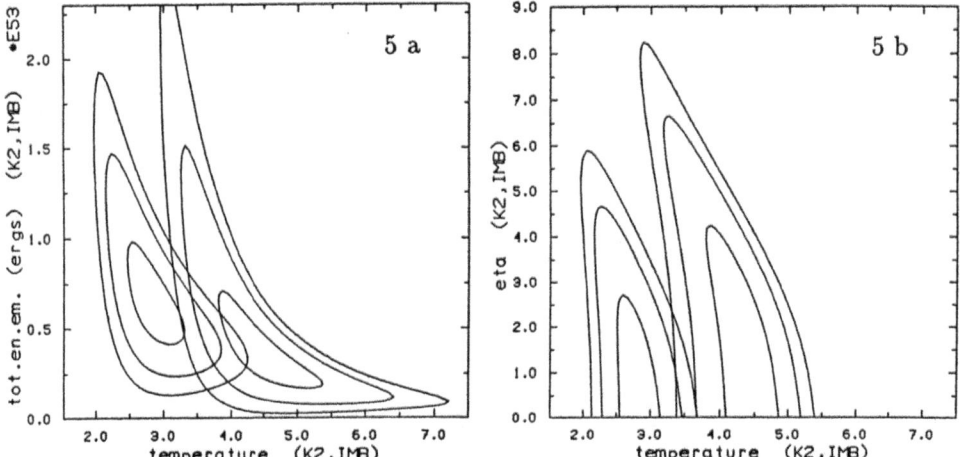

Figure 5: Maximum likelihood regions in the η_ν-T_ν-$E_{\bar{\nu}_e}^{tot}$-space. Fig. 5 a displays the T_ν-$E_{\bar{\nu}_e}^{tot}$-planes through the best fit values of η_ν for the IMB (right) and KII (left) detections. Fig. 5 b pictures the η_ν-T_ν-planes connected to the best fit values of $E_{\bar{\nu}_e}^{tot}$ (see table 4). The lines mark 68 %, 95 % and 99 % likelihood volumes. For the evaluation of the experimental data (Hirata et al., 1987, Haines, 1988) the distance to the LMC was set equal to 50 kpc, and for the detector masses we used M_{KII} = 2140 tons, M_{IMB} = 6800 tons. The efficiency curves were constructed according to the data presented by Koshiba (1988) and Secco (1988), respectively.

The attempt to extract any information about the spectral distribution from the neutrinos detected by the IMB and the Kamiokande II collaborations on Feb. 23.316 (UT) 1987 (Bionta et al., 1987, Hirata et al., 1987) of course suffers from the small numbers of events as well as from the time integrated nature of the signals. Nevertheless, a maximum likelihood test reveals the possible parameter regions, and the IMB data as well as the Kamiokande II data do not exclude a non-blackbody shape of the spectra (see figs. 5 a,b). Table 4 lists the best fit values for the parameter triplet $(\eta_\nu, T_\nu, E_{\bar{\nu}_e}^{tot})$ in both cases, where $E_{\bar{\nu}_e}^{tot} \cdot (D/50\text{kpc})^2$ is the total energy emitted in electron antineutrinos.

	η_ν	T_ν [MeV]	$E_{\bar{\nu}_e}^{tot}$ [ergs]
K II	0...2.7	$2.8^{+1.0}_{-0.6}$	$6.5^{+8.0}_{-4.0} \cdot 10^{52}$
IMB	0...4.4	$4.5^{+1.9}_{-1.2}$	$3.5^{+11.5}_{-2.5} \cdot 10^{52}$
max. consistency	0...2.6	$3.5^{+0.9}_{-0.9}$	$5.1^{+5.4}_{-2.6} \cdot 10^{52}$

Table 4 : Best fit values to the Kamioka and IMB detections in the $(\eta_\nu, T_\nu, E_{\bar{\nu}_e}^{tot})$ space. As conservative error estimates for T_ν and $E_{\bar{\nu}_e}^{tot}$ we took the total ranges of parameter values lying in the 95 % likelihood volumes. The numbers for η_ν do not bear strong statistical evidence.

The two experiments yield only a weak consistency. This need not be a point to worry about, since the different detector properties make them sensitive for different signal characteristics. So it may very well be that they did not see the same temporal window during the whole period of protoneutron star cooling. In order to get more than a rough idea about possible energy distributions of the emitted neutrinos one has, however, to wait for the next galactic supernova, accompanied by a considerably higher number of detections.

5. Summary and Conclusions

We have demonstrated that the spectra of SN 1987A are well explained if the progenitor possessed a H-rich envelope of about 10 $M_\odot$ at the time of explosion. Since its main sequence mass was close to 20 $M_\odot$ and its He-core had around 6 $M_\odot$ it follows that the progenitor has undergone moderate mass loss only.

The presence of significant overabundances of He and s-process nuclei in the H-rich envelope is strong evidence for significant mixing prior to and during or after to the explosion. Moreover, the early emission of X-rays and gamma-ray lines indicate that additional mixing has occured during the explosion. Strong IR-lines of Fe, Ni, and Co have been seen in November 1987 (Rank et al., 1988) and April 1987 (Ericson, personal communication) by the KAO with velocities up to 3000 km/s, corresponding to the velocity of the supernova photosphere in April 1987. However, we find no evidence for Fe and Co enhancement in the early optical spectra. It is likely, therefore, that the radioactive Co has been mixed inhomogenously far into the H-rich envelope, and has formed clumps there. This would also help to explain the relatively constant flux of X- and γ- rays.

All these observational facts are difficult to explain in the framework of 1-dimensional supernova models and may indicate that the progenitor of SN 1987A had been a rapidly spinning star. We have analysed, therefore, the effects of rotation on the collapse of the core of a massive star. Our main result was , however, that prompt explosions are even more unlikely, at least for the rotation laws which have been considered in the present study. But we have to note that our initial models do not fulfil the necessary condition for approximate barotropes, namely that the angular velocity is constant on cylinders. Moreover, the initial models were not in equilibrium with respect to centrifugal forces. It is not yet clear how strongly our results are affected by these inconsistencies.

We have shown, nevertheless, that rotation causes important modifications of the standard core-collapse scenario which deserve more attention. As we have shown, a rapidly spinning stellar core will bounce at fairly low densities and will achieve a quasi-stationary state on time scales of a few tens of milliseconds. Its further evolution will be governed by neutrino and angular momentum losses, and it is conceivable that it will enter a second dynamical collapse stage, which is no longer dominated by the pressure of relativistic leptons. The question, whether this second collapse can lead to a prompt explosion certainly should be investigated. Moreover, explosive O-burning in a rapidly rotating star may also help to eject the stellar envelope, as was suggested by Bodenheimer and Woosley (1983). In fact, the combined action of thermonuclear burning and rotation may be the cause of the observed large-scale mixing and the early break-up of the supernova shell. It may also explain why the amount of oxygen seen so far in the spectra of SN 1987A seems to be rather low (Danziger, this volume). Finally, from the non-sherical density and entropy distributions after core-bounce we expect significant effects on the emission of neutrinos. In particular, the neutrino spectra and, maybe, also their luminosity will be quite anisotropic, making the interpretations of the observed neutrino events more uncertain.

Our Monte Carlo simulations have shown that, in contrast to standard models, the emerging neutrino spectra are not well described by black-body spectra, but may be fitted by Fermi-Dirac distributions with non-zero chemical potentials. The typical average neutrino energies are found to be somewhat lower than those obtained from flux-limited diffusion transport schemes. Moreover, at certain times the electron-antineutrino luminosity may exceed the luminosity of all other neutrino species by a significant factor. All these effects are caused by the fact that the newly born neutron star possesses an extended atmosphere and, therefore, a "neutrino sphere" is not well defined. They should be taken into consideration if attempts are made to interpret the neutrino observations from SN 1987A.

References

Arnett, W.D. 1977, *Astrophys. J. Suppl.*, **35**, 145
Arnett, W.D. 1987, *Astrophys. J.*, **319**, 136
Arnett, W.D., Fu, A. 1988, preprint
Bionta, R.M. et al. 1987, *Phys.Rev.Lett.* **58**, 1494
Bodenheimer, P., Woosley, S.E. 1983, *Astrophys. J.* **269**, 281
Burrows, A., Lattimer, J.M. 1986, *Astrophys. J.* **307**, 178
Casatella, A., et al. 1987, *Astron. Astrophys.* **144**, 335
Catchpole, R.M. et al. 1987, *Monthly Notices Roy. Astron. Soc.*, **229**, 15p
Cooperstein,J., Van Den Horn, L.J., Baron, E.A. 1986, *Astrophys. J.* **309**, 653
De Rujula, A. 1987, *Physics Lett.* **193**, 525

Dufour, R.J. 1984, *IAU Symposium 108*, D. Reidel Publ. Comp., p. 353

Falk, S.W., Arnett, W.D. 1977, *Astrophys.J.Suppl.*, **33**, 515

Fransson, C., et al. 1988, CFA-preprint No. 2638

Haines, T. 1988, Aspen Conf. on SN 1987A

Hempe, K. 1981, *Astron. Astrophys.*, **98**, 19

Hillebrandt, W. 1987, in "High Energy Phaenomena Around Collapsed Stars", F. Pacini, ed., D. Reidel Publ. Comp., p. 73

Hillebrandt, W., Wolff, R.G. 1985, in "Nucleosynthesis", W.D. Arnett, J.W. Truran, ed., Univ. Chicago Press, p.131

Hillebrandt, W., Höflich, P., Truran, J.W., Weiss, A. 1987, *Nature* **327**, 597

Hirata, K., et al. 1987, *Phys.Rev.Lett.* **58**, 1490

Höflich, P. 1987a, in "ESO-workshop on SN1987A", I.J. Danziger, ed., ESO, Garching, p. 447

Höflich, P. 1987b, in *IAU Colloquim 108*, "Atmospheric Diagnostic of Stellar Evolution", K. Nomoto, ed., Springer Verlag, p.288

Höflich, P. 1988, in *Proceedings of Astronimical Society of Australia*, Canberra, June 1988, in press

Janka, H.-Th. 1987, in "Nuclear Astrophysics", Proc. 4th Ringberg Workshop, W. Hillebrandt et al., ed., Springer Lecture Notes **287** , p. 319

Janka, H.-Th., Hillebrandt, W. 1988, in preparation

Johnson, H.L. 1966, *Ann. Rev. Astron. Astrophys.*, **4**, 197

Koshiba, M. 1988, in Proceedings of workshop "Supernova 1987A, one year later", l.c., in press

Kudritzki, R.P.; Groth, H.G.; Butler, K.; Husfeld, D.; Becker, S.; Eber, F.; Fitzpatrick, E. 1987, in "ESO-workshop on SN1987A", I.J. Danziger, ed., ESO, Garching, p.39

Mayle, R., Wilson, J.R., Schramm, D.N. 1987, *Astrophys. J.* **318**, 288

Menzies, J.W. et al. 1987, *Monthly Notices Roy. Astron. Soc.*, **227**, 39

Myra, E.S., Bludman, S.A., Hoffman, Y., Lichtenstadt, I., Sack, N., Van Riper, K.A. 1987, *Astrophys. J.* **318**, 744

Nomoto, K., Hashimoto, M. 1988, in "Theory of Supernovae", G.E. Brown, ed., Physics Reports, in press

Nomoto, K., Shigeyama, T., Hashimoto, M. 1987, in "ESO-workshop on SN 1987A", I.J. Danziger, ed., ESO, Garching, p. 325

Sato, K., Suzuki, H. 1987, *Phys.Rev.Lett.* **58**, 2722

Secco, J.L. 1988 , in Proceedings of workshop "Supernova 1987A, one year later", l.c., in press

Sedov, L.I. 1959, *Similarity and Dimensional Methods in Mechanics*, Academic Press, New York, p. 260

Tassoul, J.-L. 1978, "Theory of Rotating Stars", Princeton Univ. Press, New Jersy (see Ch. 14 and references therein)

Weaver, T.A., Woosley, S.E., Fuller, G.M. 1982, Bull.Am. Astron. Soc. **14**, 957

Weiss, A., Hillebrandt, W., Truran, J.W. 1988, *Astron. Astrophys.* **197**, L11

Woosley, S.E. 1988, preprint

Woosley, S.E., Pinto, P.A., Ensman, L. 1987, *Astrophys. J.* **324**, 466

DISCUSSION.

Gallino: What about the observation of r-isotopes from SN1987a ?

Hillebrandt: We have looked at the possibility to detect r-process elements from SN 1987a. Unfortunately at present there is very little hope for detections because either the abundances are too low or oscillator strengths are too small or lines are blended by strong lines. Nevertheless I would expect overabundances of r-nuclei. If our interpretation of the Ba-lines is correct, ^{13}C should have been present in the He-shell by the time the supernova exploded. An r- or n-process, therefore, is very likely.

X-RAY OBSERVATION OF SN1987A FROM GINGA

Yasuo Tanaka
Institute of Space and Astronautical Science
3-1-1 Yoshinodai, Sagamihara
Kanagawa Prefecture 229, Japan

ABSTRACT. X-rays from SN1987A emerged in July 1987, much earlier than predicted. The energy spectrum is quite unusual for any of the known classes of X-ray source, indicating the presence of two separate components; a soft and a hard components. The soft component is significantly time-variable, and exhibited a remarkable flare-like increase in January, 1988. The intense soft component was quite unexpected, and its origin is yet unidentified. Whereas, the hard component has remained relatively unchanged for more than 300 days, much longer than originally expected for the ^{56}Co origin.

1. OBSERVATION OF SN1987A

Search for X-rays from the supernova 1987A in the Large Magellanic Cloud (LMC) with the X-ray astronomy satellite Ginga started from February 25, 1987, right after the optical discovery (Shelton 1987; Nelson and Jones 1987), and the observation has been continuing after the first detection of X-rays from the SN (Makino et al. 1987). Main X-ray detector of Ginga is a set of proportional counters of a total area of 4000 cm^2 with a full field of view of 2° x 4° (for more detail see Maikno 1987). SN1987A is about 0.6° away from LMC X-1, a bright and variable X-ray source in the LMC. In order to observe the SN separated from LMC X-1, two different modes of observation have been employed; (1) slow scans along a path through the SN and LMC X-1, and (2) pointing observations at a position about 1° offset from LMC X-1 on the side of the SN as shown in Fig. 1, which gives an exposure to the SN at approximately half the maximum sensitivity but with little contribution from LMC X-1.

2. RESULTS

2.1. Detection of X-Rays from SN1987A

In Fig. 2, we show the result obtained from the scanning observations on September 2 and 3 in the form of histograms of the count rate with time in different energy bands. In the energy range below 10 keV, the counts

341

M. Caffo et al. (eds.), Astronomy, Cosmology and Fundamental Physics, 341–350.

342

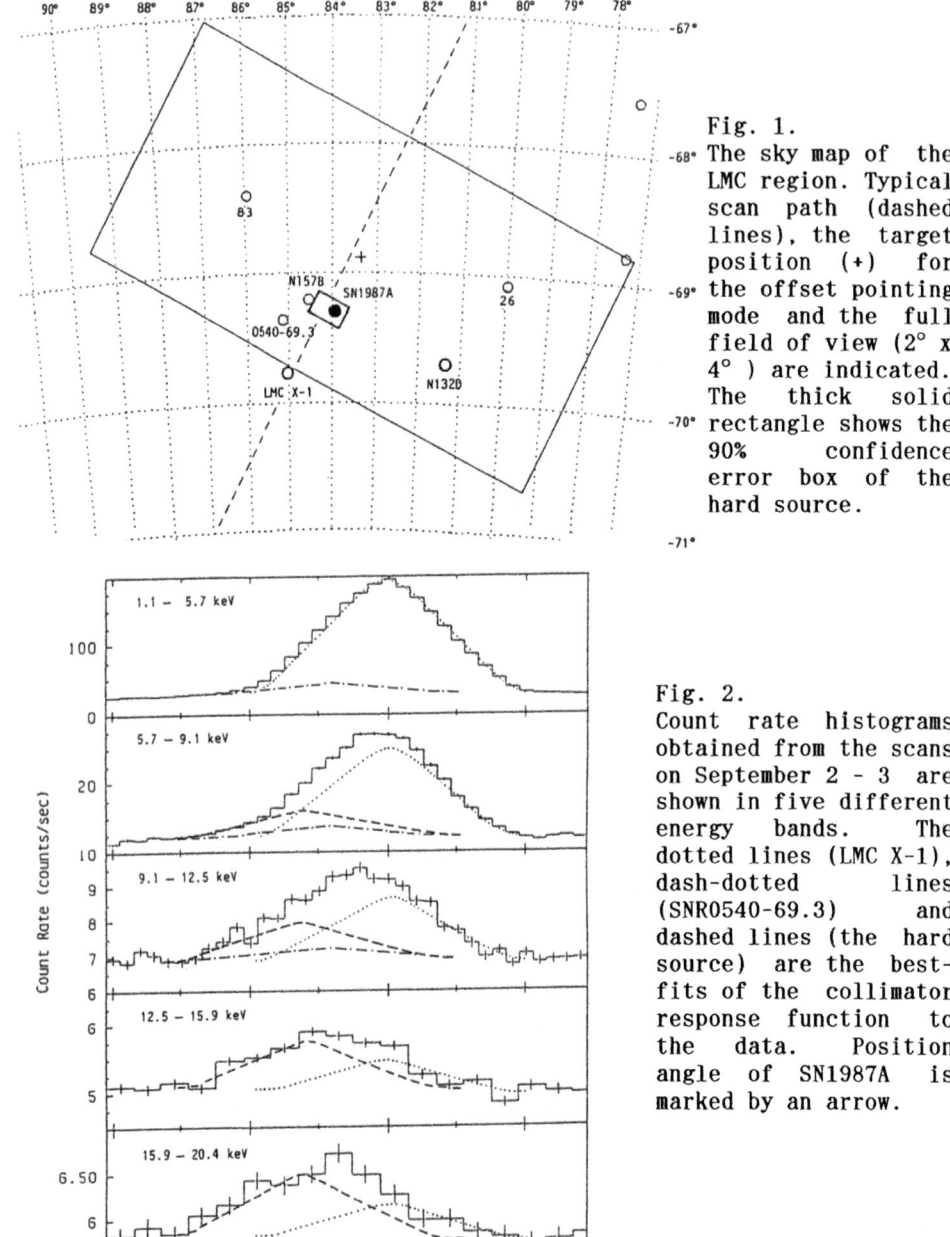

Fig. 1.
The sky map of the LMC region. Typical scan path (dashed lines), the target position (+) for the offset pointing mode and the full field of view (2° x 4°) are indicated. The thick solid rectangle shows the 90% confidence error box of the hard source.

Fig. 2.
Count rate histograms obtained from the scans on September 2 - 3 are shown in five different energy bands. The dotted lines (LMC X-1), dash-dotted lines (SNR0540-69.3) and dashed lines (the hard source) are the best-fits of the collimator response function to the data. Position angle of SN1987A is marked by an arrow.

from LMC X-1 dominate. On the other hand, as the energy increases, the peak shifts towards the line position near the SN. The contribution of

LMC X-1 quickly diminishes with increasing energy. The histograms clearly reveal the presence of another source which has a much harder spectrum than LMC X-1. We are certain that this source did not exist in the March - April period. By fitting the well calibrated collimator response function to the observed count-rate histograms, we can determine the intensity as well as the line position of the hard X-ray source. The contribution of the nearby supernova remnant 0540-69.3 is also taken into account using the result of a separate Ginga observation of this remnant. The result of the fit is shown by the dashed lines in Fig. 2. The line position of the hard source thus determined is in agreement with that of the SN within $\pm 0.1°$ (90% confidence limit).

We then carried out separate scans along several different paths parallel to each other with separations of $0.3°$ to $05°$. From the peak count rates observed on these scan paths, we determined the 90% confidence error box of the hard source with the size $0.2°$ x $0.3°$ as shown in Fig. 1. This error box indeed included the SN. This source was not present until April, 1987, and emerged some time between May and July, 1987. Furthermore, the spectrum of this source was found to be quite unusual for any of the known classes of X-ray source. From these facts, we considered this source to be identified with SN1987A. The first results were published elsewhere (Dotani et al. 1987).

We report here the latest results from the continued observations. We employ only the pointing mode after determining the error box, since pointng observations give the source flux and energy spectrum with much better statistics than scanning observations.

2.2 X-Ray Light Curve

The region around SN1987A is densely populated with many X-ray sources, though most of them except LMC X-1 are faint. In order to determine the flux from SN1987A, the contributions from nearby sources must be carefully corrected for. We attempted to keep LMC X-1, by far the brightest and also variable, essentially outside the field of view. Cosequently, the exposure to LMC X-1 has been less than 2%, except for the observation of July 4 in which the exposure happened to be 10%. For the correction for LMC X-1, we employ the average spectrum obtained from the frequent Ginga observations of LMC X-1. Since LMC X-1 has an ultrasoft spectrum, the softest of the sources in the field of view, its contrbution is maximum towards low energy end. Therefore, although the actual intensity of LMC X-1 during a pointing SN observation is not measurable, the upper limit of its contribution is determined by the flux at the low-energy peak of the observed spectrum ($\sim$ 2.5 keV, see Fig. 3 below) before the nearby-source correction. The variation of LMC X-1 is taken care of by including a large enough systematic error of $\pm 60\%$ of the average, and its effect is insignificant above 6 keV.

In addition to LMC X-1, the contributions from five more sources, SNR0540-69.3, N132D, N157B, and the sources No. 26 and No. 83 in the Einstein survey by Long, Helfand and Grabelsky (1981), were also corrected for. The spectrum of each of these sources was determined from separate Ginga observations. The other sources in the field of view are much fainter and estimated to be negligible.

In fact, all nearby sources have softer spectra than that of the SN, hence their contributions quickly dimisnish towards higher energies, and are found to be less than 40% (less than 20% from LMC X-1) of the total counts above 6 keV, except for the data on July 4 with a larger LMC X-1 contamination (50%), and less than 10% above 16 keV. Thus, it is beyond doubt that a significant flux comes from SN1987A even in the range well below 10 keV. An example of the observed spectrum before and after the correction for the nearby source contributions is shown in Fig. 3.

A background measurement is performed immediately before or after each SN observation. The systematic error in the process of background subtraction was estimated to be about 1 % of the background and is incorporated in the error estimation of the SN flux. In fact, this is the largest source of uncertainty in the range above 10 keV.

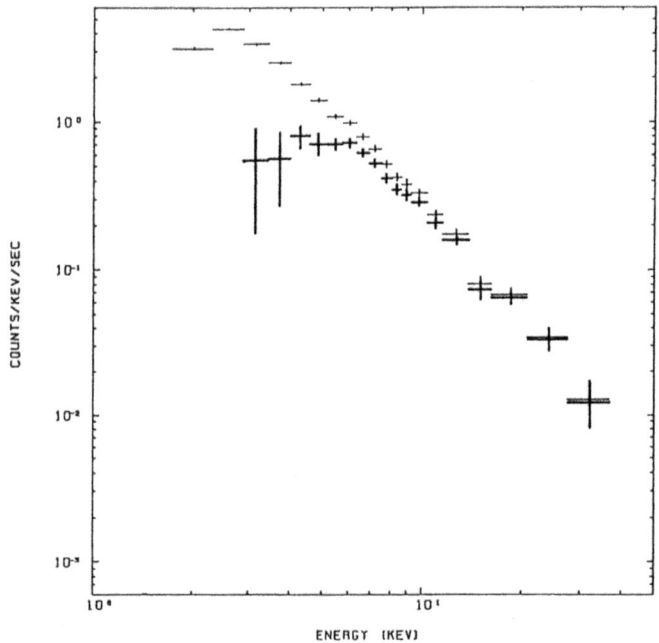

Fig. 3.   Observed spectra of SN1987A before (thin line) and after (thick line) the correction for the nearby source contributions, but not corrected for the aspect and detection efficiency.

The light curves of SN1987A so far obtained are shown for two energy bands, 6 - 16 keV and 16 - 28 keV, in Fig. 4. We chose these two bands because there seems to exist a major difference in the nature of X-rays between the two bands, as shown later. We shall call these two energy bands the soft X-ray band and the hard X-ray band, respectively. The count rates are given after subtraction of the contributions from nearby sources, and the aspect correction. For an approximate conversion of a count rate (counts/sec) to an energy flux in unit of 10^{-12} ergs/cm^{2}sec, multiply 4.6 for the soft X-ray band and 32 for the hard X-ray band.

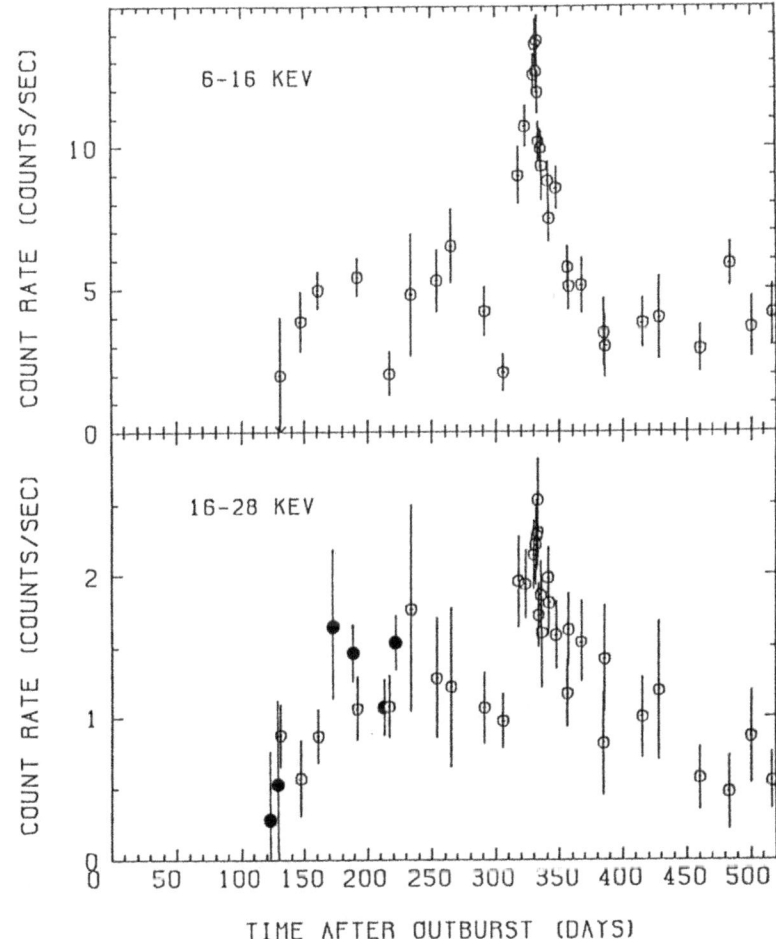

Fig. 4.   X-ray light curves in two energy bands obtained by
scanning (filled) and pointing (open) observations.

The first positive detection of X-rays was July, 1987, both in the soft
and hard X-ray bands. The actual epoch of the X-ray emergence above the
Ginga   detection   limit   remains   somewhat   unceratin.   However,   the
emergences in the soft and hard X-ray bands appear to be essentially
simultaneous; the difference is less than 50 days.
        The intensity in the soft X-ray band is found to vary significantly
by a factor of two to three until the end of 1987. The intensity on
December 26 was the lowest so far observed. After twelve days, on
January 7, 1988, the source exhibited a dramatic intensity increase.
The intensity increased further and stayed at the maximum level from
January 19 through 22. Then, on January 23, it dropped by 30 %, which
is statistically significant.   Since then, it decreased gradually

346

through March 14 and returned to the general level of 1987. We shall
hereafter call this event the "Januray flare". Figure 5 shows the light
curve of the soft X-ray band during the January flare, in which a rapid
intensity change is noticeable.

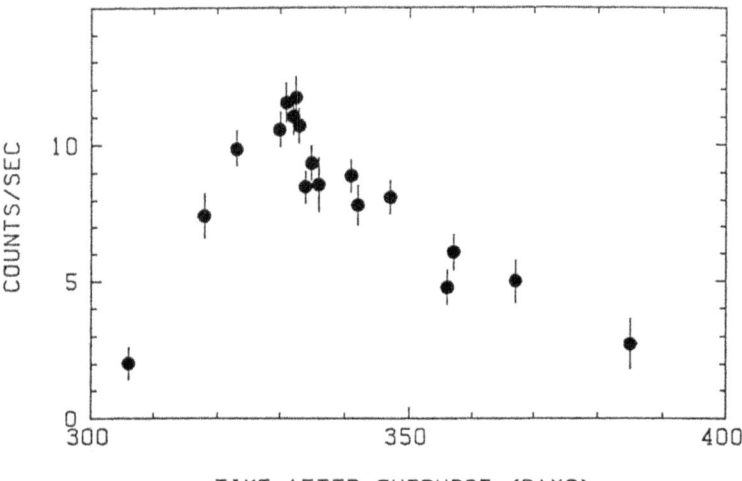

TIME AFTER OUTBURST (DAYS)

Fig. 5. Light curve in the soft X-ray band
during the January flare.

On the other hand, the X-ray intensity in the hard X-ray band (16 - 28
keV) exhibits much less change than those in the soft X-ray band. Since
July, 1987, the intensity in the hard X-ray band has remained
essentially constant within the statistical uncertainties through
December, 1987. In the January flare, the hard X-ray intensity
increased simultaneously by a factor of two, which was however much
smaller than the factor of increase in the soft X-ray band. After the
flare, the intensity returned to the pre-flare level. The intensity in
the hard X-ray band has remained fairly stationary for more than 300
days, excluding the time of the January flare which is discussed later.

2.3. Energy Spectrum

The energy spectrum of SN1987A has a unique shape, suggesting the
presence of two separate components (Dotani et al. 1987). Fig. 6 shows
the average spectra for three different intensity levels. This
averaging is justified because the spectra observed on different days
but at the same intensity level are consistent to be the same within
statistical uncertainties. Spectrum A shows the average of September 28
and December 26 for the lowest intensity group, and Spectrum B that of
August 3, September 3 and November 4. Spectrum C is the average from
January 19 through 22 when the intensity was highest during the January
flare. Spectrum D is that of February 14 and 26 after the flare. These
spectra are corrected for the energy-dependent detection efficiency.

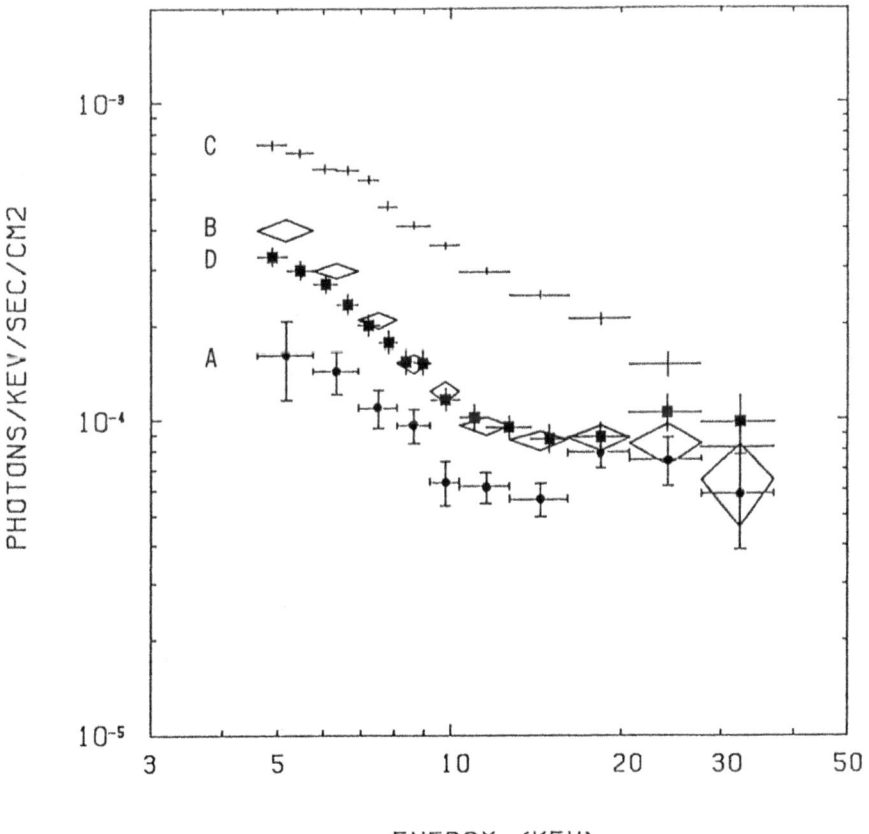

ENERGY (KEV)

Fig. 6.   Average spectra of SN1987A for three different intensity
levels, corrected for the detection efficiency.

Obviously, these spectra cannot be expressed by any of the simple models
such as thermal bremsstrahlung, power-law, and blackbody spectra.   In
particular, Spectra A and B (also D) are quite different in intensity
from each other below 15 keV. Whereas, they are essentially the same
above 15 keV. This fact as well as the spectral shape suggest that
there are two separate components; a variable soft component and a
fairly stationary hard component.   Furthermore, the observed flux
minimum in the 10 - 15 keV range in Spectrum A is an indication that the
hard component is cut off below 20 keV.

We attempt to determine the spectra of the hard and soft components
by employing the following model. For the hard component, we assume the
form $I_h \times E^{-1}\exp(-\sigma N_h)$, where N_h designates the "absorption column".
This form was simply chosen for a qualitative representation of the
result of the Kvant observations in a higher energy range (Sunyaev et
al. 1987) and also the numerical results of the Compton degradation
model (for references see Section 3.). Therefore, N_h is a parameter
introduced for simulating a cut off, hence does not represent the actual

absorption column. The assumed power-law is not critical for the present range limited below 40 keV. The soft component is assumed to be expressed by a thermal bremsstrahlung spectrum of temperature T_s including the effect of absorption by a column density N_s.

This assumed model spectrum gives satisfactory fits to all of the average spectra. As the result, Spectra A, B and D gave the N_h-values all approximately equal to 10^{25} H atoms/cm^2 for the cosmic abundances of element. For these three spectra, the hard component intensity I_h are found to be the same within the statistical errors, and kT_s-values are around 10 keV.

However, Spectrum C is not very sensitive to determine the N_h-value uniquely. As expected from the shape of Spectrum C, three parameters, N_h, T_s and I_h are strongly coupled to each other. For the range of N_h allowed for Spactra A, B and D, a wide range of T_s vs. I_h is acceptable for Spectrum C. This range includes the case in which the intensity of the hard component I_h did not change but the temperature of the soft component kT_s rose higher than 30 keV during the flare. In this case, the count rate increase in the hard X-ray band (16 - 28 keV) during the January flare (see Fig. 4) is interpreted as due to an enhanced contribution of the increased and hardened soft component. On the other hand, the acceptable range also allows the case in which kT_s did not increase much higher than 10 keV, whereas I_h increased significantly at the same time as the soft component increased. If the latter were the case, it would imply that the soft and hard components are not independent but connected generically with each other. In either case, the luminosity of the soft component reached 10^{38} ergs/sec at the flare peak, for the assumed distance of 50 kpc.

For example, the result of the fitting to the spectra B and C for the case of $N_h = 10^{25}$ H atoms/cm^2 are shown in Fig. 7. In this figure, the spectra are those as observed and not corrected for the energy-dependent detection efficiency. The structure of the hard component below 10 keV is an instrumental artifact caused by the "K X-ray escape" events of Xe gas filled in the counters. Note taht the iron line is clearly visible during the January flare (Spectrum C). The energy of the iron line observed during the flare is determined to be 6.8 ± 0.2 keV with an equivalent width of about 200 eV. Therefore, the line is emitted most likely from helium-like or hydrogen-like iron ions. This may be a supportive, if not compelling, evidence for the thermal origin of the soft component.

The absorption column of the soft component, N_s, is generally found to be about 10^{23} H atoms/cm^2 or less. The lower bound of N_s is difficult to estimate because of the increasing systematic errors in the corrections for the nearby sources towards lower energies. For the same reason, whether or not N_s changes with time remains uncertain.

3. DISCUSSIONS.

What are the origings of X-rays from SN1987A? The spectral shape of the hard component is qualitatively in agreement with that expected from the model of the down-Comptonized gamma-ray lines from ^{56}Co (e.g., Itoh et al. 1987; Xu et al. 1988; Ebisuzaki and Shibazaki 1988). However, the

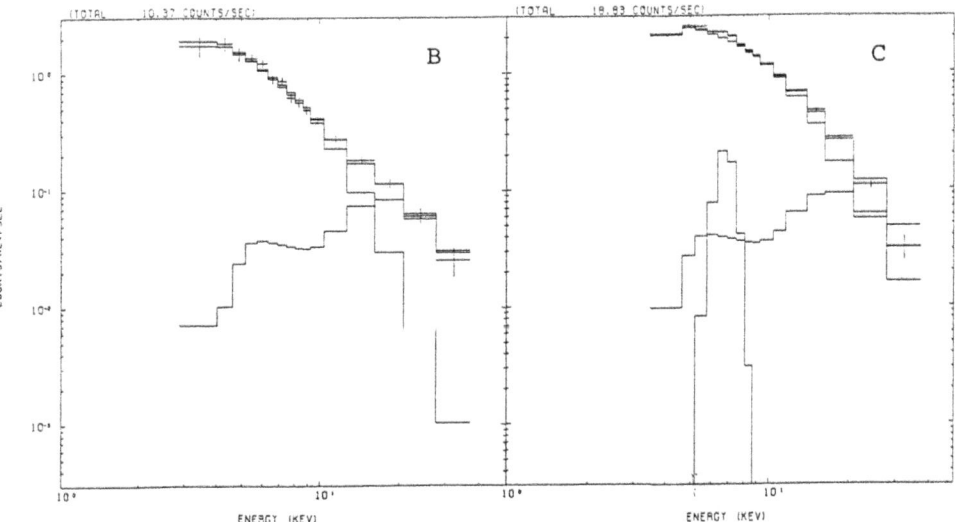

Fig. 7. Spectral fits of the soft (bremsstrahlung) component and
the hard component to the average spectra B (left) and C
including the iron line (right). These spectra are not
corrected for the energy-dependent detection efficiency.

model must explain the observed facts that X-rays emerged much earlier
than expected (e.g., McCray et al. 1987) and the intensity has essentially been constant over 300 days. For a possible explanation, a large-
scale mixing of ^{56}Co into the outer layers has been proposed (e.g., Itoh
et al. 1987; Sunyaev et al. 1987; Ebisuzaki and Shibazaki 1988; Woosley
and Pinto 1988). However, this model would require a fine tuning of the
^{56}Co distribution in order to reproduce the constancy of the hard X-ray
component for such a long period of time. While X-rays of ^{56}Co origin
must exist, whether this accounts for all hard X-rays is a big qustion.
    The presence of an intense soft component was quite unexpected. A
possible scenario for generating soft X-rays below 10 keV is that the
expanding ejecta hits fairly dense circumstellar matter formed near the
SN by the mass loss of the progenitor (Masai et al. 1987), and that the
reverse shock-heated ejecta emits thermal X-rays. For this scenario, the
January flare is due to the collision between the ejecta and a
particularly dense shell or cloud. Masai et al. (1988) were able to
reproduce the general shape of the soft X-ray light curve. However, the
rapid intensity drop on a time scale of 1 - 2 days occurred during the
January flare remains to be explained. Also, since the soft component
of this origin is entirely independent of the hard component, the hard
component is expected to be unchanged during the flare. As discussed
earlier, the soft component temperature during the flare should then be
higher than 30 keV. Such a high temperature would be difficult to
explain, since X-rays from the high-density ejecta dominate.
    Besides, the near simultaneity of the emergences of the soft and
hard components within an uncertainty of less than 50 days is a mere

accidental coincidence, if the soft component is generated outside.

As an entirely different possibility, the neutron star possibly born at the center could be the power source of the observed X-rays. Bandiera, Paccini and Salvati (1988) propose that a synchrotron nebula formed around a young pulsar may be visible in both soft and hard X-ray bands through the gaps, if the ejecta were fragmented. Apart from this, a possibility exists that the central neutron star is accreting mass and producing thermal emission. Fabian and Rees (1988) consider the case of mass accretion from a binary companion. In this relation, it may be of interest that the luminosity at the peak of the January flare is roughly equal to the Eddington limit for a 1Mo neutron star. Search for regular pulsations is obviously of great importance. So far, no pulsation has been detected from Ginga. Search still continues.

ACKNOWLEDGEMENT

The author is indebted to the Ginga Team for the strong support. This work was mainly carried out with Drs. H. Inoue and M. Itoh.

REFERENCES

Bandiera,R. Paccini,F. & Salvati,M. (1988), Nature 332, 418.
Dotani, T. et al. (1987), Nature 330, 230.
Ebisuzaki,T. & Shibazaki,N. (1988), Astrophys.J. Letters 327, L5.
Fabian,A & Rees,M.J. (1988), to be published in Nature.
Itoh,M., Kumagai,S., Shigeyama,T., Nomoto,K. & Nishimura,J. (1987), Nature 330, 233.
Long, K.S., Helfand, D.J. & Grabelsky, D.A. (1981), Astrophys.J. 248, 925.
Makino, F. (1987), Astrophys. Lett. & Communications. 25, 223.
Makino, F. et al. (1987), IAU Circular No.4447.
Masai,K., Hayakawa,S., Itoh,H. & Nomoto,K. (1987), Nature 330, 235.
Masai, K., Hayakawa, S., Inoue,H., Itoh,H., & Nomoto,K. (1988), preprint, to be published in Nature.
McCray,R., Shull,J.M. & Sutherland,P. (1987), Astrophys.J. Letters 317, L73.
Shelton, I. (reported by W. Kunkel and B. Madore); Jones, A., Nelson (reported by F.M. Bateson) (1987), IAU Circular No.4316.
Sunyaev, R. et al. (1987), Nature 330, 227.
Woosley,S. & Pinto,P. (1988), Gamma Ray Spectroscopy in Astrophys., American Inst. of Physics, in press.
Xu,Y., Sutherland,P., McCray,R. & Ross,R.R. (1988), Astrophys.J. 327, 197.

MODELS FOR THE X-RAY EMISSION FROM SUPERNOVA 1987 A

F. Pacini
Arcetri Astrophysical Observatory and
University of Florence (Italy)
Largo E. Fermi 5,
50125 Florence (Italy)

An X-ray source coinciding in position with Supernova 1987 A was discovered in July 1987 by the group operating the Ginga satellite (Dotani et al. 1987). Its existence has been confirmed by measurements made on board the Mir-Kvant Observatory (Sunyaev et al. 1987), as well as by several balloon experiments. The emission extends at least from 6 KeV up to the MeV range (see the reviews by Tanaka and Trümper in this Conference).

A possible interpretation of this phenomenon involves the Compton diffusion of γ-ray photons emitted by radioactive Cobalt (see, e.g., Mc Cray et al. 1987; Sunyaev et al. 1987). This model is based upon the behavior of the bolometric light curve of the Supernova whose luminosity has been decaying with the same half-life of Co^{56}. Spectral lines which can be attributed to the presence of Co^{56} have been detected in the infrared and in the γ-ray spectrum. It is then assumed that the original γ-ray photons emitted in the radioactive decay lose energy in the SN envelope when they are scattered by electrons. They escape as X-rays only if they can reach the outer regions of the envelope before their energy is decreased too much, otherwise they are inevitably absorbed by heavy elements (photoelectric opacity). At early times, all photons are absorbed while later on low and medium energy X-rays are no longer produced because of the reduced number of scatterings. The main predictions of this model are the following:

1. an X-ray luminosity around 10^{39} ergs s^{-1}

2. a low energy cut-off below about 20 KeV, due to photoelectric absorbtion.

3. a progressive hardening of the X-ray spectrum.

M. Caffo et al. (eds.), Astronomy, Cosmology and Fundamental Physics, 351–354.
© 1989 by Kluwer Academic Publishers.

4. a gradual disappearance of the X-ray emission in the energy
range $\leq$ 30 KeV, beginning around the end of 1987.

The observed properties of the source are not in complete agreement with the predictions. In particular, apart from some temporary
fluctuations, the flux has remained nearly constant from September
1987 until Spring 1988.

If one wants to explain the persistency of the X-ray emission one
must postulate an inhomogeneous mixing of Co^{56} inside the star. Some
sort of Maxwell's demon may then have arranged the distribution of Co^{56}
in such a way as to maintain a constant X-ray luminosity. The amount
of Co^{56} which was initially produced can be determined from the bolometric light curve and it turns out to be about 0.07 $M_{\odot}$. Provided that
this material is suitably distributed, the X-ray luminosity of SN 1987
A can be maintained at the present level by radioactivity until late
1988 (Bandiera and Salvati, private communication). If this ultimate
deadline will be violated, one will be forced to conclude that the X-ray
photons produced by radioactivity do not contribute appreciably to the
total flux and that an additional mechanism is also involved.

In any case, the existence of a flux below $\sim$ 20 KeV cannot be explained by the radioactivity model and requires an additional mechanism.
A definite possibility stems from the likely presence of a newly born
neutron star inside SN 1987 A.

The burst of neutrinos detected from SN 1987 A at the time of the
explosion is usually regarded as the signature certifying the birth of
such a neutron star which is likely to have become a pulsar. Of particular importance would then be the properties of the non-thermal nebula
produced in the central region of the Supernova by the relativistic
pulsar wind and the ensuing synchrotron radiation.

This radiation should be visible if the SN envelope has already
undergone fragmentation because of internal instabilities (Rayleigh-
Taylor?). We will assume that this has happened within a few months
after the explosion. In order to justify this assumption we recall the
filamentary nature of the Crab Nebula (where internal fragmentation has
occurred spontaneously, before any substantial interaction with the
interstellar medium) and also the observational evidence for asymmetry,
clumpiness and non-spherical motions in SN 1987 A (see the talk by
Hillebrandt in this Conference). We also stress that the formation of
filaments in a Supernova shell is likely to be an early event since it
requires that the thermal velocities in the gas be larger than the
general speed of expansion in the envelope.

In order to compute the observational properties of the pulsar
nebula we have followed the methods discussed in previous papers (Pacini
and Salvati 1973; Bandiera et al. 1984). As long as the pulsar energy

input remains constant ($t < \tau$) and the lifetime of the emitting electrons is less than the age of the nebula (break frequency $\nu_b < 10^{18}$ Hz), the evolution of the X-ray flux from a nebula expanding with constant velocities is given by

$$S_\nu \alpha \ t^\beta \qquad \beta = 0.5 \text{ for } \gamma \leq 1; \ \beta = 1-\gamma/2 \text{ for } \gamma > 1$$

The spectrum should be

$$S_\nu \alpha \nu^{-\alpha} \qquad \alpha = 0.5 \text{ for } \gamma \leq 1; \ \alpha = \gamma/2 \text{ for } \gamma > 1$$

The observed X-ray flux $\geq 10^{38}$ ergs s^{-1} requires a pulsar with a period $P \leq 18$ ms if $B \sim 4 \times 10^{12}$ gauss on the stellar surface. Correspondingly, the magnetic field in the nebula would be ≥ 0.5 gauss and the X-ray emitting electrons would be confined by radiation losses within a distance $\leq 6 \times 10^{13}$ cm.

If the X-ray emission observed by Ginga is explained by this mechanism, we predict that its intensity should be nearly constant for a time $\sim \tau$ (probably decades or centuries).

In our model the long term constancy is naturally accompanied by short term variabilities due to temporary occultations of the pulsar nebula by fragments of the envelope. Since these variations have been observed in the X-ray flux (see the talk by Tanaka), their time scale Δt determines the typical age of the fragments $d\Delta \sim D\Delta t/t$ ($D \sim 10^{16}$ cm is the present age of the envelope, t the age of the Supernova). Variabilities have been reported with typical $\Delta t \sim$ weeks, which entail a size of the fragments $\sim 10^{14} - 10^{15}$ cm. The influence of these fragments upon the visibility of the central source should be more important in the low than in the high energy range. This is in agreement with the X-ray measurements by Ginga which show that the variability in the 6-16 KeV range is much more marked than in the 16-28 KeV band.

Future observations will be able to discriminate between the relative importance of the radioactivity and pulsar nebula model as far as the X-ray emission from SN 1987 A is concerned. We note explicitly that the pulsar model naturally predicts that the radioemission from SN 1987 A should resurrect as soon as the surrounding medium becomes optically thin to radiowaves. A radio flux around 25 mJy from SN 1987 A could then be expected to arise in the coming months (Bandiera et al. 1988). Also, a contribution to the X-ray emission could arise directly in the magnetosphere of the neutron star and be pulsed. The evidence from other similar sources suggests that the pulsed flux is likely to be a relatively small percentage of the total (in the range 1-10%).

I am greatly indebted to R. Bandiera and M. Salvati for our collaboration in investigating the origin of the X-ray emission of SN 1987 A.

References

Bandiera, R., Pacini, F. and Salvati, M., Ap. J. 285, 134 (1984).
Bandiera, R., Pacini, F. and Salvati, M., Nature 332, 418 (1988).
Dotani et al, Nature, 330, 230 (1988).
McCray, R., Shull, J.M. and Sutherland, P., Ap. J. (Letters) 317, L73 (1987).
Pacini, F. and Salvati, M., Ap. J. 86, 249 (1973).
Sunyaev, R. et al., Nature, 330, 226 (1987).

DISCUSSION.

Hillebrandt: If, as we know from the IR-observations, the radioactive isotopes were mixed into the hydrogen-rich envelope and if the shell broke up due to Rayleigh-Taylor instabilities, I would expect all kind of scales on which clumpy structures might exist. So in my opinion the hard X- ray flux can either increase or decrease or stay roughly constant for a considerable time, depending on the distribution of clumps.

BEYOND THE STANDARD MODEL

D.V. NANOPOULOS
University of Wisconsin–Madison
Department of Physics
1150 University Avenue
Madison, WI 53706
USA

ABSTRACT. This talk is divided in five parts. Part 1 is dealing with the problems of the standard model, while part 2 covers possible ways to extend the standard model. After discussing Grand Unified Theories (GUTs) in part 3, supersymmetric theories are analyzed in part 4. Finally, part 5 provides a lengthy discussion of a recently derived, genuine 4–dimensional superstring model, flipped $SU(5) \times U(1)$.

1. The Problems With The Standard Model

We believe today that the standard $SU(3)_C \times SU(2)_L \times U(1)$ gauge model provides a fair description of all presently established Low Energy Phenomenology (LEP).[1] The identification card for the standard model reads as follows:

1. Framework: relativistic quantum field theory

2. Dimensions: $4 = 3$–space $+1$–time

3. Basic symmetries:

   (a) Poincaré invariance: "gauge group" $0(1,3)$

   (b) Gauge invariance: gauge group $SU(3)_C \times SU(2)_L \times U(1)$ \hfill (1)

4. Building blocks:

$$\text{Fundamental fields :} \begin{cases} \text{a) quarks-leptons (FERMIONS)} \\ \text{b) gauge-Higgs fields (BOSONS)} \end{cases}$$

The credentials of the standard model have been established after many years of hard theoretical and experimental work.[1] There is a set of very crucial tests that any model has to pass in order to be taken seriously. Among others, these tests include:

1. Absence (to the appropriate order) of Flavour Changing Neutral Currents (FCNC) both in the real part[2] ($\Delta m_{K_L - K_S}$, $K_L \to \mu\mu$...) and in the imaginary part[3] (CP–violation, ...)

355

M. Caffo et al. (eds.), Astronomy, Cosmology and Fundamental Physics, 355–382.

2. Appropriate dipole moments either magnetic $[(g-2)_{\mu,e}]$ or electric $(d_{n,e})$. The present experimental values are extremely precise

$$(g - 2)_{\mu}^{4)} = (2.331.848^{+52}_{-40}) \cdot 10^{-9} \tag{2}$$

$$d_n^{5)} \leq 4 \cdot 10^{-25} e.cm \tag{3}$$

3. Validity of the weak isospin $I_W = \frac{1}{2}$ rule[6], implying $\rho = 1$, where the ρ-parameter is defined by

$$\rho \equiv \left(\frac{M_W}{M_Z \cos \theta_W} \right)^2 \tag{4}$$

in a self–explanatory notation. The present experimental value[7]

$$\rho = 0.998 \pm 0.0086 \tag{5}$$

already leaves no room for large deviations from unity.

Sure enough, the standard model passes these tests with flying colors. I have concentrated on these tests because they are very sensitive to the low–energy structure (particle as well as interaction content), and they put under severe scrutiny any extension of, or alternative to, the standard model. Needless to say, the major discoveries[8] at CERN of the intermediate vector bosons $W^{\pm}$, Z^o with characteristics exactly as predicted by the standard model,[1] signal once more that we are on the right track. Then why do we discuss extensions or alternatives to the standard model?

Theorists always felt uneasy with the standard model. For sure it gives a fair description of what has been observed, but true enough, it leaves much to be desired. The heart of the matter is the arbitrariness and uncorrelation of far too many parameters. Either on the mass front (fermion, Higgs, gauge boson masses) or on the coupling constant front (gauge, scalar couplings) everything is at random. This is not exactly what one expects from a fundamental theory. It is convenient to look at the problems raised by the three sectors of the theory, namely the fermion, Higgs and gauge sector.

1. <u>Fermion sector</u>:

    (a) quarks–leptons: incommunicado → masses; quantum numbers uncorrelated

    (b) superfluous family repetition

    (c) Mass source: arbitrary Yukawa couplings

2. <u>Higgs sector</u>: Arbitrary structure

(a) Yukawas (h_i)

(b) Self–couplings (λ_i)

(c) $-\mu^2$: "funny" mass terms

triggers
↓

3. <u>Gauge sector</u>: The most constrained, but SPONTANEOUS BREAKDOWN (SB) arbitrary and problematic as it leads to the

4. Gauge (scale) hierarchy problem:

As is well known, the raison d'etre of the Higgs particles is their ability to get vacuum expectation values (v.e.v) different from zero and thus cause spontaneous breakdown of some gauge symmetry. Clearly, the mass of the gauge bosons, m_G, will be proportional to the v.e.v., v: $m_G \sim v$. On the other hand, v will depend on the parameters of the Higgs potential and it is not surprising that v is proportional to the Higgs mass: $v \sim m_H$ and thus finally reaching the conclusion that

$$m_G \simeq < H > \equiv v \sim m_H \tag{6}$$

And here starts the problem. The weak gauge boson $(W^{\pm}, Z^0)$ masses are much lighter than other known or conjectured mass scales in physics, such as the grand unification[9] (GUT) mass scale $M_X (\geq 10^{15} GeV)$ or the Planck scale $M_P (\simeq 10^{19} GeV)$:

$$\frac{M_W}{M_X} \leq 0(10^{-13}), \quad \frac{M_W}{M_P} = 0\left(10^{-17}\right) . \tag{7}$$

Scalar boson masses are notoriously unstable with a strong tendency to rise to the largest available mass scale. Elementary scalar bosons propagating through space–time foam[10] at the Planck length scale would acquire masses $0(M_P)$, or propagating through the GUT vacuum also pick up masses $O(M_X)$. That is certainly catastrophic, since (6) would imply that $M_W \sim 0(M_X$ or $M_P)$ in violent contradiction with (7)! Even if one tries to cancel out these contributions at the tree level, so as to start off with some "light" Higgs fields with masses $0(M_W)$, then radiative corrections will generate mass shifts $O(\alpha^n M_X$ or $\alpha^n M_P)$ still unbearable. Even if one is willing to forget about space–time foam[10] or grand unification[9], there are also contributions to scalar boson masses of similar magnitude as before, from the quadratically divergent diagrams if one cuts off the loop moments at $Q = O(M_P)$, namely

$$\delta m_H^2 (\sim \delta M_W^2) \simeq \Lambda^2 \simeq M_P^2 \tag{8}$$

a rather disastrous result. This is the gauge hierarchy problem. Clearly, somethings goes wrong with our standard model! We are badly in need of some possible mechanism which replaces (8) by something like

$$\delta m_H^2 (\sim \delta M_W^2) \simeq \Lambda^2 \leq 0(M_W^2) \tag{9}$$

There are two obvious ways that we can satisfy (9). One scenario is to cut the integration off at momenta $0(1)$ TeV by <u>dissolving</u> the Higgs boson at that scale.

Then, Λ is identified with the compositeness scaled Λ_c

$$\Lambda_c \leq 0(TeV) \tag{10}$$

So, Higgs bosons are not elementary and then by induction, and why not, fermions or even gauge bosons may be composite. This is, in general terms, the <u>COMPOSITENES framework</u>. An alternative strategy[11] relies on the observation that the boson and fermion loop diagrams have opposite signs. Perhaps they could be made to cancel so that the effective cut–off would be determined by the boson and fermion masses:

$$\Lambda^2 \simeq |m_B^2 - m_F^2| \tag{11}$$

This cancellation persists through the many orders of perturbation theory required to reduce (8) to an acceptable magnitude (9) if the bosons and fermions have identical couplings and similar masses[11,12]

$$|m_B^2 - m_F^2| \leq 0(M_W^2) \tag{12}$$

This requires approximate supersymmetry. This is the SUPERSYMMETRY (SUSY)[1] framework.

In either case, the standard model needs some modification, and the very exciting thing is that in either framework, COMPOSITE or SUPERSYMMETRIC, one is bound to find a lot of new "stuff" around M_W, if the gauge hierarchy problem is to be resolved. If we go composite, clearly "excited" states of the known particles cannot be far off because of (10), or if we go supersymmetric, the SUSY partners of the known particles should satisfy (12).

2. Extension(s) Of The Standard Model

The easiest way to classify the possible extensions of the standard model is to look back to its identification card (1) and try to change its content. So, we are naturally led to the following possibilities:

1. <u>Framework</u>: maybe at very short distances like the Planck scale ($\sim 10^{-33}$ cm), quantum gravitational fluctuations cause space–time to become foamy[10] or even fractal, so that the usual notions of relativisitic quantum field theory do not apply.[14] One may need modification[14,15] of quantum mechanics, and such schemes sometimes called[15] unquantum mechanics (UQM) have been already proposed. One of the relevant main characteristics of such schemes[15] is the fact that symmetries do not necessarily and automatically imply conservation laws; a notion that may eventually be of tantalizing importance.[15,16,17] Recently, such ideas have been used to provide a possible explanation of the vanishing of the cosmological constant.[18]

2. <u>Dimensions</u>: maybe at very short distances like the Planck scale ($\sim 10^{-33}$ cm) space–time dimensions suffer a proliferation. It is not inconceivable that at distances smaller than

the Planck scale, there are $(d+3)$-space + 1-time dimensions, while the extra d-dimensions curl up at the Planck scale so that the "observable" dimensions at large distances are the standard 3+1. Such ideas are realized in the framework of Kaluza–Klein theories, which try to unify gravity with all other fundamental interactions.[19]

3. Basic symmetries: here, one may distinguish two basic philosophies: either at short distances $(<< M_W^{-1})$, the basic symmetries are reduced or they are extended. In the case of symmetry reduction, one hopes that despite the chaotic behavior at very short distances, at large distances Poincaré or gauge invariance are progressively established.[20] This is the anti–unification program.[20] Also here, conservation laws are not valid at very short distances, because there are simply no symmetries present at such small distances. In the more conventional case of symmetry extension, one may distinguish:

a) Extension of the Poincaré algebra, and that brings us uniquely to SUSY;[13]

b) Extension of the $SU(3)_C \times SU(2)_L \times U(1)$ gauge symmetry, either by sticking in extra factors like $SU(2)_R, \ldots$, or by going to larger gauge groups containing $SU(3)_C \times SU(2)_L \times U(1)$ like SU(5), 0(10), $E_6, \ldots$, i.e., moving to GUTs.

4. Building blocks: maybe at short distances $\leq M_W^{-1}$ all basic fields are:

i) Composite fields: a) fermions (quarks–leptons)

                             b) vector, scalar bosons (gauge–Higgs).

or

ii) Extended objects: in the simplest case of 1–dimensional objects we are talking about superstrings while in the case of 2–dimensional objects about supermembranes.

I believe that the above classification covers more or less all possible attempts for extension of the standard model.

I take it for granted that we all agree on the excellent performance of the standard model at large distances $(\geq M_W^{-1})$, so that we all want to keep it valid, at least as an effective theory, at such large distances. Anything that dissolves part or all the standard model content at distances close to M_W^{-1} is lumped into the COMPOSITENESS framework. Anything that keeps the standard model fundamental, but adds extra group factors or extra elementary particles even at scales close to M_W is lumped into the SUPERSYMMETRY framework. Such a major division emerged naturally from attempts to solve the gauge hierarchy problem. Despite many efforts for many years no major breakthrough has occured in the COMPOSITENESS framework, and thus I'll concentrate on the SUPERSYMMETRY framework.

3. Grand Unified Theories (GUTs)

The spectacular success of the electroweak unification[1] opened the way to more ambitious projects. GUTs[9,21,22,23] that unify electroweak and strong interactions are such a project. The general strategy is expressed schematically by

$$G \xrightarrow{\;\;M_X\;\;} SU(3)_C \times SU(2)_L \times U(1) \xrightarrow{\;\;M_W\;\;} SU(3)_C \times U(1)_{E-M} \qquad (13)$$

One assumes the existence of a large group G [like $SU(5)$, $0(10)$, E_6 ...] which at some superhigh energy scale M_X suffers spontaneous breakdown (SB) to the "standard" group which eventually is spontaneously broken at M_W to $SU(3)_C \times U(1)_{E-M}$. The consequences of such a natural and simple assumption are rather dramatic:

1) At energy scales above $0(M_X)$, there is only one gauge coupling constant g_G, which at lower energies, because of the renormalization effects[21] which are different for strong and electroweak interactions, give rise to the three observed gauge coupling contants g_3, g_2, g_1. In such a case, one is able to predict the value of the electroweak mixing angle θ_W at low energies[24] in simplest models

$$\sin^2 \hat{\theta}_W (M_W) = 0.214^{+0.004}_{-0.003} \tag{14}$$

in poor agreement with the experimental value[7]

$$\sin^2 \hat{\theta}_W (M_W) = 0.228 \pm 0.0044 \tag{15}$$

Also, one is able to calculate[21] M_X, which is the mass of the superheavy gauge bosons $[G/SU(3) \times SU(2) \times U(1)]$ in simplest models[9]

$$M_X = (1.5 \pm 0.5) \cdot 10^{15} \Lambda_{\overline{MS}} \tag{16}$$

with $\Lambda_{\overline{MS}} = 0.1$ to 0.2 GeV.

2) Quarks and leptons lie necessarily in the same multiplet(s). That means:

    i) relations between their internal quantum numbers, e.g., commensurate electric charge (charge quantization) or common weak isospin, etc.

    ii) Relations between their masses, which start identical or very similar at superhigh energies but because of renormalization effects become different at low energies. One is able to predict[25,26], for example

$$\frac{m_b}{m_\tau} \simeq 2.8\text{--}2.9 \tag{17}$$

in excellent agreement with experiment.

    iii) Baryon and lepton number violating interactions. A clear manifestation of lepton number violating interactions will be the possible existence neutrino masses. One predicts[27] in GUTs

$$m_\nu \simeq \frac{m^2}{M} \simeq (10^{-5} \rightarrow 100)eV \tag{18}$$

where m indicates some low energy mass while M stands for some superhigh energy scale $(M_X, M_P?)$. In principle, the neutrino mass spectrum, is such that neutrino oscillations are allowed. A dramatic manifestation of the baryon number violating interactions will be the nucleon instability. The minimal SU(5) model[9] or other minimal GUTs do predict proton decay mainly to $e^+\pi^o$, with a rather "short" lifetime[28]

$$\tau(p \to e^+\pi^o) = \left(\frac{M_X}{4 \cdot 10^{14} GeV}\right)^4 \cdot (1-2) \cdot 10^{29} \text{ years} \tag{19}$$

which by using (16) reads as

$$\tau(p \to e^+\pi^o) \simeq 10^{29\pm1} \text{ years} \tag{20}$$

in rather sharp contradiction with the presently available experimental lower bound[24,28]

$$\tau(p \to e^+\pi^o) > 4 \cdot 10^{32} \text{ years} \tag{21}$$

There are several ways out from these grave difficulties suffered by the minimal SU(5)[9] or other minimal GUTs. The best way is for sure

4. Supersymmetry

Supersymmetry is a new kind of symmetry[13] relating bosons to fermions:

$$Q|F> = |B>, Q|B> = |F> . \tag{22}$$

It is generated by spinorial charges Q_α (α a spinor index) which, as you might expect for fermionic operators, obey an anticommutation algebra:

$$\left\{Q^i_\alpha, Q^{+\dot\alpha}_j\right\} = -2(\sigma^\mu)^{\dot\alpha}_\alpha P_\mu \delta^i_j : i = 1, ..., N. \tag{23}$$

It has been proven[29] that this is the only possible way of combining internal symmetry [the index i in Eq. (23)] with Lorentz invariance. A theory is said to have simple supersymmetry if $N = 1$ and extended supersymmetry if $N > 1$. Particles in renormalizable gauge theories can only have helicities between +1 and -1, and since each charge Q changes helicity by 1/2 a unit:

$$h = +1 \xrightarrow{Q} + 1/2 \xrightarrow{Q} 0 \xrightarrow{Q} -1/2 \xrightarrow{Q} -1 \tag{24}$$

there can be at most four extended supersymmetries in a SUSY gauge theory. The analogous argument applied to a theory including gravity, in which case the allowed helicities range from +2 to -2, tells us that up to 8 extended supersymmetries are possible in a supergravity theory.[30] Since general relativity expresses invariance under general coordinate transformations which allow different local translations at each point, the supersymmetry transformations must also be local in supergravity theory, just as gauge theories embody local phase transformations. In what follows, we will mainly restrict ourselves to simple N

= 1 SUSY, in which case the permissible supermultiplets are the graviton smultiplet which includes a spin–3/2 gravitino, and

$$\text{gauge} : \begin{pmatrix} 1 \\ 1/2 \end{pmatrix} \quad \text{and chiral} : \begin{pmatrix} 1/2 \\ 0 \end{pmatrix} \tag{25}$$

smultiplets. Conventional gauge interactions of spin–1/2 fermions such as quarks are accompanied by interactions involving the supersymmetric partners of fermions and gauge bosons:

$$g(\bar{f}\gamma^\mu f G_\mu) \leftrightarrow \sqrt{2}g(\bar{f}\tilde{f}\tilde{G} + h.c.) \tag{26}$$

while Higgs Yukawa interactions are accompanied by interactions involving the fermionic partners of Higgses:

$$\lambda(ffH) \leftrightarrow \lambda(f\tilde{f}\tilde{H}) \tag{27}$$

Such simple relations like (26) and (27) between different sets of coupling constants, as well as similar relations between fermion and boson masses, lie behind the magic properties of SUSY field theories. SUSY field theories are extremely ultraviolet convergent field theories, thanks to the cancellations between suitably "dressed" fermion and boson loops. A mathematical expression of these cancellations is given by the no–renormalization theorems[31], which tell us that if a term is absent from the superpotential at the tree level, then it cannot be generated in any order in perturbation theory. This "set it and forget it" principle is the reason that SUSY has been employed to solve all types of hierarchy problems in particle phyics. I discussed in detail before how SUSY may stabilize "light" Higgs masses and therefore solve[11], at least in part, the cumbersome gauge hierarchy problem. Another serious hierarchy problem, the strong CP hierarchy problem[32] may also find its solution[33] in the SUSY framework. Simply recall that the non–perturbatively created $\theta_{QCD} F^\alpha_{\mu\nu} \tilde{F}^\alpha_{\mu\nu}$ term, where θ_{QCD} is a free parameter and $F^\alpha_{\mu\nu}$ the gluon field strength, violates CP and P and thus contributes to DEMON (d_n) and thus θ_{QCD} should be[34]

$$\theta_{QCD} \leq 10^{-9} \tag{28}$$

if the bound (3) has to be satisfied. This is a rather small number! Once more, SUSY may rescue[33] the situation. Starting with $\theta_{QCD} = 0$, the radiative corrections to quark masses, which eventually create catastrophically large contributions to θ_{QCD}, behave themselves and thanks to the fermion–boson loop cancellations any possible $\delta\theta_{QCD}$ always satisfies naturally Eq. (28). The principle is the same as in the gauge hierarchy problem: "set it and forget it". SUSY respects and stabilizes hierarchies[11,33].

A remarkable implication of the magic convergent propergies of SUSY theories is the existence of nontrivial finite theories. Yes, there are four–dimensional SUSY Yang–Mills (Y–M) field theories, like $N = 4$ or $N = 2$ with vanishing one–loop β–function(s) that are finite[35,36]. The implications, especially for quantum gravity, may be rather dramatic. The following table (Table 1) shows the convergent properties of SUSY theories.

Having established the credentials of SUSY field theories, I move next to discuss their physics applications.

	Y–M Field Theor.	SUSY (Y–M) Field Theor.			Supergravity
		N = 1	N = $2^{36)}$	N = $4^{35)}$	N = 8
			$(\beta)_{1-\text{loop}} = 0$		
Divergences	$\ell n\Lambda$, Λ^2, ...	$\ell n\Lambda$	FINITE		?

Table 1

Divergent(less) field theories

In minimal SUSY GUTs[12] where the only new particles with masses $0(M_W) << M_X$ are the spartners of the conventional $SU(3)_C \times SU(2)_L \times U(1)$ particles, one finds[37,38,39]

$$\sin^2 \hat{\theta}_W (M_W) = 0.236 \pm 0.003 \tag{29}$$

and

$$M_X = 6.10^{16}\Lambda_{\overline{MS}} \simeq 10^{16} GeV \tag{30}$$

while m_b/m_τ remains numerically unchanged[39] from its value (17) in conventional GUTs. The large value (30) of M_X would yield a very long nucleon lifetime if the decay amplitude were $\propto 1/M_X^2$ and the lifetime $\propto M_X^4$ as in conventional GUTs, (19). However, it has been realized[39,40,41] that in minimal SUSY GUTs, the exchange of a heavy Higgs color triplet H_3 smultiplet can give $\triangle B = 1 = \triangle L$ interactions which are $\propto 1/M_{H_3} = 0(1/M_X)$, suggesting a dependence $\propto M_X^2$ of the baryon lifetime. The tree level H_3 exchange gives an interaction of dimension 5 involving two quarks or leptons and two of their spin–zero spartners. This interaction must be dressed mainly by $\widetilde{W}$ exchange[39] [$\tilde{g}$ or $\tilde{\gamma}$ exchanges[41] are usually suppressed[42]] to give a $\triangle B = 1 = \triangle L$ four–fermion interaction. Color symmetry and quark mass factors in the H_3 Yukawa couplings favor the participation of second and third generation particles, so that the dominant nucleon decay modes in minimal SUSY GUTs are[39,41]

$$p \to \bar{\nu} K^+, n \to \bar{\nu} \bar{K}^o \tag{31}$$

Taking the experimental lower limit $\tau(n \to \bar{\nu} K^0) > 0(10^{31}$ years), one finds that if $m_{\tilde{q}} \simeq m_{\tilde{l}} \simeq m_{\widetilde{W}} \simeq 100$ GeV then

$$m_{H_3} \geq 0(10^{16}-10^{17}) GeV . \tag{32}$$

It does appear that baryon decay should be expected soon in the context of minimal SUSY GUTs.

One may wonder if there are at all realistic SUSY models encompassing all different phenomenological constraints previously mentioned. Indeed, realistic SUSY model building is not an easy task. However, any effort is worthwhile since SUSY models are left as the only candidates for a physical description of the world, at least up to enegies of the Planck scale M_P. Any high standard(s) SUSY model should satisfy the following two **SUSY golden rules:**

1) It should provide naturally an acceptable form of SUSY breaking such that

$$0((20GeV)^2) \leq m_B^2 - m_F^2 \equiv \tilde{m}^2 \leq 0(M_W^2) \tag{33}$$

where $\tilde{m}^2$ is a typical boson–fermion mass splitting of supermultiplet. The sparticle mass spectrum should be such that not only all types of low energy constraints are satisified, but in addition some possible potential problems of the standard model should find a satisfactory resolution.

2) It should provide a complete solution to the three–fold gauge (scale) hierarchy problem: create, stabilize and **dynamically** explain the scale hierarchy. All (small) mass scales should be determined **dynamically** in terms of one fundamental one, the super–Planck scale $M \equiv (M_P/\sqrt{8\pi}) = 2.4.10^{18}$ GeV:

$$\frac{\tilde{m}}{M} \simeq \frac{M_W}{M} \simeq 0(10^{-16}) \tag{34}$$

Suprisingly enough, such models have been constructed, and understandably enough have been called No–scale models.[43,44] Actually, one may argue on general grounds that the success of such a program is guaranteed only in the framework of local supersymmetry or supergravity. It is quite interesting that phenomenological as well as theoretical constraints have led us <u>uniquely</u> into the supergravity framework, something highly desirable because it involves only <u>local</u> symmetries and provides <u>automatically</u> unification with gravity. It is beyond the scope of this talk to discuss no–scale supergravity models and its phenomenological applications[45] It suffices to say here that the No–scale supergravity framework is the one that emerges[46] as the low energy (infrared) limit of superstring theories[47]!

Until now, starting from the standard model and trying to resolve its enigmas we reached the no–scale supergravity framework,[45] which considered to be the best effective theory to describe physics below the Planck scale. But what about the fundamental theory? It seems that point–like field theories do not have the power to provide a consistent quantum theory of gravity, thus we have to do something drastic, move to

5. Superstrings and Flipped SU(5).

The simplest way to understand the recent progress is to remind ourselves that the strings are 1–dimensional objects, sweeping out a 2–dimensional world–sheet, and thus a lot of the string physics is a reflection of (two–dimensional) world–sheet physics. For example the string coordinates $X^\mu, \mu = 0, 1, \ldots$ correspond to two–dimensional scalar bosons. A fundamental property that one needs for a consistent string theory is conformal symmetry in two–dimensions.[48] But unfortunately, even if you implement conformal symmetry at the classical level, quantum (world–sheet) corrections may break it, thus creating a conformal anomaly. In the case of superstrings what we need to know is that the value of the anomaly is $c = -15$ and thus we need to have a system on the world–sheet that has $c = 15$ and thus $c_{tot} = 0$. In the case of the bosonic string the 15 becomes 26. For example, if we consider free world–sheet scalars or free world–sheet fermions, each one of these contribute $c_b = 1$, or $c_f = 1/2$ respectively to the anomaly. In the case of explicitly realized $N = 1$ world sheet supersymmetry (needed for the consistency of the superstring e.g., no tachyons ...) the number of bosons (n_b) is equal to the number of fermions (n_f) and thus

$$c = n_b \cdot 1 + n_f \cdot 1/2 = n_b \cdot 3/2 = 15 \tag{35}$$

implying $n_b = 10$, which people rushed to interpret as the critical number of dimensions $D_c = 10$, in which a consistent superstring has to live in. It turned out that this interpretation was not the only correct one, thus opening the way for the construction of genuine, consistent, 4–dimensional (4–D) superstring theories.[49,50,51] It is conceptually easy to see what is going on. Let us concentrate on the, physically relevant, closed strings and especially on the heterotic strings.[52] The heterotic string may be defined as a closed string whose left(L) and right sectors (R) have different world–sheet (symmetry) properties. This is quite legal, because the L and R modes on a closed string are completely decoupled except for the Virasoro condition of mass–matching at each string level. Thus we are entitled, if we wish so, to make the L–sector to have $N = 1$ world–sheet SUSY while leaving the R–sector (world–sheet) nonsupersymmetric. Thus in the case of the R–sector we have to satisfy the conformal anomaly–cancellation equation

$$c_R = n_b \cdot 1 + n_f \cdot 1/2 = 26 \tag{36}$$

with n_b in general different from n_f, and thus for $D = 10$ (i.e., $n_b = 10$), equation (36) would imply $n_f = 32$.

Incidentally, an equal treatment of all these world–sheet real fermions, from the point of view of boundary conditions, gives an apparent $SO(32)$ global symmetry which is gauged when acted upon suitably by the left–sector. This is the famous[47] $SO(32)$ gauge group in $D = 10$. In the case of "unequal" treatment, in terms of the boundary conditions of the world sheet fermions (say in two groups of 16–fermions each) we may derive similarly the $O(16) \times O(16)$ or $E_8 \times E_8$ gauge groups[47]. But, until now we have tacitly assumed that $N = 1$ (world–sheet) SUSY implies *necessarily* $n_b = n_f$, which is only true in the case of a linear (or explicit) realization of $N = 1$ (world–sheet) SUSY. Remarkably enough, it has been found recently[50] that non–linear realizations of $N = 1$ (world–sheet) SUSY are possible, under certain easily satisfiable constraints, and thus the constraint $n_b = n_f$ used in equation (35) to derive $n_b = 10$ and thus $D_c = 10$ is naturally evaded! We are now free, to start with $(n_b)_L = 4$, (i.e., $D = 4$) and thus get, using eq. (35), $(n_f)_L = 22$, while $(n_f)_R = 44$ from eq. (36). The magic of $D_c = 10$ has been lost and any D between 4 and 10 is acceptable. It is better to call $D_m = 10$ the maximal number of allowed space–time dimensions rather than the critical number. Clearly, there is nothing special for $D = 10$ as there is nothing special for $D = 4$! Because we happen to live in 4–dimensions it makes sense to concentrate on genuine 4–dimensional superstrings. In this case, since there are 44 world sheet real–fermions on the right sector, the starting gauge group is $SO(44)$, big enough to encompass all "known" physics and even more.

The rules for constructing consistent, 4–dimensional superstring models have been worked out in detail,[49,50,51] and need not be repeated here. The "uniqueness" of $N = 1$, $D = 10$, $E_8 \times E_8$ (or $SO(32)$) is translated to $N = 4$, $D = 4$, $SO(44)$. As $D = 10$, or $E_8 \times E_8$ do not sound like $D = 4$ or $SU(3)_C \times SU(2)_L \times U(1)_Y$, similarly here $N = 4$ or $SO(44)$ do not sound like $N = 1$ or the standard model. As in the case of compactified schemes, where the uniqueness is traded in for "reality", the same thing happens here but in a more sophisticated and clearer way. In the case of compactified schemes,[47] the rules are rather loose so we need a lot of "outside" information, while in the case of 4–D superstrings the rules are well–defined and we have to select physically from a plethora of models. The arbitrariness in this case, lies in the selection of the boundary conditions for the 22 (44) left–(right)–sector world sheet fermions and the GSO conditions[53] that eliminate a lot of, otherwise acceptable, (massless) states. But once you decide on this selection, a consistent string theory has been defined and you just have to work out the low–energy physics implications. But how are we going to select the "right" boundary/GSO conditions? For the moment, this "top–bottom" approach does not shed more light than I described above so we have to make a suitable melange of the "top–bottom" with the "bottom–top" approach for the construction of a realistic model. There is a very *valuable clue*, a generic characteristic of 4–D superstring theories: the absence of adjoint (or higher dimension) real scalar representations.[54] The origin of this fact lies in the fact that when we "come" down from $(N = 4, SO(44))$ to $(N = 1$, some generic GUT G) the GSO projections needed to get only $N = 1$ SUSY (and/or chiral fermions) kill the adjoint or other higher dimensional real scalar representations. Actually, this was also a characteristic feature[55] in the $D = 10$, $E_8 \times E_8$ superstring theory, so we should not be profoundly shocked. What is stunning are the far reaching consequences of such a generic feature. Let me remind you, that in grand unified theories (GUTs) like $G \equiv SU(5)$, $SO(10)$, $E_6, \ldots$ you *always need* at least the Higgs adjoint (among other things) to "come" down to the standard model. Again this should not come as a surprise because the $\frac{G(\equiv SU(5), SO(10), E_6 \ldots)}{SU(3) \times SU(2) \times U(1)}$ gauge bosons, in order to be massive need to "eat" Goldstone bosons coming from the Higgs representations. Because the gauge bosons belong to the adjoint of G it is not difficult to understand that we may generically need Higgs adjoints or higher representations to do the job. But then we are leading to a catastrophy, because since the 4–D superstrings do not contain scalar adjoints or higher representations (reps),[54] who is going to provide the "realistic" breaking of the GUT group G down to the standard model? In the case of

$D = 10$, $E_8 \times E_8$ superstring theory the problem was resolved by invoking the so called Hosotani–Wilson–loop breaking.[56,55] By using a multiply connected compactified space needed[55] also for a realistic number of generations: (3, ...) one has a kind of non–Abelian Bohm–Aharanov effect, thus being able to "get rid" of the states that do not satisfy the "quantization" rules. In such a way one gets a restricted pattern of symmetry breaking[55,57] that in many cases leads to realistic models. But, in 4–D superstring theories there is no compactified space, simple or multiple and thus no way of explicitly employing the Hosotani–Wilson loop mechanism.[56] We are stuck! One way out of this impasse is to reject the whole idea of grand unification. But let me remind you that grand unification, i e , the gauge unification of strong and electroweak interactions has turned out to be of fundamental importance, at least as an Organizing Principle. It seems extremely difficult to extend the standard model and still keep intact all its beauty, like natural absence of all kind of flavour changing neutral currents (FCNC), natural conservation of baryon and lepton number, etc. Two famous examples that failed badly are, the otherwise appealing ideas, of (low–scale) compositeness and technicolour. In sharp contrast, a GUT–extension of the standard model automatically keeps the above natural properties and it makes them look even better!

In physical terms it seems difficult to start from $(N = 4, SO(44))$ and get the standard model directly. What you usually get are many extra $U(1)'s$ (or bigger gauge groups), many extra representations at low energies that spoil everything (FCNC, proton decay, neutrino masses, unification of gauge coupling constants, ...), in other words a rather complicated and messy situation[58,59]! I do believe that the absence of grand unification, as an organizing principle, creates grave problems. But then, rejecting the idea of a "direct" derivation of the standard model, i.e., accepting grand unification as a fundamental principle, we are back to the problem of absence of Higgs adjoints or higher real reps. What is going on?

Have we excluded 4–D superstring theories as able to provide realistic models? No! We have just made apparent the unique features of

Flipped SU(5)[60−64] among all known GUTs has the unique property of being able to break down to the standard model, by small Higgs reps like the 10. The Higgs adjoint or higher reps are not needed. So, that is the one! But let us take things from the beginning. We are learning from the books that the fifteen states (per generation) of the standard model (numbers in the parentheses indicate the transformation properties under $SU(3)_C \times SU(2)_L$)

$$\begin{pmatrix} u \\ d \end{pmatrix}_L \sim (3,2)\,; \; u_L^c\,; \; d_L^c \sim (\bar{3},1)$$
$$\begin{pmatrix} \nu_e \\ e \end{pmatrix}_L \sim (1,2)\,; \; e_L^c \sim (1,1)$$

(37)

may be reshuffled to fit in the $\bar{5}$ and $\underline{10}$ reps of SU(5) (colour indices are suppressed)

$$\bar{5} \equiv \begin{pmatrix} d^c \\ d^c \\ d^c \\ e \\ \nu \end{pmatrix} , \quad \underline{10} \equiv \begin{pmatrix} 0 & u^c & -u^c & u & d \\ -u^c & 0 & u^c & u & d \\ u^c & -u^c & 0 & u & d \\ -u & -u & -u & 0 & e^c \\ -d & -d & -d & -e^c & 0 \end{pmatrix} \tag{38}$$

Since $\bar{5} \equiv (\bar{3},1) + (1,2)$ and $\underline{10} \equiv (3,2) + (\bar{3},1) + (1,1)$, we have no problems of finding the correct places for $(3,2)$, $(1,2)$ or $(1,1)$, as there is only a way. But for the $(\bar{3}, 1)$ we have a dichotomy problem: is it d^c or u^c that belongs to $\bar{5}$ (and thus u^c or d^c that belongs to $\underline{10}$)? Georgi and Glashow (G–G)[9] provided a convincing solution: $d^c \epsilon \bar{5}$ and $u^c \epsilon \underline{10}$. The reason being charge quantization! Because SU(5) is a simple group that contains $SU(3)_C \times SU(2)_L \times U(1)_Y$, the electric charge operator has to be traceless. Thus

$$TrQ_{\bar{5}} = 0 \rightarrow Q_x = \frac{1}{3}Q_{e-} \tag{39}$$

i.e., $d^c \epsilon \bar{5}$ (and thus $u^c \epsilon \underline{10}$). Notice that necessarily all the members of the $\underline{10}$ carry electric charge, thus we cannot use the $\underline{10}$ as a Higgs multiplet candidate, due to electric charge conservation. Furthermore, the chain

$$SU(5) \xrightarrow[<\Phi>]{M_X} SU(3)_C \times SU(2)_L \times U(1)_Y \xrightarrow[<h>]{M_W} SU(3)_C \times U(1)_{EM} \tag{40}$$

needs (at least) a Higgs adjoint $\Phi \sim \underline{24}$ for the GUT breaking at $M_X (\geq 10^{14} GeV)$ and a Higgs $h \sim \underline{5}$ for the electroweak breaking at $M_W (\sim 100\, GeV)$.

The use of these reps create some (at least) technical problems. In the case of the Higgs adjoint Φ (GUT breaking), the $SU(3) \times SU(2) \times U(1)$ phase is not a unique one and one may get other non–physical phases (e.g., SU(4) $\times$ U(1)), actually all degenerate in the SUSY SU(5) case,[65] which may create a lot of cosmological embarrassments. In the very early universe who is telling the universe its proper "phase route" and who is keeping it from sticking into the wrong vacuum? In the cases of the Higgs pentaplet h (electroweak breaking) we are facing a, maybe, more embarassing problem. As eq. (38) indicates, the h contains together with the $(1,2)$ part $(\equiv h^2)$ responsible for the electroweak breaking and fermion masses, another part $(\bar{3},1)(\equiv h^3)$ coupled to fermions with the same strength (Yukawa couplings) as the $(1,2)$ part and efficient to mediate proton decay. Recent limits on proton lifetime imply $m_{h^3} \geq 10^{13} GeV$ [66] while $m_{h^2} \leq O(1 TeV)$ if h^2 is the electroweak Higgs. This is the gauge hierarchy problem. We may invoke low–energy, space–time supersymmetry to ensure that once fixed at their classical (tree) level values m_{h^3}, m_{h^2}, quantum corrections will not upset them. But who is responsible for such a huge mass splitting in the first place (i.e., classical level)?

There is coupling $24 \cdot 5 \cdot \bar{5}$ in SU(5), i.e., $\Phi h \bar{h}$ but when Φ gets a big vacuum expectation value (v.e.v) $\sim M_X$, both (1,2) and $(\bar{3},1)$ get M_X masses. Then we have, say, to introduce a bare mass term $O(M_X)h\bar{h}$ and ask for a cancellation of the big M_X mass for the (1,2) down to the $O(M_W)$ accuracy level. This is an extraneous fine–tuning. It would be nicer, if for group theoretical reasons when Φ gets a v.e.v. only the $(\bar{3},1)$ part of h gets a mass while the (1,2) remains automatically massless. This is the missing partner mechanism proposed a few years ago.[67] It was put to work in the Georgi–Glashow SU(5) model by using $\Phi \sim \underline{75}$ and further introducing $\underline{50}$ and $\underline{\bar{50}}$ reps. The idea was interesting but the price for its implementation was rather high! Clearly in our 4–D superstring theories such an implementation is impossible since we don't even have[54] the 24, and we are talking about $\underline{75}$'s and $\underline{50}$'s. Since we see that G–G SU(5) is failing badly as superstring derived model, let us go back to step one and try the other solution to the dichotomy problem, namely: $u^c \epsilon \bar{5}$ and $d^c \epsilon \underline{10}$. In this case the $\bar{5}$ and $\underline{10}$ multiplets become

$$\bar{5} \equiv \begin{pmatrix} u^c \\ u^c \\ u^c \\ \nu \\ e \end{pmatrix}, \quad \underline{10} = \begin{pmatrix} 0 & d^c & -d^c & d & u \\ -d^c & 0 & d^c & d & u \\ d^c & -d^c & 0 & d & u \\ -d & -d & -d & 0 & \nu^c \\ -u & -u & -u & -\nu^c & 0 \end{pmatrix}. \tag{41}$$

We notice immediately that we go from (38) $\leftrightarrow$ (41) by replacing $u \leftrightarrow d$, $u^c \leftrightarrow d^c$, $\nu \leftrightarrow e$, $\nu^c \leftrightarrow e^c$, thus the name flipped SU(5). There are certain important points that need to be made before going further: 1) Clearly now $TrQ_{\bar{5}} \neq 0$! What's going on is that we need to augment $SU(5) \to SU(5) \times U(1)$ such that the electric charge operator is not any more inside the simple SU(5) group but is a combination of an U(1) inside SU(5) and the external U(1). Since SU(5) $\times$ U(1) is not a (semi) simple group, TrQ need not be zero. But what about charge quantization? Well, there is no problem if primordially $SU(5) \times U(1)$ was embedded inside a simple group ($SO(10)$, E_6 or SO(44) as in the 4–D superstring case). Actually, in the superstring case where $SU(5) \times U(1)$ "emerges" from bigger groups by GSO projections ("cuts"), and not real Higgs mechanism it may even be possible to eliminate the "magnetic monopole" problem.[60] There is no Higgs–induced phase transition, between a phase where all of $U(1)_{E.M}$ was inside a big group and a phase where the $U(1)_{E.M}$ has been liberated, necessary for the creation of "magnetic monopoles". Of course it remains to be seen if more complicated topological structures,[68] monopole–like, with "funny" charges, survive. We note from eq. (41), that the $\underline{10}$ contains now, a colorless, chargeless member, the right–handed neutrino ν^c. So, not only the lepton–quark symmetry has been full re-established for both L– and R–handed components, but the scalar 10 has been promoted to a very serious candidate for a Higgs representation.

Indeed the chain of breaking

$$SU(5) \times U(1) \xrightarrow[\langle H \rangle \sim \langle \underline{10} \rangle]{M_X} SU(3)_C \times SU(2)_L \times U(1)_Y \xrightarrow[\langle h \rangle \sim \langle \underline{5} \rangle]{M_W} SU(3)_C \times U(1)_{E.M.} \tag{42}$$

is now possible by using the $\underline{10}$ for the big (GUT) breaking and the $\underline{5}$ (of eq. (38)) as before, for the small (electroweak) breaking. Furthermore, beyond the no–need of Higgs-adjoints anymore, we resolve $\underline{naturally}$ the G–G SU(5) problems discussed previously: i) The $SU(3)_C \times SU(2)_L \times U(1)_Y$ phase is a $\underline{unique\ one}$. There are no other degenerate (or not) competing phases and so we are free from the cosmological embarrassments discussed above.[61] In addition, we resolve another tough problem plaguing superstring inspired models starting with $G \equiv SU(3) \times SU(3) \times SU(3)$. It has been noticed, for some time now,[69] that in such models the universe will be trapped by thermal effects in the "origin" of the Higgs–field space for a long time, namely untill $T \sim M_W$. But, then in order for the Higgs field to roll to its real global minimum $\sim (M_X)$ it takes a really long time and makes the universe full of coherent Higgs–field energy instead of "normal" radiation. When the coherent Higgs field decay, the liberated huge amount of entropy dilutes everything. Thus, unless $M_X < 0(10^{10} GeV)$,[69] which on the other hand is fatal for proton decay,[58] there is a severe cosmological problem.[69] In our case of flipped SU(5), this problem is easily avoided because SU(5) is going through a strong coupling phase early enough ($\Lambda_{SU(5)} \sim 10^{15}$ GeV) which shifts upwards the origin and thus the Higgs–field reaches its global minimum very quickly.[70] In other words, SU(5) is a big enough group to have $\Lambda_{SU(5)} \sim M_X$ (and thus no huge entropy release), while SU(3) is not. ii) The missing partner mechanism is $\underline{automatic}$ here.[61] Simply notice that there is a $10_H \cdot 10_H \cdot 5_h$ coupling and that the common partner between a $\underline{10}$ and an $\underline{5}$ is the $(\bar{3}, 1)$! Once $\langle 10_H \rangle \sim M_X$ we get $M_X 10_H 5_h \sim M_X (\bar{3}, 1)_H \cdot (3, 1)_h$ as the only allowable $SU(3) \times SU(2) \times U(1)$ invariant combination.

Thus, our dream has been materialized: naturally supermassive color Higgs triplets and massless $(O(M_W))$ electroweak Higgs doublets. And all this, by avoiding the use of the Higgs–adjoint or higher real reps and make use only of $\underline{5}$, $\underline{10}$ of Higgs reps, as the 4–D superstring theories ordered.[54] But before going to the string flipped SU(5), it is illuminating to consider the simpler case of

The Ordinary Supersymmetric Flipped $SU(5) \times U(1)$ has the following (minimal) particle content:

$$F_i \equiv (\underline{10}, 1/2), \quad \bar{f}_i \equiv (\bar{5}, -3/2), \qquad \ell_i^c \equiv (1, 5/2), \quad i = 1, 2, 3 \quad \text{generation inde}$$
$$H_1 \equiv (\underline{10}, 1/2); \quad \overline{H}_1 \equiv (\overline{10}, -1/2): \qquad \text{"Heavy" Higgs}$$
$$h_1 \equiv (5, -1); \quad \bar{h}_1 \equiv (\bar{5}, 1): \qquad \text{"light" Higgs}$$
$$\phi_m \equiv (1, 0), \quad m = 1, 2, 3, 4 (\equiv N_g + 1): \qquad \text{"singlets"} \tag{43}$$

where now, the numbers in the parentheses indicate the SU(5) reps and the U(1) charges. Notice that $\bar{f} \sim \bar{5}$ and $F(H) \sim \underline{10}$ as in eq. (41) while $h_1 \sim 5$ as in eq. (38), the reason being that $F_i(H), \bar{f}_i, \ell_i^c$ fill in a $\underline{16}$ rep. of SO(10) ($M_i \equiv F_i + \bar{f}_i + \ell_i^c$), while $h_1 + \bar{h}_1$ come from a $\underline{10}$ rep. of SO(10).

The most general superpotential, consistent with $SU(5) \times U(1) \times (Z_2 : H_1 \to -H_1)$ is

$$\begin{aligned} W = & \lambda_1^{ij} F_i F_j h_1 + \lambda_2^{ij} F_i \bar{f}_j \bar{h}_1 + \lambda_3^{ij} \bar{f}_i \ell_j^c h_1 + \lambda_4 H_1 H_1 h_1 \\ & + \lambda_5 \bar{H}_1 \bar{H}_1 \bar{h}_1 + \lambda_6^{im} F_i \bar{H}_1 \phi_m + \lambda_7^m h_1 \bar{h}_1 \phi_m + \lambda_8^{mnp} \phi_m \phi_n \phi_p \end{aligned} \tag{44}$$

The reason for the extra (R–symmetry like) $Z_2 : H_1 \rightarrow -H_1$ should be apparent, it avoids disastrous couplings e.g., $F_i H_j h_1$ which would make d^c, s^c, b^c superheavy, etc.

Each and every term in the superpotential $W(44)$ has an important physical role: λ_1 is responsible for down quark masses, λ_2 gives up quark masses, λ_3 gives charge fermion masses, while λ_4, λ_5 realize, as explained above, the missing partner mechanism. The role of λ_6 is very important, it realizes the see–saw mechanism as originally conceived:[71] use an extra singlet to couple to ν_i^c and make it (Dirac) superheavy, while leaving ν_i massless, and not by using suitable big reps give to ν_i^c a Majorana–type superheavy mass.[72]

The see saw mass matrix for the neutrinos is of the form

$$
(\nu_i, \nu_i^c, \phi_m)
\begin{pmatrix}
0 & m_u^{ij} & 0 \\
m_u^{ji} & 0 & \lambda_6^{im}\langle \bar{H}_1 \rangle \\
0 & \lambda_6^{mi}\langle \bar{H}_1 \rangle & O(M_W)
\end{pmatrix}
\begin{pmatrix}
\nu_i \\
\nu_i^c \\
\phi_m
\end{pmatrix}
\tag{45}
$$

where we have explicitly indicated the $Q = 2/3$ quark masses $m_u^{ij} (\equiv \lambda_2^{ij} < h_1 >)$, and which gives light eigenstates $\nu_i + O(\frac{m_u}{M_X})\phi_m$ with masses $O(\frac{M_W^3}{M_X^2})$ and Dirac eigenstates $\frac{(\nu_i^c \pm \phi_i)}{\sqrt{2}} + O(\frac{m_u}{M_X})\nu_i$ with masses $O(M_X)$.

The role of λ_7 is to provide the needed electroweak scale mixing in SUSY theories between h_1 and $\bar{h}_1$, by allowing some $\langle \phi_m \rangle \sim O(M_W)$, while λ_8 makes it sure that such a case is feasible.

It should be noticed, that because of the new structure of the $\underline{5}$ and $\underline{10}$ (eq. (41)), we do not automatically get[22] $(m_i)_{Q=-1/3} = (m_i)_{Q=-1}$ at M_X, as in the G-G SU(5) model,[9] since here in principle λ_1 and λ_3 are unrelated. At least for the heaviest generation, $m_b = m_\tau|_{M_X}$ is a very successful relation, because when renormalized down to low energies it gives $m_b \sim 3m_\tau$, the right relation.[22] But don't panic, we will recover it later in the stringy flipped SU(5)![63,64] Also notice, that the see–saw mechanism eq. (45) is rather imperative in flipped SU(5), otherwise we would end up with $(m_i)_{Q=0} = (m_i)_{Q=2/3}$ at M_X, as implied by the λ_2 term only, and in the absence of a λ_6 term! A rather embarrassing situation. So, λ_6-type terms are useful.

Recently, a detailed study[73] of this SUSY flipped SU(5) model has produced some interesting results: 1) It is possible to realize a "double" radiative breaking mechanism.[74] In other words, both breakings indicated in the chain (40) may be triggered by quantum effects thus determining dynamically that:[73]

$$
\frac{< H_1 >}{M_{SU}} \sim \frac{M_X}{M_{SU}} \sim 10^{-1-2}; \frac{\overset{(-)}{< h_1 >}}{M_{SU}} \sim \frac{M_W}{M_{SU}} \sim 10^{-16}
\tag{46}
$$

where M_X indicates the "meeting point" (unification scale) of the SU(3)$_C$ and SU(2)$_L$ gauge coupling constants g_3 and g_2, while M_{SU} indicates the super (or eventually string) unification scale where g_3, g_2 and g_1 meet. Remarkably enough, one gets[73] excellent predictions for $\sin^2 \theta_W$ and Λ_{QCD} (or α_3) for $M_X \in O(10^{15-16} GeV)$ and $M_{SU} \in O(10^{17} GeV)$. Eventually, M_{SU} is going to be identified with the superstring unification scale, calculated to be in the $O(10^{17} GeV)$ range! One also should notice that the relevant scale for proton decay is M_X and we finally predict[73]

$$\tau_p(p \to e^+ \pi^0) \simeq 10^{35 \pm 2} \text{ years.} \tag{47}$$

2) Concerning the "small" (electroweak) radiative breaking we remark that consistency with low–energy phenomenology demands that[73]

$$60 \, GeV \leq m_{top} \leq 90 \, GeV, \tag{48}$$

and that the $\frac{<h_1>}{<\bar{h}_1>}$ ratio is dynamically determined to be[73]

$$\frac{<h_1>}{<\bar{h}_1>} \sim O(1/5) \tag{49}$$

for most of the phenomenologically (e.g., $m_{\text{squarks,gluinos}} > O(60\text{--}80 \text{ GeV})$) allowed region. We will later see the great importance of eq. (49) when we discuss hierarchical fermion masses in the stringy flipped SU(5) model.

While the ordinary SUSY flipped SU(5) model[61] has several remarkable properties, still it leaves a lot of space for improvement. Consistency with the programme of grand unification (e.g., charge quantization), makes it desirable to embed flipped SU(5) into some bigger (simple) group like SO(10), E_6... in the framework of point–like quantum field theory. Here we are running into trouble. In the case of SO(10), the λ_4 coupling of eq. (44) would become $\underline{16}_{H_1} \cdot \underline{16}_{H_1} \cdot \underline{10}_{h_1}$ which decomposes under SU(5) × U(1) as $(\underline{10} + \bar{\underline{5}} + 1)_{H_1}(\underline{10} + \bar{\underline{5}} + 1)_{H_1}(\underline{5} + \bar{\underline{5}})_{h_1}$ and thus we don't get[62] only superheavy h_1 triplets ($\langle \underline{10}\rangle_{H_1} \underline{10}_{H_1} \underline{5}_{h_1}$) as before, but also superheavy h_1–doublets ($\langle \underline{10}\rangle_{H_1} \underline{5}_{H_1} \bar{\underline{5}}_{h_1}$)! All initial motivation has been lost, the partners finally found each other. In the case of E_6, each generation ($F_i, \bar{f}_i, \ell_i^c$) belongs to a $\underline{27}$ as is the case for H_1, h_1. The Yukawa couplings should come from $\underline{27}^3$ or $\overline{\underline{27}}^3$ and here starts the trouble. The λ_2 couplings of eq. (44) imply that $\bar{h}_1$ belongs to a $\underline{27}$ while the λ_5 coupling suggests that $\bar{h}_1$ belongs to a $\overline{\underline{27}}$: the $\bar{h}_1$ becomes schizophrenic.[62] Further extension to bigger groups is still more problematic. Another, central issue that it is not addressed by the ordinary SUSY flipped SU(5) is the generation (or family) problem. The superpotential of eq. (44), consistent with all available symmetries cannot distinguish between different generations, so the Yukawa couplings $\lambda_1^{ij}, \lambda_2^{ij}, \lambda_3^{ij}$, should be arbitrarily chosen, e.g., $\frac{\lambda^e}{\lambda^t} \sim O(10^{-5})$. We expect a little better from an ambitious model and thus we turned to

Stringy Flipped SU(5)[63,64] has been suggested to overcome all the above problems, and provide a realistic model directly derived from 4-D superstrings. We use the fermionic construction[51] of 4-D heterotic superstrings, which entails the following field content: (the three numbers in the subscript parenthesis refer to the $U(1)_1 \times U(1)_2 \times U(1)_3$ quantum numbers of the state.)

$$M_i = [F_i \equiv (10, \tfrac{1}{2}), \ \bar{f}_i \equiv (\bar{5}, -\tfrac{3}{2}), \ l_i^c \equiv (1, \tfrac{5}{2})]_{(\tfrac{1}{2}, 0, 0)} \text{ for } i = 1$$

$$(0, \tfrac{1}{2}, 0) \text{ for } i = 2$$

$$(0, 0, -\tfrac{1}{2}) \text{ for } i = 3$$

$H_1 \equiv (10, \tfrac{1}{2})_{(\tfrac{1}{2}, 0, 0)}$	;	$\bar{H}_1 \equiv (\overline{10}, -\tfrac{1}{2})_{(-\tfrac{1}{2}, 0, 0)}$
$H_2 \equiv (10, \tfrac{1}{2})_{(0, \tfrac{1}{2}, 0)}$	;	$\bar{H}_2 \equiv (\overline{10}, -\tfrac{1}{2})_{(0, -\tfrac{1}{2}, 0)}$

(50)

"Heavy" Higgs

$h_i \equiv (5, -1)_{(-1, 0, 0)}$ for $i = 1$; $\bar{h}_i \equiv (\bar{5}, 1)_{(1, 0, 0)}$ for $i = 1$

$(0, -1, 0)$ for $i = 2$; $(0, 1, 0)$ for $i = 2$

$(0, 0, 1)$ for $i = 3$; $(0, 0, -1)$ for $i = 3$

$h_{12} \equiv (5, -1)_{(\tfrac{1}{2}, \tfrac{1}{2}, 0)}$; $\bar{h}_{12} \equiv (\bar{5}, 1)_{(-\tfrac{1}{2}, -\tfrac{1}{2}, 0)}$: "Light" Higgs

$\phi_{23} \equiv (1, 0)_{(0, 1, 1)}$; $\bar{\phi}_{23} \equiv (1, 0)_{(0, -1, -1)}$

$\phi_{13} \equiv (1, 0)_{(1, 0, 1)}$; $\bar{\phi}_{13} \equiv (1, 0)_{(-1, 0, -1)}$

$\phi_{12} \equiv (1, 0)_{(1, -1, 0)}$; $\bar{\phi}_{12} \equiv (1, 0)_{(-1, 1, 0)}$: "Singlets"

Clearly, the resemblance of the string particle spectrum (Eq. (50)) to the minimal (desired) spectrum (43) is stunning. Naturally as it is a dynamically derived particle spectrum Eq. (50) is far richer than Eq. (43) as is also the case now for our dynamically derived flipped $SU(5) \times U(1)$ which is extended to $[SU(5) \times U(1)] \times U(1)^3$. We will show next that this dynamical "augmentation" of the particle spectrum/gauge group leads to far better physics than the ordinary SUSY flipped $SU(5)$ model.

The most general renormalizable (trilinear) superpotential which is allowed by the $SU(5) \times U(1)^4$ symmetry is[64]

$$\begin{aligned} W_R = &\lambda_1 \ (F_1 F_1 h_1, \ F_2 F_2 h_2, \ F_3 F_3 h_3) + \lambda_2 \ (F_1 \bar{f}_2 \bar{h}_{12}, \ F_2 \bar{f}_1 \bar{h}_{12}) \\ &+ \lambda_3 \ (\bar{f}_1 l_1^c h_1, \ \bar{f}_2 l_2^c h_2, \ \bar{f}_3 l_3^c h_3) + \lambda_4 \ (H_1 H_1 h_1, \ H_2 H_2 h_2) \\ &+ \lambda_5 \ (\bar{H}_1 \bar{H}_1 \bar{h}_1, \ \bar{H}_2 \bar{H}_2 \bar{h}_2) \\ &+ \lambda_7 \ (h_1 \bar{h}_2 \phi_{12}, \ h_1 \bar{h}_3 \bar{\phi}_{12}, h_2 \bar{h}_3 \phi_{23}, \ \bar{h}_1 h_2 \bar{\phi}_{12}, \ \bar{h}_1 h_3 \bar{\phi}_{13}, \ \bar{h}_2 h_3 \bar{\phi}_{23}) \\ &+ \lambda_8 \ (\phi_{12} \phi_{23} \bar{\phi}_{13}, \ \bar{\phi}_{12} \bar{\phi}_{23} \phi_{13}) \end{aligned}$$

(51)

It should be noticed that even if, as we discuss later, one combination of the $U(1)$'s is "anomalous",[75,76] the other $U(1)$'s alone suffice to fix the superpotential. Moreover, the "anomalous" $U(1)$ remains as a global symmetry of the theory. The other a priori possible

$SU(5) \times U(1)^4$-invariant superpotential terms of the forms $F_i H_i h_i$ and $H_1 \bar{f}_2 \bar{h}_{12}$, $H_2 \bar{f}_1 \bar{h}_{12}$ are forbidden by a generalized R-parity $\overset{(-)}{H}_{1,2} \leftrightarrow \overset{(-)}{H}_{1,2}$. Such an R-symmetry is a necessary feature[77] of $N = 1$, $D = 4$ supergravity with a $U(1)_{F-I}$ factor, such as is in this model, and it is an <u>automatic</u> symmetry of the massless modes coming from an underlying string theory.[78] Furthermore, even if this R-parity were absent, it would be possible to find a basis for the fields in which there is manifestly no lepton-Higgs mixing, and the light spectrum contains the three desired charges $+\frac{1}{3}$ antiquark triplets. Having explained and justified the structure of W_R (51) we proceed now to discuss it physical implications.

i) Gauge Symmetry Breaking:

The model (51) has a $\overset{(-)}{\phi}_{23}$ D- and F-flat direction: $\langle \phi_{23} \rangle = \langle \bar{\phi}_{23} \rangle \equiv V_{23} \neq 0$. There are other flat directions for the GUT-Higgses: $\langle H_1 \rangle = \langle \bar{H}_1 \rangle \equiv V_1 \neq 0$ and $\langle H_2 \rangle = \langle \bar{H}_2 \rangle \equiv V_2 \neq 0$. Values of $V_{1,2,23}$ very similar to those of Eq. (46) can be generated by a first stage of radiative corrections.[73] There is a unique way of breaking $SU(5) \times U(1)^4$ down to $SU(3)_C \times SU(2)_L \times U(1)_Y$ at the GUT scale. This is because one combination of the generation-dependent $U(1)$'s is "anomalous": $U(1)_1 + U(1)_2 - U(1)_3$ and thus automatically broken at M_{Pl} to render a consistent theory,[76] one combination is broken by V_{23}: $U(1)_2 + U(1)_3$, one combination is broken by V_1: $U(1) + U(1)_1$ and one combination is broken by V_2: $U(1) + U(1)_2$. Notice also that despite the fact that $V_{23} \neq 0$ gives large masses to the $\bar{h}_{2,3}$ (as well as to $\overset{(-)}{\phi}_{12}$, $\overset{(-)}{\phi}_{13}$) electroweak Higgses are available (h_1, $\bar{h}_{12}$) to give masses both to the charge $-\frac{1}{3}$ quarks and charged leptons: $\langle h_1 \rangle = v_1 \neq 0$, and to the charge $+\frac{2}{3}$ quarks: $\langle \bar{h}_{12} \rangle = \bar{v}_{12} \neq 0$. These v.e.v.'s are generated[73] by a second stage of radiative corrections and they are very similar to those of Eqs. (46) and (49).

Amazingly enough, while we started with a more complicated particle spectrum (51) and gauge group $SU(5) \times U(1)^4$ than those (43) of the ordinary flipped $SU(5) \times U(1)$ model, we succeeded to get rid (automatically) of the surplus, not needed massless states (e.g., $\bar{h}_{2,3}$) and still get a unique chain breaking

$$SU(5) \times U(1)^4 \xrightarrow[\substack{\langle H_{1,2} \rangle \equiv V_{1,2} \\ \langle \phi_{23} \rangle \equiv V_{23}}]{M_X} SU(3)_C \times SU(2)_L \times U(1)_Y \xrightarrow[\substack{\langle h_1 \rangle \equiv v_1 \\ \langle h_{12} \rangle \equiv \bar{v}_{12}}]{M_W} SU(3)_C \times U(1)_{E.M.} \quad (52)$$

Furthermore, a phenomenological analysis[73] of this stringy model renders even more satisfactory results concerning the values of $\sin^2 \theta_W$, Λ_{QCD}, than the ordinary SUSY flipped $SU(5)$. Especially the value of M_{SU}, which was identified before with the superunification scale where g_3, g_2, and g_1 meet, literally coincides with the value of the superstring unification scale derived from first principles, as it should be the case for a consistent string-derived model! Notice also that there is no disaster with stringy proton decay, especially there are no dangerous light color Higgs triplets, so it follows the same pattern discussed above[73] (see Eq. (47)). Having found an acceptable (and phenomenologically interesting) gauge symmetry breaking pattern down to the standard model, makes it possible to study the low-energy physics implications of our stringy flipped $SU(5)$ model.

To start with we notice that the superpotential W_R (51) has to be reduced in compliance with our findings above. Several states (h_2, h_3, ϕ_{12}, ϕ_{13}) are getting superheavy masses and thus they are effectively ostracized from low-energy physics. Things become very constrained indeed. Clearly, the "reduced" W_R hardly contains all the terms necessary for "good" low energy physics. Even, the "unreduced" W_R of Eq. (51) does not contain any λ_6-type see-saw couplings, which are of fundamental importance as discussed previously. What is going on? Is the model dead? No! There are non-renormalizable superpotential terms to be added to W_R (51). These are generic features of string theory, and many can be identified as originating from the exchanges of specific massive modes. They are restricted not only by $SU(5) \times U(1)^4$ invariance but also by R-invariance[70,78] and by string theory arguments[79] which identify flat directions at the string tree level. The arguments strongly suggest[79] that the above mentioned flat directions of the trilinear superpotential (W_R) remain flat in the presence of the non-renormalizable terms. These suggestions, not only support the stability of the above discussed gauge symmetry breaking pattern, but at the same time can be used as a "selection rule" for allowable non-renormalizable terms.

ii) Hierarchical Fermion Masses:[64]

The renormalizable, trilinear superpotential W_R (51), when suitable "reduced" in order to expel irrelevant superheavy states, yields

$$W'_R \ni \lambda_1 F_1 F_1 h_1 + \lambda_2 (F_1 \bar{f}_2 \bar{h}_{12} + F_2 \bar{f}_1 \bar{h}_{12}) + \lambda_3 (\bar{f}_1 l_1^c h_1) \tag{53}$$

as the only terms relevant for fermion masses. At first sight, such an output does not look very impressive! We get only one charge $-\frac{1}{3}$ quark mass; two charge $\frac{2}{3}$ quark masses and one charge -1 lepton mass, different from zero. But remember, quarks or leptons have a hierarchical mass structure: $m_t \geq O$ (50 GeV) and $m_{u,d} \sim O$ (MeV) while $m_e \simeq 0.5$ MeV and $m_\tau \simeq 1.7$ GeV. So, maybe we are getting something that we should get from a correct theory. At tree level only t, b, τ and c are getting direct masses and then we have to use non-renormalizable terms to give masses to the other fermions naturally suppressed by inverse powers of M_{SU} and thus hierarchical. A mechanism proposed many years ago,[80] finally is finding its proper place. Notice that, for historical reasons, we have to identify (F_1, $\bar{f}_1$, l_1^c) with (t, b, τ, ν_τ), (F_2, $\bar{f}_2$, l_2^c) with (c, s, μ, ν_μ) and (F_3, $\bar{f}_3$, l_3^c) with (u, d, e, ν_e). At this level it is interesting to stress that m_t has a tendency to be heavier than m_b or m_τ, because $m_t \propto < \bar{h}_{12} >$ and $m_{b,\tau} \propto < h_1 >$ and $\frac{<h_1>}{<\bar{h}_{12}>} \sim \frac{1}{5}$ according to Eq. (49). Thus, the disparity between Yukawa couplings (λ_1, λ_2, λ_3) need not be as big as in more conventional approaches, and is in full accordance with explicit calculations of certain Yukawa couplings in 4-D superstring theories.[81] Furthermore, as a residue of the intermediate $SO(10)$ symmetry, we recover,[64] for the first time in superstring theories, the relation $m_b = m_\tau$ at M_{SU}, which implies $m_b \simeq 3m_\tau$ at low-energies, a rather successful prediction. The same $SO(10)$ symmetry implies that $F_1 \bar{f}_2 \bar{h}_{12}$ and $F_2 \bar{f}_1 \bar{h}_{12}$ couplings (λ_2) are equal at the superunification scale, as is shown in Eqs. (53) and (55). This does not mean that $m_c = m_t$. One reason is that the $F_{1,2}$ fields have as yet undetermined mixings with the $H_{1,2}$ fields. Since the (F_1, H_1) and (F_2, H_2) diagonalizations are independent, the couplings of the physical t and c quarks may be different. Moreover, the renormalization group equations for the couplings of F_1 and F_2 are different because of the large $\lambda_1 F_1 F_1 h_1$ coupling, implying $m_c \neq m_t$ after renormalization. Next, putting together all the lowest order non-trivial contributions to the fermion mass matrices including the ones coming from non-renormalizable terms, as explained before, we get (here I have set $M_{SU} \equiv 1$):

$$M_{-\frac{1}{3}} = v_1 \begin{pmatrix} \lambda_1 & \lambda_{11}V_1V_2 & O(\epsilon) \\ \lambda_{11}V_1V_2 & \lambda_{12}V_1^2V_2^2 & O(\epsilon) \\ O(\epsilon) & O(\epsilon) & \lambda_{13}V_1^2V_2^2V_{23} \end{pmatrix} \tag{54}$$

for the charge $-\frac{1}{3}$ quark mass matrix,

$$M_{\frac{2}{3}} = \bar{v}_{12} \begin{pmatrix} \lambda_{21}V_1V_2 & \lambda_2 & O(\epsilon) \\ \lambda_2 & \lambda_{22}V_1V_2 & O(\epsilon) \\ O(\epsilon) & O(\epsilon) & \lambda_{23}V_1V_2V_3 \end{pmatrix} \tag{55}$$

for the charge $+\frac{2}{3}$ quark mass matrix, and

$$M_{-1} = v_1 \begin{pmatrix} \lambda_3 & \lambda_{31}V_1V_2 & O(\epsilon^2) \\ \lambda_{32}V_1V_2 & \lambda_{33}V_1^2V_2^2 & O(\epsilon^2) \\ O(\epsilon^2) & O(\epsilon^2) & \lambda_{34}V_1^2V_2^2V_{23} \end{pmatrix} \tag{56}$$

for the charge -1 lepton mass matrix. Interestingly enough, since we expect, as discussed above, V_1, V_2 and V_{23} to be large, but smaller than M_{SU} (see also Eq. (46)), the non-renormalizable-term contributions to $M_{-\frac{1}{3}, \frac{2}{3}, -1}$ are hierarchically smaller than the renormalizable terms $\lambda_1, \lambda_2, \lambda_3$ in (53). Thus, a hierarchical mass-matrix with small Calibbo-Kobayashi-Maskawa mixing angles is finally emerging.[64,82] Notice though, that at this stage there is still no mixing between $(t, b; c, s)$ with (u, d)! This is because of a discrete symmetry P_3: $M_3 \rightarrow -M_3$ which is respected by all interaction terms in the low-energy effective theory. A mechanism, involving the low-energy supersymmetry breaking scale $\tilde{m}$, has been suggested[83] to break this discrete symmetry and allow all types of mixing. In this case,[83] $\epsilon \simeq \frac{\lambda^2}{32\pi^2}\tilde{m}$ in Eqs. (54) - (56), which produces reasonable mixings for $\tilde{m} \in [100 \text{ GeV}$ - 1 TeV], the favorable range of $\tilde{m}$, suggested by the solution of the gauge hierarchy problem. Similarly, in the case of the neutrino mass matrix, the allowable non-renormalizable terms are able to provide all needed effective λ_6-like couplings and thus the see-saw mechanism is recovered![64] Actually, because of the P_3 discrete symmetry, discussed above, there is hardly any mixing(s) between (see Eq. (34)) (ν_τ, ν_μ) and (ν_e) while we get very different masses for the three neutrinos:[64,82)

$$m_{\nu_\tau,\nu_\mu} \simeq O(eV)$$
$$m_{\nu_e} \simeq O(10^{-6} - 10^{-9}\ eV)$$

(57)

It is rather amazing how very different, dynamically derived, properties of the model conspire to produce a hierarchical fermion mass spectrum. The extra gauge group $U(1)^3$, which came out of the string theory when we asked only for a $SU(5) \times U(1)$ gauge group, plays an important role in restricting the tree-level allowable Yukawa couplings, as well as the specific form of non-renormalizable terms. Clearly, the $U(1)^3$ group plays the role of the family group, but here it is dynamically derived, local and understandable! Then, the existence of flat directions in W_R (51), supported by string-derived arguments, produces dynamically determined values for V_1, V_2, V_{23} in the range shown in Eq. (46), which is exactly what is needed to make the mass-matrices (54), (55) and (56) hierarchical. Meanwhile Eq. (49) makes it plausible that m_t is bigger than m_b or m_τ etc. I really find it hard to believe that all this is mere coincidence. With respect to the ordinary SUSY flipped $SU(5)$ model, the stringy flipped $SU(5)$ is clearly winning in the following points:

(i) The $SU(5) \times U(1)$ is primordially embedded in a simple group $SO(44)$, thus automatic recovery of charge quantization without destroying the missing partner mechanism.

(ii) The needed Z_2 (or $Z_2 \times Z_2$) symmetry(ies) is of dynamical origin,[77,78) as it is the $U(1)^3$ accompanying family group mainly responsible for the observed hierarchical fermion mass spectrum.

I believe these points are strong enough to go the superstring way, without even invoking the usual "consistency of Quantum Gravity" argument(s) as supporting evidence.

Acknowledgements

I would like to express my sincere and heartfelt thanks to the Organizing Committee of the 3rd ESO–CERN symposium for their warm and generous hospitality. Their excellent choice of place (Bologna) and time (900th anniversary of Bologna University) was impecable, as it was the atmosphere created by the local organizers and authorities. This work was supported in part by U.S. Department of Energy Contract No. AC02–76ER–00881 and in part by the University of Wisconsin Alumni Research Foundation.

378

References

1. For a review of the present status of the standard model see: R. Peccei, contribution in these proceedings.

2. S.L. Glashow, J. Iliopoulos and L. Maiani, Phys. Rev. D2 (1970) 1285.

3. M. Kobayashi and T. Maskawa, Progr. Th. Phys. 49 (1973) 652.

4. J. Bailey et al., Phys. Lett. 68B (1977) 191; F. Combley, F.J.M. Farley and E. Picasso, Phys. Rep. 68 (1981) 93.

5. For a recent review see: J.H. Pendlebury, in the proceedings of the Ninth Workshop on Grand Unification, Aix–les–Bains (Savoie), France, April 1988.

6. D.A. Ross and M. Veltman, Nucl. Phys. B95 (1975) 135.

7. G. Costa et al, Nucl. Phys. **B297** (1988) 244; U. Amaldi et al, Phys. Rev. **D36** (1987) 2191.

8. UA1 Collaboration, G. Arnison et al., Phys. Lett. 122B (1983) 103; ibid. 126B (1983) 398; UA2 Collaboration, M. Banner et al., Phys. Lett. 122B (1983) 476; P. Bagnaia et al., Phys. Lett. 129B (1983) 130.

9. H. Georgi and S.L. Glashow, Phys. Rev. Lett. **32** (1974) 438.

10. J.A. Wheeler, Relativity, Groups and Topology, eds. B.S. and C.M. de Witt, (Gordon and Breach, New York, 1963); S. Hawking, D.N. Page and C.N. Pope, Phys. Lett. 86B (1979) 175 and Nucl. Phys. B170 (FS1) (1980)283.

11. E. Witten, Nucl. Phys. B188 (1981) 513; M. Dine, W. Fischler and M. Srednicki, Nucl. Phys. **B189** (1981) 575; S. Dimopoulos and S. Raby, Nucl. Phys. **B192** (1981) 353;
For earlier attemtps, see: P. Fayet, in Unification of the Fundamental Particle Interactions, ed. S. Ferrara, J. Ellis and P. Van Nieuwenhuizen, (Plenum Press, New York, 1980), p. 587.

12. S. Dimopoulos and H. Georgi, Nucl. Phys. B193 (1981) 150; N. Sakai, Zeit für Phys. C11 (1982) 153.

13. Y.A. Gol'fand and E.P. Likhtman, Pis'ma Zh. Eksp. Theor. Fiz. 13 (1971) 323; D. Volkov and V.P. Akulov, Phys. Lett. 46B (1973) 109; J. Wess and B. Zumino, Nucl. Phys. B70 (1974) 39.
For a review see: P. Fayet and S. Ferrara, Phys. Rep. 32C (1977) 249.

14. S. Hawking, Commun. Math. Phys. 87 (1982) 395.

15. J. Ellis, J.S. Hagelin, D.V. Nanopoulos and M. Srednicki, Nucl. Phys. B241 (1984) 381.

16. D. Gross, Nucl. Phys. B236 (1984) 349.

17. T. Banks, L. Susskind and M. Peskin, Nucl. Phys. B244 (1984) 125; S. Hawking, Nucl. Phys. B244 (1984) 135; G. Parisi, Phys. Lett. 136B (1984) 392.

18. S. Hawking, Phys. Lett 134B (1984) 403; S. Coleman, Nucl. Phys. B307 (1988) 867 and to be published.

19. For a review see: M.J. Duff, B.E.W. Nilsson and C.N. Pope, Phys. Rep. 130 (1986) 1.

20. D. Förster, M. Ninomiya and H.B. Nielsen, Phys. Lett. 94B (1980) 135; J. Iliopoulos, D.V. Nanopoulos and T. Tomaras, Phys. Lett. 94B (1980) 141.

21. H. Georgi, H.R. Quinn and S. Weinberg, Phys. Rev. Lett. 33 (1974) 451.

22. A.J. Buras, J. Ellis, M.K. Gaillard and D.V. Nanopoulos, Nucl. Phys. B135 (1978) 66.

23. For GUT reviews, see: P. Langacker, Phys. Rep. 72C (1981) 185; "Grand Unification with and without Supersymmetry and Cosmological Implications" by C. Kounnas, A. Masiero, D.V. Nanopoulos and K.A. Olive in International School for Advanced Studies Lecture Series No. 2, World Scient. Publ. Comp. (1984).

24. For a recent review see: P. Langacker, DESY preprint DESY 88-076 (1988).

25. D.V. Nanopoulos and D.A. Ross, Nucl. Phys. B157 (1979) 273 and Phys. Lett. 108B (1982) 351.

26. C.T. Hill, Phys. Rev. D24 (1981) 691; M.A. Machacek and M.T. Vaughan, Phys. Lett. 103B (1981) 427.

27. Folklore of the late 70's.

28. For a review see: K. Enqvist and D.V. Nanopoulos, Progr. in Particle and Nucl. Phys. 16 (1986) 1.

29. R. Haag, J.T. Lopuszanski and M. Sohnius, Nucl. Phys. B88 (1975) 257.

30. D.Z. Freedman, P. van Nieuwenhuizen and S. Ferrara, Phys. Rev. D13 (1976) 3214; S. Deser and B. Zumino, Phys. Lett. 62B (1976) 335.

31. J. Wess and B. Zumino, Phys. Lett. 49B (1974) 52; J. Iliopoulos and B. Zumino, Nucl. Phys. B76 (1974) 310; S. Ferrara, J. Iliopoulos and B. Zumino, Nucl. Phys. B77 (1974) 413; M.T. Grisaru, W. Siegel and M. Rocek, Nucl. Phys. B159 (1979) 420.

32. G. 't Hooft, Phys. Rev. Lett. 37 (1976) 8; Phys. Rev. D14 (1976) 3432; R. Jackiw and C. Rebbi, Phys. Rev. Lett. 37 (1976) 172; C.G. Callan, R.F. Dashen and D.J. Gross, Phys. Lett. 63B (1976) 334.

33. J. Ellis, S. Ferrara and D.V. Nanopoulos, Phys. Lett. 114B (1982) 231.

34. V. Baluni, Phys. Rev. D19 (1979) 2227; R.J. Crewther et al., Phys. Lett 88B (1979) 123; [Erratum, 91B (1980) 487].

35. S. Mandelstam, Phys. Lett. 121B (1983) 30; Nucl. Phys. B213 (1983) 149; P.S. Howe, K.S. Stelle and P.K. Townsend, Nucl. Phys. B236 (1984) 125.

36. P.S. Howe, K.S. Stelle and P.C. West, Phys. Lett. 124B (1983) 55.

37. S. Dimopoulos, S. Raby and F.A. Wilczek, Phys. Rev. D24 (1981) 1681; L.E. Ibánez and G.G. Ross, Phys. Lett. 105B (1982) 439.

38. M.B. Einhorn and D.R.T. Jones, Nucl. Phys. B196 (1982) 475.

39. J. Ellis, D.V. Nanopoulos and S. Rudaz, Nucl. Phys. B202 (1982) 43.

40. S. Weinberg, Phys. Rev. D26 (1982) 287; N. Sakai and T. Yanagida, Nucl. Phys. B197 (1982) 533.

41. S. Dimopoulos, S. Raby and F.A. Wilczek, Phys. Lett. 112B (1982) 133.

42. J. Ellis, J.S. Hagelin, D.V. Nanopoulos and K. Tamvakis, Phys. Lett. 124B (1983); See also: V.M. Belyaev and M.I. Vysotsky, Phys. Lett. 127B (1983) 215.

43. E. Cremmer, S. Ferrara, C. Kounnas and D.V. Nanopoulos, Phys. Lett 133B (1983) 61.

44. J. Ellis, A.B. Lahanas, D.V. Nanopoulos and K. Tamvakis, Phys. Lett 134B (1984) 429; J. Ellis, C. Kounnas and D.V. Nanopoulos, Nucl. Phys. B241 (1984) 406, ibid B247 (1984) 373 and Phys. Lett 143B (1984) 410; J. Ellis, K. Enqvist and D.V. Nanopoulos, Phys. Lett. 147B (1984) 27.

45. For a review see: A.B. Lahanas and D.V. Nanopoulos, Phys. Rep. 145 (1987) 1.

46. E. Witten, Phys. Lett. 155B (1985) 151; J. Ellis, C. Gómez and D.V. Nanopoulos, Phys. Lett. 171B (1986) 203; S. Ferrara, C. Kounnas and M. Porati, Phys. Lett. 181B (1986) 263.

47. M.B. Green, J.H. Schwarz and E. Witten, "Superstring Theory", Cambridge Univ. Press, Cambridge (1987).

48. S. Ferrara, R. Gatto and A.F. Grillo, Nuov. Cim. 12A (1972) 959; F. Mansouri and Y. Nambu, Phys. Lett. 39B (1972) 375; A.M. Polyakov, Phys. Lett. 103B (1981) 207, 211.

49. K.S. Narain, Phys. Lett. 169B (1986) 41; W. Lerche, D. Lüst and A.N. Schellekens, Nucl. Phys. 287 (1987) 477; H. Kawai, D.C. Lewellen and S.H.H. Tye, Phys. Rev. Lett. 57 (1986) 1832 and Nucl. Phys. 288 (1987) 1.

50. I. Antoniadis, C. Bachas, C. Kounnas and P. Windey, Phys. Lett. 171B (1986) 51.

51. I. Antoniadis, C. Bachas and C. Kounnas, Nucl. Phys. B289 (1987) 87; I. Antoniadis and C. Bachas, Nucl. Phys. B298 (1988) 586.

52. D.J. Gross, J.A. Harvey, E. Martinec and R. Rohm, Nucl. Phys. B256 (1985) 468 and 267 (1986) 75.

53. F. Gliozzi, J. Scherk and D. Olive, Phys. Lett. **65B** (1976) 282 and Nucl. Phys. **122B** (1977) 253.

54. H. Dreiner, J. Lopez, D.V. Nanopoulos and D. Reiss, Madison preprint MAD/TH/88–17.

55. P. Candelas, G. Horowitz, A. Strominger and E. Witten, Nucl. Phys. **B258** (1985) 46; E. Witten, Nucl. Phys. **258** (1985) 75.

56. Y. Hosotani, Phys. Lett. **129B** (1983) 193.

57. S. Cecotti, J.P. Derendinger, S. Ferrara, L. Girardello and M. Roncadelli, Phys. Lett. **156B** (1985) 318; M. Dine, V. Kaplunovsky, M. Mangeno, C. Nappi and N. Seiberg, Nucl. Phys. **259B** (1986) 519; J.D. Breit, B.A. Ovrut and G. Segré, Phys. Lett. **188B** (1987) 415.

58. J. Ellis, K. Enqvist, D.V. Nanopoulos and K. Olive, Phys. Lett. **188B** (1987) 415; and Nucl. Phys. **297B** (1988) 103.

59. B. Greene, K.H. Kirklin, P.J. Miron and G.G. Ross, Phys. Lett. **180B** (1986) 69; and Nucl. Phys. **B274** (1986) 574 and **B278** (1986) 667; L. Ibanez, H.P. Nilles and F. Quevedo, Phys. Lett. **187B** (1987) 25 and **192B** (1987) 332; D. Bailin, A. Love and S. Thomas, Phys. Lett. **194B** (1987) 385 and Nucl. Phys. **298B** (1988) 75.

60. S.M. Barr, Phys. Lett. **112B** (1982) 219; J.P. Derendinger, J.E. Kim and D.V. Nanopoulos, Phys. Lett. **139B** (1984) 170.

61. I. Antoniadis, J. Ellis, J.S. Hagelin and D.V. Nanopoulos, Phys. Lett. **194B** (1987) 231.

62. B. Campbell, J. Ellis, J.S. Hagelin, D.V. Nanopoulos and R. Ticciati, Phys. Lett. **198B** (1987) 200.

63. I. Antoniadis, J. Ellis, J.S. Hagelin and D.V. Nanopoulos, Phys. Lett. **205B** (1988) 459.

64. I. Antoniadis, J. Ellis, J.S. Hagelin and D.V. Nanopoulos, Phys. Lett **208B** (1988) 209.

65. T. Hübsch, S. Meljanac, S. Pallua and G.G. Ross, Phys. Lett. **161B** (1985) 122.

66. J. Ellis, M.K. Gaillard and D.V. Nanopoulos, Phys. Lett. **80B** (1979) 360; B. Campbell, J. Ellis and D.V. Nanopoulos, Phys. Lett. **141B** (1984) 229.

67. A. Masiero, D.V. Nanopoulos, K. Tamvakis and T. Yanagida, Phys. Lett. **115B** (1982) 380; B. Grinstein, Nucl. Phys. **206B** (1982) 387; C. Kounnas, D.V. Nanopoulos, M. Srednicki and M. Quiros, Phys. Lett. **127B** (1983) 82.

68. X.G. Wen and E. Witten, Nucl. Phys. **261B** (1985) 651.

69. K. Yamamoto, Phys. Lett. **168B** (1986) 341; K. Enqvist, D.V. Nanopoulos and M. Quiros, Phys. Lett. **169B** (1986) 343.

70. B. Campbell, J. Ellis, J.S. Hagelin, D.V. Nanopoulos and K. Olive, Phys. Lett. **197B** (1987) 355.

71. H. Georgi and D.V. Nanopoulos, Nucl. Phys. **155B** (1979) 52.

72. M. Gell–Mann, P. Ramond and R. Slansky, in "Supergravity". Proceedings of the Supergravity workshop, Brookhaven, 1979, ed. by P. Van Nieuwenhuizen and D.Z. Freedman (Amsterdam, 1979), p. 315.

73. J. Ellis, J.S. Hagelin, S. Kelley and D.V. Nanopoulos, Madison–CERN prepint MAD/TH/88–8 and CERN–TH.4990 (1988).

74. J. Ellis, J.S. Hagelin, D.V. Nanopoulos and K. Tamvakis, Phys. Lett. **125B** (1983) 275.

75. R. Barbieri, S. Ferrara and D.V. Nanopoulos, Zeit. für Phys. **C13** (1982) 267 and Phys. Lett. **116B** (1982) 16.

76. M. Dine, N. Seiberg and E. Witten, Nucl. Phys. **289B** (1987) 589.

77. R. Barbieri, S. Ferrara, D.V. Nanopoulos and K.S. Stelle, Phys. Lett. **113B** (1982) 219.

78. S. Cecotti, S. Ferrara and M. Villasante, Int. J. Mod. Phys. **A2** (1987) 1839.

79. I. Antoniadis, J. Ellis, C. Kounnas, and D.V. Nanopoulos, MAD/TH/88–24 (1988).

80. J. Ellis and M.K. Gaillard, Phys. Lett. **88B** (1979) 315; D.V. Nanopoulos and M. Srednicki, Phys. Lett. **124B** (1983) 37.

81. I. Antoniadis, J. Ellis, E. Floratos, D.V. Nanopoulos and T. Tomaras, Phys. Lett. **191B** (1987) 96; S. Ferrara, L. Girardello, C. Kounnas and M. Porrati, Phys. Lett. **192B** (1987) 368 and **194B** (1987) 358.

82. G. Leontaris and D.V. Nanopoulos, Phys. Lett **212B** (1988) 327.

83. G. Leontaris, Phys. Lett. **207B** (1988) 447.

HIGH REDSHIFT OBJECTS

Richard G. Kron
Yerkes Observatory
P.O. Box 0258
Williams Bay, WI 53191
USA

ABSTRACT. This contribution reviews some topics in the field of high-redshift objects. Attention is directed to particular current issues, namely the redshift distribution for QSOs at the highest redshifts; the redshift distribution for field galaxies and its application for studies of cosmology and galaxy evolution; and the nature of the radio galaxies of highest redshift.

1. Introduction

In the recent past the highest known redshift for quasars (QSOs) and for radio galaxies has been substantially increased, partly due to the development of search techniques and partly due to improved instrumentation. These new observations naturally lead to a number of new questions related to the nature, origin, and evolution of the sources, both individually and as populations.

The discussion is organized by the categories of QSOs, normal galaxies, and radio galaxies. One of the interesting questions is the relationship between QSOs and radio galaxies, and the relationship between ordinary galaxies and both radio galaxies and QSOs. The conventional categorization of sources adopted here is not intended to imply that the distinctions drawn from objects at low redshift necessarily apply at larger redshifts.

Not discussed here are a number of other important topics in the field of high-redshift objects, such as absorption lines seen in QSOs, radio structures at large redshifts, gravitational lenses, and spatial correlations of the various sources. The review is thus far from being comprehensive, but the selection of topics is intended to be at least representative of problems in the field.

The review is restricted to those objects with spectroscopically-measured redshifts, as opposed to redshifts inferred from photometric quantities. In this same spirit, source counts without spectroscopic redshift information are not considered.

2. QSOs

2.1. REDSHIFT CUTOFF

2.1.1. *Background.* A feature in the redshift distribution would have profound cosmological significance, as it would suggest that at least some objects act as clocks and

M. Caffo et al. (eds.), Astronomy, Cosmology and Fundamental Physics, 383–398.
© 1989 by Kluwer Academic Publishers.

point to a particular cosmic epoch as being special. The establishment of a feature in the redshift distribution would be a unique accomplishment in observational cosmology. It is of the greatest importance to limit the epoch of condensation of mass on various scales in the universe, because in a variety of models structures of galactic mass are expected to form late (Blumenthal *et al.* 1984). Efstathiou and Rees (1988) have discussed how the number of $z > 4$ QSOs can be used as a powerful constraint on the conventional Cold Dark Matter picture. QSOs are a natural probe because they can be seen to enormous redshifts and they are expected to be associated with galaxies of stars (e.g. Boroson, Persson, and Oke 1985). The appearance of C, N, and O lines in the spectra of QSOs at $z > 4$ provides further constraints on the history of nucleosynthesis.

The term "redshift cutoff" implies a sharp decline in density with increasing redshift. It is normally understood that the comoving densities are evaluated over some particular range in rest-frame luminosity. (Both densities and luminosities depend on the adopted cosmological model, but the choice of the cosmological model cannot obscure the presence of a real feature in the redshift distribution.)

The idea of a cutoff in the QSO redshift distribution was proposed by Sandage (1972). Osmer (1982) later concluded that his non-detection of $z > 3.7$ QSOs was statistically significant, and this result has strongly influenced subsequent discussions (e.g. Schmidt and Green 1983). However there is no consensus on the exact redshift or character of the cutoff (cf. Véron 1986). Schmidt, Schneider, and Gunn (1988) concluded on the basis of new data that the comoving density was lower at $z = 3.3$ by a factor of seven with respect to the density at $z = 2.2$. Koo and Kron (1988) developed an alternative picture that not only had no sharp features in the redshift distribution, but had no decline in densities, up to redshifts much higher than so far sampled. Others have adopted a middle ground, proposing that there is a decline in density of QSOs beyond some redshift (e.g. Hazard, McMahon, and Sargent 1986; Warren, Hewett, and Osmer, 1988), without necessarily implying a feature in the redshift distribution.

2.1.2. *Sources of Uncertainty.* There is room for skepticism that a real feature in the redshift distribution has been positively identified, *viz*:

> The Osmer (1982) and the Schmidt, Schneider, and Gunn (1988) studies both compared results from different surveys in different redshift intervals.
> If the luminosity function evolves in shape, the choice of the luminosity range in the comparison of different redshift intervals involves some arbitrariness.
> To address this possibility one can instead trace the luminosity function down in luminosity to a limit such that the emissivity converges. Koo and Kron did this (1988) and found no decline in the emissivity with increasing redshift (out to the limit of the available data, $z \sim 3$).
> The statistics are very poor — often results are claimed based on samples with only a dozen or even fewer QSOs in each redshift interval.

These points and the following discussion are based only on published results, and may be substantially changed once new surveys are completed.

2.2. QSO SURVEYS

To avoid the dangers involved with comparing different surveys, one can instead evaluate individual surveys that each have a broad redshift range, a redshift selection function that can be evaluated reliably, and a large number of objects. The limiting magnitude should be as faint as possible, because the steep slope at bright magnitudes can make the results sensitive to calibration errors. With these survey criteria in mind I have selected for discussion three large-telescope slitless spectroscopic surveys and two multicolor surveys.

Crampton, Cowley, and Hartwick (1987) have published a IIIa-J + blue grens survey with the Canada-France-Hawaii Telescope, 1000 Å mm^{-1}, with follow-up slit spectroscopic confirmation. The effective wavelength range is $\lambda\lambda\,3500$ 5400. I will restrict the discussion to the 4 deg^2 subset. The B magnitude *versus* redshift diagram for all $z > 1.0$ objects in this restricted area is shown in Figure 1a.

Vaucher and Weedman (1980) undertook another IIIa-J survey (at 1600 Å mm^{-1}) with the Kitt Peak grism. Magnitudes and redshifts are given for a restricted area of 2.7 deg^2, and these points have been added to Figure 1a. In general their redshifts are obtained directly from the grism plates, but many of the points plotted, especially those near $z \sim 2.5$, have slit spectroscopic redshifts also.

Osmer (1980) published slit spectroscopic results for the Hoag and Smith (1977) IIIa-F survey at CTIO (also 1600 Å mm^{-1}). The area covered is 5.1 deg^2. Individual fields were centered on previously known QSOs; these QSOs are not considered in the following analysis. The data for $z > 1.0$ are plotted in Figure 1a, the magnitudes having been corrected to the observer's-frame B waveband. Although the IIIa-F emulsion can in principle detect QSOs with Lyman α up to $z = 4.7$, the net wavelength sensitivity favors lower redshifts because the grism was blazed for the violet (see Carswell and Smith 1978 for a detailed discussion).

The Koo and Kron (1988) color-selected survey in 0.29 deg^2 is spectroscopically complete to $B = 21$. We claim that the redshift selection should be reasonably uniform, although as the redshift exceeds 3 there may be a greater likelihood that some QSOs could be overlooked as stars. The data for $B < 21$ are plotted in Figure 1b.

Marano, Zamorani, and Zitelli (1988) have conducted a survey in 0.69 deg^2 that is very similar to that of Koo and Kron (1988) and which is also spectroscopically complete to $B = 21$. The data for $z > 1.0$ are plotted in Figure 1b.

2.2.1. *Redshift-Magnitude Diagrams*. A reasonable estimate of the completeness limit for the continuum flux for the Osmer (1980) and Vaucher and Weedman (1980) surveys is $B = 19.8$. It is apparent from Figure 1a that the Crampton, Cowley, and Hartwick (1987) survey is complete to a deeper limit, which is expected from differences in seeing and spectral dispersion, but for convenience I adopt $B = 19.8$ generally. Note also that the Crampton, Cowley, and Hartwick (1987) survey is less incomplete for $z < 1.8$ than the other two surveys because lines weaker than Lyman α can be more readily detected with their higher dispersion.

Figures 1ab have luminosity limits drawn in according to the K-correction for a "standard" QSO given in Koo and Kron (1988), and assuming $H_0 = 50$ km sec^{-1} Mpc^{-1} and $q_0 = 0.5$ (these values are adopted throughout the following discussion). Too high a luminosity threshold would yield very poor statistics, and too low a threshold would make the higher redshifts incomplete (for example, in Figure 1b the chosen limit will evidently select against some high-redshift objects).

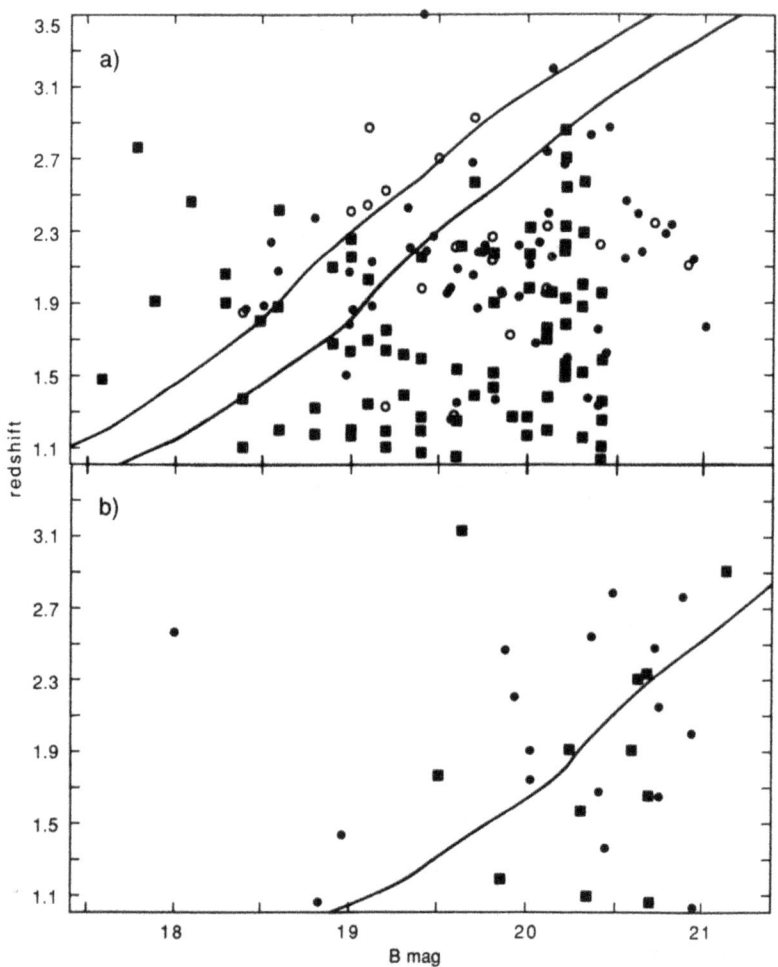

Figure 1a. — Redshift *vs.* apparent B magnitude for three QSO grism surveys complete to $B = 19.8$. Open circles: Vaucher and Weedman (1980); squares: Crampton, Cowley, and Hartwick (1987); filled circles: Osmer (1980). The loci are lines of constant restframe absolute B magnitude of -27.0 and -26.5, as described in the text.

Figure 1b. — Redshift *vs.* apparent B magnitude for two QSO color-selected surveys complete to $B = 21$. Filled circles: Marano, Zamorani, and Zitelli (1988); squares Koo and Kron (1988). In both cases the published J magnitudes have been transformed to B by $B = J + 0.1$. The absolute magnitude locus is $M_B = -25.3$.

2.2.2. *Redshift Distributions*. Figure 2a shows the derived redshift distribution for the QSOs selected by the grens and grism surveys according to a rest-frame absolute magnitude criterion $M_B < -26.5$. Figure 2b shows the redshift distribution for all QSOs in these same surveys to a slightly brighter limit, $M_B < -27.0$. Figure 2c shows the redshift distribution for the two color-selected surveys, extending to $M_B = -25.3$. In the range of redshifts around $z \sim 2$ the volume element dV/dz is nearly constant with redshift, and the plotted histograms are therefore a representation of the comoving densities.

Figure 2a does show a decline with increasing redshift, though not necessarily by a factor of seven between $z = 2.2$ and $z = 3.3$ (note the absence of QSOs in the bin $1.5 \leq z < 1.7$). Schmidt, Schneider, and Gunn (1988) also adopted the Crampton, Cowley, and Hartwick (1987) and Osmer (1980) surveys to establish their density point at $z = 2.2$, but in their analysis the line flux, rather than the continuum flux, was considered.

An apparent decline in Figure 2ab is qualitatively expected on the basis of the photographic emulsion wavelength sensitivity function and other factors (Carswell and Smith 1978). It must be shown that the redshift selection is properly understood before strong statements can be made that there is a cutoff, or even a decline, in QSO densities at large redshifts, especially so since a selection by another absolute magnitude limit, as in Figure 2b, does not show the effect nearly as strongly. The fact that Figure 2c shows no convincing decline either could be taken as evidence for incompleteness in the slitless surveys, or it could mean that the luminosity function has changed in shape. The absence of QSOs at very large z is presumably due to the combined limits in luminosity and in apparent magnitude. In any case the statistical precision is clearly not great.

In Figure 1 the most luminous QSOs are among the highest-redshift QSOs. Thus if Figure 2 were weighted according to luminosity rather than numbers, the decline would be less apparent.

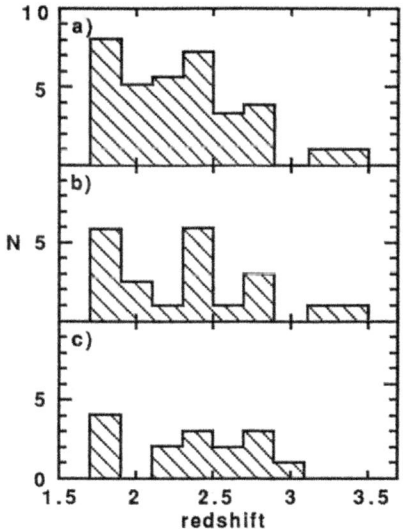

Figure 2a. — Redshift distribution, $\Delta z = 0.2$, derived from Figure 1a for all QSOs with $M_B < -26.5$. Objects falling near the limit in luminosity are counted as one-half.

Figure 2b. — Same as for Figure 2a, except the lower limit in absolute magnitude is 0.5 mag brighter.

Figure 2c. — Redshift distribution derived from Figure 1b for all QSOs with $M_B < -25.3$.

2.2.3. *Surveys to Higher Redshifts.* The previous comments do not directly address the results at still higher redshifts of Osmer (1982) and Schmidt, Schneider, and Gunn (1988). Osmer (1982) found no QSO in 5.1 deg^2 with $3.7 < z < 4.7$ and $m_{6100} < 20$, whereas he expected 10 or so with the assumption of no evolution between $z \sim 2.5$ and $z \sim 4$ (see also Schmidt and Green 1983). On the other hand, an extrapolation of the luminosity function of Koo and Kron (1988) to these high redshifts would predict about one QSO to have appeared in Osmer's window. This comparison is however difficult because Koo and Kron's standard band (photographic J, $\lambda \sim 4650$) is far from m_{6100} and experiences large K-correction at large z. Moreover, Osmer analysed line strengths whereas Koo and Kron analysed continuum fluxes.

A more direct comparison comes from the important new survey by Warren *et al.* (1987), which is elaborated in Warren, Hewett, and Osmer (1988). In 30 deg^2 they found 3 QSOs with $m_R < 20$ in the interval $4 \le z < 4.5$, and 21 QSOs in the interval $3 \le z < 4$. An approximate scaling of these results to 5.1 deg^2 would suggest one or possibly two QSOs in Osmer's range $3.7 < z < 4.7$. [It is also well to keep in mind the remarkable serendipitous discovery of a QSO with $z = 4.39$ at $R = 21.5$ (McCarthy *et al.* 1988), currently the object with the second highest known redshift.] Taken at face value, these numbers indicate comoving densities that are lower than the densities at $z = 2$ by a factor of about 4 for $3 \le z < 4$ and a factor of about 6 for $4 \le z < 4.5$. The completeness of their survey is currently being evaluated. Warren, Hewett, and Osmer (1988) also remark on the apparent flatness of the luminosity function at high redshifts, consistent with the tentative proposals by Hazard, McMahon, and Sargent (1986) and Koo and Kron (1988). If this change in the shape of the luminosity function is confirmed, then arguments for a cutoff that are based on a particular range in absolute magnitude would be luminosity-dependent.

According to Schmidt, Schneider, and Gunn (1988), 44 QSOs have been found in 14.3 deg^2 in their new 4-shooter survey. Of these, 15 have detected Lyman α, implying $2.6 < z < 4.3$, which suggests a total QSO surface density of about 8 deg^{-2} (using the redshift distribution of Koo and Kron 1988). If the survey were limited by continuum flux, this surface density would correspond to a completeness limit in the continuum of only $R_{lim} \sim 19$, but most of the QSOs in Schmidt, Schneider, and Gunn (1987) are fainter than this. With an imperfect knowledge of the line strength distribution at fixed continuum luminosity at high redshift, it is not possible to make reliable transformations between surveys limited by line flux and those limited by continuum flux. In principle counts to a limit in line flux may show a density decline that is not followed by counts to a limit in continuum flux.

Despite these uncertainties, the prospect for the future are bright. Surveys for QSOs will proceed at an accelerated rate thanks to efficient fiber-coupled spectrographs. The new surveys will comprise thousands, rather than hundreds, of QSOs, and this greater statistical weight will allow the detailed shape of the luminosity function to be established as a function of redshift. Techniques for enhancing the selection of QSOs at the highest redshifts are being refined, and some future surveys will use these advances to concentrate on the problem of the shape of the high-redshift tail.

3. Normal Galaxies

3.1. COSMOLOGICAL TESTS

Observable properties of galaxies allow them to be used as tracers of the geometry of the universe on very large scales, assuming that the evolutionary effects are properly understood. The classical cosmological tests involve some indicator of proper distance, the measurement of which is independent of the measurement of redshift. For a uniformly-distributed, non-evolving population of sources, this distance indicator could be the apparent radius θ, the apparent brightness $\mathcal{L}$, the apparent number of sources per solid angle A, and, if the sources can be used as clocks, the look-back time τ. (Other physical properties can also be measured at large distances under special conditions.) The surface brightness $I = \mathcal{L}/\pi\theta^2 = I_0(1 + z)^{-4}$. Since this relation is independent of the cosmological model, it follows that $\mathcal{L}$ and θ^2 are equivalent in terms of cosmological information; together they provide a crucial consistency check. An especially powerful test is the combination $A(\mathcal{L}, z)$ or $A(\theta^2, z)$. These relations are computed for a specific cosmological model, and they can be compared with measurements once the statistical distributions of $\mathcal{L}$ and θ^2 are specified. This "classical" approach leads to a direct measure of the deceleration parameter q_0, and in particular the method is sensitive to the cosmological consequences of uniformly-distributed matter. For practical application of the $A(\mathcal{L}, z)$ and $A(\theta^2, z)$ tests, the following are required:

The statistical distribution functions of intrinsic luminosities and sizes for galaxies, as derived within a local volume that is large enough to be representative.

Techniques for measuring A, $\mathcal{L}$, and θ^2 to large distances that a) optimize the signal-to-noise ratio, and b) control for redshift-dependent selection effects.

Some way to control for evolution of the observable properties of the sources. Populations of sources may become luminous, or may coalesce, over the time span represented by the observations, and number evolution as well as luminosity evolution must be considered.

No definitive measurement of q_0 as derived from studies of galaxies at large redshift has yet been accomplished, largely because these conditions have not been satisfied.

3.2. FAINT GALAXY SPECTRA

To illustrate the quality of information that is currently being achieved for distant galaxies, I show in Figure 3 examples of spectra of galaxies from the field redshift survey that I have been working on with D. Koo, A. Szalay, and J. Munn (Koo and Kron 1986). (By "field" galaxy I mean a galaxy chosen for study without regard for its environment.) The galaxies illustrated all have $R \sim 21$, high redshift, and a spectrum of reasonable quality. The spectra in the survey are generally unexciting, but there are cases of strong [O II] λ 3727 (e.g. Figure 3a), enhanced Balmer absorption lines (e.g. Figure 3b), and cold, old-looking energy distributions (e.g. Figure 3c).

A selection of the high-redshift tail, as in Figure 3, by definition yields objects that are uncommon. For a given luminosity, the objects so selected will be those with relatively small K-corrections or relatively strong spectral features. These effects, which must be present at some level, have to be very well understood before unexpected differences in samples at large redshifts can be taken to be evidence for galaxy evolution.

3.3. COMPLETENESS LIMIT

3.3.1. *Signal-to-Noise Ratio.* Catalogues of galaxies are usually stated as being complete to some limit in magnitude, apparent diameter, or surface brightness. In practice observations are actually limited by the signal-to-noise ratio achieved in a certain integration time. This applies both to the initial selection of catalogued galaxies, and to subsequent spectroscopic studies. The signal-to-noise ratio, S/N, may itself be considered

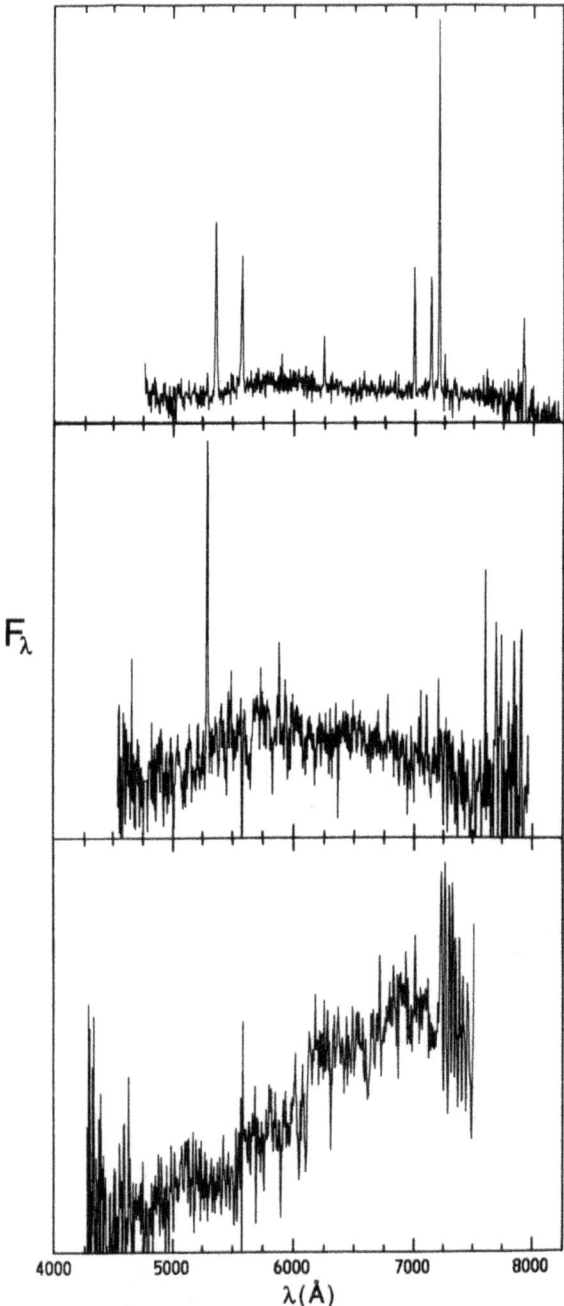

Figure 3abc. — Top: Galaxy a10601, $z = 0.4384$, 4000 sec exposure; middle: galaxy a04920, $z = 0.4186$, 12000 sec exposure; bottom: galaxy b12564, $z = 0.5374$, 37000 sec exposure. The ordinate is linear in flux per unit wavelength, with 0 at the bottom. All spectra were obtained with multislit masks and the KPNO Cryogenic Camera.

as the extrinsic quantity that is appropriate for sample definition.

As the measurement aperture for a galaxy is enlarged, the S/N of the measurement of the enclosed light $L(\theta)$ increases at first and then declines, according to the surface brightness of the galaxy with respect to the sky background, and according to the shape of the image profile. Galaxies of the same apparent total magnitude can thus have a wide range of S/N, which also depends on the size of the galaxy image compared to the size of the seeing disk. When the sky noise dominates within the optimum radius θ, the S/N $\sim \theta I$, where I is the average surface brightness interior to the optimum radius.

If effects of profile blur due to seeing are ignored, the sensitivity of the S/N to the cosmological model can be easily derived as follows.

Let D = source proper radius, and $rR_0 = f(q_0, H_0, z)$, the cosmological expansion parameter. Then since
$\theta \sim D(1 + z)/rR_0$,
$I \sim I_0(1 + z)^{-4}$, and
S/N $\sim \theta I$,
it follows that
$(S/N)^2 \sim D^2 I_0^2 (1 + z)^{-6} (rR_0)^{-2}$.

This result can be compared to the sensitivity of the apparent magnitude L to the cosmological model, which differs from the above by a factor of $I_0(1 + z)^{-4}$, or simply I. The practical effect of selecting according to S/N depends on the statistical distribution of θI for the galaxies in a sample. Different types of galaxies have different characteristic surface brightnesses: blue galaxies have lower characteristic surface brightnesses than redder galaxies (Sandage 1983), and there is also a dependence on total luminosity (Binggeli, Sandage, and Tarenghi 1984).

3.3.2. Selection Effects and the Redshift Distribution. Figure 4 shows that the effect of selection by $(S/N)^2$, to be compared to the selection by magnitude. To calculate this I have assumed the gaussian luminosity function for Virgo cluster spirals of Sandage, Binggeli, and Tammann (1985), and the dependence of I_e and θ_e on M_B of Binggeli, Sandage, and Tarenghi (1984). The result is schematic because the calculation presumes a direct relation between the characteristic surface brightness I_e and the average surface brightness within the optimum radius. The magnitude intervals of Figure 4 and the S/N limits were chosen to match approximately the survey of Broadhurst, Ellis, and Shanks (1988). Confinement to only a particular galaxy class (Sb) is for convenience, and does not affect the generality of the argument.

Both Koo and Kron (1986) and Broadhurst, Ellis and Shanks (1988) found no extended tail to high redshifts in their comparison of data to models for $A(z)$ at fixed magnitude, and concluded that the expected signature of higher luminosities at higher redshifts was not detected. Both groups also remarked on the lack of large numbers of dwarf galaxies in their respective surveys. If one considers that the measured redshifts are limited instead by S/N, then according to Figure 4 the conclusions could be substantially changed, because of the selection in favor of especially large or especially high surface-brightness galaxies. Once again it appears that very careful modelling of the selection process at each redshift is required before we can evaluate either evolutionary or cosmological effects.

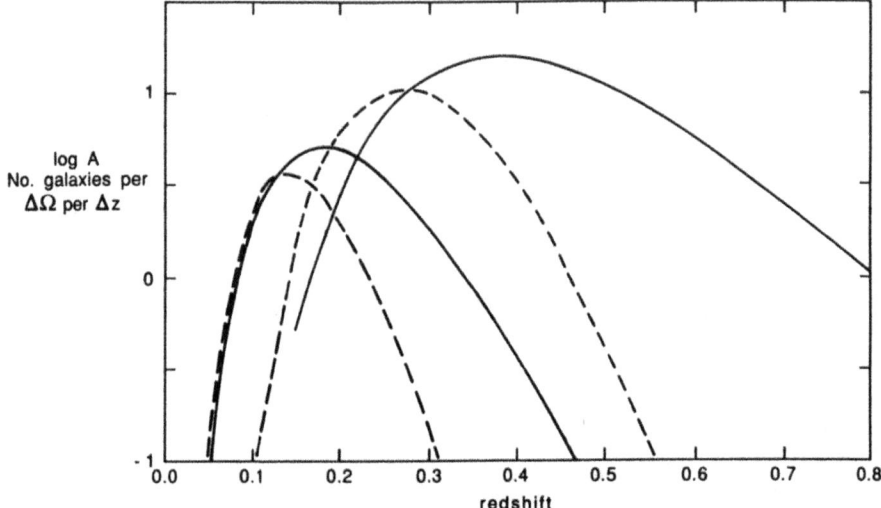

Figure 4. — Schematic relative differential redshift distributions according to different selection criteria. The solid curves give the log number of galaxies selected by magnitude in one-magnitude intervals at $B = 20.0$ and $B = 22.5$. The assumptions are a gaussian luminosity function with $\mu = -20$ and $\sigma = 1.1$, $q_0 = 0.5$, a K-correction appropriate to Sb galaxies, and no evolution. The dashed curves give the redshift distribution for two intervals in the quantity $(S/N)^2$ that correspond approximately to the two magnitude intervals.

3.4. INTEGRATION TIME AS A FUNCTION OF REDSHIFT

The detection problem is actually worse, and the selection effects are correspondingly larger, when the spectral character of the background sky noise and the character of typical energy distributions for galaxies are considered explicitly. In order better to evaluate selection effects, it would be desirable to observe distant galaxies with the same S/N per angular and velocity resolution element as for nearby galaxies. This may seem like an overly ambitious goal, but the following exercise is nevertheless worthwhile to show the loss in information as redshift increases.

One approach is to consider a galaxy observed in some fixed observed band $\Delta\lambda_{obs}$. In this case the velocity resolution in the galaxy rest frame is independent of redshift. The favorable case is for a galaxy with an energy distribution $f_\lambda \sim$ const. The surface brightness, specifically the number of photons sec^{-1} arcsec^{-2} $\Delta\lambda_{obs}^{-1}$, depends on z as $(1 + z)^{-5}$, and for fixed S/N in the background-limited case, the required integration time goes as $(1 + z)^{10}$! Thus an 8 meter telescope would have to integrate 6 times as long at $z = 0.5$ as a 4 meter telescope at $z = 0.1$, at the same angular and spectral resolution, surface brightness, and S/N, assuming no spread of information due to seeing.

The alternative is to observe the same spectral region by an appropriate shift $\lambda_{obs} = (1 + z)\lambda_{em}$ and $\Delta\lambda_{obs} = (1 + z)\Delta\lambda_{em}$. In this case the number of photons sec^{-1} arcsec^{-2} $\Delta\lambda_{obs}^{-1}$ goes as $(1 + z)^{-3}$. However, in the optical range the sky continuum background increases towards longer wavelengths very approximately as $f_\lambda \sim \lambda$ (this again leads to a favorable conclusion), which means that the number of photons in the sky background within $\Delta\lambda_{obs}$ increases as $(1 + z)^3$. The result is that the required exposure time increases as $(1 + z)^9$, which is not much better than before.

The conclusion is that either a substantial loss of information as the redshift increases has to be anticipated in the analysis of distant galaxy data, or very long exposures must be acquired, even with the largest telescopes. This explains why the largest absorption-line redshifts so far obtained for normal galaxies still do not exceed $z \sim 1$.

4. Radio Galaxies

4.1. PHENOMENOLOGY

The identification of strong radio sources with optically faint extended objects has led to the discovery of galaxies of the highest known redshifts. The sources typically have a double-lobed, edge-brightened radio morphology with steep radio spectra. The optical spectra show strong (sometimes very strong) emission of generally lower ionization and velocity width than is typical for active nuclei. There appears to be a correlation of the strength of the [O II] λ 3727 line and radio power (Spinrad 1986). Lyman α is visible for all objects with $z > 1.6$ (Spinrad et al. 1985). Often the emission is extended, sometimes over very large regions (~100 kpc, as in 3C 326.1, McCarthy et al. 1987a). Ratios of emission line strengths are compatible with photoionization, but the deduced energetics are more plausible if either the central ionizing source radiates anisotropically, or if the ionization occurs in situ (McCarthy et al. 1987b). The kinematical field is often coherent, but there are also a number of examples of velocity structure (e.g. 3C 368, Djorgovski et al. 1987a) with components extending beyond ~ 1000 km sec^{-1} in the rest frame (e.g. 3C 294, Djorgovski 1988). The great strength of the lines would allow in some cases higher spectral resolution to be achieved, which may in turn reveal narrow components.

The emitting gas follows the morphology of the underlying continuum, and both tend to be aligned with the principal radio axis (Chambers et al. 1987, McCarthy et al. 1987). This last result was a surprise because the alignment does not appear nearly as strongly at $z < 0.6$. Aside from this apparent evolutionary effect, the great significance is that the radio properties and the optical properties are clearly coupled, which means that the selection as a radio source cannot be ignored when evaluating the optical data. In some ways this development could have been anticipated, since there is growing awareness that the presence of an active nucleus (in these cases revealed by the radio source) is somehow correlated with greatly enhanced star formation over extended regions (Sanders et al. 1988, and references therein).

4.2. COMPARISON WITH NEARBY COUNTERPARTS

Powerful radio galaxies at large redshift thus have a rich phenomenology, and the challenge is to place the observations into a context that can be related to events observed nearby. One of the reasons that this comparison is difficult is simply the extremes encountered: the high-redshift objects are the most luminous radio sources within the observable horizon (the 3CR catalogue has been essentially completely identified), and the space density is sufficiently low that there can be no local comparison sample, even in principle: Cygnus A notwithstanding, statistically one would need to look to $z > 0.3$ in order to find one object of radio power comparable to the 3CR galaxies in the range $1 < z < 2$.

Extended low-level line emission is a reasonably common event in nearby radio galaxies (Heckman et al. 1986; Robinson et al. 1987), but in the distant samples the emission is far more spectacular, exceeding by orders of magnitude the luminosities seen locally. The high-power Heckman et al. sample has log $P_{408\,MHz} \sim 26.7$ W Hz^{-1}, whereas the distant steep-spectrum 3CR sources have log $P_{175\ MHz} \sim 29.4$ W Hz^{-1} (Chambers, Miley, and van Breugel 1988). On the other hand Dunlop and Longair (1988) have shown that distant radio sources that are ~ 100 times fainter than the 3CR radio galaxies nevertheless share many of the optical properties of the 3CR galaxies.

The complex optical morphology often observed in the distant radio galaxies is suggestive of galaxies merging or at least interacting (Djorgovski *et al.* 1987). This view is popular because of the analogy with nearby active nuclei and the evidence for such galaxy interaction associated with them. However, the evidence for disturbances in the nearby radio galaxies studied by Heckman *et al.* (1986) and Hutchings (1987) are on scales and at surface brightnesses such that they would often be difficult to detect in the distant objects. Moreover as noted earlier the strong correlation of the alignment of the optical and radio elongation axes only appears at high redshift. Collimation of the jets and subsequent interaction with the ambient gas may provide a trigger for large-scale star formation (e.g. van Breugel 1988), thus accounting for the morphological alignment. A secular change in the density of gas or the number of clouds might account for the redshift dependence. Alternatively, the evolution of the optical properties may be a reflection of the evolution of the radio source properties. The merger hypothesis may describe the morphology only indirectly, for example if the nuclear source were activated by the merger.

4.3. NATURE OF THE OPTICAL CONTINUUM LIGHT

Since the radio galaxies are the most distant galaxies known (in fact spanning the range of redshifts of QSOs), it is natural that they have been used as probes of early galaxy formation and evolution. Not a great deal is known about the continuum emission in view of its great faintness at high redshift (§ 3.4). Since stellar absorption lines have yet to be revealed for $z > 1$, arguments that the continuum light is actually of stellar origin are indirect:

> The continuum is extended.
> The line-emitting region follows the continuum, as expected for a mixture of gas and stars.
> The emission lines are often like those from H II regions, i.e. consistent with *in situ* photoionization from hot stars.
> The continuum luminosities are consistent with bright galaxies of stars.

A stellar agglomeration with normal surface brightness would not be visible at large z because of the strong relativistic dimming (e.g. Weedman and Williams 1987), and a scaling to account for the enhanced luminosity would probably yield an image that is more compact than some of those observed. Thus it is clearly of great interest to explore more fully the nature of the continuum light from these objects. For example, much of Lilly's (1988) argument about the age of the 0902+34 system at $z = 3.4$ depends on the shape of the energy distribution as observed in broad bands.

4.3.1. *Color-Color Diagrams as a Diagnostic.* If the continuum from distant radio galaxies really can be understood as light from ordinary stellar populations, then for each galaxy it should be possible to find a Bruzual model that matches the broad-band colors. By selecting a homogeneous photometric data base for galaxies all at about the same redshift, the data can be easily compared to the models. The photometry should cover a wide range in wavelength in order better to reveal curvature in the spectral energy distributions.

Lilly and Longair (1984) and Eisenhardt and Lebofsky (1987) have published photometry suitable for this purpose. Eisenhardt and Lebofsky's data were obtained with the same aperture size for each galaxy in all colors; the basic results from Lilly and Longair's data are essentially the same. Figure 5 shows $V_B - H$, $H-K$ colors from Eisenhardt and Lebofsky (1987) for seven galaxies in the range $0.89 < z < 1.13$. (Three galaxies are common to both studies.) Figure 5 shows that $H-K \sim 1$, and the dispersion around this value may in fact be largely observational error. This value for $H-K$ is indeed what is expected for composite stellar populations at $z \sim 1$. [QSOs at this redshift tend to have $H-K \sim 0.7$ (Hyland and Allen 1982).] The color $V_B - H$, however, shows a large range. The redder galaxies correspond reasonably closely to a stellar population with little or no recent star formation, but the bluer galaxies have colors that are not matched by Bruzual models.

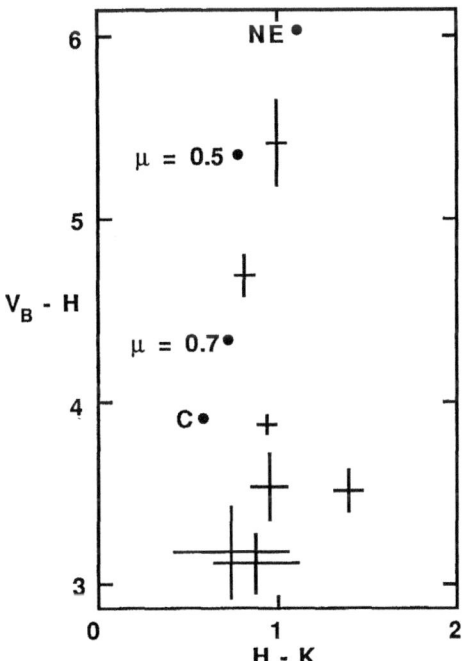

Figure 5. — Color - color diagram for seven 3CR galaxies in the interval $0.89 < z < 1.13$. Photometric data are from Eisenhardt and Lebofsky (1987), using their "P.C." values. The four Bruzual model points assume $z = 1$ and are taken from their Figure 3, the caption for which gives details.

An interesting question is whether the colors are correlated with the luminosity of the galaxy, as might be expected if a recent burst of star formation has both elevated the luminosity and made the colors bluer. No such correlation of H band luminosity and optical-infrared color is apparent in these data. Dunlop and Longair (1988) have however commented on a possible correlation in their sample of galaxies between degree of blueness and the strength of [O II] λ 3727.

A long-standing question is why distant luminous galaxies have such a small dispersion in luminosity. The dispersion in K-band luminosity of 3CR sources is only $\sigma_K \sim 0.4$ mag (Lilly and Longair 1984; Lebofsky and Eisenhardt 1986). This property was originally ascribed to some process that truncated the upper end of the giant elliptical luminosity function in rich clusters. However it now appears, as the previous discussion has emphasized, that the powerful radio galaxies at large redshift bear little resemblance to the quiescent bright ellipticals seen locally, and it also appears that the radio source is directly related to the optical emission (and to the near-infrared emission: see Chambers, Miley, and Joyce 1988). The continued exploitation of near-infrared imaging will permit both much higher photometric precision, and the possibility of studying color gradients.

This review was supported by NSF AST-8705517. I thank those who sent preprints of current work. The figures were prepared by Richard Dreiser, and I thank also Jeff Munn for assistance with Figure 3.

References

Binggeli, B., Sandage, A., and Tarenghi, M. 1984, *A.J.*, **89**, 64.

Blumenthal, G.R., Faber, S.M., Primack, J.R., and Rees, M.J. 1984, *Nature*, **311**, 517.

Boroson, T.A., Persson, S.E., and Oke, J.B. 1985, *Ap.J.*, **293**, 120.

Broadhurst, T.J., Ellis, R.S., and Shanks, T. 1988, preprint.

Carswell, R.F., and Smith, M.G. 1978, *M.N.R.A.S.*, **185**, 381.

Chambers, K.C., Miley, G.K., and van Breugel, W. 1987, *Nature*, **329**, 604.

Chambers, K.C., Miley, G.K., and Joyce, R.R. 1988, *Ap.J. (Letters)*, **329**, L75.

Chambers, K.C., Miley, G.K., and van Breugel, W.J.M. 1988, *Ap.J. (Letters)*, **327**, L47.

Crampton, D., Cowley, A.P., and Hartwick, F.D.A. 1987, *Ap.J.*, **314**, 129.

Djorgovski, S. 1988, in *Towards Understanding Galaxies at Large Redshift*, eds. R.G. Kron and A. Renzini (Kluwer:Dordrecht), p. 259.

Djorgovski, S., Spinrad, H., Pedelty, J., Rudnick, L., and Stockton, A. 1987, *A.J.*, **93**, 1307.

Dunlop, J.S. and Longair, M.S. 1988, preprint.

Eisenhardt, P.R.M. and Lebofsky, M.J. 1987, *Ap.J.*, **316**, 70.

Efstathiou, G. and Rees, M.J. 1988, *M.N.R.A.S.*, **230**, 5P.

Hazard, C., McMahon, R., and Sargent, W.L.W. 1986, *Nature*, **322**, 38.

Heckman, T.M., Smith, E.P., Baum, S.A., van Breugel, W.J.M., Miley, G.K., Illingworth, G.D., Bothun, G.D., and Balick, B. 1986, *Ap.J.*, **311**, 526.

Hoag, A.A. and Smith, M.G. 1977, *Ap.J.*, **217**, 362.

Hutchings, J.B. 1987, *Ap.J.*, **320**, 122.

Hyland, A.R. and Allen, D.A. 1982, *M.N.R.A.S.*, **199**, 943.

Koo, D.C. and Kron, R.G. 1988, in *Observational Cosmology, IAU Symp. 124*, eds. A. Hewitt, G. Burbridge, and L.Z. Fang (Reidel:Dordrecht), p.383

Koo, D.C. and Kron, R.G., 1988, *Ap.J.*, **325**, 92.

Lebofsky, M.J. and Eisenhardt, P.R.M. 1986, *Ap.J.*, **300**, 151.

Lilly, S.J. 1988, *Ap.J.*, **333**, 161.

Lilly, S.J. and Longair, M.S. 1984, *M.N.R.A.S.*, **211**, 833.

Marano, B., Zamorani, G., and Zitelli, V. 1988, *M.N.R.A.S.*, **232**, 111.

McCarthy, P.J., Spinrad, H., Djorgovski, S., Strauss, M.A., van Breugel, W., and Liebert, J. 1987a, *Ap.J. (Letters)*, **319**, L39.

McCarthy, P.J., van Breugel, W., Spinrad, H., and Djorgovski, S. 1987b, *Ap.J. (Letters)*, **321**, L29.

McCarthy, P.J., Dickinson, M., Filippenko, A.V., Spinrad, H., and van Breugel, W. 1988, *Ap.J. (Letters)*, **328**, L28.

Osmer, P.S. 1980, *Ap.J. Suppl.*, **42**, 333.

Osmer, P.S. 1982, *Ap.J.*, **253**, 28.

Robinson, A., Binette, L., Fosbury, R.A.E., and Tadhunter, C.N. 1987, *M.N.R.A.S.*, **227**, 97.

Sandage, A. 1972, *Ap.J.*, **178**, 25.

Sandage, A. 1983, in *Internal Kinematics and Dynamics of Galaxies, IAU Symp. 100*, ed. E. Athanassoula (Reidel:Dordrecht), p. 367.

Sandage, A., Binggeli, B., and Tammann, G.A. 1985, *A.J.*, **90**, 1759.

Sanders, D.B., Soifer, B.T., Elias, J.H., Madore, B.F., Matthews, K., Neugebauer, G., Scoville, N.Z. 1988, *Ap.J.*, **325**, 74.

Schmidt, M. and Green, R. 1983, *Ap.J.*, **269**, 352.

Schmidt, M., Schneider, D.P., and Gunn, J.E. 1987, *Ap.J. (Letters)*, **316**, L1.

Schmidt, M., Schneider, D.P., and Gunn, J.E. 1988, *A.S.P. Conf. Series*, **2**, 87.

Spinrad, H. 1986, *Pub. A.S.P.*, **98**, 269.

Spinrad, H., Filippenko, A.V., Wyckoff, S., Stocke, J.T., Wagner, R.M., and Lawrie, D.G. 1985, *Ap.J. (Letters)*, **299**, L7.

van Breugel, W.J.M. 1988, in *Hotspots in Extragalactic Radio Sources*, Schloss Ringberg, FRG, Feb. 1988.

Vaucher, B.G. and Weedman, D.W. 1980, *Ap.J.*, **240**, 10.

Véron, P. 1986, *Astron. Ast.*, **170**, 37.

Warren, S.J., Hewett, P.C., and Osmer, P.S. 1988, *A.S.P. Conf. Series*, **2**, 96.

Warren, S.J., Hewett, P.C., Osmer, P.S., and Irwin, M.J. 1987, *Nature*, **330**, 453.

Weedman, D.W. and Williams, K.L. 1987, *Ap.J.*, **318**, 585.

DISCUSSION.

Ulrich: In addition to the various ways of doing cosmochronology and cosmology which you summarised very nicely, there is another one based on chemical evolution. My question is: What is known of the abundances of the elements for high redshift objects from: 1) metal rich absorption line systems at $z > 3$ in quasar spectra; 2) high redshift galaxies $(z > 2)$; 3) emission lines in quasars at $z > 3$?

Kron: This is an intriguing question, but unfortunately the limits on chemical evolution are currently not too restrictive. 1) The magnesium doublet for $z > 3$ is redshifted into an instrumentally awkward part of the spectral band. See Meyer and York 1988 *Ap. J.* **319** ,L45 for a discussion of a detection of NiII at $z = 2.811$. 2) All known galaxies with $z > 2$ are radio galaxies. The emission line ratios are often consistent with HII regions, but on the other hand little is known about either the ionizing spectrum or the details of the radiation transfer. Thus, chemical abundances are highly model dependent. 3) The quasar emission spectra at $z > 3$ exhibit a wide range of types including some cases with relatively narrow lines. As far as I know, no systematic change in QSS spectra with redshift has yet been established.

Brandenberger: 1) What is known about the luminosity function of quasars? 2) Can one argue from the data that there are no galaxies (normal) with $z > 1$?

Kron: 1) The emerging picture is a luminosity function with a relatively flat slope at fainter luminosities, with a much steeper slope at higher luminosities. At $z = 2$, the change in slope occurs at roughly $M_B = $ -26.5, 500 QSOs per comoving cubic Gigaparsec per magnitude interval. These values are for the cosmological parameters $H_o = 50$ Km/sec Mpc^{-1} and $\Omega = 1$. As I discussed in my talk, the luminosity function evolves strongly with redshift. In particular, there is some evidence that the luminosity function at the bright end is flatter at very high redshifts. 2) Normal galaxies at $z > 1$ present formidable detection problems. Therefore, the absence of data for this population cannot be used to argue that such objects do not exist.

CONCLUDING LECTURE

L. Van Hove
CERN
Geneva

1. When preparing this concluding lecture, I looked back at the Proceedings of the two first ESO–CERN Symposia held at CERN in November 1983
and at ESO in March 1986, and I chose to relate what was presented in
Bologna to the evolution of the main topics in the last five years.
    While no fundamental advances can be reported, the main progress
lies in my opinion in the impressive amount of experimental, observational and interpretative work done in both disciplines. Nature did not
(yet?) fulfill any of the expectations raised by the ambitious speculations of theoretical physicists in the 70's and early 80's, which were
given much attention in the first Symposium. But, so to say in compen-
sation, came the unexpected and very fruitful gift of the supernova
of February 1987, which still goes under its ugly code SN1987A.
    In 1983, hopes were high to detect nucleon decays with lifetimes
of order 10^{30} to 10^{32} years, as predicted by the "grand unified", supersymmetry and supergravity schemes which were then popular. Such decays
were not found, but two of the huge underground nucleon decay experiments (IMB and Kamiokande) turned out to detect neutrino bursts which
are generally believed to originate from SN1987A (although one does
not yet understand all features of the data, e.g., the angular distributions). This marks the beginning of a second type of neutrino astronomy, the neutrino supernova watch, the first type concerning solar
neutrinos [M.Koshiba; names between square brackets refer to contributions to the Symposium].
    As reported at the Symposium by the Rome group, a few hours befo-
re the IMB and Kamiokande bursts, there may have been coincidences between neutrino signals and peaks in the spectra of two small gravitational wave detectors which were operating in Rome and Maryland. As far
as I can see, the evidence that these coincidences are linked with the
supernova is weak, and it is of course a great pity that none of the
existing large gravitational wave detectors was operating in February
of last year.
    As far as astronomical observations are concerned, although it
appears that SN1987A did not quite behave according to pre–existing
theoretical models for type 2 supernovae, most observations can be interpreted by adaptations of these models. The availability of an enormous amount of data ranging from visible light to gamma rays makes the

M. Caffo et al. (eds.), Astronomy, Cosmology and Fundamental Physics, 399–403.

supernova for many years to come a unique source of information on stellar collapse. The present status was discussed in the special supernova session of the Symposium [L.Woltjer, W.Hillebrandt, Y.Tanaka, J. Trumper, F.Pacini].

2. The dark matter problem, which became prominent in the early 80's, is now a very central issue for astronomers and particle physicists alike. Its nature and its abundance are still very uncertain. Efforts continue for the development of novel detectors [L.Gonzales-Mestres], and a Rochester-Brookhaven-Fermilab Collaboration has undertaken a very interesting search for cosmic axions [B.Moskowitz]. Even if no novel type of dark matter were found, the nucleon decay story might repeat itself, leading to a successful use of the new detectors for unforeseen purposes.

With one exception to be mentioned presently, the theoretical situation concerning the possible nature of dark matter has not changed much. In his report at the Symposium, M.Turner listed a great many possibilities but gave his preference to axions and to the old idea of a neutrino mass of the order of the present experimental limit for electron neutrinos.

The exception is linked with new work on the possible influence of the quark-hadron phase transition on the big bang nucleosynthesis (BBN). As was first suggested in 1985 [1], although the transition is predicted to take place at a temperature of the order of 200 MeV (i.e. $2.3 \ 10^{12}$K), it is conceivable that it created strong inhomogeneities in the space distribution of baryons, which could have survived for a minute or so till BBN time when the temperature was about 0.1 MeV. The key question turns out to be the cosmic abundance of lithium 7. If one takes the observed abundance to be primordial, it imposes interesting limits on the properties of the quark-hadron phase transition [H. Reeves]. If on the other hand one regards the primordial abundance of ^{7}Li to be uncertain [D.Lynden-Bell], the total density of baryonic matter could be much larger than the generally quoted value (which is around 3% of the critical density). This is the viewpoint developed by C.R. Alcock et al. and presented by him at the recent Bombay Conference on Physics and Astrophysics of Quark-Gluon Plasmas. They even consider the possibility that baryonic matter alone could achieve the critical density [2]!

The theoretical basis for the existence of the quark-hadron phase transition was discussed by H.Satz. From lattice Quantum Chromodynamics, there is very strong evidence for the existence of a first order transition, but its parameters are not accurately determined. Experimentally, the search for the quark-gluon plasma has recently started at CERN and Brookhaven with the beginning of operation of ion beams of up to 200 GeV per nucleon at CERN and up to 15 GeV per nucleon at Brookhaven. Satz also summarized the first results, especially the very interesting findings of the NA38 Collaboration at CERN concerning suppression of J/ψ production in central nucleus-nucleus collisions, which may indicate production of a short-lived plasma.

3. Regarding the large scale structure of the universe, the symposium
has further strengthened the contrast between the very inhomogeneous
distribution of luminous matter at the largest measured scales and the
high degree of isotropy of the microwave background radiation after
correction for the earth's motion with respect to it [M.Geller and R.
Partridge, respectively]. Geller reviewed the results of the systematic redshift surveys carried out in the last ten years or so. They are
impressive and give strong evidence for "voids" separated by "sheets".
But they do not permit the determination of the characteristic size
of the inhomogeneities, because the presently observed sizes are comparable to the dimensions of the available samples. Various theoretical
models continue to be developed. While hard conclusions cannot be reached, models involving biasing and cold dark matter seem to be
favoured.

When one corrects for the Döppler shift due to a velocity of some
600 km/sec for the local group, the cosmic background radiation (CBR)
is confirmed to be very isotropic with a temperature of 2.75 ± 0.04 K
for the wave length range 0.1-12 cm. The observed upper limits and possible sources of anisotropies for various angular ranges were discussed
in detail by Partridge, who also addressed the indications for additional radiation at wavelengths below 1 mm.

The accuracy of CBR measurements is now such that the far-infrared
background from dust becomes a possible contributor at wavelengths below 1 mm [Partridge]. In an interesting contribution linked with the
steady state theories of the expanding universe, F.Hoyle discussed the
possible formation of iron whiskers in interstellar space. They could
be a source of thermalization of electromagnetic radiation.

The two foregoing items really belong to a newly emerging discipline involving physics and astrophysics, the study of condensed matter
in outer space, i.e., under conditions of microgravity. It attracts
growing attention because of the experimental opportunities offered
by space research. One may also recall in this respect one of the most
remarkable results of the 1986 observations of Halley's comet, to wit,
the extraordinarily low density of its nucleus, of order 0.2 g/cm^3.

4. Regarding the status of particle physics, the most important developments since the discovery of the weak bosons W and Z in 1983 have
been a long series of experimental advances within the framework of
the Standard Model, based on quarks, leptons and the SU(3)xSU(2)xU(1)
theories of the strong and electroweak interactions. The theoretical
attempts to find a satisfactory unified theory for these interactions
and gravity were at the level of supergravity at the time of the first
symposium and of superstrings at the time of the second. Since then,
many superstrings theories were invented, including schemes no longer
requiring higher dimensionality of spacetime [A.Salam], but little progress was made to relate them to the real world. A very recent version
was presented at the meeting by D. Nanopoulos; it will probably take
a few years to assess its value.

R.Peccei had the heavy task of reviewing the Standard Model, which
continues to be amazingly successful despite its shortcomings of principle. He concentrated on the electroweak sector where perturbation
theory is generally applicable and where the most significant tests

of the theory are made. The experimental precision is now such, especially for the masses of the weak bosons, that radiative corrections
are tested quantitatively. The sixth or top quark appears to be too
heavy to have been found at CERN; the next opportunity is at the Tevatron Collider at Fermilab. The study of the overall consistency of present data with theory leads to the tentative conclusion that the top
quark should exist and have a mass somewhere in the range 50–200 GeV.

There are two new experimental developments relevant for CP violation, the discovery of a large $B\bar{B}$ mixing and the direct measurement
of CP violation in $\Delta S=1$ transition amplitudes of K decay. In contrast,
no progress is reported on the Higgs mechanism.

We add a few remarks on aspects of the Standard Model which were
not covered by Peccei. They concern the non-perturbative properties
which are of particular importance for Quantum Chromodynamics (QCD),
the theory of strong interactions, now that accurate work is needed
to compare theory and experiment. The most important developments concern the understanding of particle production processes, which proceed
through a quark-gluon phase followed by the non-perturbative transformation of these "partons" into hadrons. The physics of the quark-gluon
plasma mentioned in section 2 belongs to this category. But also as
"simple" a question as the quark-gluon structure of the proton ground-
state still reveals surprises; the European Muon Collaboration recently measured deep inelastic muon scattering off polarized protons and
found results which contradict the conventional picture for the spin
structure of the proton in the framework of QCD [3].

Also the non-perturbative structure of the vacuum continues to
raise unresolved problems in the Standard Model. One which deserves
to be quoted in this meeting is the debate on whether the very weak
baryon-number violation predicted by the model could have been much
larger at temperatures of the order of 1000 GeV, implying the possibility of strong effects in the evolution of the net baryon number of
the expanding universe [4].

5. As a last topic I mention very briefly the domain of cosmological
speculations. It is some nine years ago that the idea of cosmological
inflation was invented to solve the puzzles of the early expansion of
the universe by invoking a phase transition of the matter fields at
a temperature around 10^{15} GeV. This general idea remains extremely popular, and one now knows that (exponential or non-exponential) inflation can be arranged to occur in many theoretical schemes due to the
coupling of gravitational and matter fields. Many aspects of inflation
were discussed by A.Salam, A.Starobinski, M.Turner and D.Nanopoulos.
Other inflationary models would have been presented if S.Hawking,
A.Linde or others had participated in the symposium. The field remains
wide open for further speculations. One notes an increasing trend toward inhomogeneous scenarios where the universe beyond our observational horizon is different from the part we see.

6. In conclusion, as far as I can see, the main change in the relationship between astrophysics and particle physics since 1983 is the emergence of a well organized and well equipped particle astronomy, concen-

trating mostly on neutrinos and on the search for dark matter, but concerned for example also with the high energy point sources [4]. On the other hand, despite much work and many advances, the two dominant problems discussed five years ago are still with us today; the particle physicists are still looking for a satisfactory theory unifying the strong, electroweak and gravitational interactions, while in astronomy one is still very far from a consensus on the origin of the large scale structures in the universe.

It should be recognized that these are formidable problems, very different from the usual problems of natural science. Cosmology in particular must try to reconstruct the deep past from the few relevant relics which are available for present observation, and the phenomena in question do not repeat themselves. The particle physicists are better off because they deal with reproducible experiments, but for them the relevant time scales are far too short for observation. For each high energy reaction they study, all they can observe are the (meta)-stable relics of the particle creation mechanisms and the shortlived particle decays. Even worse, the building blocks of all strongly interacting matter, quarks and gluons, never occur in free states.

In very different ways, astronomy and particle physics are confronted with the fact that the most fundamental processes are not observable and must be reconstructed from "relics". This reconstruction is practically impossible in a model-independent way and it requires much guesswork in the form of theoretical assumptions. These are apt to be influenced by a priori choices or convictions which are not necessarily correct. The "theory of everything" is not yet here, and if it were we would be naive to believe that it is unique. But meanwhile our two disciplines progress steadily through experimentation, observation and theoretical interpretation, each one being helped by the advances of the other. This is why meetings such as the present one are of great interest and value.

It was a privilege for CERN and ESO to be associated through our 3rd symposium with the 900th anniversary of the University of Bologna, and I speak on behalf of the two organizations and of all participants when I express our deep gratitude to the authorities of the University and the City for their generous hospitality in the magnificent setting of the Palazzo Re Enzo. Our thanks go also to Professors G. Giacomelli and A. Renzini and to their colleagues for their excellent organizational work which helped to make our meeting such a pleasant experience. Finally, I wish to express to Professor Giancarlo Setti my deepest personal appreciation for our friendly collaboration in organizing the ESO-CERN Symposia since their inception.

REFERENCES
[1] J.H.Applegate and C.J.Hogan, Phys. Rev. D31 (1985) 3037;
    S.A.Bonometto et al., Phys. Lett. B157 (1985) 216.
[2] Proceedings of the International Conference on Physics and Astrophysics of Quark Gluon Plasma, Bombay, 8-12 February 1988, ed. B. Sinha, to be published by World Scientific, Singapore.
[3] P.Arnold and L.McLerran, Phys. Rev. D36 (1987) 581;
    J.Ellis et al., Phys. Lett. B194 (1987) 241;
    P.Arnold and L.McLerran, Phys. Rev. D37 (1988) 1020.
[4] G.B.Yodh in ref.[2].

POSTERS

HARD X-RAY OBSERVATION OF THE SUPERNOVA 1987A

A. Bazzano, P. Ubertini, C. La Padula
Istituto di Astrofisica Spaziale, CNR
C.P. 67
00044 - Frascati
Italy

THE OBSERVATION

The Supernova 1987A has been observed during the period 1988 April 5.00 - 5.42 UT with a combined hard X-ray (15-150 KeV) and a high energy gamma ray (50-500 MeV) payload.
The experiment was carried out in the framework of an international collaboration among the Istituto di Astrofisica Spaziale, Italy, the University of New South Wales, Australia, the Case Western Reserve University, USA, the Imperial College, London, UK and the Astronomisches Institut of Tubingen University, Germany. The payload was flown on a stratospheric balloon launched from the Balloon Station of Alice Springs, Australia and floated at an altitude corresponding to 3.0 mbar for about 10 hours.

THE HARD X-RAY DETECTOR

The hard X-ray detector (shown in Fig. 1) was a high pressure Xenon-filled MWPC with a sensitive area of 500 cm^2 in the energy range 15-150 KeV, and a spectral resolution of $\Delta E(\%)=93/E^{1/2}$ corresponding to 8% at 130 KeV. The field of view of the detector was limited by an hexagonal copper collimator with an aperture of 4 deg FWHM. The observation was performed continously tracking the SN in azimuth while the elevation angle of the detector was tilted every 10 minutes off-source for the background determination.

THE PRELIMINARY RESULT

The preliminary analysis of seven hours of data obtained from the hard X-Ray detector, performed just after the flight, resulted in a 4.8 sigma flux detection in the

M. Caffo et al. (eds.), Astronomy, Cosmology and Fundamental Physics, 407–408.
© *1989 by Kluwer Academic Publishers.*

energy range 30-110 KeV, from a region of 4 degrees centered on the SN position in the sky.
This flux, provisionally assuming a continuum Crab-like spectrum, corresponds to an emission of $(1.4 \pm .4) \times 10^{-4}$ ph/cm^2 s KeV at 50 KeV, as shown in fig. 2, compared with other previously reported data.

CONCLUSIONS

Our data is consistent with a general increase of the SN flux in the hard x-ray range as also suggested by the MIR station data.

ACKNOWLEDGMENTS

We thanks all the scientists and technicians of the Institutes partecipating to this observation for their contribution.

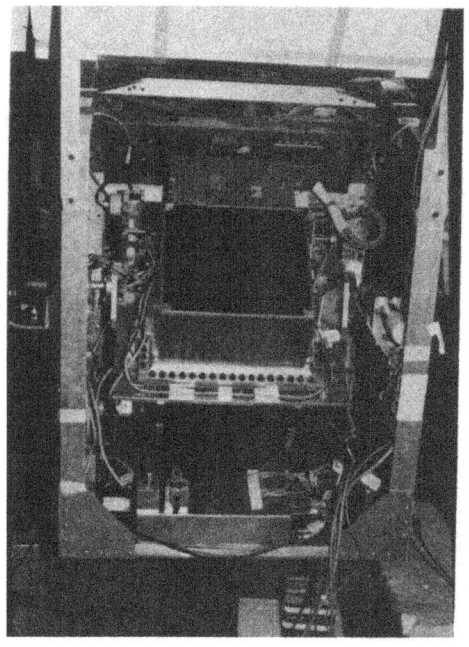

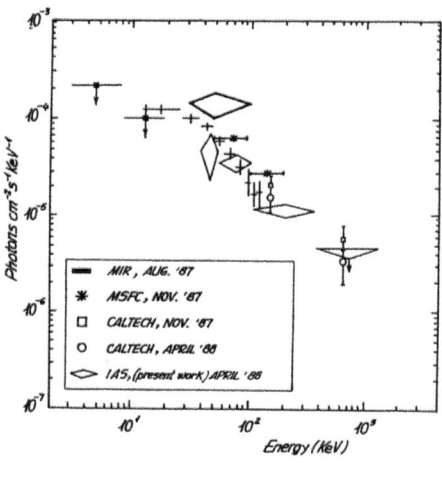

Figure 1. The X-Ray detector   Figure 2. SN87a spectral data

The UA6 experiment at the CERN Collider

CERN-Lausanne-Michigan-Rockefeller collaboration

Presented by P. Giacomelli
The Rockefeller University

The UA6[1] experiment uses an internal hydrogen jet target in the CERN SPS main ring to study electron pairs, π^o, η and direct γ production in the central region as well as low t elastic scattering and diffraction dissociation in pp and ppbar collisions at $\sqrt{s}$ = 24.3 GeV. For direct photon analysis we study a transverse momentum range $3 < p_t < 7$ GeV/c, which at the $\sqrt{s}$ of the experiment corresponds to the x_t ($x_t = 2p_t/\sqrt{s}$) range 0.25-0.58. Direct photons are produced in hadronic collisions through gluon Compton scattering, quark-antiquark annihilations and bremsstrahlung off a quark line. In pp collisions the main source of direct photons is gluon Compton scattering[2,3], while in ppbar collisions, in the UA6 x_t domain, 40% to 80% of the cross section is expected to be due to quark-antiquark annihilation[4]. Studies of pp reactions therefore allow a determination of the gluon structure function, whereas in ppbar reactions gluon fragmentation can also be studied; finally the reactions allow a measurement of Λ_{QCD}.

The experimental set-up, shown in fig. 1, is located in a 12 m long straight section of the CERN Super Proton Synchrotron (SPS). The detector consists of a H_2 cluster jet used as an internal target, followed by a double-arm magnetic spectrometer. The angular coverage of each arm in the laboratory system is 20 to 100 mrad in polar angle and 70° in azimuth; this corresponds to a solid angle of 1.8 sr in the ppbar centre-of-mass system. An arm consists of: five multiwire proportional chambers (MWPCs), two in front and three behind a 2.3 T·m dipole magnet; an ionization chamber (dE/dx); a Li/Xe transition radiation detector and an electromagnetic (EM) calorimeter.

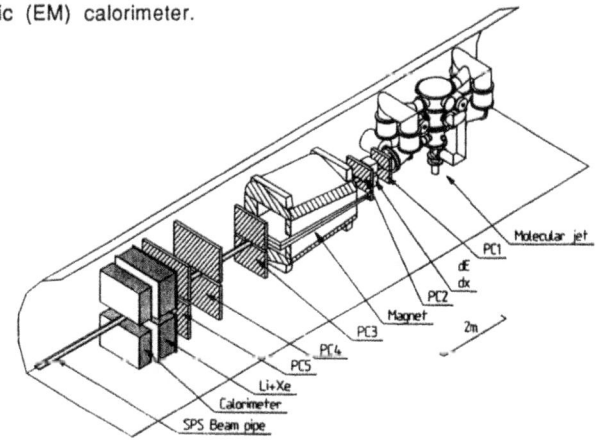

Fig. 1: The UA6 Experimental apparatus in ppbar configuration

M. Caffo et al. (eds.), Astronomy, Cosmology and Fundamental Physics, 409–412.
© *1989 by Kluwer Academic Publishers.*

The jet target[5] consists of a vertical stream of pure hydrogen clusters, each cluster containing 10^5 molecules, which traverses the SPS beam. The jet density can be varied up to 4×10^4 atoms per cm^3. Until 1987 the SPS Collider operated with 3 antiproton and 3 protons bunches, containing typically 1.3×10^{10} antiprotons and 10×10^{10} protons per bunch, giving instantaneous luminosities of 5.5×10^{29} cm^{-2} s^{-1} and 4.0×10^{30} cm^{-2} s^{-1}, respectively, at full jet density. This is because each particle bunch traverses the jet at the SPS revolution frequency of 43.4 kHz. With the new Antiproton Collector (ACOL) we expect to increase by a factor of 5 our ppbar luminosity. The luminosity is determined, to within ±4%, from the rate of elastic ppbar (or pp) events. This rate is monitored by a set of solid-state counters, located at 90^0 in the laboratory system, which detect recoil protons. The dimensions of the interaction volume are given along the beam by the length of the jet , which is 8 mm, and transverse to it by the height of the SPS beam, usually less than 1 mm, and by the overlap of the jet and beam, both of width 3 mm. The experiment is performed during Collider operation of the SPS when beams are particularly clean, thus eliminating beam-halo problems common to intense secondary beams.

The two electromagnetic calorimeters[6] are of the lead/proportional-tube type. The calorimeter is divided into three identical modules of eight radiation lengths each. Located between the first two modules is a hodoscope of seven horizontal scintillator counters used for triggering. The position resolution for showers was found to be 3.5 mm (RMS) at 10 GeV, improving to 1.5 mm at 75 GeV. Test-beam results show that the calorimeter response to electrons in the energy range 10-100 GeV is linear to better than 0.6%, and that the energy resolution is given by $\sigma(E)/E = 0.33/\sqrt{E}$ (E in GeV). The overall energy scale of the calorimeter is determined and adjusted off line on an hourly basis, by centering the π^0 mass peak at its known value. Before adjustment, the energy scale was found to vary, over periods of days, within ±8%. The events which pass the pretrigger requirements are then analysed by a hardware processors which is used to enhance the selection of events with high-p_t localized EM energy deposition.

In the off-line analysis, individual electromagnetic clusters are identified in the following way. Clusters of adjacent tubes containing more than 30 MeV are formed around any seed tube containing more than E_{min} GeV. In this way, clusters are found in the horizontal and vertical views of each module. The clusters are then associated module to module in depth when a line through the cluster centroids extrapolates to the target. The small size of the interaction volume is an advantage at this stage of the clustering.

The distribution of the invariant mass of two clusters in the same arm shows clear π^0 and η peaks. Reconstructed π^0 and η's are taken to be any two cluster combination with an invariant mass of less than 0.2 GeV/c^2 and between 0.4 and 0.66 GeV/c^2, respectively. The uncorrelated background under the π^0 and η peaks is 14% and 50%, respectively. Results on the production of these mesons have already been published[7].

An EM shower is identified as a direct photon candidate if (i) it is located inside a proper fiducial area of the calorimeter, taking well into account the magnet aperture; (ii) it does not reconstruct to a π^0 or a η with any other shower in the same calorimeter; (iii) no charged particle points to it within a radius of 1 cm; (iv) its RMS width in the first module is less than 1.35 cm.

The number of direct photon candidates has been corrected by subtracting single-photon candidates due to π^0's and η's. This was performed using a Monte Carlo simulation program which generates these mesons according to the parametrization of Bourquin and Gaillard[8]. The generated events have then been passed through the same analysis chain used for the real data. The Monte Carlo reproduces the measured π^0's very well. The final γ/π^0 ratio after background subtraction and correlation for acceptances is shown in fig. 2. The invariant cross section for direct γ production in ppbar collisions at $\sqrt{s}$ = 24.3 GeV[9], evaluated at an average y of 0.4, is shown in fig. 3. Also shown in fig. 3 are predictions by Aurenche et al.[4] using four quark flavours and Q^2 scales obtained via a "principle of minimal sensitivity"[10], and by Contogouris et al.[11] using a "natural scale" of $Q^2 = p_t^2$. Both calculations consider two sets of proton structure functions[12], Duke and Owens (DO) sets I and II. Set I corresponds to a soft gluon distribution of the form $(1-x)^6$ with Λ=0.2 GeV/c, whereas set II uses a harder gluon distribution, $(1-x)^4$ with Λ=0.4 GeV/c. The data seem to prefer the soft gluon distribution set.

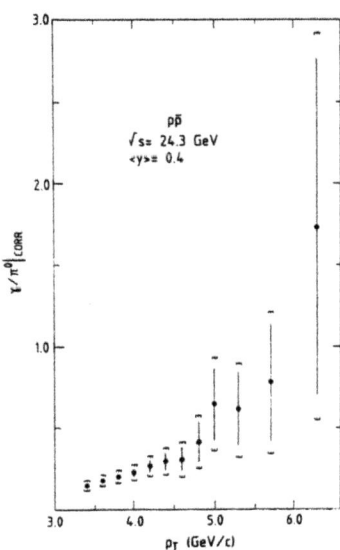

Fig. 2: The corrected γ/π^0 ratio

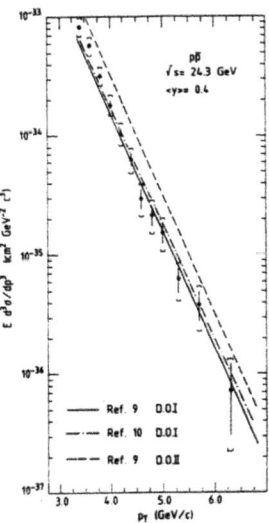

Fig. 3: The invariant cross section for direct photon production, evaluated at an average y=0.4.

References

1) The members of the UA6 Collaboration are A. Bernasconi, R.E. Breedon, L. Camilleri, R.L. Cool, P.T. Cox, L. Dick, E.C. Dukes, M. Duro, B. Gabioud, F. Gaille, P. Giacomelli, J.B. Jeanneret, C. Joseph, W. Kubishta, J.F. Loude, E. Malamud, C. Morel, O.E. Overseth, J.L. Pages, J.P. Perroud, P. Petersen, D. Ruegger, R.W. Rusack, G.R. Snow, G. Sozzi, M.T. Tran, A. Vacchi, G. Valenti and G. Von Dardel.
2) M. Diakonou et al., Phys. Lett. B 91 (1980)301.
3) CCOR Collab., A.L.S. Angelis et al., Phys. Lett. B98 (1981)115.
4)   P. Aurenche et al., Phys. Lett. B140 (1984)87, Nucl. Phys. B286 (1987)509, Nucl. Phys. B297 (1988)661.
5) L. Dick and W. Kubishta, Hadronic Physics at intermediate energy (1986)209.
6) G. Snow (for Expt. UA6), in Proc. Gas Sampling Calorimetry Workshop II, Fermilab, 1985 (U.S. Government Printing Office, 1986)174.
7) J. Antille et al., Phys. Lett. B194 (1987)568.
8) M. Bourquin and J.M. Gaillard, Nucl. Phys. B114 (1976)334.
9) A. Bernasconi et al., Phys. Lett. B206 (1988)163.
10) P.M. Stevenson, Phys. Rev. D23 (1981)2916, Nucl. Phys. B203 (1982)472.
11) A.P. Contogouris, in Proc. Advanced Research Workshop on QCD hard hadronic processes, St. Croix (virgin Islands) 1987
     N. Mebarki, Ph.D. thesis, McGill Univ. (1987)
     A.P Contogouris and N. Mebarki, private communication
12) D.W Duke and J.F. Owens, Phys. Rev. D30 (1984)49.

PRECISE DETERMINATION OF THE BACKGROUND RADIATION TEMPERATURE

P. Crane
European Southern Observatory, Karl-Schwarzschild-Str. 2
D-8046 Garching bei München, West Germany

D.J. Hegyi
Dept. of Physics, University of Michigan,
Ann Arbor, MI 48109, USA

M. Kutner
Rennselaer Polytechnic Institute,
Troy, N.Y. 12181, USA

N. Mandolesi
Istituto TE.S.R.E./CNR, via d'Castagnoli,
I-40126 Bologna, Italy

ABSTRACT. A new analysis technique has reduced the uncertainty in the CN equivalent widths used to measure the cosmic background radiation temperature at 2.64 mm., and a careful search for CN emission at 2.64 mm has effectively eliminated the possibility of local mechanisms contributing to the observed CN excitation. The CN molecules towards ζ Oph are accurately in thermal equilibrium with the microwave background at a temperature of T_{CBR} = 2.796 (+0.019, -0.041) K.

INTRODUCTION

Measurements of the temperature of the Cosmic Background Radiation at several wavelengths are one of the few ways we have to investigate the thermal history of the Universe. Since recent measurements at $\lambda < 1$ mm. (Matsumoto et al., 1988) have shown a departure from a pure blackbody, it is now more important to make careful measurements of the CBR temperature at or near the peak in the blackbody curve. This paper reports such a measurement.
    The rotational levels of the CN molecule are a sensitive probe of the CBR temperature in the submillimeter region of the spectrum which is difficult to study with ground based radiometers or with rocket-borne bolometers. With the single assumption that the CN molecules in diffuse interstellar clouds are in thermal equilibrium with the background radiation, the relative populations of the two lowest rotational levels can be used directly to determine T_{CBR} at 2.64 mm. For the weak interstellar lines studied here, the equivalent widths of

413

M. Caffo et al. (eds.), Astronomy, Cosmology and Fundamental Physics, 413–415.

the optical absorption lines provide accurate measurements of the rotational state populations after making small corrections for saturation.

This paper reports improved precision in the determination of the optical equivalent widths using a new analysis technique and a direct test of the assumption that the CN molecules are in thermal equilibrium. This latter test involved a careful search for possible emission from CN at millimeter wavelengths from the interstellar cloud in the direction of ζ Oph.

OPTICAL ABSORPTION LINES

The data discussed here were obtained during two observing runs in 1984 and 1985 at the European Southern Observatory's 1.4m Coudé Auxilliary Telescope and has been discussed previously (Crane et al., 1986).

By analyzing each of the 61 20 minute runs individually, a better understanding of the uncertainties in the equivalent width determinations was possible. Essentially the histogram of the equivalent widths for each line from the 61 runs was formed, and the center of the histogram was used to determine the most probable value for each equivalent width. A detailed discussion of the technique is found in Crane and Hegyi (1988). Table 1 summarizes the equivalent widths of interest for this discussion.

Table 1: CN Equivalent Widths

Line	Equivalent Width mÅ
R(0)	7.746 ± 0.041
R(1)	2.454 ± 0.024
P(1)	1.255 ± 0.024

Using the above equivalent widths, and an intrinsic line width of the CN lines towards ζ Oph of 1.47 km s^{-1} for the saturation correction (see Crane et al., 1986 for details) the temperatures given in Table 2 have been determined. The entry ΔT_{local} given in Table 2 is discussed in the next section.

Table 2: T_{CBR} at 2.64 mm

Line Ratio	T_{raw}	T_{corr}
R(1)/R(0)	2.9578±0.0242	2.7956±0.0228
P(1)/R(0)	2.9944±0.0402	2.7965±0.0375
T_{ave}		2.7958±0.0195
ΔT_{local}		< 0.0365
T_{CBR}		2.7958(+0.0194,-0.0414)

We note that the extremely good agreement between the saturation corrected temperatures gives us confidence that our technique is correct.

MILLIMETER OBSERVATIONS

In order to test the assumption that the CN molecule is in thermal equilibrium with the cosmic background radiation, a search for emission at 2.64 mm. from the interstellar cloud was performed using the NRAO 12 meter telescope on Kitt Peak.

The final spectrum obtained during roughly 30 hours of on source integration time had an rms uncertainty of 6.9 mK. The data show a negative dip or absorption at the point where CN emission is expected. This implies that emission is rather unlikely. The rms fluctuations in the data averaged over the expected CN line width of 1.47 km/s is 4.8 mK. Using this and the fact that absorption is not physically reasonable, the observations constrain $T_{ANT} < 4.8$ mK. This can be turned into a limit on the local excitation temperature of CN after taking account of optical depth effects. This limit is

$$T_{local} < 0.0356 \text{ K}$$

SUMMARY

The correction for local excitation has been reduced to $\Delta T_{local} < 0.0356$ K from -0.06 ± 0.04 K, and the uncertainty in T_{CBR} due to the errors in the optical equivalent width measurements is now ±0.0195 K. The final value of T_{CBR} at 2.64 mm. is

$$T_{CBR} = 2.796 \ (+0.019, -0.041) \text{ K}.$$

It is gratifying that our result agrees so well with the value of Johnson and Wilkinson (1986) which has slightly higher precision than ours. Their value of of $T_{CBR} = 2.783\pm0.025$ at 1.2 cm. was determined using a balloon radiometer.

Averaging the measurements for $\lambda \geq 2$ mm., we determine $T_{CBR} = 2.755\pm0.018$ K with a χ^2 of 12.8 for 8 degrees of freedom.

REFERENCES

Crane, P., and Hegyi, D.J., 1988, Astrophys. J. (Letters), **326**, L35.
Crane, P., Hegyi, D.J., Mandolesi, N., and Danks, A.C., 1986, Astrophys. J., **309**, 822.
Matsumoto et al., 1988, Astrophys. J., **329**, 567.
Johnson, D.G., and Wilkinson, D.T., 1986, Astrophys. J. (Letters), **313**, L1.

AN AGE ESTIMATE FOR THE NGC456, 460 AND 465 SMC CONSTELLATION.

A. Dapergolas[1], E. Kontizas[1], M. Kontizas[2] F. Pasian[3], M. Pucillo[3] and P. Santin[3]

(1) National Observatory of Athens, Astronomical Institute, P.O. Box 20048, GR-Athens 11810, Greece.
(2) University of Athens, Athens, Greece.
(3) Osservatorio Astronomico di Trieste, Trieste, Italy.

The aim of this paper is to study the distribution of spectral types of the stellar content in an outlying region (wing) of the Small Magellanic Cloud (SMC) which contains the brightest early type stars and therefore the most massive in this galaxy.

The studied field here is an extended area (1189 arcmin^2) where three associations forming a constellation are included (Westerlund, 1961).

High quality UJ film copies of objective prism plates taken with the 1.2m U.K. Schmidt telescope were used. The dispersion of the spectra is 830 Å/mm for the high resolution plates and 2440 Å/mm for the low resolution plates, at H_γ line.

The data presented here contain stellar spectral types down to B=18.5mag and the whole studied field is shown in Fig. 1, where the dashed line gives the limits of the adopted area of the constellation (Westerlund, 1961). The spectral classification criteria are the same described previously (Dapergolas et al., 1986).

The limiting apparent visual magnitude of the data sets has very little colour dependence (see Kontizas et al., 1987a).

The high resolution B and A spectra (830 Å/mm) were scanned with a PDS at Trieste Observatory for more accurate spectral classification. The measurements were carried out with a slit 20X20μm and a step 10μm. The effective resolution of the measurements is ≈20 Å.

The distribution of spectral types for the field stars outside the constellation was subtracted from the distribution of the constellation area after been normalized and the results are shown in Fig. 2. A completeness factor (that is a parameter designed to take account the overlapping images) has been taken into consideration.

Discussion

The so derived distribution of spectral types (Fig. 2) free of foreground and background stars shows that the majority of the constellation area stars are early type, O, B and A. A small number but not negligeable of all other spectral types is also present in this area. The ratio of early to late type stars $R=N_{(O+B+A)}/N_{(K+M)}$ being an age indicator (Kontizas et al., 1987b) allowed us to give an age

M. Caffo et al. (eds.), Astronomy, Cosmology and Fundamental Physics, 416–417.

estimate $\approx 6 \times 10^7$yr. The theoretical models for this age predict a
similar distribution of spectral types of stars. Finally it is also
worth emphasizing that the O stars are fainter than the evolved bright B
and A type stars.

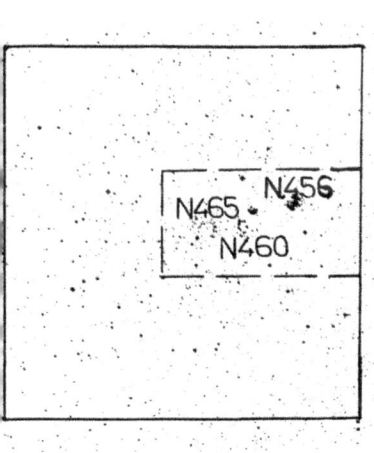

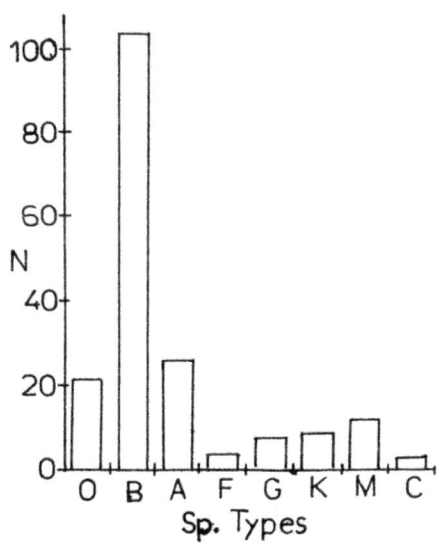

Fig. 1. Map of the studied
field. The dashed line indicates the constellation
area.

Fig. 2. Distribution of spectral types
of the constellation field. The
background and foreground stars are
excluded.

Acknowledgements

The authors would like to express their sincere thanks to the 1.2m
U.K. Schmidt Telescope Unit for loan of the observational material. We
also acknowledge the financial support by the Greek General Secretariat
of Research and Technology and the CNR of Italy.

References

Dapergolas, A., Kontizas, E. and Kontizas, M.: 1986, Astron. Astrophys.
Suppl. Ser., 65, 283.
Kontizas, E., Morgan, D.H., Dapergolas, A. and Kontizas, M.: 1987a,
Astron. Astrophys. Suppl. Ser., 70, 1.
Kontizas, E., Kontizas, M., Dapergolas, A. and Hadzidimitriou, D.: 1987b
IAU Symp. No. 126, (in press).
Westerlund, B.: 1961, Uppsala Astronomiska Observatoriums Annaler, Bond
5, No. 2., p.3.

ON THE ORIENTATION OF SPIRAL GALAXIES IN THE PERSEUS SUPERCLUSTER

Piotr FLIN and Włodzimierz GODŁOWSKI
Jagiellonian University Observatory, Kraków, Poland.

Various scenarios of galaxy origin predict different types of galaxy-supercluster alignment (Shandarin 1974, Dekel 1985). In the framework of classic scenarios the lack of alignment, i.e. the random distribution of galaxy rotation axes is expected in the hierarchical clustering picture, whereas the existence of an alignment is expected in fragmentation scenario. The perpendicularity of rotation axes of galaxies to the supercluster main plane (hereafter smp) is consistent with the prediction of turbulence model, while the alignment of rotation axes of galaxies with smp supports the adiabatic model of galaxy origin. The true spins of galaxies are known for small numbers of objects. So, in the present study the distribution of normals to galaxy planes is investigated. Two possible normals to galaxy plane, along which real galaxy rotation axis should be situated are determined (FG). Both solutions are taken into consideration, which blurs anisotropy, but does not introduce the spurious one. In order to better determine the spatial orientation of galaxy disc both parameters the position angle of the major axis and the axial ratio of the galaxy image were considered.

In the present study we investigate the orientation of spiral galaxies within radial velocity range 4000 km/s and 8500 km/s, situated on the sky in the region $22^h \leqslant \alpha \leqslant 4^h$ and $21^\circ.5 \leqslant \delta \leqslant 45^\circ$. The radial velocity data were collected from the literature as well as the equatorial coordinates (α, δ), morphological types and diameters a and b of the major and minor axes of galaxies. The position angles came from UGC or GTT. Some additional measurements of diameters and position angles were performed by us. The search of alignment was made transforming coordinates and position angles of galaxies (α, δ, p) into coordinate system connected with the Perseus supercluster (l,b,P). For galaxies seen "face-on" it was accepted that P=0. The spatial geometry of the Perseus supercluster is known (Chincarini et al. 1983), which allows us to assume that b=0 corresponds to smp. The

M. Caffo et al. (eds.), Astronomy, Cosmology and Fundamental Physics, 418–420.
© *1989 by Kluwer Academic Publishers.*

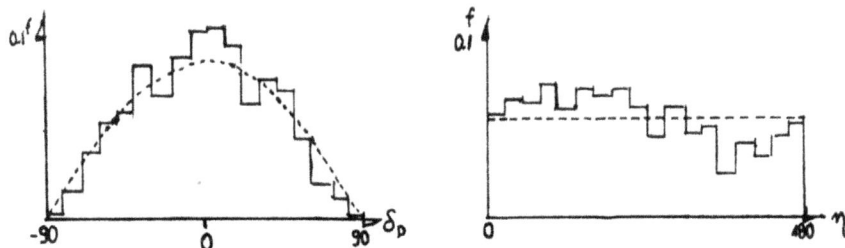

Fig.1. The distribution of the angle δ_D and the angle η (broken lines illustrate expected isotropic distribution).

galaxy tilt i was calculated assuming the oblate spheroid and using the Holmberg's formula. The sample containing 686 galaxies was divided into four subsamples, according to galaxy membership to the main ridge of the supercluster ($\alpha \in (20^o, 50^o)$) and the distance from smp.

Two angles dealing with the orientation of galaxies in supercluster were determined: the polar angle δ_D between the direction normal to galaxy plane and smp, and the azimuthal angle η between the projection of the normal on smp and the origin of coordinates. The formula for calculating δ_D and η as derrived in FG are:

$$\sin \delta_D = -\cos i \sin b \pm \sin i \sin P \cos b$$

$$\sin \eta = (-\cos i \cos b \sin l + \sin i (\mp \sin P \sin b \sin l \mp \cos P \cos l)) / \cos \delta_D$$

There is an excess of angles in bins close to the value 0° and a deficit in bins $|60^o-90^o|$. The quotient the observed number of solution falling into the bins $|0^o-30^o|$ to that falling into bins $|60^o-90^o|$ is 5.9, whereas the value 3.7 is expected for a random distribution. Such distribution shows that rotation axes of galaxies are aligned with smp.

The isotropy of the angular distribution was checked using statistical analysis proposed by Hawley and Peebles (1975), as given in FG. The Table 1 presents the result of the statistical analysis. N denotes the number of galaxies, $P(\geqslant \chi^2)$ and $P(\geqslant \triangle)$ are the probabilities that departure from isotropy are caused by random fluctuations in the χ^2-test and the wave model respectively, C is the value of the autocorrelation test (for isotropic distribution C=0 with δ_C =6), F-coefficient describes the direction of the departure from isotropy (F<0 when rotation axes are aligned with smp), the index $W=(N_\| - N_\perp)/(N_\| + N_\perp)$, where $N_\|$ and $N_\perp$ are number of rotation axes parallel ($|\delta_D| < 45^o$) and perpendicular to smp. From Table 1 it follows that the observed distribution in both angles is highly anisotropic. The rotation axes of galaxies are aligned with smp. Moreover their projections onto smp are pointing toward the main structure, i.e. the Perseus supercluster, covering the zone of avoidance and the western

Table 1. The result of statistical analysis

	N	$P(\geqslant x^2)$	$P(\geqslant \Delta)$	C	F-coef	W	$P(\geqslant x^2)$	$P(\geqslant \Delta)$	C	F-coef
		the angle δ_D					the angle η			
all	686	0.00	0.00	114.3	-.132	.197	0.00	0.00	61.2	-.213
a	113	0.01	0.00	12.8	+.523	.198	0.00	0.27	25.7	-.410
b	150	0.00	0.00	-4.4	-.425	.198	0.00	0.00	32.0	-.294
c	140	0.00	0.00	48.4	-.157	.232	0.00	0.00	30.5	-.691
d	283	0.02	0.00	65.0	-.214	.181	0.00	0.00	38.1	+.146

a-galaxies with $\alpha \in (20°-50°)$ and $|b| \leqslant 3°$; b-galaxies with $\alpha \in (20°-50°)$ and $|b| > 3°$; c-galaxies with $\alpha \notin (20°-50°)$ and $|b| \leqslant 3°$; d-galaxies with $\alpha \notin (20°-50°)$ and $|b| > 3°$.

part of the Lynx-Ursa Major Supercluster (Giovanelli and Haynes 1982). The division into subsamples does not change this general picture. Once $F > 0$ is appearing, but this is not accompanied by the change of the sign of the W-index, and moreover the latter index shows that almost 20 per cent of solutions have an excess of rotation axes in the region $|\delta_D| < 45°$. The similar situation of the change of the sign of the F-coefficient in the case of η angle is observed for the one subsample.

The result of the present study is that rotation axes of spiral galaxies in the Perseus supercluster are aligned with smp. Moreover, the projections of rotation axes onto smp are also non-random. They point toward the structure and it possible extention. This two effects are mutually connected, but the present analysis does not permit us to state which effect is the principal one. The existence of alignment is noticable in both high- and low-density regions of the supercluster. Our result constitutes good evidence in favour of adiabatic model of galaxy origin.

Acknowledgments
PF thanks the Institute of Astronomy of the University of Rome "La Sapienza" and the Vatican Observatory where during his stay there the paper has been prepared.

References
Chincarini,G.L.,Giovanelli,R.,Haynes,M.P.,1983,Astr.Astroph. 121,5
Dekel,A.,1985,Astrophys.J.,298,461
Flin,P.,Godlowski,W.,1986,Mon.Not.R.Astr.Soc.222,525 (FG)
Giovanelli,R.,Haynes,M.P.,1982,Astr.J.87,1355
Gregory,S.A.,Thompson,L.A.,Tifft,W.G.,1981,Ap.J.243,411(GTT)
Hawley,D.L.,Peebles,P.J.E.,1975,Astr.J.80,477
Nilson,P.,1973,Uppsala Astr.Obs.Ann.V,vol.I (UGC)
Shandarin,S.F.,1974,Sov.Astr.18,392

Thermal coupling of the CMBR and the primordial gas in the post–recombination epoch ($z \leq 1000$)

Daniele Galli and Francesco Palla
Osservatorio Astrofisico di Arcetri
L.go E. Fermi 5, 50125 Firenze
Italy

ABSTRACT. The thermal evolution of the Universe after the recombination era is followed by studying in detail the coupling of the radiation field with the primordial gas, due to the presence of a minor, but finite, amount of trace molecules. It is shown that, under some circumstances, the formation of H_2 and LiH molecules can lead to an appreciable heating of the gas, that would otherwise cool adiabatically as the Universe expands. We also present results for the expected distortions of the CMBR, due to the formation of H_2 in highly excited states and the subsequent radiative cascade to the ground state.

In the framework of the standard Big Bang cosmology, the thermal evolution of the baryonic matter is governed by its interaction with the cosmic background radiation field. While the radiation temperature, T_{rad}, at any redshift z is simply given by $T_{rad} = T_0(1 + z)K$, where $T_0 = 2.76K$ is its present value, the evolution of the gas temperature is more complicated and must be obtained by solving the kinetic equation (cf. Peebles, 1968):

$$\dot{T}_{gas} = -2T_{gas}\frac{\dot{R}}{R} + \frac{8}{3}\sigma_T\frac{aT_{rad}^4(T_{rad} - T_{gas})x_e}{m_e c} + \frac{2}{3}\frac{\Lambda}{K}, \tag{1}$$

where R is the Universe scale factor, x_e is the ionization fraction, σ_T is the Thomson cross-section, Λ is the heating function, a and K the blackbody and Boltzmann constants. The gas temperature is thus determined by the balance between the adiabatic cooling and two sources of heat transfer from the radiation field to the gas via Compton scattering on the residual free electrons and via excitation of the internal degrees of freedom of the molecules. In particular, the last term on the r.h.s. of eq.(1) can act as an effective heating source for the gas if the rotovibrational levels are radiatively excited and then collisionally de-excited (e.g. Khersonskii 1986).

The main molecular species that can be formed after the recombination epoch are H_2, HD, HeH, LiH, and their abundance as a function of redshift for various cosmological models has been computed by several authors (Shchekinov and Entel 1984; Lepp and Shull 1984; see Dalgarno and Lepp 1987 and Palla 1988 for a

M. Caffo et al. (eds.), Astronomy, Cosmology and Fundamental Physics, 421–423.

review). Of particular interest here are the results of Shchekinov and Entel (1984) and Izotov and Kolesnik (1984) that found an asymptotic relative H_2 abundance of $\simeq 1 - 3 \times 10^{-4}$ at $z \leq 1000$ with a weak dependence of the final value of H_2 on the density parameter Ω. To study in detail the energy transfer from the radiation field to the molecules we have therefore modelled H_2 as a system consisting of the first 3 vibrational levels and 20 rotational levels; for LiH we included 20 rotational levels of the ground vibrational level, and for HD we considered only the first two rotational levels. The heating function Λ is then computed by solving for each molecular species with abundance x_i the system of equations:

$$\Lambda(z) = \sum_i x_i(z) \sum_{v,j} \left(n_{v,j} R_{v,j;v',j'} - n_{v,j'} R_{v',j';v,j} \right) h\nu_{v',j';v,j}, \qquad (2)$$

The resulting evolution of the gas temperature is shown in Fig. 1, where we plot the run of the difference between T_{rad} and T_{gas} as a function of z.

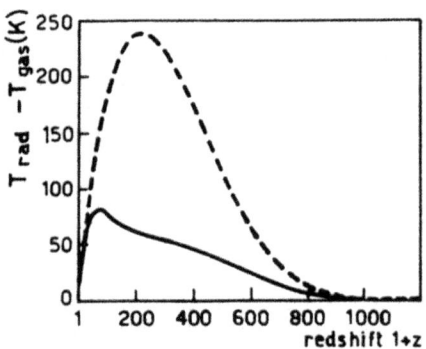

Figure 1

The dashed curve is the evolution in the case where only Compton heating is included in eq.(1), and no molecules are taken into account. The effect of including molecular heat transfer in the calculation is shown by the straight line and is twofold: a) the loss of thermal contact is more gradual, the maximum difference between T_{rad} and T_{gas} being reduced by a factor of ~ 4; and b) the peak is shifted towards lower redshifts. Notice that, due to the choice of a linear abscissa, the effect of LiH at low z ($z \leq 50$), does not show up clearly in the Figure, but it is such that $T_{gas} \simeq T_{rad}$. Although the thermal evolution is greatly affected, no opacity effects are expected, τ_{H_2} being always $\ll 1$, due to the still minor fraction of molecules present in the gas. The most important outcome of these calculations is that the growth of the isothermal density fluctuations that survived the recombination epoch will be greatly delayed due to the strong dependence of the Jeans mass on the matter temperature.

As a final result, in Table 1 we present the calculations of the expected distortion of the CMBR, $\Delta I/I$, due to the sudden formation of H_2 in the gas. For the example given in Table 1 we have assumed a conservative value of the H_2 relative concentration of 10^{-6} and taken the redshift of H_2 formation, z_f, as a free parameter. In the fifth column we give the redshifted wavelength at present time of the (v, j) transition with a given Δv for which $\Delta I/I$ is maximum.

Table I

Δv	v	j	z_f	λ_{max} (μ)	λ_{obs} (μ)	$\Delta I/I$
0	3	15	50	4.55	232	1.5
1	2	11	100	1.79	181	2.3
2	2	11	100	1.50	152	91
3	4	11	200	1.11	223	97

References

Dalgarno, A., and Lepp, S. 1987, in IAU Symposium 118, Astrochemistry, eds. M.S. Vardya and S.P. Tarafdar, 109.

Izotov, Yu.I., and Kolesnik, I.G. 1984, Sov. Astr., **28**, 15.

Khersonskii, V.K. 1986, Sov. Astr., **24**, 114.

Lepp, S., and Shull, J.M. 1984, Ap. J., **280**, 465.

Palla, F. 1988, in NATO-ASI "Galactic and Extragalactic Star Formation", R. Pudritz and M. Fich eds., 519.

Peebles, P.J.E. 1968, Ap. J., **153**, 1.

Shchekinov, Yu.A., and Entel, M.B. 1984, Sov. Astr., **27**, 622.

NEARBY SPIRAL GALAXIES:
OPTICAL AND RADIO PROPERTIES

G. Giuricin, F. Mardirossian, M. Mezzetti

Dipartimento di Astronomia
Universita' degli Studi di Trieste
via Tiepolo, 11
34131 Trieste, Italy

The study of the relations between the radio continuum emission of spirals and other global properties (e.g.HI content, colours, Hα emission strengths, infrared emission, arm classification), which are know to be probably different in different environments, can be of aid in understanding the nature of the radio continuum emission. In the present study we examine a sample of nearby spirals ($V_o < 3000 Km/s$).

As a sample of nearby (non-cluster) galaxies with known HI and radio continuum data, we have considered: i) the spiral and irregular members of Geller and Huchra's (1983) nearby galaxy group and the nearby ($V_o < 3000 km/s$) spirals of the list of 324 "isolated" galaxies whose HI properties have been surveyed by Haynes and Giovanelli (1984).We have taken the HI fluxes from Hayes and Giovanelli (1984) and from the compilation of Giuricin et al. (1985). We have characterized the HI content of a galaxy by evaluating the parameters M_H/L (ratio of the HI mas to the corrected total blue luminosity expressed in the RC2 catalogue system (de Vaucouleurs et al.,1976) and σ_H (mean apparent surface density) defined as

$$\sigma_H = 4M_H/(\pi D^2)$$

where D is the absolute linear diameter (reduced to face-on value at the galactic pole) at the 25 $mag \cdot arcsec^{-2}$ brightness level in the RC2 system. The HI deficiency (or excess) of a galaxy is then evaluated by means of the residuals $\Delta Log\sigma_H = Log\sigma_H - < Log\sigma_H >$ and $\Delta LogM_H/L = LogM_H/L - < LogM_H/L >$, where the standard values $< Log\sigma_H >$ and $< LogM_H/L >$ for each galaxy morphological type refer to a reference sample, as tabulated in Guiderdoni and Rocca-Volmerange (1985). The radio continuum fluxes (or their upper limits) of our galaxies have been mostly taken from Dressel and Condon (1978), Hummel (1980), Kotanyi (1980), Hummel et al. (1985). We have transformed the fluxes observed

M. Caffo et al. (eds.), Astronomy, Cosmology and Fundamental Physics, 424–426.
© 1989 by Kluwer Academic Publishers.

at different frequencies to 2.4 GHz using the power-law spectrum $f\nu \sim \nu^{-0.8}$. We have evaluated the logarithm of the ratio R between the radio (at a frequency of 2.4 GHz) and the optical luminosity defined as

$$Log R = Log f_\nu + 0.4 \cdot (B_T^0 - 12.5)$$

where B_T^0 is the corrected blue total magnitude in the RC2 system and f_ν is the flux density at 2.4 GHz in mJy.

There appear to be no significant correlations between $Log R$ and the HI deficiency parameters $\Delta Log\sigma_H$ and $\Delta Log M_H/L$. On the other hand, $Log R$ turns out to correlate negatively with the corrected colour index $(U-B)_0^T$. But $Log R$ does not correlate (significantly) with the corrected colour index $(B-V)_0^T$. For our galaxies we have generaly taken the colour indices from the RC2 catalogue. We have found a good positive correlation between $Log R$ and $Log W$ (logarithm of the $H\alpha$ + [NII] emission-line equivalent widths), which are taken directly from Kennicutt and Kent (1983).

Hence, our analysis of a large sample of non-cluster spirals, indicates that the radio continuum emission of spirals is a little linked to their integrated star formation histories (to which the colours $(U-B)_0^T$ are related); it is more closely linked to their current star formation rates (which are represented by $H\alpha$ emission strengths). The absence of a relation between the HI content and the radio emission in our sample may be related to the fact that star formation activity may be more associated with molecular gas rather than with HI gas.

There appears to be no dependence of the radio power of our group spirals either on the compactness of the group to which they belong or on the projected distance from their closest neighbour as listed in Geller and Huchra's (1983) catalogue of groups. An excess of overall radio emission in interacting galaxies when compared to relatively isolated objects has often been reported in the literature. In this respect, examining the radio properties of Dahari's (1985) sample of interacting and asymmetric galaxies, we have found that the radio power (per light unit) correlates positively with the interaction strength IAC as defined by Dahari (1985). A similar $Log R - IAC$ correlation has been evidenced by Giuricin et al. (1987) for the central sources of the interacting spirals of Dahari's (1985) list. This correlation makes it clear that the strength of interaction, evidenced from the degree of asymmetry and distortion of a galaxy rather than just the presence of a companion appears to be the cause for the observed radio radio power enhancement. Enhanced star formation triggered by interaction can account for the radio emission excess.

Furthemore, we have examined the radio emission of Elmegreen and Elmegreen (1987) sample of $\approx$ 760 spiral galaxies with know arm classes AC ranging from

$AC = 1$ (i.e., galaxies having fragmented arms with no symmetry) to $AC = 12$ (i.e., galaxies characterized by two long, sharply defined arms). We have found no difference in the radio power (per light unit) of flocculent $(AC = 1 - 4)$ and grand design spirals $(AC = 5 - 12)$. This finding can be taken as a further piece of evidence that star formation activity in spirals does not appreciably depend on the presence of density wave modes. Star formation might be controlled by local processes which operate independently of a wave (see Elmegreen and Elmegreen, 1986 for other lines of evidence, which are based on colours, metallicities, $H\alpha$ and ultraviolet fluxes, blue and infrared surface brightnesses).

In general, our results give further support to the (controversial) view that the radio emission of spiral galaxies is connected with their young stellar population, although a contribution from the old disk population can not be excluded.

References:

Dahari, O.:1985, Astrophys. J. Suppl.57,643

de Vaucouleurs, G., de Vaucouleurs, A., Corwin, H.:1976, Second Reference Catalogue of Bright Galaxies, Austin, University of Texas Press.

Dressel, L.L. and Condon, J.J.:1978, Astrophys. J. Suppl. 36,53.

Elmegreen, B.G. and Elmegreen, D.M.:1986, Astrophys. J. 311,554.

Elmegreen, D.M. and Elmegreen, B.G.:1987, Astrophys. J. 314,3.

Geller, M. and Huchra, J.P.:1983, Astrophys. J. Suppl. 52,61.

Giuricin, G., Mardirossian, F., Mezzetti, M.: 1985 Astron. Astrophys. 176, 175.

Giuricin, G., Mardirossian, F., Mezzetti, M.: 1987 Astron. Astrophys. 176, 175

Guiderdoni, B. and Rocca-Volmerange, B.: 1985, Astron. Astrophys. 151, 108.

Haynes, M.P. and Giovanelli, R.:1984, Astron. J. 89,758

Hummel, E.:1980, Astron.Astrophys. Suppl. 41,151.

Hummel, E., Pedlar, A., van der Hulst, Y.M., Davies, R.D.: 1985, Astron. Astrophys. Suppl. 60,293.

Kennicutt, R.C. and Kent, S.M.:1983, Astron. J. 88,1094.

Kotanyi, C.G.:1980, Astron. Astrophys. Suppl. 41,421.

RECENT RESULTS ON DETECTOR DEVELOPMENTS FOR LOW ENERGY SOLAR NEUTRINOS AND DARK MATTER: SSG and DSC DEVICES

L. GONZALEZ-MESTRES and D. PERRET-GALLIX

L.A.P.P.
B.P. 909, 74019 Annecy-le-Vieux Cedex, France

Abstract: We briefly discuss the main problems related to the detection of low energy solar neutrinos and cold dark matter, and report on the status of current developments, mainly Superheated Superconducting Granules (SSG) and Dedicated Scintillating Crystals (DSC).

1. INTRODUCTION

No really convincing techniques exist by now for real time detection of low energy solar neutrinos (E < 1 MeV) or galactic dark matter. In the first case, all proposed detectors face extremely severe difficulties to reject the various possible backgrounds. In the case of cold dark matter, a very high sensitivity (often below 1 keV) appears to be required, in addition to exceptional low background performances.

Among the proposed real time techniques for solar neutrino detection one may mention:

1) $\nu - e^-$ scattering experiments, where the electron recoil energy is of the same order as that of the incoming neutrino, but unfortunately directionality is lost at such low energies. Due to the low event rate (< 1/ton.day), and to the lack of a specific signature, background rejection becomes extremely difficult. Liquid scintillators, bolometers or superfluid ^{4}He have, however, been proposed as possible detectors.

2) ν-nucleus scattering benefits of coherent cross-sections leading to event rates of the order of 0.1/kg.day, but the deposited recoil energy is very low (1-10 eV), and again no distinct signature would be provided. Therefore, even if a detector sensitive to E $\approx$ 1 eV where feasible, background rejection remains an unsolved dilemma. Among the proposed techniques are: SSG and bolometers.

3) Inverse β reactions, such as ^{115}In, have for a while attracted the interest of several groups, due to the clean signature expected from the reaction:

$$\nu_e \, (E > 128 \text{ keV}) + {}^{115}\text{In} \rightarrow {}^{115}\text{Sn}^{**} + \beta^- \, (E - 128 \text{ keV})$$

where the ^{115}Sn** excited state decays with a lifetime of 3.3 μs emitting simultaneously two γ rays of energies 496 and 116 keV. The possibility of using this delayed coincidence, and even separate the 116 keV γ from the 496 keV one, offers in principle a way to reject very harmful backgrounds. But the natural radioactivity of ^{115}In (E_β < 486 keV, event rate = 0.22/g.s) generates fortuitous events in delayed coincidence with itself or with erratic γ's. High performances are, therefore, required in energy and time resolution, as well as detector segmentation. Several techniques are being considered: liquid scintillators, superconducting tunneling junctions, SSG, InP , crystal scintillators,...

Dark matter detection must be discussed in terms of the specific candidate(s) one is searching for. Dedicated detectors for monopoles or cosmic axions already exist, but other particles such as WIMPs or solar axions set rather severe problems. Again, specific cases must be dis-

427

M. Caffo et al. (eds.), Astronomy, Cosmology and Fundamental Physics, 427–430.
© 1989 by Kluwer Academic Publishers.

cussed, mainly in terms of event rates for nucleus recoil. If such a rate is higher than the best existing backgrounds ($\approx$ 1/keV.kg.day), a recoil experiment (energy deposition: 10 eV < E < 50 keV) may be feasible, although background below E $\approx$ 1 keV has never been explored and a sharp rise is observed below E = 20 keV in existing double β germanium detectors. Among the particles that may eventually be detected through nucleus recoil, due to the large expected event rates (10-6000/kg.day) are: Dirac neutrinos, s-neutrinos and cosmions. For particles that do not exhibit full strength coherent scattering (e.g. the neutralino), it may be useful to consider inelastic scattering of the form: WIMP + N $\rightarrow$ WIMP + N*, where the excited nucleus N* decays by emitting a γ ray. Such a technique may be particularly interesting if the N* lifetime is long enough to allow for a delayed time coincidence. Event rates lie at best in the range 0.1-1/ton.day.

2. SOME RECENT DEVELOPMENTS

SSG detectors [1] are potentially the best suited cryogenic device for large mass experiments, due to the simple X-Y read-out system, which allows for a very strong segmentation with a reasonable number of electronic channels. As explained elsewhere in these Proceedings [2]. Several basic problems must be solved before SSG become an operating detector. In the micro-avalanche scenario [1, 3], and initial energy deposition E produces, after the primary interaction, the flip of several extra granules due to heat propagation inside the detector. We can then write the equation:

$$E + V_{flip} q^\ell = V \int_T^{T+\Delta T} c_{det} dT' \qquad \{1\}$$

where V_{flip} is the number of granules having changed state in a isothermal hot sphere of volume V reached by heat propagation (ΔT increase in temperature) from the interaction point. q^ℓ is the released latent heat per unit volume, and c_{det} the average specific heat of the SSG colloid. Then, by purely dimensional arguments (write: $x = V/E$, $y = V_{flip}/E$), the solution of {1} brings: $V_{flip} (V = \infty) \propto E$, which restores linearity since the signal in magnetic flux is proportional to the flipping volume.

Waiting for a clear-cut experimental confirmation of the possibility to implement such a scenario (requiring fast heat transfers between the dielectric and the granules), we present here irradiation results obtained since the last ESO-CERN Symposium, where dark matter detection was discussed by P.F. Smith [4]. Fig. 1 shows irradiation results at T$\approx$1.5 K with several low energy sources for two collections of granules (998 pure tin and a $Sn_{99}Sb_1$ alloy) of size 10 μm < ϕ(diameter) < 25 μm obtained by sieving from $\phi_{mean} \simeq$ 25 μm batches prepared by EXTRAMET [5]. $(\Delta H/H)_{min}$ is the relative threshold in applied magnetic field [1] set by performing a small sweep above the value (H_{test}) of the applied magnetic field (H_0) at which the irradiation takes place. The observed sensitivity is in all cases better than the one expected from global heating calculations, although local heating is less and less important as the grain size decreases [1]. The existence of local heating effects at such small sizes is demonstrated, however, by comparing the ^{55}Fe 6 keV γ results for both samples, which differ only in purity (i.e. normal state resistivity). The most resistive metal leads to more sensitive granules, as predicted by the local heating model [1].

Results at T $\simeq$ 400 mK give even better results [6], and a clear increase in sensitivity is observed as the operating temperature is lowered (more than 9% of grains sensitive to 6 keV γ's , ΔH/H up to 0.04). Fig. 2 shows the superimposed irradiated (full line) and non-irradiated (dashed line) differential superheating curves (number of counts per unit increase in H_0). The irradiated curve was obtained increasing H_0 to H_{test}, then staying for 10 min at this value of the applied magnetic field, and $\Delta H = 0$ (no sweep), the source being permanently implanted inside the detector. The non-irradiated curve corresponds to an uninterrupted increase of H_0 up to its highest value. Again, the observed sensitivity is better than that predicted by global heating calculations.

Dedicated scintillating crystals are also a potentially interesting development, especially for solar neutrino detection if a high quality scintillator can be made of some indium compound [7]. Similar performance to NaI(Tl) should be required for a serious candidate. $InBO_3$ is now being studied, as its powder doped with Tb^{3+} or Eu^{3+} is known to provide a high light yield [8], and transparent crystals have already been grown [9, 10]. Fluorides, such as $LiInF_4$ and $LiBaF_5$ are also being considered [10], and silicates or germanates may also be interesting [7] as $GdSi_2O_5:Ce^{3+}$ is known to be a rather efficient scintillator.

The use of high quality scintillators for dark matter detection would pose several basic problems. The first one concerns the light yield from a recoiling nucleus, if a recoil experiment is to be performed. A comparatively weak luminescence is expected, but experimental work on the subject should be pursued further. If inelastic scattering is used for neutralino detection, there is in addition the difficulty to get the required signature. The isotopes with predicted event rates in the range 0.05-1/ton.day and excited state lifetime of more than 1 ns are [11]: ^{169}Tm, ^{187}Os, ^{119}Sn, ^{83}Kr, ^{57}Fe, ^{151}Eu. ^{57}Fe (E_γ = 14.4 keV, τ = 98 ns, event rate $\simeq$ 0.2/ton.day) seems particularly well suited for a scintillator experiment, because of the comparatively long lifetime, and transparent crystals incorporating Fe^{2+} or Fe^{3+} have already been grown. ^{183}W has an interesting event rate (0.6/ton.day for a neutralino of mass m = 50 GeV), and WO_4Cd scintillators are already on the market. However, the lifetime of $^{183}W^*$ is too short to allow for a delayed time coincidence.

^{119}Sn (E_γ = 23.9 keV, τ = 18 ns, event rate 0.1/ton.day) may be an interesting material for a SSG detector, as fast enough electronics can in principle be built [12] and tin is commonly used in current SSG studies.

Another interesting technique for the study of rare events could be some combination of bolometric and scintillator techniques, where materials exhibiting luminescence at very low temperature would provide simultaneously time information through the emitted light and good energy resolution through the thermistor signal. Such an idea requires the development of new photosensitive devices, able to operate at very low temperature and with a high quantum detection efficiency. Perhaps such devices can already be found in the field of cryogenic detectors [2]. Furthermore, the presence of a nucleus recoil may manifest itself in an anomalous phonon/light ratio, as compared to electrons and photons.

References

[1]   For an updated introduction to SSG detectors, see: L. Gonzalez-Mestres and D. Perret-Gallix, in "Superconductive Particle Detectors", Ed. A. Barone, World Scientific Pub. 1987; and K. Pretzl, same reference.

[2]   L. Gonzalez-Mestres and D. Perret-Gallix, these Proceedings.

[3]   L. Gonzalez-Mestres and D. Perret-Gallix, Proceedings of the Moriond Meetings on " Neutrinos and Exotic Phenomena" (January 1988) and "Dark Matter" (March 1988) Ed. Frontieres.

[4]   P.F. Smith, Proceedings of the III ESO−CERN Symposium, Garching March 1986, Ed. G. Setti and L. Van Hove, ESO report (1986).

[5]   EXTRAMET, Zone Industrielle, rue de la resistance, 74-Annemasse, France.

[6]   More details can be found on: J. Boniface, J. Ditta, L. Gonzalez-Mestres, L. Massonnet, D. Meylan and D. Perret-Gallix, Contribution to the International Conference on Advanced Technology and Particle Physics, Como 13-17 June 1988.

[7]   L. Gonzalez-Mestres and D. Perret-Gallix, Preprint LAPP-EXP-87-03, presented at the Rencontre sur la Masse Cachée dans l'Univers et la Matière Noire, Annecy (LAPP), July 1987 (Annales de Physique, France).

430

[8]   F.J. Avella, O.J. Sovers and C.S. Wiggins J. Electrochem. Soc. Solid State Science
        114, 613 (1968); V.P. Dotsenko, I.V. Berezovskaya, E.A. Zhikhareva and N.P.
        Efryushina, Izvestiya Akademii Nauk USSR, Neorg. Mater. 20, 1942 (1984).
[9]   K. Oka and H. Unoki, Journal of Crystal Growth 64, 385 (1983).
[10]   J.-P. Chaminade, private communication, and N. Legros, rapport de stage de DEA,
        Lab. de Chimie du Solide du CNRS Talence (June 1988).
[11]   J. Ellis, R. Flores and J.D. Lewin, CERN preprint TH-5040/88 (1988).
[12]   A. de Bellefon et al., in Proceedings of the II European Workshop on Low Tempera-
        ture Devices for the Detection of Solar Neutrinos and Dark Matter, Annecy May
        2−6 1988, Ed. L. Gonzalez-Mestres and D. Perret-Gallix, Frontieres (in preparation).

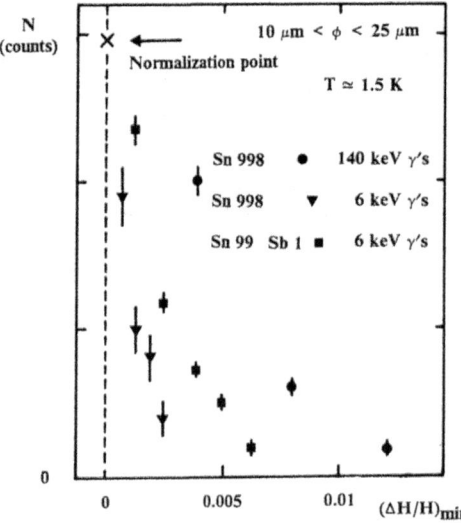

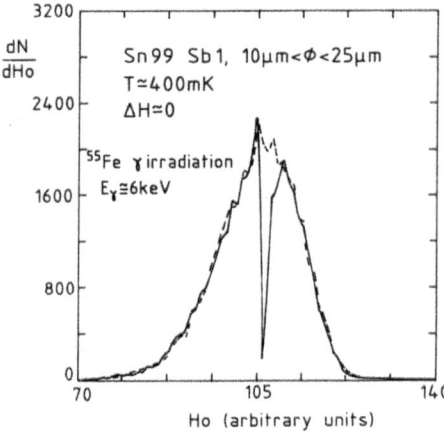

Fig.1 : Number of flips for a fixed time
         period versus $(\Delta H/H)_{min}$ for differ-
         ent irradiations. The superheating
         curve has a FWHM ≈ 20%. The
         sources used were ^{99}Tc and ^{55}Fe.

Fig.2 : Low temperature (0.4 K) irradiation
         of SSG with ^{55}Fe, showing a large
         increase in sensitivity as compared to
         1.5 K data. The irradiated curve (full
         line) exhibits the gap due to missing
         granules having changed state during
         irradiation.

NEW ESTIMATES FOR THE C/M RATIO

M. Kontizas[1], E. Kontizas[2] and A. Dapergolas[2]

(1) University of Athens, Laboratory of Astrophysics,
Panepistimiopolis, GR-Athens 151-71, Greece.
(2) National Observatory of Athens, Astronomical Intitute,
P.O. Box 20048, GR-Athens 11810, Greece.

The intermediate mass stars in their final stage of evolution as red giants can exhibit the effects of interesting astrophysical processes like the mixing to the stellar photosphere of products of the nucleosynthesis in their interior i.e carbon stars. In recent years observational studies of late type giants i.e M and C type stars in the Magellanic Clouds and theoretical investigations of the asymptotic giant branch (AGB) have allowed a valuable comparison between theoretical predictions and observations for this final evolutionary stage of low and intermediate mass stars.

Blanco et al. (1980) and Blanco and McCarthy (1983) have found that the ratio of the number of C to M type stars (later than M5) exhibits enormous variations in going from the galactic bulge C/M≈0.002 to the solar neighbourhood (≈0.01) to the LMC (≈1.7) and the SMC (≈25).

C/M ratio in the Magellanic Clouds

Extended areas of the SMC and LMC (Fig. 1) have been examined in order to search for C and M type stars down to the limit of detection B≈18.5mag (for the used material).

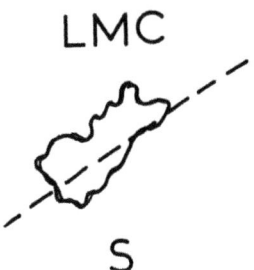

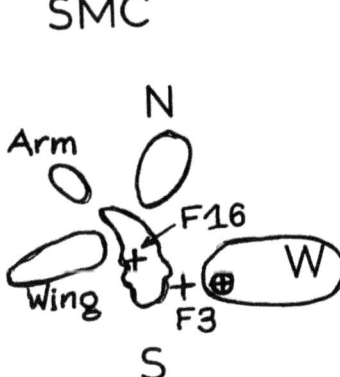

Fig. 1 Location of the examined areas for deriving the C/M ratios in LMC and SMC.

M. Caffo et al. (eds.), Astronomy, Cosmology and Fundamental Physics, 431–432.
© 1989 by Kluwer Academic Publishers.

High quality film copies of objective prism plates taken with the
1.2m UK Schmidt telescope were used (dispersion 2440 and 830 Å/mm at
H_γ) in the wavelength range 3200 – 5400 Å. The classification criteria
and the effectiveness of low and medium dispersion spectra of the 1.2m
UK Schmidt plates are discussed in previous paper by Dapergolas et al.
(1986). The derived values of the C/M ratios are given in Table I.

Table I

Location of the studied area	Searched area arcmin^2	C/M Present	Previous
SMC West	4132	0.04	
SMC North	450	0.05	
SMC Arm	474	0.04	
SMC Wing	4352	0.04	
SMC F3[*]	440	0.07	≈0.79[*]
SMC F16[*]	489	0.10	≈0.89[*]
LMC North	841	0.04	
LMC South	1453	0.05	
Galactic bulge			≈0.002[*]

[*] Blanco and McCarthy (1983).

For the field 3 all the carbon stars of Blanco were identified
whereas for field 16 only 75% of them.

Discussion

Assuming an error ≈15% in the derived number of stars per spectral
type, from Table I it can be seen that when all M stars are included the
derived C/M ratios a) are 10 times smaller for LMC and SMC than those
derived previously b) are 10 times larger for LMC and SMC than those of
our Galaxy c) are of the same order for the LMC and SMC.

Metallicities and age variations have been assumed to be responsible for the large observed differences in C/M derived previously. In
fact theory indicates that for decreasing metallicity the fraction of
AGB stars becoming C type stars increases, and for decreasing metallicity the AGB tracks shift to higher temperatures and the fraction of AGB
stars later than M5 rapidly drops. (i.e all AGB stars in galactic
globular clusters are K-type and all oxygen stars in the Magellanic
Cloud clusters containing C stars are earlier than M5).

From the values derived here where all M type stars are included it
can be suggested that the ratio C/M is not so strongly dependant on
metal abundance at least for the metallicity difference known for the
LMC and SMC.

Acknowledgements
The authors thank the U.K. Schmidt Unit for loan of the material.

References
Blanco, V.M., McCarthy, M.F. and Blanco, B.M.: 1980, Ap. J, 242, 938.
Blanco, V.M. and McCarthy, M.F.: 1983, Astron. J., 88, 1442.
Dapergolas, A., Kontizas, E. and Kontizas, M.: 1986, Astron. Astrophys.
    Suppl. Ser., 65, 283.

SMALL SCALE ISOTROPY OF THE COSMIC MICROWAVE BACKGROUND
AT 230 GHz

E. Kreysa, R. Chini
Max-Planck-Institut für Radioastronomie
Auf dem Hügel 69
5300 Bonn 1

ABSTRACT. We observed a sample of 25 radio-quiet quasars at 230 GHz
with the IRAM 30m-telescope. Most of the sources show no emission
down to a flux density level of 1 mJy. The implications of this result for
radio-quiet quasars will be discussed in a forthcoming paper [Chini et
al. 1988]. Here we use the same data in order to derive upper limits for
the small scale anisotropy of the cosmic microwave background radiation
at 230 GHz.

The IRAM 30m mm-telescope is situated at 2870 m altitude on Pico
Veleta in southern Spain. With an aperture efficiency of ~ 25 % at 230
GHz (λ = 1.3 mm) it is the largest existing telescope for this frequency
[Baars et al. 1987]. We used our own ^{3}He-cooled bolometer system
[Kreysa 1985] which is equipped with diffraction-limited field optics, resulting in a beam diameter of 11" (FWHM). Beamswitching was performed
by a focal plane chopper, using a square waveform with a frequency of
8.5 Hz and 30" amplitude. The bandwidth of the filters in front of the
bolometer is about 50 GHz and, under perfect conditions, the system-
limited sensitivity is 50 mJy sec$^{1/2}$. These conditions are rare however,
and most of the time we are limited by atmospheric fluctuations ("sky-
noise").

For high-sensitivity observations it is essential that the base line
is stable and not affected by variable ground-pickup in the sidelobes.
This could easily lead to spurious "detections" in the ON-ON observing
mode. The stability of the baseline was checked by slewing the telescope
in both elevation and azimuth. Analysis of these scans leads to the result that spurious signals, due to these effects, are below the 0.1 mJy
level.

For our program we chose all radio-quiet active galactic nuclei from
a table published by Neugebauer et al. (1986) that were detected by
IRAS in at least 3 wavelengths. The resulting list of 29 sources represents a complete sample in this sense. We omit 3 southern sources and
the relatively bright active galaxy Mk 231 from further discussion.

Each source was observed in one beam for 10 sec and in the other
beam for the same time. Ten of these ON-ON integrations represent one
measurement of 200 sec integration time. Up to 10 measurements were

M. Caffo et al. (eds.), Astronomy, Cosmology and Fundamental Physics, 433–435.

done in one measurement sequence. We aimed at a final statistical error of about 1 mJy. Depending on atmospheric conditions, from one to five measurement sequences were required to reach that goal. To determine atmospheric opacity, sky dips were done frequently and calibration was provided by observations of planets (preferably Uranus and Neptune).

The data from two observing runs in November 1987 and April 1988 are shown in Fig. 1.

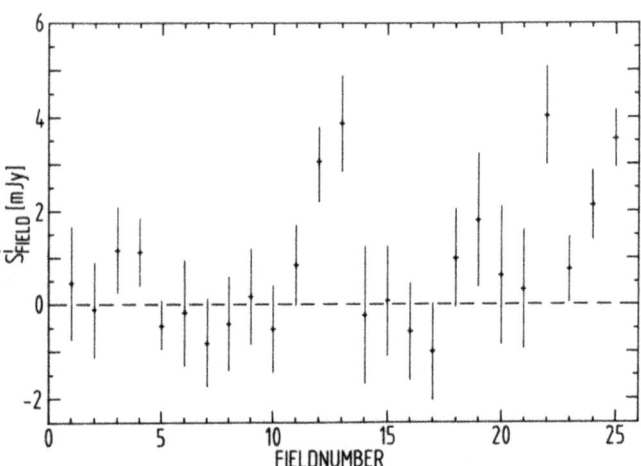

Figure 1. Final averages for all 25 field observed

The large majority of the sources could not be detected. This is not surprising if one interprets the FIR-data as emission from heated dust. Indeed, if one calculates the expected 230 GHz flux density from the IRAS data on the basis of this model, none of the 25 sources should have been detected (Most of the predicted flux densities are below 1 mJy). The reason for the four positive detections is probably that we picked up synchrotron radiation from the compact sources. The case is quite clear for sources no. 12, 13 and 22 where radio data at longer wavelengths fit very well with our measured flux at 1.3 mm. The situation is less clear-cut in the case of source 25. It has been detected at cm wavelengths but there is not enough spectral information to extrapolate as far as 230 GHz. We also exclude this source from the analysis. The data from the remaining 21 fields show a scatter that is reasonably consistent with normally distributed errors as shown in Fig. 2. Applying Boynton-Partridge statistics [Boynton and Partridge, 1973], we can test our measured distribution for the existence of a underlying anisotropy of the cosmic microwave background. At the 95 % confidence level we can say that the statistical signal is less than 0.94 mJy. ON-ON measurements test the statistical sky fluctuations δS_{rms} with an efficiency of $(1.5)^{1/2}$ [Uson and Wilkinson, 1984] at the separation of the two beams. Therefore $\delta S_{rms} = 0.77$ mJy. For a blackbody of 2.75 K observed at 230 GHz the flux scale can be converted to a temperature scale

$$\frac{\delta T}{T} = 0.245 \frac{\delta S}{S}$$

The flux S per beam of the CMB is 0.733 Jy. Our final result is

$$\frac{\delta T}{T} \le 2.6 \ E-4$$

at a 95 % significance and for a scale of 30".

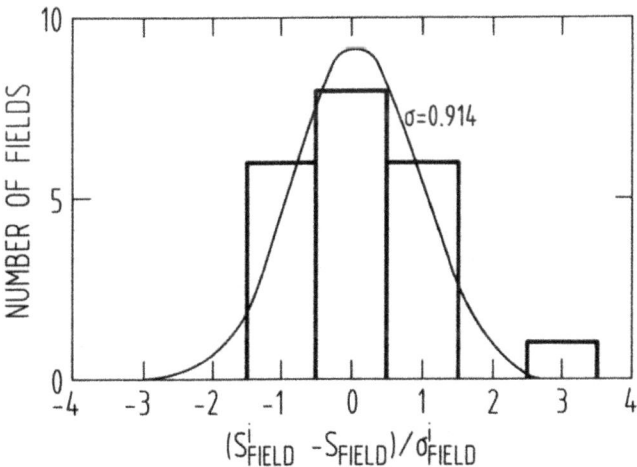

Figure 2. Scatter of results for 21 fields.

Improvements are possible. A total integration of 35 h time was required for these data. From the statistics of our data we derive an average sensitivity of 70 mJy $\sec^{1/2}$. This means that some of the data have been obtained under less than ideal conditions. The situation would have been much better if we had used empty fields at high declinations as was done by Uson and Wilkinson. Apart from more stable atmospheric conditions and constant sidelobe contributions, this would also avoid ambiguities due to source emission. The biggest improvement, however, could be achieved with bolometer arrays.

REFERENCES

Baars, J.W.M., et al. 1987, Astron. Astrophys. 175, 319
Boynton, P.E. and Partridge, R.B. 1973, Ap. J. 181, 243
Chini, R., et al. 1988, Astron. Astrophys., submitted
Kreysa, E. 1985, Proc. URSI Symp. 'MM- and Submm Wave Radio Astronomy', Granada, p. 153
Neugebauer, G., et al. 1986, Ap. J. 308, 815
Uson, M.J. and Wilkinson, D.T., 1984, Ap. J. 283, 471

PREDICTABLE CHEMISTRY OF SYSTEMS CONTAINING UNSATURATED QUARKS

Christian K. JØRGENSEN
Section de Chimie, Université de Genève
CH 1211 Geneva 4, Switzerland

Much like the neutrino (Pauli, 1930; Fermi, 1934), the *Quarks* (Gell-Mann, 1964) rationalize a huge manifold of properties of "elementary" particles. These quarks carry the electric charges u(u +2e/3; d(down) -e/3; s(strange) -e/3; c(charm) +2e/3; b(beauty) -e/3,.. The (8+10) lightest baryo "consist" of three quarks selected from the "flavours" u, d and s. Only u and d quarks play a re in conventional nuclei $u^{Z+A} d^{2A-Z}$ (with charge $+Ze$ and atomic weight in the unit 1 amu = 0.93 GeV quite close to the positive integer A) and though strong inter-quark correlations cluster quar 3 and 3, the 1932-1964 model of Z protons and $(A - Z)$ neutrons is no longer appropriate [1].

It took quarter of a century to detect neutrinos explicity. Several theorists believe quarks be unconditionally confined, though at kT above 0.2 GeV (prevailing up to 20 microseconds aft the Big Bang) a quark plasma is now supposed. When this quark soup coagulated to baryons (cooling below 2 terakelvin) a few unsaturated quark-systems (difference between number of quarks a number of anti-quarks not divisible by 3) may have survived (even if our accelerators cannot sma baryons; a lasting source may be protons passing close to black holes) and Zel'dovich estimated the concentration to one per 10^{10} amu (far more than found in Earth and Moon samples). Later, valu [2] of one per 10^{20} amu (i.e. 6000/g) were suggested. This happens to be the amount of fractio charge in 10^{-4} g niobium spheres reported by the group of Fairbank at Stanford University since 197 There is no need [3] for Z of these "nuclei" to be close to 41; chemical properties are compatible wi Z close to 73, 105, 143,.. since the propensity toward fission is attenuated. High-density uds-mat [4-7] with a comparable number of u, d and s quarks may compensate the usual mass excess 0.2 an per s by decreased (Pauli exclusion) kinetic energy and be metastable (or, what is a scary thoug more stable) and be a major constituent of heavy, non-luminous "dark matter". Then, fractiona charged systems may occur [8] $u^{A+\delta-\epsilon} d^{A+\epsilon} s^{A-\delta-1}$ with $(3A - 1)$ quarks (probably atomic weig 100 to 10^6 amu above A) and a moderate charge $Z = (\delta - \epsilon + 1/3)$. If Z_0 is an integer, all $(3A -$ systems have $Z = Z_0 + 1/3$ but adducts with an anti-di-quark [1] have $Z = Z_0 + 2/3$, and in gener $(3A-2)$ systems may be surrounded by a meson cloud containing one more quark than (a fluctuati or invariant) number of anti-quarks, provided by the surrounding vacuum.

Glashow pointed out that many physical discoveries are unscheduled surprises: one percent the paradoxical element argon found in the air 1894; all data of nuclear physics obtained (befc 1932) from radioactive isotopes extracted from uranium and thorium minerals; and the cosmic ra being worthy precursors of the super-colliders. Chemical properties of fractionally charged syste are almost exclusively determined by the $Z = (Z_0 + 1/3)$ or $(Z_0 + 2/3)$ [here written $Z.33$ and $Z.6$

436

M. Caffo et al. (eds.), Astronomy, Cosmology and Fundamental Physics, 436–437.
© 1989 by Kluwer Academic Publishers.

intercalating 200 new "elements" in the Periodic Table up to $Z_0 = 100$. Any negatively charged systems very rapidly zooms down (like a negative muon) on a conventional nucleus (mainly Z=1 and 2 in Cosmos) decreasing its positive Z. Any central system (with radius below 10^{-11} cm) shows chemical properties characterized by the total amount Z of positive charge. A survey [8] of expected chemistry (as a function of Z) includes the quite mobile carriers with $Z = 0.33$ of fractional charge stabilized by the electronic density in metals and low-gap semiconductors, like also 0.66, able to form a strongly polarizing, tiny cation (like protons). The anions $1.33^{-0.66}$, $1.66^{-0.33}$, and $9.66^{-0.33}$ are quite large and polarizable (like iodide) and $9.33^{-0.66}$ more electrovalent (like chloride). The cations $2.33^{+0.33}$ and $2.66^{+0.66}$ are large but quite electrovalent, and $10.33^{+0.33}$ and $10.66^{+0.66}$ highly polarizable, as Ag^+ and Tl^+. The "elements" $Z = 5.33, 5.66, 6.33, 6.66, ..$ should readily be involved in quark-organic chemistry. Comparison are made [8] between 24.33, 24.66 and the known chemistry of chromium and manganese. Lackner and Zweig [9] proposed a rationalization of the expected chemistry of fractionally charged systems with Mulliken electronegativities χ_M. The relations between χ_M, the Pauling χ_P, and the Sanderson χ_S electronegativities are subtle (and rather loose), but it can be demonstrated [8] by comparison with the known lanthanide chemistry that χ_M is inconsistent, and largely irrelevant to the chemical behavior between $Z = 58.33$ and 70.66 (of which traces may occur in rare-earth minerals).

De Rújula, Giles and Jaffe [10] concluded their paper by the conceivable analogy between quark mines [a rich stroke would be one per 4×10^{15} amu, the average radium concentration in the Earth's crust] and gold mines. The geochemical fractionation, and suggestions for pre-concentrated, quite readily available materials have been reviewed [8] as well as detection methods after further enrichment. The most pratical way may be droplet-jet deflection in electric fields, recently applied [11] to tetraethylene glycol with a limit of one $(Z_0 + 1/3)$ per 2×10^{19} amu, and of one $(Z_0 + 2/3)$ per 4×10^{19} amu. It would be helpful if solutions of compounds involving Z above 8, or finely divided powders (like pigments in a paint jet) could be studied. Since the atomic weights may be high (above 300) or falling in the "windows" between $(A + 0.05)$ and $(A + 0.90)$, it is also worthwhile to continue mass-spectrometric searches (e.g. including a tandem-accelerator). Though Z may be an integer, there is a fair chance that a substantial part of such species have fractional Z values. This would be particularly interesting for any future cometary, Jovian or asteroid samples. As already discussed previously [8], even rather scarce species may significantly influence stellar nucleosynthesis [12], and in particular, act as WIMP (weakly interacting massive particles) in the solar center. For chemists, it is important to note that naked neutral species heavier than 10^8 rapidly fall into the center of the Earth (because the Brownian motion cannot compete with the vertical fall), but that rest-masses as high as 10^{12} amu, when accompanied by an electronic cloud at least 50 pm, fall very slowly, and would tend to accumulate in the Ocean between the manganese nodules at the sea-floor.

1] C.K.Jørgensen: Naturwissenshaften 69 (1982) 420; 71 (1984) 151 & 418.
2] R.V.Wagoner and G.Steigman: Phys.Rev. D 20 (1979) 825.
3] C.K.Jørgensen: Nature 305 (1983) 787.
4] E.Witten: Phys.Rev. D 30 (1984) 272.
5] A.DeRújula and S.L.Glashow: Nature 312 (1984) 734.
6] E.Farhi and R.L.Jaffe: Phys.Rev. D 30 (1984) 2379.
7] F.C.Michel: Phys.Rev.Lett. 60 (1988) 677.
8] C.K.Jørgensen: in "Molecular Structure and Energetics" (J.F.Liebman and A.Greenberg Ed.), in press.
9] K.S.Lackner and G.Zweig: Phys.Rev. D 28 (1983) 1671; D 36 (1987) 1562.
10] A.DeRújula, R.C.Giles and R.L.Jaffe: Phys.Rev. D 17 (1978) 285.
11] J.Van Polen, R.T.Hagstrom and G.Hirsch: Phys.Rev. D 36 (1987) 1983.
12] V.Trimble: Rev.Mod.Phys. 47 (1975) 877.

THE OPAL DETECTOR AT LEP

The OPAL collaboration [*]

OPAL, one of the 4 experiments being installed at the Large Electron-Positron (Lep) collider at CERN, will study the physics of e^+e^- interactions in the 100-200 GeV energy range. In particular it will perform precision measurements of the Standard Model parameters, like the mass of the vector bosons W and Z and the Weinberg mixing angle, as well as search for new particles (like the Higgs bosons and supersymmetric particles) and new phenomena. For these reasons a general purpose detector has been designed.

The interaction region is surrounded by a sequence of detectors, all with a cylindrical structure, with a barrel and two end caps. Tracking is performed by a central drift chamber; electron and photon energy measurements are done by a high resolution electromagnetic calorimeter using lead glass Çerenkov counters. The magnet iron joke, placed outside the electromagnetic calorimeter, is instrumented as a hadron calorimeter with plastic streamer tubes. Muon detection is provided by a set of drift chambers in the barrel region, and plastic streamer tubes in the end caps. The detector (fig. 1) is being installed, and it will be fully operational in July 1989.

The Central Detector consists of 3 different tracking chambers: the vertex detector, the jet chamber, and the Z-chambers. The vertex detector is a high precision drift chamber which will provide correlated measurements of r-ϕ and z coordinates. Precision of $\sigma_{r\text{-}\phi}$ = 50 μm and σ_z = 500 μm will be achieved. The jet chamber is a pictorial drift chamber which will measure the coordinates of charged particle over a solid angle of 97% of 4π. The space coordinates are measured through wire position, drift time and charge division measurements.   Energy loss measurements, dE/dx, in the chamber gas are also performed, to identify low momentum particles. The Z-chambers are 24 drift chambers (one for each sector of the jet chamber) which measure track coordinates along the beam direction with an accuracy of 300 μm.

The magnet is a warm coil; it will produce a central maximum field of 0.4 T.

The Time of Flight Counters consist of 160 scintillators, 7 m. long and 4.5 cm thick, mounted outside the magnet coil. They will improve the identification of pions, kaons and protons in the low momentum region (1.0 - 2.5 GeV/c).

438

M. Caffo et al. (eds.), Astronomy, Cosmology and Fundamental Physics, 438–440.
© 1989 by Kluwer Academic Publishers.

The Electromagnetic Presampler detects electromagnetic showers originating in the magnet coil and thus measures the position of photon conversion. It consists of two layers of aluminum limited streamer tubes read-out by 1 cm wide cathode strips.

The Electromagnetic calorimeter consists of systems of lead glass Cerenkov counters. It provides electron and photon detection with excellent energy resolution, 99% coverage of solid angle and good rejection of hadrons. The barrel consists of an array of 9440 lead glass counters, each equipped with a photomultiplier, and pointing toward the interaction region. It has a thickness of 22 radiation lengths. The achieved energy resolution combining the data from the presampler is $\sigma/E = 0.08/\sqrt{E}$. The end caps consist of 2300 lead glass blocks, instrumented with single stage phototubes (located in a magnetic field) used in conjunction with a high gain, low noise amplifier.

The Hadron calorimeter is composed of the Barrel, of two End Caps and of two Pole Tips. The Barrel and End Caps are sandwiches of 10 cm. iron and plastic streamer tubes. The calorimetry is achieved by reading out induced signals on external pads and strips. From tests in the energy range of 5 - 50 GeV, the energy resolution of the bare hadron calorimeter is $\sigma/E = 1.2/\sqrt{E}$. The Pole Tips are built out of sandwiches of 8 cm iron and thin proportional chambers of new design, operating in high gain mode. The prototype calorimeter showed linear respose up to 90 GeV hadrons with the same energy resolution of the Hadron Barrel and End Caps.

The Muon detector identifies muons by their penetration through the lead glass and iron, and by requiring the agreement in the position of the external track with the extrapolated one from the central detector. The barrel consists of 4 layers of drift chambers, 10.5 m. long. This structure will allow simultaneous determination of the position and direction of each penetrating track, with a space resolution of about 1 mm in the transverse drift direction. The End Caps are 4 planes of resistive limited streamer tubes, with two dimensional cathode strip read-out for each layer of tubes so that the x-y coordinates can be measured.

The Forward Detector completes the electromagnetic coverage for the OPAL detector by tagging small angle electrons and photons; it will also be used as a luminosity monitor. The Forward Detector consists of lead-scintillators calorimeters, tube chambers and scintillation counters. The energy resolution of the lead-scintillator calorimeter is $\sigma/E = 0.20/\sqrt{E}$.

OPAL

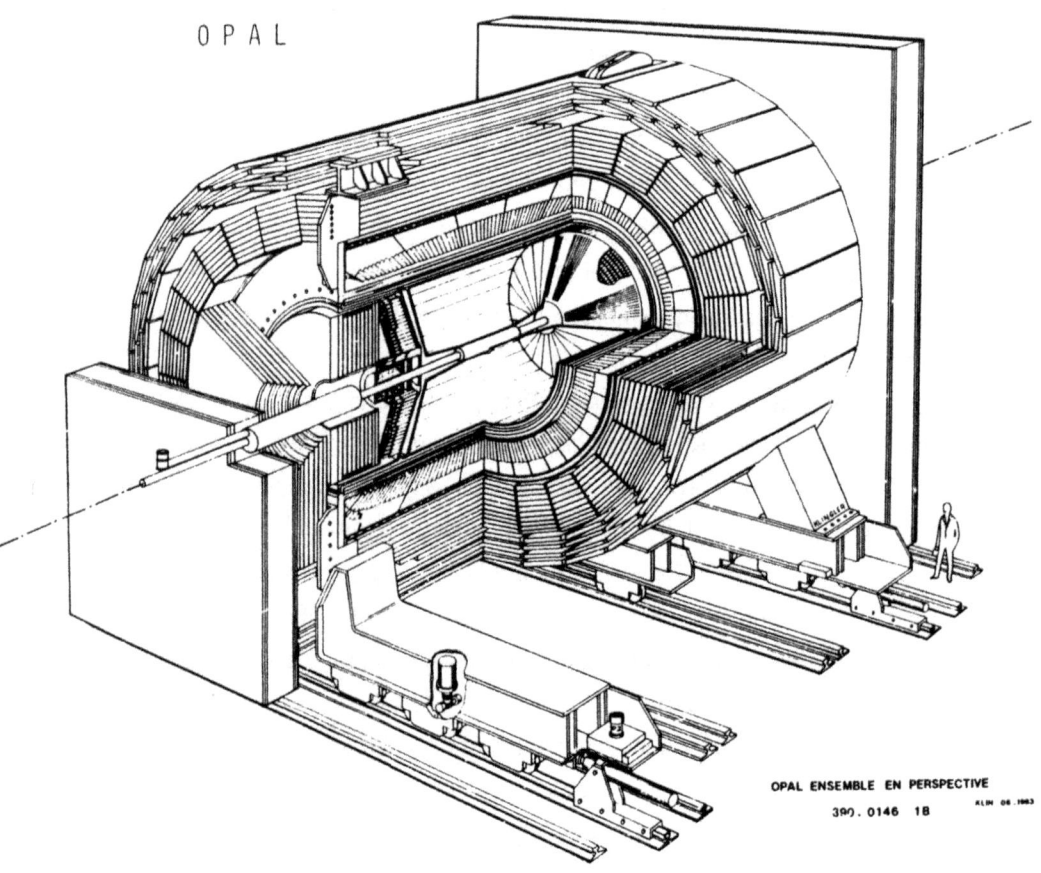

Fig. 1. *Schematic layout of the OPAL detector at LEP*

* Poster presented by the Bologna group of the OPAL collaboration (P. Capiluppi, M. Cuffiani, I. D' Antone, M. M. Deninno, F. Fabbri, G. Giacomelli, G. Mandrioli, S. Marcellini, A. Montanari, F. Odorici, B. Poli, A. M. Rossi and G. P. Siroli)

MEASUREMENTS OF T$_{CBR}$ AT 1.3 MM

Eliana Palazzi and Nazzareno Mandolesi
Istituto TE.S.R.E.-CNR, Bologna, Italy
Philippe Crane
European Southern Observatory, Munich,BRD
Dennis J. Hegyi
University of Michigan Ann Arbor, Michigan,USA
J. Christopher Blades
Space Telescope Science Institute, Baltimora,USA

1. INTRODUCTION

The determination of T$_{CBR}$ at 1.3 mm is particularly interesting because it lies just at the peak of the Black-Body spectrum, where deviations can be most likely (Matsumoto et al., 1988). However, measurements at this wavelength are extremely difficult from the ground.

The CN molecule provides a fortuitous thermometer to determine the CBR temperature at 1.3 mm and also at 2.6 mm. The tecnique involves measuring weak absorption lines in interstellar clouds (Crane et al., 1986). These absorption lines provide a direct measure of the rotational state populations provided the saturation correction is not too large.

We present here observations of the star HD154368 which has a high column density of CN and is particular suitable for determing T$_{CBR}$ at 1.3 mm . At 2.6 mm the errors in saturation correction are in this case so large that the resulting uncertainty in T$_{CBR}$ is larger that in other sources (i.e. ζ Ophiuchi). At 1.3 mm the saturation correction is not important, and we present a preliminary value of T$_{CBR}$. This result is limited primarily by statistical uncertainties.

2. OBSERVATIONS AND DATA REDUCTION

The data discussed here were obtained during an 11 day observing run in June 1987 at the European Southern Observatory 's 1.4 m Coude' Auxilliary Telescope and associated Coude' Echelle Spectrograph. During 9 of the 11 nights, a 1024 x 512 elements CCD detector was used

M. Caffo et al. (eds.), Astronomy, Cosmology and Fundamental Physics, 441–444.
© *1989 by Kluwer Academic Publishers.*

producing a resolution of 75,000 and sampling the spectrum at 32.3 mÅ per pixel.

In the remaining 2 nights, we used a Reticon detector with a resolution of 150,000 at λ = 3874 Å ; each individual diode of the Reticon sampled 17.4 mÅ of the spectrum. The higher resolution of the Reticon was necessary to determine the intrinsic linewidth of CN. From the 9 spectra obtained we measured an intrinsic width (FWHM) of 14.28 ± 2.32 mÅ.

During the 9 CCD nights we obtained 89 spectra of HD154368. A typical sum-spectrum of one night before normalization (fig. 1) shows clearly the absorbtion CN lines R(0), R(1), P(1), R(2) and P(2) of the optical transition $B^2\Sigma - X^2\Sigma$.

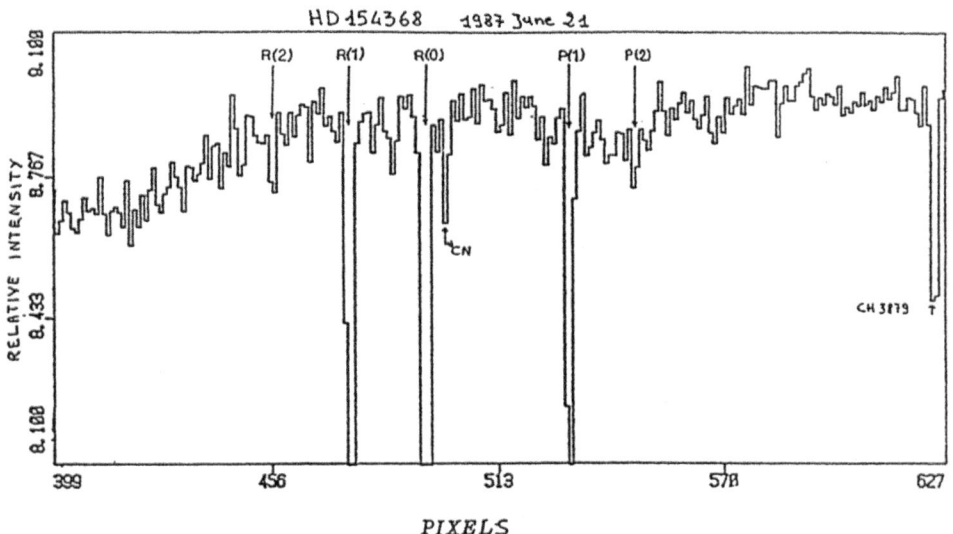

Fig1. A typical sum spectrum of one night before normalization is shown. The absorption CN lines R(0), R(1), P(1), R(2) and P(2) are clearly visible.

The determination of the equivalent widths of the lines provides, after a saturation correction, accurate measurements of the rotational state populations, which can be used directly to find the CBR temperature T_{CBR} at 2.64 mm and 1.32 mm.

A detailed description of the method used in the data
analysis is given in Crane et al.(1986) , Crane and Hegyi
(1988).
    Table I summarizes the equivalent widths obtained.

TABLE I: CN EQUIVALENT WIDTHS

LINE	EQUIVALENT WIDTHS (mÅ)
R(0)	25.10 ± 0.15
R(1)	15.02 ± 0.13
P(1)	9.80 ± 0.14
R(2)	0.576 ± 0.108
P(2)	0.450 ± 0.077

Using the equivalent widths above and making the saturation
correction using the linewidth we have determined, we find
the values of T_{CBR} given below in Table II.

TABLE II: MEASUREMENTS OF THE T_{CBR} AT 1.32 mm

Lines ratio	$T_{uncorrected}$ (K)	$T_{corrected}$ (K)
R(2)/R(1)	2.973 ± 0.166	2.631 ± 0.130
R(2)/P(1)	2.771 ± 0.148	2.599 ± 0.130
P(2)/R(1)	3.106 ± 0.166	2.733 ± 0.130
P(2)/P(1)	2.887 ± 0.148	2.699 ± 0.130

    We note that the uncorrected values of T_{CBR} are a firm
upper-limit since the saturation correction as well as any
local excitation correction (we have not included any
since we expect it to be less then 10 mK) would reduce
the values of T_{CBR}.
    As mentioned T_{CBR} at 2.64 mm has a relatively large
uncertainty. We find

$$T_{CBR} (2.64 \text{ mm}) = 2.82 ± 0.12 \text{ K}$$

which agrees with the more precise value reported in the
work of Crane, Hegyi, Kutner and Mandolesi at this
conference.

3. CONCLUSIONS

Our preliminary value for T$_{CBR}$ at 1.32 mm is :

$$T_{CBR} (1.32 \text{ mm}) = 2.68 \pm 0.10 \text{ K}$$

The quoted uncertainty is mainly due to the errors in determining the equivalent widths but also includes an allowance for uncertainty in the saturation correction. This value already indicates no departure from a pure 2.75 K blackbody at $\lambda > 1.3$ mm.

REFERENCES

Crane, P, and Hegyi, D.J., 1988, *Astrophys. J.(Letters)*, L35, 326.

Crane, P., Hegyi, D.J., Mandolesi, N., and Danks, A.C.D., 1986, *Astrophys. J.*, 309, 822.

Matsumoto, T., Hayakawa, S., Matsuo, H., Murakami, H., Sata, S., Lange, A.E., and Richards, P.L., 1988, *Astrophys. J.*, 329, 567.

CORRELATIONS BETWEEN THE DATA RECORDED BY THE MONT BLANC NEUTRINO DETECTOR AND BY THE MARYLAND AND ROME GRAVITATIONAL WAVE DETECTORS DURING SN1987A

Giovanni V. Pallottino
Department of Physics
University of Rome "La Sapienza"
P.le A. Moro, 2 - 00185 Roma, Italy

ABSTRACT The author reports the analysis of the detectors mentioned in the title over a period of 18 hours that includes the Mont Blanc 5 neutrino burst time. There is statistical evidence for correlations during a period of about two hours, centered on the 5 neutrino burst. The effect is mainly due to a dozen of large amplitude Maryland and Rome events during the above 2 hour period. The interpretation of these events as due to gravitational radiation from SN1987A should be rejected, according to the standard theory of the cross-section of g.w. detectors, since it would imply an emission of energy too large by a factor of about 10^4.

1. Introduction

I present here part of the results of the analysis of the data recorded by the neutrino detector of Mont Blanc and by two gravitational wave antennas, one in Rome the other in Maryland, during SN1987A, that were reported by Guido Pizzella to "Les Rencontres de Physique de la Vallée d'Aoste", La Thuile, on February 29, 1988 [1]. The authors belong to the following institutions:
the Departments of Physics of the two Universities of Rome
the Department of Physics and Astronomy of the University of Maryland
the Institute of Cosmo-Geophysics of CNR, Turin
the Institute of Nuclear Research of the Academy of Sciences of USSR, Moscow

I recall that the gravitational wave (g.w.) antennas considered here, that were in operation during the SN time, are low sensitivity detectors, but their performance, as well as that of the auxiliary instrumentation (accelerometers, electromagnetic antennas, etc. [2]), was very satisfactory. We examined carefully the data provided by these instruments and arrived to the conclusion that it is not possible that the g.w. events considered in the analysis reported in what follows may be

445

M. Caffo et al. (eds.), Astronomy, Cosmology and Fundamental Physics, 445–451.
© *1989 by Kluwer Academic Publishers.*

446

due to seismic, acoustic or electrical disturbances.

As regards the neutrino detector, we have used in the analysis all the events recorded by the apparatus (at a rate of about 40 per hour): usually they are considered as noise when occurring with Poisson statistics, as signal when occurring in bunches with low probability of being due to chance. They are, in other words, discriminated on the basis of statistics and not of physics and therefore any of them may convey useful physical information.

Since, however, we do not dispose of an interpretation of the correlations here reported, in the following I shall talk of "neutrinos" and of "g.w." events in inverted commas referring to the events recorded by the corresponding detectors, without nor excluding nor presuming that these events are actually due to, respectively, physical neutrinos and physical g.w.

2. The gravitational wave detectors

The basic features of the two antennas are the following.

	Rome	Maryland
mass (Kg)	2300	3100
resonance frequency (Hz)	858	1600
orientation	29^0 to E-W in S-W quadrant	E-W

The accuracy of the recorded times is ±0.1 s for the g.w. antennas, ±1 ms for the Mont Blanc detector. The data of the two antennas are filtered to form two time series, with sampling time of 1 second, that represent energy innovations, that is variations of the energy of the fundamental vibration mode (as due to a short burst of gravitational radiation or to the effect of the noise).

These quantities, denoted as E, are expressed in kelvin degrees. Their value in joule is obtained by multiplying by the Boltzmann constant 1.38×10^{-23} J/K; their value in eV, by dividing by the charge of the electron, that is multiplying by 0.86×10^{-4} eV/K.

Since the cross section of a resonant detector is proportional to the mass, prior to the analysis we have normalized the data of the antennas by multiplying the Maryland data by the ratio of the masses (2300/3100). Even after this normalization, however, we cannot expect that the same gravitational wave burst produces equal signals in the two antennas because of two effects: (a) the different orientations of the detectors; (b) their different resonance frequencies (the input burst may have different spectral content at these frequencies).

This explains why we have combined the two streams of g.w. data in a single time series, to be used in the analysis, by summing the corresponding energy innovations:

$$E(t) = E_M(t) + E_R(t) \qquad\qquad (1)$$

where the subscripts M and R stay for Maryland and Rome. We also note that the light distance between Rome and Maryland is about 30 ms, much smaller than the spacing of the samples (1 s).

3. The first observations

Shortly after SN1987A took place, we noticed (fig. 1) that our antenna had recorded a large (135 K) energy innovation 1.5 s before the first of the 5 neutrinos detected by the Mont Blanc collaboration [3], with a probability of 3% to be observed in 3 seconds before the M. B. event [4]. This is a reasonable range for the delay if the neutrino mass fulfils the relation: $m_e\, c^2 \lesssim 10$ eV.

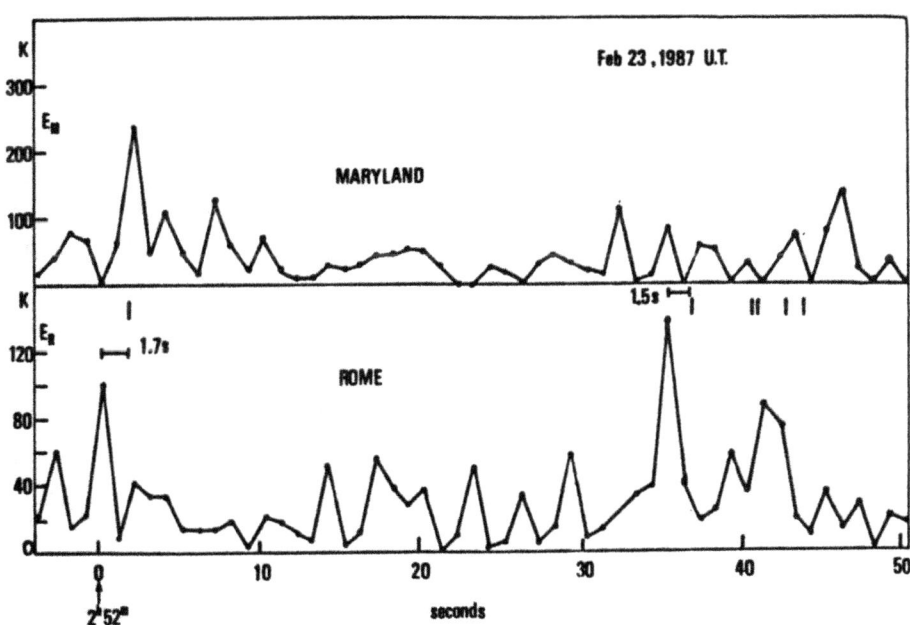

Fig. 1 The Maryland and Rome data plotted versus time. The five neutrinos of the Mont Blanc burst are also indicated together with the first background neutrino preceding them.

Later, examining the data of the Maryland antenna, we found at the same time a peak (with probability of 5.6%): the fact that they are in coincidence improves of one order of magnitude the previous estimate for the Rome observation. We are, of course, aware that attributing these observations to g.w. requires an energy emitted by the supernova too large by a factor 10^4.

4. The analysis of the data

According to the above experimental observations, we decided to perform a systematic search for correlations between all the events of the Mont Blanc detector and the data of the g.w. antennas. I only report, in what follows, the search for coincidences between the neutrino events and large amplitude g.w. signals occurring in a 1 second interval between 0.7 and 1.7 seconds before the neutrino arrival times. I consider, in particular, the case of energy innovations, as given by eq.(1), larger than 150 K.

In a period of 135 minutes centered at 2:45 of February 23 we have N_e=102 neutrinos and N_{gw}=187 g.w. energy innovations above 150 K. In a window of 1 second we expect

$$\bar{n} = \frac{N_e \ N_{gw} \ 1 \ s}{135 \ \text{minutes}} = 2.355 \qquad (2)$$

accidental coincidence events. When we count how many times a g.w. event precedes a neutrino in the interval 0.7-1.7 seconds we find 14 events (fig. 2). Assuming a Poisson distribution the probability for the 14 events to be due to chance is:

$$P(n \geq 14)_{2.355} = 2.1 \times 10^{-7} \qquad (3)$$

We have performed the same calculations for a large number (20,000) of values of delay (±10,000 s), and we have counted how many times we have zero coincidences, one coincidence, two coincidences and so on: the distribution is in excellent agreement with the Poisson theoretical curve (computed for n=2.355). We conclude that the above probability estimate is correct.

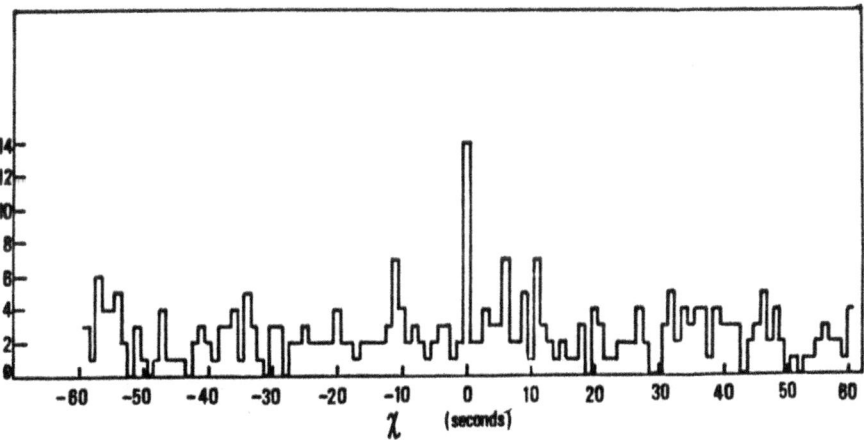

Fig. 2   Observed coincidences between g.w. events larger than 150 K and neutrinos during 135 minutes centered at 2:45 of February 23, for delays in the range -60, +60 s. The abscissa is displaced so that the peak at zero corresponds to a delay of 1.2±0.5 s from g.w. events to neutrino arrival times.

The above analysis has been repeated for different values of threshold (from 100 K to 200 K, in steps of 20 K) for the g.w. signals. The results are summarized in fig. 3, where the probability to have accidentally the observed number of coincidences (in the window 0.7-1.7 s) is shown versus the threshold (in parenthesis we give the integer closest to the expected number of accidentals). We notice that, starting from 100 K, the difference between the number of observed and expected coincidences is about 10, up to about 170 K, where the possible real signals begin to be eliminated. Therefore we infer that something like 10-12 neutrinos out of the observed 102 (in the 135 minutes interval) significantly contribute to the correlation effect, that is they could be signals instead of noise.

When a similar analysis is performed on other periods (ranging from 12:00 U.T. of February 22 to 6:00 U.T. of February 23, when there was a power line interruption in Maryland), the results show neglegible deviations from the statistical expectations.

450

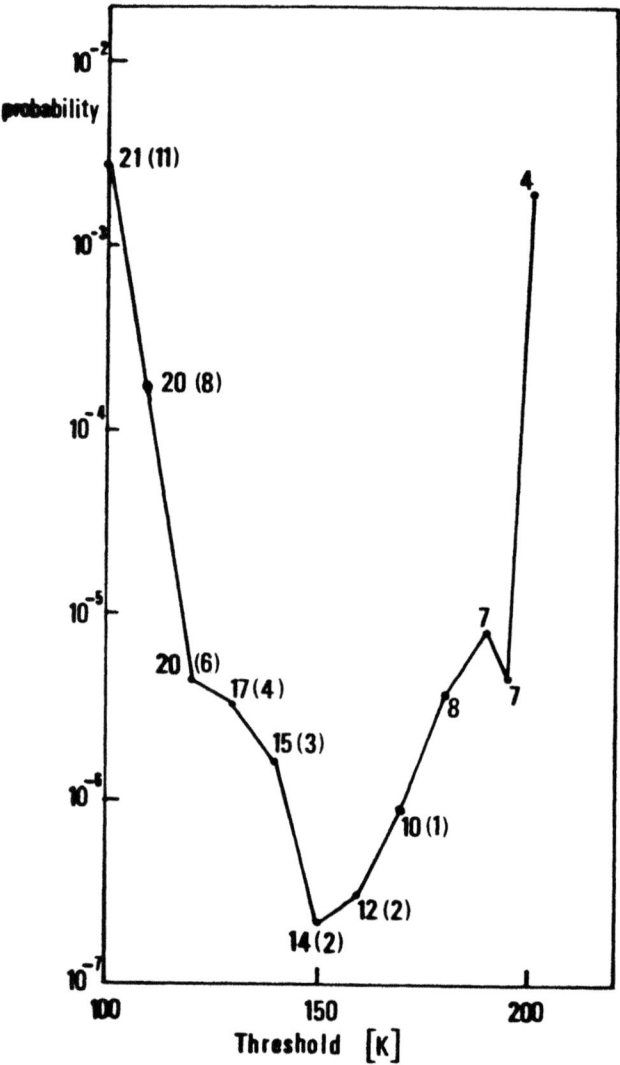

Fig. 3   *The Poisson probability computed for the coincidences (-1.2 ±0.5 s) for various thresholds of the g.w. events during the period of 135 minutes. The numbers indicate the number of the observed coincidences, those in parenthesis the expected numbers of the corresponding accidentals.*

451

5. Final remarks

We do not dispose, at present, of a physical interpretation of the correlations here reported. We plan to check the statistical results by extending the analysis over a much longer period, using the data of the detector during the month of February 1987. We also intend to investigate whether using the data of other neutrino detectors (in particular those of Kamiokande) the above results are confirmed or not.

REFERENCES

(1) M. Aglietta, E. Amaldi, M.Badino, M. Bassan, G. Bologna, P. Bonifazi, C. Castagnoli, M.G. Castellano, A. Castellina, E. Coccia, C. Cosmelli, V.L. Dadykin, S. Frasca, W. Fulgione, P. Galeotti, D. Gretz, A.S. Malguin, I. Modena, G.V. Pallottino, G. Pizzella, P. Rapagnani, F. Ricci, V.G. Ryassny, O.G. Ryazhskaya, O. Saavedra, V.P. Talochkin, G. Trinchero, G. Vannaroni, S. Vernetto, J. Weber, G. Wilmot, G.T. Zatsepin, V.F. Yakushev: 'Analysis of the data recorded by the Mont Blanc neutrino detector and by the Maryland and Rome gravitational wave detectors during SN1987a' presented to *Les Rencontres de Physique de la Vallée d'Aoste*, La Thuile, February 1988.

(2) F. Bronzini, S. Frasca, G.V. Pallottino, G. Pizzella, G. Vannaroni: *Nuovo Cimento C* **8**, 300-319 (1985).

(3) M. Aglietta, M.Badino, G. Bologna, C. Castagnoli, A. Castellina, V.L. Dadykin, W. Fulgione, P. Galeotti, F.F. Kalchukov, B. Kortchaguin, P.V. Kortchaguin, A.S. Malguin, V.G. Ryassny, O.G. Ryazhskaya, O. Saavedra, V.P. Talochkin, G. Trinchero, S. Vernetto, G.T. Zatsepin, V.F. Yakushev: *Europhys. Lett.* **3**, 1315-1320 (1987).

(4) E. Amaldi, P. Bonifazi, M.G. Castellano, E. Coccia, C. Cosmelli, S. Frasca, M. Gabellieri, I. Modena, G.V. Pallottino, G. Pizzella, P. Rapagnani, F. Ricci, G. Vannaroni: *Europhys. Lett.* **3**, 1325-1330 (1987).

CBR FLUCTUATIONS AT DIFFERENT WAVELENGTHS

Mirosław Panek[1], Bronisław Rudak[2]
Copernicus Astronomical Center
[1] 00-716 Warszawa, Bartycka 18
[2] 87-100 Toruń, Chopina 12/18

The analysis of the CBR fluctuation amplitudes detected simultaneously on both sides of the $2.74K$ peak provides an interesting cosmological test. The relation of amplitudes is different for considered models producing the fluctuations: nonrelativistic comptonization on hot gas (A), emission from cosmic dust at high redshifts (B), and primordial density perturbations (C).

The homogeneous part of A or B can give the CBR spectrum as measured by Matsumoto et al. (1988). Fluctuations are generated by perturbations of the Compton y parameter (A), by the dust density or temperature perturbations (B), or by primordial effects (C) observed on the background given by A or B.

Figures 1 and 2 (following page) show values of the relative intensity perturbations R_λ:

$$R_\lambda = \left(\frac{\delta I}{I}\right)_\lambda \left(\frac{\delta I}{I}\right)_{\lambda_0}^{-1}$$

To use R_λ values to identify or constrain models of origin of fluctuations we suggest the following procedure:
- Look for the CBR fluctuations at two wavelengths - one in the RJ region ($\lambda > 2mm$) and the other in the Wien region ($400\mu m < \lambda < 1.1mm$). Angular scales of both measurements should be identical.
- Find amplitudes of fluctuations at both wavelengths and investigate their correlation for the same locations on the sky.
- If they are correlated then it means that fluctuations at both wavelengths are generated by the same process:
   a) anticorrelation indicates the comptonization perturbations
   b) correlation indicates the dust or primordial perturbations
- Find the value of R_λ and use it to constrain the fluctuation generating process. An additional measurement at $\sim 100\mu m$ is necessary to correct for the interstellar dust radiation fluctuations.

References

Matsumoto, T., et al., 1988. *Astrophys.J.*, **329**, 567.
Panek, M., Rudak, B., 1988. *MNRAS*, in press.

M. Caffo et al. (eds.), Astronomy, Cosmology and Fundamental Physics, 452–453.
© *1989 by Kluwer Academic Publishers.*

453

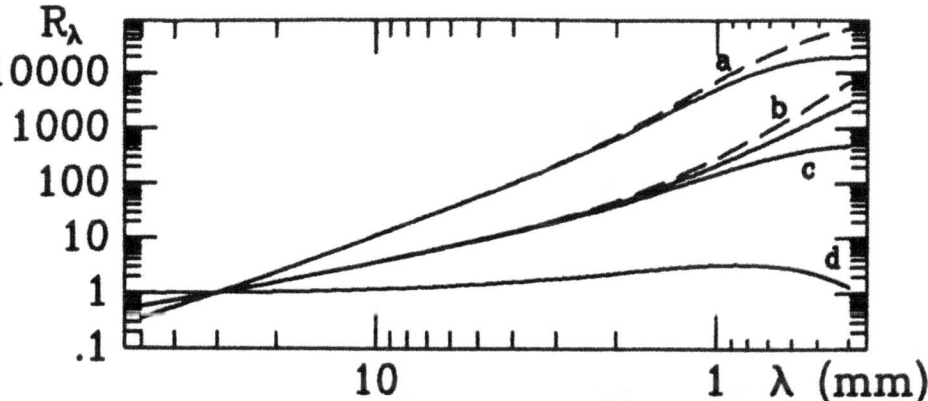

Figure 1. The amplitudes of the CBR fluctuations normalized to fluctuations at $\lambda_0 = 3cm$, R_λ, for the models with cosmic dust. The temperatures of the relic radiation and of the dust are $T_B = 2.74K$ and $T_d = 3.55K$, respectively.
(a) (dashed line) - fluctuations of dust temperature, (continuous) - perturbations of dust density. The dust emissivity index is $\alpha = 2$, and the optical depth of the dust is $\tau(700\mu m) = 0.2$.
(b) (dashed and continuous) - same as (a) but $\alpha = 1$, $\tau(700\mu m) = 0.02$,
(c) (continuous) - dust density perturbations, $\alpha = 1$, $\tau(700\mu m) = 0.2$.
(d) primordial perturbations imposed on the CBR spectrum, the dust emissivity index is $\alpha = 1$, and $\tau(700\mu m) = 0.2$.

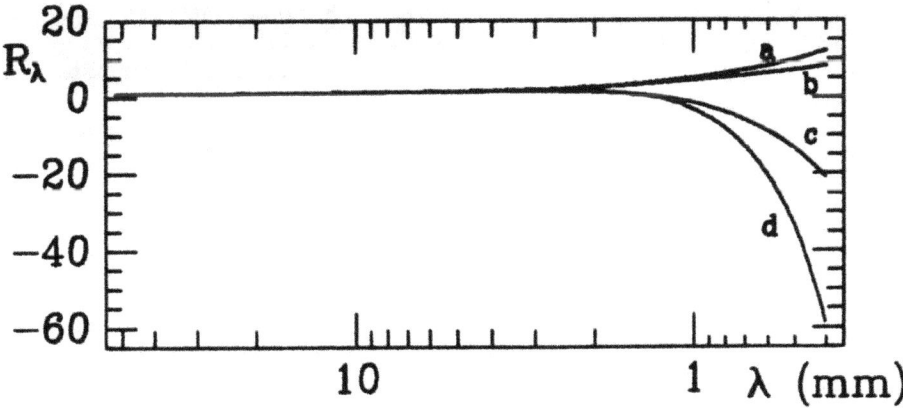

Figure 2. R_λ for the models with nonrelativistic comptonization. The CBR spectrum is described by $T_B = 2.74K$ and the Compton parameter y.
(a) primordial perturbations imposed on the CBR spectrum with $y = 10^{-4}$,
(b) same as (a) but $y = 0.028$,
(c) y-perturbations imposed on the CBR spectrum with $y = 0.028$,
(d) same as (c), $y = 10^{-4}$.

Temperature relation at a deflagration hadronization front

Ornella Pantano* and John Miller*†

*International School for Advanced Studies (S.I.S.S.A.), Trieste, Italy.
†Department of Astrophysics, University of Oxford, England.

At high density, strongly interacting matter is believed to take the form of a quark-gluon plasma in which the quarks and gluons are not confined within hadrons and chiral symmetry is maintained. The transition from the quark-gluon plasma state to the low density hadronic phase is of interest in connection with early universe cosmology, heavy ion collisions and, possibly, neutron star interiors. We will consider here the situation if the transition is of first order and bubbles of the low density phase are nucleated within a supercooled plasma. A transition front can move either supersonically or subsonically relative to the medium ahead (corresponding to detonation or deflagration solutions respectively)[1] and we have previously presented a comparison of these two types of front based on an analysis of their associated characteristic structure[2]. Deflagrations require smaller supercooling and it seems likely to be these which occur in practice. An important feature of deflagrations is that the rate at which energy crosses the interface (the "transition rate") cannot be calculated directly from the hydrodynamic equations but must be supplied as a separate condition derived from considerations of the transformation process. This further condition places constraints on the possible values of fluid parameters on either side of the interface. We will here describe the situation for planar symmetry and vanishingly small baryon number density.

The energy-momentum conservation equations in the rest frame of the interface (in the limit in which surface contributions can be neglected) are

$$F_H = (e_h + p_h)\, \gamma_h^2 v_h = (e_q + p_q)\, \gamma_q^2 v_q, \tag{1}$$

$$p_h + (e_h + p_h)\, \gamma_h^2 v_h^2 = p_q + (e_q + p_q)\, \gamma_q^2 v_q^2, \tag{2}$$

where the subscripts h and q refer to the hadron and quark phases respectively, F_H is the hydrodynamical energy flux across the interface, e is the energy density, p is the pressure, v is the fluid three velocity in the rest frame of the interface and $\gamma = 1/\sqrt{1 - v^2}$.

The required additional condition is obtained by setting F_H equal to F_T, the rate at which energy crosses the interface as derived from considerations of the transition process. We use here a simplified expression for F_T (which is analogous to the one commonly used in classical theory[3]):–

$$F_T = \alpha \left[\Phi\left(T_q\right) - \Phi\left(T_h\right) \right], \tag{3}$$

454

M. Caffo et al. (eds.), Astronomy, Cosmology and Fundamental Physics, 454–455.
© 1989 by Kluwer Academic Publishers.

where $\Phi(T_q)$ is the thermal flux away from the interface (at temperature T_q) into the hadron phase, $\Phi(T_h)$ is the corresponding flux from the hadronic matter towards the interface and α is an accommodation coefficient ($0 \le \alpha \le 1$) which takes account of deviations away from the ideal situation. The limit $\alpha = 0$ corresponds to zero hydrodynamical flow across the interface and pressure equilibrium between the two phases. If equations of state are now specified for the two phases, one can then calculate the relationship between the temperatures on the two sides of the interface. Figure 1a shows the relation between T_q and T_h, for various values of α, in the case where the hadron phase is represented as an ideal gas of massless, point-like pions and the MIT bag model is used for the quark phase. Figure 1b shows the corresponding behaviour of the energy flux. We note two important features: a) for $T_q = T_c$ (where T_c is the critical temperature for the transition), T_h must also be equal to T_c, giving zero hydrodynamical flux across the interface; b) for a range of temperatures T_q above the minimum of the curve $T_q(T_h)$ there are two possible values of T_h consistent with each value of T_q, a fact which may be associated with the appearance of instabilities. Use of a more sophisticated equation of state for the quark phase does not cause any qualitative change in these curves.

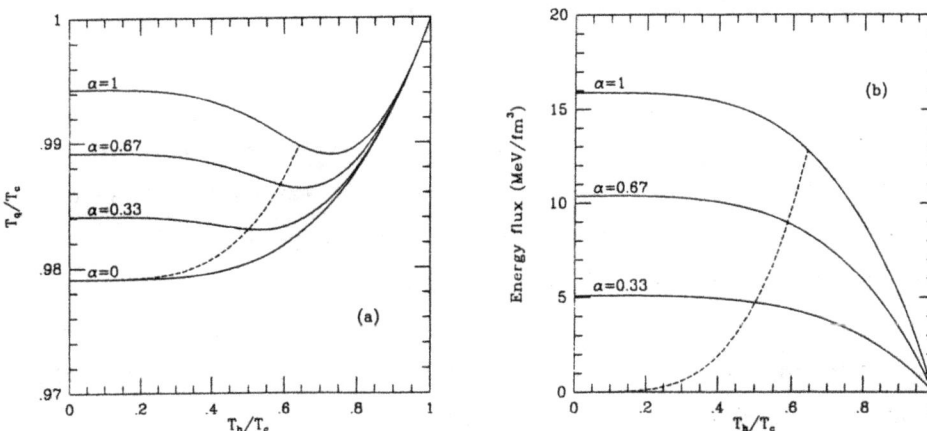

Fig. 1a. Temperature relation between the two phases. The dotted line is the locus of the Chapman-Jouguet points (where v_h is equal to the sound speed). Only points to the right of this correspond to stable deflagrations.

Fig. 1b. Energy flux across the interface as a function of T_h/T_c.

References

1) M. Gyulassy, K. Kajantie, H. Kurki-Suonio and L. McLerran, *Nucl. Phys.* **B237**, 477 (1984).

2) J.C. Miller and O. Pantano, *SISSA preprint, Astro 15* (1988).

3) T.G. Theofanus, L. Biasi, H.S. Isbin and H.K. Fauske, *Chem. Eng. Sci.* **24**, 885 (1984).

SEARCH FOR CBR ANISOTROPIES AND OBSERVATIONS OF THE MAGELLANIC CLOUDS AT MILLIMETRIC WAVELENGTHS

L. Piccirillo[1,2], P. Andreani[1], G. Dall'Oglio[1]

L. Martinis[3], L. Pizzo[1], L. Rossi[4], C. Venturino[1]

(1) – Dipartimento di Fisica, Università di Roma "La Sapienza", Roma
(2) – Istituto Superiore PP.TT., Roma
(3) – E.N.E.A. Tib. Frascati
(4) – Istituto Astrofisica Spaziale, Frascati

ABSTRACT. We present preliminary results from a ground-based experiment carried out in Antarctica and devoted to the search for the Cosmic Background Radiation (CBR) anisotropies. We find an upper limit of $\frac{\Delta T}{T} < 2 \cdot 10^{-4}$ at 2.5° angular scale and at 2 mm wavelength.

The millimetric emission of the Large and Small Magellanic Clouds has been also detected. Their spectra present an excess probably due to a very cold dust component coexisting with the warm dust observed by IRAS.

INTRODUCTION

A fundamental problem in modern Cosmology consists of the understanding of the evolution of the universe from a highly homogeneous and isotropic epoch, as it is inferred from the CBR observations, to the present structure.

Cosmologists think that the primordial universe must be sprinkled with some "seeds" which will become the actual structures. These seeds produce the spatial fluctuations (anisotropies) in the radiation coming from the last-scattering surface

456

M. Caffo et al. (eds.), Astronomy, Cosmology and Fundamental Physics, 456–460.

: the detection of the CBR anisotropies is then essential to provide an explanation to the present matter distribution.

Millimetric and submillimetric observations of the CBR from the ground are contaminated by the atmospheric emission and fluctuations, mainly due to the water vapour content. For this reason very cold and dry sites are preferred. Since Antarctica matches these features, we performed an experiment at the Italian Base in Terra Nova Bay, devoted to the detection of the millimetric diffuse radiation.

THE EXPERIMENTAL SET-UP

The instrument consists of two distinct detectors coupled to a telescope. The first detector is sensitive to the CBR while the second one is sensitive both to the CBR and to the atmospheric emission. This latter is used either to subtract the atmospheric contaminations in the first channel and to detect sources which are well above the atmospheric noise. Their main characteristics are listed in table 1.

Table 1

Detectors characteristics

	λ_{eff} (μm)	T_{oper} (K)	R $\mu V / K$	$A \cdot \Omega$ $cm^2 sr^{-1}$
chann. 1	2090	0.35	10	~ 0.6
chann. 2	1160	1	50	~ 0.6

The optics (see fig.1) consists of a 1 metre diametre off-axis paraboloid, which defines a field of view (FOV) of about 1°. The radiation collected by the primary is

458

reflected into the detectors by a secondary plane mirror. The paraboloid wobbles at a frequency of 8 Hz performing a beam-switching in the sky of 2.5°.

The experiment points to the zenith and the scanning of the sky is performed by the Earth's rotation.

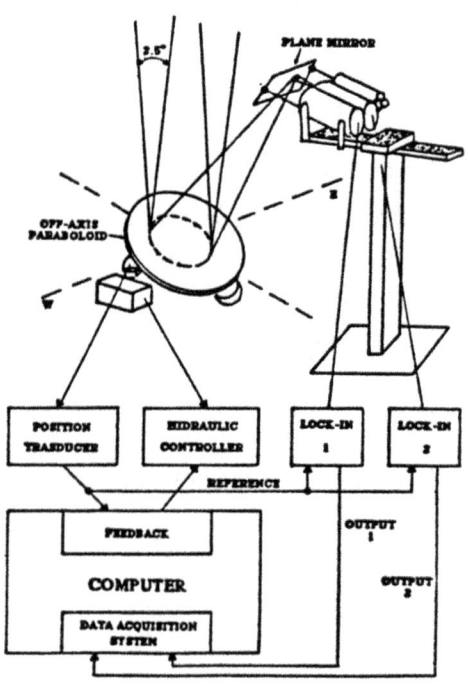

fig.1 - the experimental set-up (from Andreani et al., 1988a)

OBSERVATIONS AND DATA ANALYSIS

A complete revolution of the Earth produces the scanning of ∼ 70 independent regions at the latitude of the Terra Nova Bay (= −74.7°). An independent region in the sky, corresponding to one FOV, is obtained by averaging the data over about 1000 s.

Because of the uncertainty of our poynting system, we compare all our measurements with the IRAS (Infrared Astronomical Satellite) 100 μm data (see Andreani et al., 1988a). As far as the observation of the Magellanic Clouds is concerned, we find, in spite of the huge difference in wavelength, a striking correlation between our and the IRAS data (see Andreani et al., 1988b). As it is shown in fig.2, the 60 μm, 100 μm, 1 mm and 2 mm intensities do not match the same temperature dust distribution. Our measurements show a strong millimetric emission of these galaxies. This could be a signature of a very cold dust well mixed with the IRAS dust (Cox and Mezger, 1986).

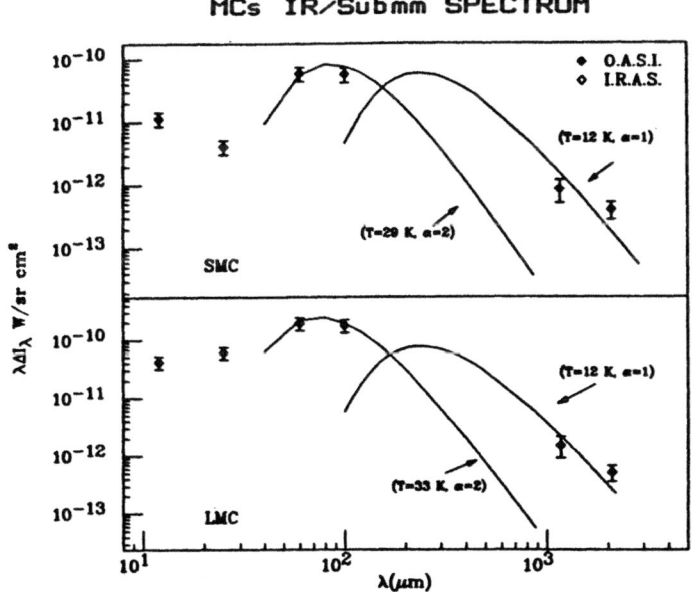

fig.2 - the IR/Submm spectrum of the Magellanic Clouds obtained from the IRAS data and our measurements (from Andreani et al., 1988b)

Finally, in order to give an upper limit to the CBR anisotropies, we analyse those regions of the IRAS sky which are free from dust emission. We have evaluated the upper limit to the CBR anisotropies by testing 87 independent points with the usual statistic procedures (see Lasenby and Davies, 1983). The result turns out to be :

$$\frac{\Delta T}{T} < 2 \cdot 10^{-4}$$

at 2.5° angular scale and at 95 % confidence level.

REFERENCES

P. Andreani, G. Dall'Oglio, L. Martinis, L. Piccirillo, L. Pizzo, L. Rossi, C. Venturino, submitted to the Astroph. Journal (1988a)

P. Andreani, C. Ceccarelli, G. Dall'Oglio, L. Martinis, L. Piccirillo, L. Pizzo, L. Rossi, C. Venturino, submitted to the Astroph. Journal (1988b)

P. Cox and P.G. Mezger, in "Light on Dark Matter", Dordrecht, Reidel (1986)

A.N. Lasenby and R.D. Davies, Mon.Not.Roy.astr.Soc., 203, 1137 (1983)

PRIMORDIAL NUCLEOSYNTHESIS:
THE EFFECTS OF INJECTING HADRONS

M.H. Reno

Fermi National Accelerator Laboratory, Batavia, Illinois 60510

The standard big bang nucleosynthesis model (BBN) successfully predicts the light element abundances.[1] By including extra exotic particles that decay into hadrons during the time of nucleosynthesis, the standard scenario is changed. Limits on the light element abundances therefore restrict the abundance of exotic particles as a function of their lifetime. The limits described here rely on work done with D. Seckel and are described in more detail in Ref. 2. The essential ingredient in our analysis is that the hadronic decay products interact with the ambient neutrons and protons during the nucleosynthesis era, inducing a net increase in the number of neutrons and therefore increase the ^{4}He, ^{3}He and deuterium (D) abundances.

The results presented below are for exotic particle X with lifetimes $\tau_X = 0.1 - 10^4$ sec. The standard nucleosynthesis parameters are set to: neutron lifetime, $\tau_n = 900$ sec, the number of neutrinos equals three, and the baryon-to-photon ratio, $\eta > 3 \cdot 10^{-10}$.

To describe the effect, consider X particles with mass of 100 GeV. In general, at least some of the time, X decays into p, $\bar{p}$, n, $\bar{n}$, mesons and short-lived baryons. For the moment, consider just the nucleon-anti-nucleon "injections" from X decays with $\tau_X = 1$ sec, that is, the temperature of the Universe of about 1 MeV. At this temperature, the Universe is radiation dominated, so injected nucleons thermalize from scattering with photons, and thermalization times are essentially instantaneous. The stopped antibaryons then annihilate with ambient neutrons and protons:

$$(n\bar{n}) + n \rightarrow n + \pi + \ldots \rightarrow n$$
$$(n\bar{n}) + p \rightarrow n + \ldots$$
$$(p\bar{p}) + n \rightarrow p + \ldots$$
$$(p\bar{p}) + p \rightarrow p + \ldots$$

461

M. Caffo et al. (eds.), Astronomy, Cosmology and Fundamental Physics, 461–465.
© *1989 by Kluwer Academic Publishers.*

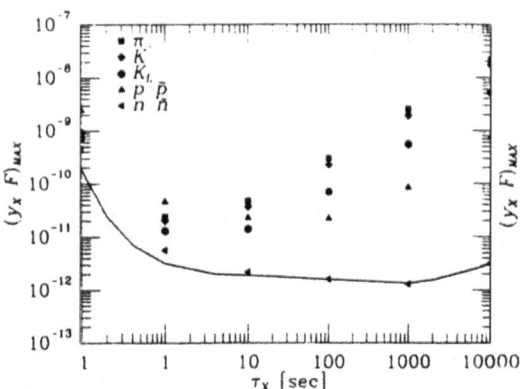

1. Maximum values of $y_X F$ for $\eta = 3 \cdot 10^{-10}$ assuming that X decays into a final state with no net baryon number. The limits coming from just $\pi^{\pm}$, K^-, K_L, $p\bar{p}$ or $n\bar{n}$ are shown separately. The factor F is defined by $F = (N_{quarks} B_{hadronic})/2 \cdot \langle n_{ch}(E_{quark}) \rangle / \langle n_{ch}(E = 33 \text{ GeV}) \rangle$, and $y_X = n(X)/s$ for entropy density s. The charged particle multiplicity n_{ch} as a function of the quark energy is used to scale the particle multiplicities of hadrons in X decay final states.

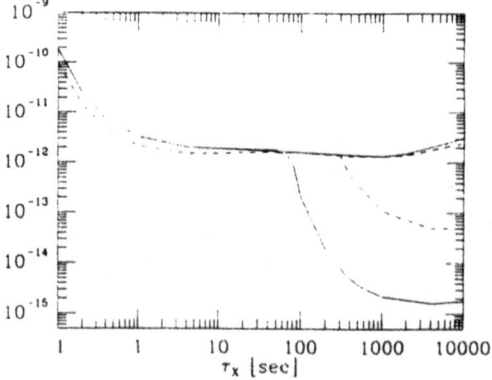

2. Maximum values of $y_X F$ for $\eta = 3 \cdot 10^{-10}$ (solid line) and $\eta = 10^{-9}$ (dashed line) for final states with no net baryon number. The limits come from $Y \leq 0.26$ and $X(D) + X(^3He) \leq 10^{-4}$.

where the injected pairs are indicated by parentheses. Taking equal cross sections for nucleon-anti-nucleon annihilation, regardless of charge, the rate for $n \to p$ conversion and $p \to n$ conversion depends on the target particle densities. Because at $T < 1$ MeV, $n(n) < n(p)$ the rates $\Gamma(n \to p) < \Gamma(p \to n)$, so $(n/p) > (n/p)^0$, where the superscript indicates the ratio of number densities in the standard BBN scenario. As the density of X increases, for X decays in the allowed lifetime range, the neutron to proton ratio increases, eventually leading to an overproduction of ^{4}He, ^{3}He, and D.

To limit the X abundance, we assumed the primordial light element abundances[3] of Y, the relative ^{4}He abundance by mass, of $Y < 0.26$, and $X(\mathrm{D}) > 10^{-5}$ with $X(\mathrm{D}) + X(^3\mathrm{He}) \leq 10^{-4}$.

We use the Wagoner nucleosynthesis code[4] with appropriate modifications. In addition to including the energy and entropy density due to X and its decay, we add terms to the rate for n to p conversion and p to n conversion that depend on the $\bar{p}$, $\bar{n}$ and injected meson interaction rates with nucleons. They also depend on the particle multiplicities from the X decays.[5] We assume that injected hadrons are stopped before they interact with nucleons, and we neglect hadrons produced from annihilations of injected particles with ambient nucleons. Photofission of light elements is unimportant before 10^4 sec, because D is efficiently reformed, so we do not explicitly include photofission processes. Furthermore, we do not include ^{4}He fission induced by fast neutrons or antiparticles.

Fig. 1 shows our results for the limit on the X abundance scaled by the entropy density s and a factor F (which equals unity for a 100 GeV X particle which decays into 2 quarks and one lepton with a branching ratio of one) as a function of lifetime for $\eta = 3 \cdot 10^{-10}$ using just the ^{4}He limits. The D limits are included in Fig. 2 for $\eta = (3, 10) \cdot 10^{-10}$.

We can apply these results to the case of decaying gravitinos (Fig. 3) and decaying photinos in theories where R-parity is violated (Fig. 4). In Fig. 3, our results appear as the lower solid line. The work of Dominguez-Tenreiro[6] is in contradiction with ours at short lifetimes, even when we rescale our limits to account for different input parameters (dashed line). For decaying photinos, we can exclude a large region of $(m_{\tilde{\gamma}} - \tau_{\tilde{\gamma}})$ parameter space. The limits are dependent on $m_{\tilde{f}}$, the mass of the particle exchanged in $\tilde{\gamma} - \tilde{\gamma}$ annihilation.

We have also examined the case where there is a net baryon number in the final state of X decays, and we find that there are special circumstances in which $\Omega_B = 1$, however, they require that X always decay into a proton (plus the usual hadrons above). The consequences of cold dark matter annihations for the neutron-to-proton ratio and the light element abundances are also discussed in Ref. 2.

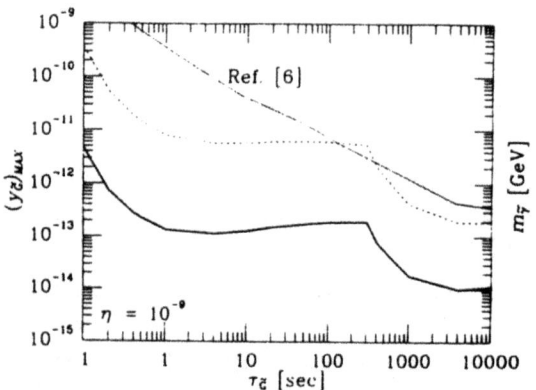

3. Maximum gravitino abundance $y_{\tilde{G}}$ versus lifetime. The bold line shows our result, and the other solid line indicates the limits from Dominguez-Tenreiro[6]. The dotted line is a rescaling of our limits to compare with the Dominguez-Tenreiro limits.

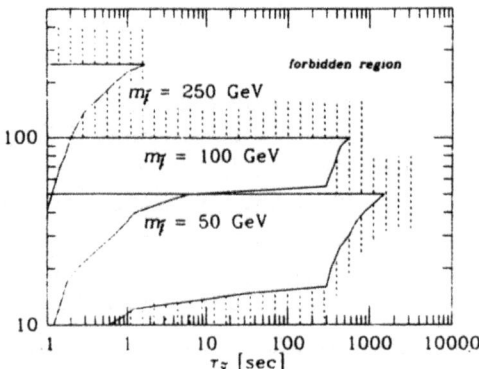

4. Mass versus lifetime for a decaying photino when $m_{\tilde{f}} = 50, 100, 250$ GeV.

REFERENCES

1. R. V. Wagoner, W. A. Fowler and F. Hoyle, *Ap. J.* **148**, 3 (1967);
   R. V. Wagoner, Ap. J. Suppl. **18**, 247 (1969), *Ap. J.* **179**, 343 (1973).

2. M. H. Reno and D. Seckel, *Phys. Rev.* **D38**, 3441 (1988).

3. For a review of the limits, see, e.g., J. Yang, M. S. Turner, G. Steigman, D. N. Schramm and K. A. Olive, *Ap. J.* **281**, 493 (1984);
   A. M. Boesgaard and G. Steigman, *Ann. Rev. Astron. Astrophys.* **23**, 319 (1985).

4. We have used L. Kawano's version of the Wagoner nucleosynthesis code. Details about the solution of the coupled differential equations appear in Ref. [1].

5. See Ref. [2] for details. The strong interaction cross sections come from low energy scattering experiments, and particle multiplicities are calculated in analogy with multiplicities in $e^+e^- \rightarrow 2$ jets.

6. R. Dominguez-Tenreiro, *Ap. J.* **313**, 523 (1987).

Power–law spectra and the large scale pattern of the CMB

Roberto Scaramella*   &   Nicola Vittorio[†]

* *International School for Advanced Studies (S.I.S.S.A.), Trieste, Italy.*
† *Dipartimento di Fisica, Università dell'Aquila, L'Aquila, Italy.*

In the framework of the gravitational instability theories, the primordial density fluctuations are commonly described as a 3-D random, isotropic gaussian field. On large angular scales these density fluctuations induce fluctuation in the temperature of the Cosmic Microwave Background (CMB) through the Sachs–Wolfe effect. Under this assumption, the CMB temperature distribution is itself a 2-D random gaussian field. The statistics of the CMB temperature field is then completely described by the two point angular a.c.f. $C(\alpha) \equiv \langle \Delta T/T(\hat{\mathbf{n}}_1) \cdot \Delta T/T(\hat{\mathbf{n}}_2) \rangle$, where $\alpha = \cos^{-1}(\hat{\mathbf{n}}_1 \cdot \hat{\mathbf{n}}_2)$ and the average is taken over the whole sky and over the ensemble constituted by all the possible realizations of last scattering surfaces. We take into account the finite resolution of the antenna, by modeling the beam with a gaussian of dispersion σ. Then, the smoothed a.c.f. is given by (Scaramella and Vittorio, 1988): $C(\alpha, \sigma) = (1/4\pi) \sum_{l=2}^{\infty} a_l^2 \, (2l+1) \, P_l(\cos \alpha) \, \exp\{-[(l+(1/2))\sigma]^2\}$. The effect of the beam, as expected, acts as a low band pass filter, which severely attenuates harmonics of order $l \gg 1/\sigma$. We study density fluctuations in a flat universe which have a scale free power spectrum, $|\delta_k^{(in)}|^2 = A \cdot k^n$, where A is the normalization constant, n is the spectral index. We consider quite a wide range of values $(-3 \lesssim n \leq 3)$ to study the trend of the considered quantities.

With the above information on the a.c.f. we evaluate the large scale pattern of the CMB temperature distribution, as observed with an antenna of resolution σ, for different primordial spectral indeces. All we need (see Vittorio and Juszkiewicz, 1986,VJ) is, besides $C(0, \sigma)$, the values of the second and fourth derivative of the a.c.f. with respect to α, evaluated at zero lag. Then we can evaluate the average total number of maxima in the CMB temperature distribution, expected in all the sky. This number is plotted in Fig. (a) as a function of the antenna beam σ, for different primordial spectral index n. For $n < -2$ the quadrupole is the dominant harmonic and the number of maxima is slightly above 2 as it should exactly be for a pure quadrupolar pattern. For $-1 < n < 3$, the total number of detectable hot spots is to a very good approximation proportional to σ^{-2}, the resolution of the antenna. For these values of n, the temperature distribution is scale free and the only characteristic scale involved is the antenna beam size. As noted in VJ, this implies that the temperature fluctuations in the CMB due to the Sachs-Wolfe effect, mimick unresolved sources: infact their number continuosly increases improving the antenna angular resolution. From an observational point of view, however, one is interested in the number of regions in the sky where the temperature fluctuation is higher than ν times the rms temperature fluctuations [i.e., $C^{1/2}(0, \sigma)$]. If these regions are sufficiently abundant and large, one could look for rare but very hot spots in the sky. The number of hot spots is well

M. Caffo et al. (eds.), Astronomy, Cosmology and Fundamental Physics, 466–467.

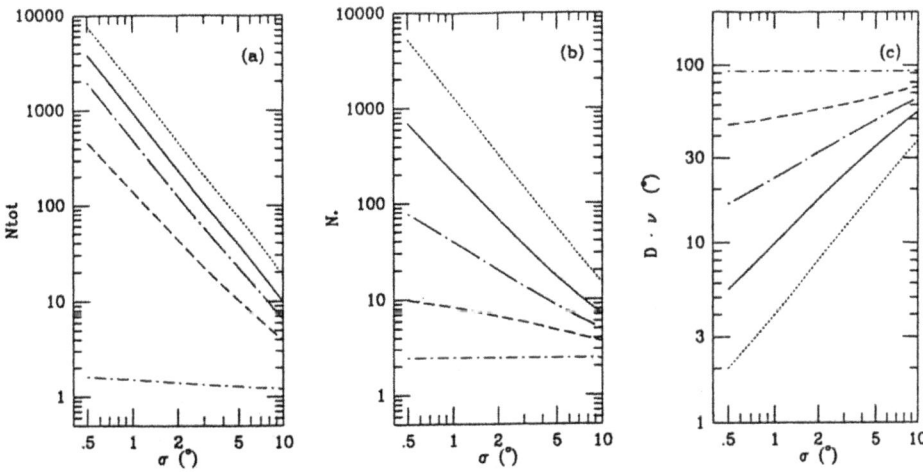

Figure. The total number of upcrossing regions N_{tot} (a), the quantity N_* (b), and the average diameter D times the threshold ν (c) are plotted as a function of the beamsize σ, for different primordial spectral index n; $n = 3$, dotted line; $n = 1$, solid line; $n = 0$, dotted–long dashed line; $n = -1$, dashed line; $n = -2.9$, dotted–dashed line. Here $\sigma = 3°.5$

approximated, for $\nu \gg 1$, by (see, VJ): $N_{up} = N_*(\sigma,n) \cdot \nu \, \exp[-(1/2)\nu^2]$. The function $N_*(\sigma,n)$ is plotted in Fig. (b), as a function of σ, for different values of the spectral index n. For $n > 2$, $N_*(\sigma,n) \propto \sigma^{-2}$. For smaller values of n, the dependence on σ is weaker and weaker. In fact, lowering n reduces the small scale (relative to the large scale) power and having a finite antenna beam becomes irrelevant. Eventually, only the quadrupole matters and N_* tends to 2. If only the quadrupole is present, however, $C^{1/2} = (5/4\pi)^{(1/2)} a_2$. Then looking for regions with $\nu \gg 1$, implies looking for very improbable values for the quadrupole anisotropy. Then the actual number, N_{up} of hot spots tends to zero. For moderate ν's, the temperature profile of the upcrossing regions is slightly steeper than the shape of the correlation function. If the temperature field is very correlated the profile is very flat and hence to beam switch at an angular scale less than the typical hot spot angular diameter leads to the risk of a strong reduction in any detectable anisotropy, because of the differential technique. The knowledge of the typical hot spot angular diameter can at least guide the design of the observational configuration, taking of course into account other trade–offs, such as atmospheric and ground emission. The expected angular diameter is inversely proportional to the number of hot spots and depends on the threshold as ν^{-1} (for $\nu \gg 1$; see VJ). The quantity $D \cdot \nu$ is plotted in Fig. (c). For fixed ν, D increases linearly with σ for $n \simeq 3$, but flattens out to $90°$ (formally for $\nu = 1$) when only the quadrupole is dominant (i.e., $n < -2$). In this limit, however, the same comment applies which we made before: a quadrupole with amplitude much higher than $C^{1/2}(0,\sigma)$ is improbable.

References

Scaramella, R., and Vittorio, N., 1988, *Ap.J.Letters* in press.
Vittorio, N., and Juszkiewicz, R., 1987, Ap.J. *Letters* **314**, L29.

OBSERVATIONS OF COSMIC FLUCTUATIONS

P. Schuecker, H.-A. Ott, H. Horstmann, V. Gericke, W.C. Seitter
Astronomisches Institut
Westfälische Wilhelms-Universität
4400 Münster
F.R.Germany

ABSTRACT. The scales and amplitudes of fluctuations in cosmic matter distribution are expected to show evolution from the homogeneity postulated for the early phases of the standard universe to the partially hierarchical clustering observed today. The present communication presents observational data which appear to trace the evolution between $z = 1000$ and the present.

1. MEASURES OF INHOMOGENEITY

Following the concepts of gravithermodynamics, Poisson distribution is used for reference in galaxy and quasar statistics. The quantity β measures the deviation from Poisson distribution, averaged over a given volume of space V_{tot}. β is defined as the ratio of the mean number counts $< N >$ and their variances $< \delta N^2 >$ within (ideally) non-overlapping volume elements V of equal size (counting cells), distributed randomly or by tesselation in V_{tot}. The fluctuation parameter $b = 1 - \sqrt{\beta}$ is zero for the exact Poisson case and $-\infty$ for complete homogeneity. In the presence of non-random inhomogeneities b is larger than zero, reaching the limiting value $b = 1$ for infinite hierarchical systems in virial equilibrium. A theory of gravitational non-linear clustering (Saslaw and Hamilton 1984) defines b as the ratio of gravitational correlation energy (cluster surplus potential energy relative to the Poisson case) to kinetic energy within a system. In the following the parameters b and β are used to characterize the fluctuations.

2. RESULTS FROM THE MRSP

We present fluctuation values obtained from galaxy and quasar samples from the Muenster Redshift Project (MRSP, Horstmann 1988, Schuecker 1988, Gericke 1988). To date 140 000 galaxies up to the completeness limit $m_J = 21^m.0$ are identified covering an area of 135 square degrees. 23 700 galaxies up to $m_J = 19^m.7$ in 108 square degrees have measured redshifts $z \leq 0.3$. 372 Ly-α quasars with $18^m.0 \leq m_J \leq 19^m.0$, $1.8 \leq z \leq 3.0$ were found so far in 85 square degrees.

In *Fig. 1a* the fluctuation parameter b is plotted versus cell volume for a volume-limited subsample of 1 421 galaxies in the ESO/SRC field No. 411 (dots with error bars) and for 10 simulated samples with Poisson distributions (dots). The galaxies have absolute magnitudes $-18^m.5 \geq M_J \geq -19^m.5$ and redshifts $0.04 \leq z \leq 0.14$ ($H_0 = 100\, km\, s^{-1}\, Mpc^{-1}$, $q_0 = 0.5$). Because of the low spatial resolution in z, introduced by the redshift error $\delta z \leq 0.008$, the galaxies are counted in ellipsoidal cells with the major axis oriented in the direction of z. The major axis is varied from 2.5 to 100 Mpc, while the minor axes assume values between 2.5 and 5.0 Mpc. The cells are arranged in a body-centered cubic lattice where volumes partially exceeding the observed region are rejected. In order to reduce the wasted space and to increase the S/N ratio in the counted frequency distributions the degree of overlapping for neighbouring cells is increased with increasing volume.

The distribution of b-values for the simulated random samples in *Fig. 1a* indicates that the different degrees of overlapping for the different volume sizes introduce no significant systematic error. Random errors, however, are large for large volumes because of the small number of volumes sampled.

The b-values for galaxies in *Fig. 1a* show noticeable deviations from zero, suggesting large-scale inhomogeneities, i.e. galaxy clustering and superclustering. The systematic increase of b with volume

468

M. Caffo et al. (eds.), Astronomy, Cosmology and Fundamental Physics, 468–470.
© 1989 by Kluwer Academic Publishers.

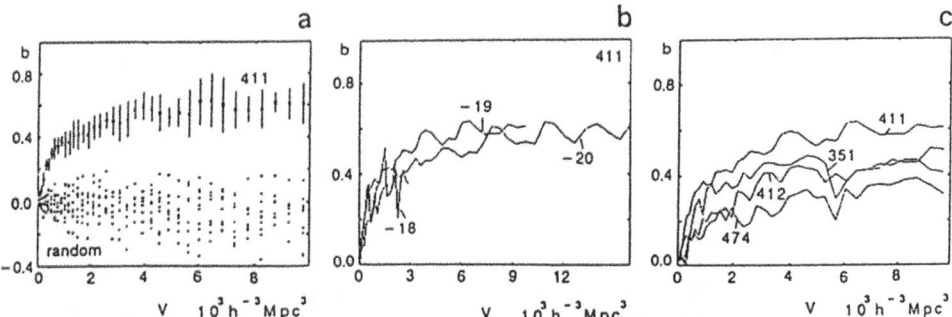

Fig. 1. Fluctuation parameter b versus volume V of the counting cell for
(a) galaxies from plate No. 411 (dots with error bars) and 10 simulated random samples (dots);
(b) galaxies with $-18^m.5 \leq M \leq -17^m.5$ (−18), $-19^m.5 \leq M \leq -18^m.5$ (−19) and
$-20^m.5 \leq M \leq -19^m.5$ (−20);
(c) galaxies of ESO/SRC Atlas fields Nos. 351, 411, 412 and 474.

size is in general agreement with N-body simulations by Saslaw (1985). The deviations between the b-values of absolutely bright and faint galaxies shown in *Fig. 1b* lie within the error bars. The constantly high values of b, even for the maximum cell volumes of $V = 15\,500\,Mpc^3$, show that these are still too small to include inhomogeneities of all sizes, and that homogenization does not (yet) set in.

In *Fig. 1c* the fluctuation parameter b is plotted versus cell volume for volume-limited subsamples from four different fields. The number of galaxies are 612 (field No. 351), 1 421 (field No. 411), 1 319 (field No. 412) and 574 (field No. 474). All $b(V)$-curves shown have similar shapes. However, the degree of inhomogeneity is different for the different fields: field No. 411 has the largest fluctuations because of the presence of the **Sculptor supercluster** (Schuecker *et al.* 1988). Some contribution to the differences in the b-values is due to varying plate quality as found from the correlation between quality parameters and maximum b-values. The latter effect will have to be corrected for to derive the true differences of clustering properties in different fields.

For a comparison of the clustering properties derived from two- and three-dimensional surveys, two-dimensional MRSP data were also used to derive b-values. Here, the counts were averaged over conical volumes with different bases extending from the observer to a distance given by the assumption of the fixed representative absolute magnitude $M = -21^m.0$ and different limiting apparent magnitudes, starting with $m_J = 16^m.0$. The b-curves calculated for fields Nos. 411 and 412 indicate somewhat higher clustering than that obtained from the three-dimensional data. The interpretation is not as straightforward as in the three-dimensional case, it is, however, conceiveable that chance superpositions contribute significantly to the apparent higher clustering values.

3. EVOLUTION OF CLUSTERING

In *Fig. 2* the fluctuation parameter b is plotted versus logarithmic cell volume for volume-limited MRSP galaxy and quasar (QSOs) subsamples from field No. 411. The data are supplemented by the b-value from current upper limits of the cosmic microwave background (CMB) fluctuations for volumes subtending 6", 1' and 6'. The subsample of 97 quasars includes all quasars in the magnitude and redshift ranges $18^m.0 \leq m_J \leq 19^m.0$ and $2.0 \leq z \leq 2.4$ from a total of 158 Ly-α quasars. The errors of the b-values for the quasar sample are obtained from simulations using four random samples. The quasar sample shows significantly smaller fluctuations than the galaxy sample, however, the b-values suggest some weak quasar clustering. For the CMB the fluctuation parameter b is far below zero.

470

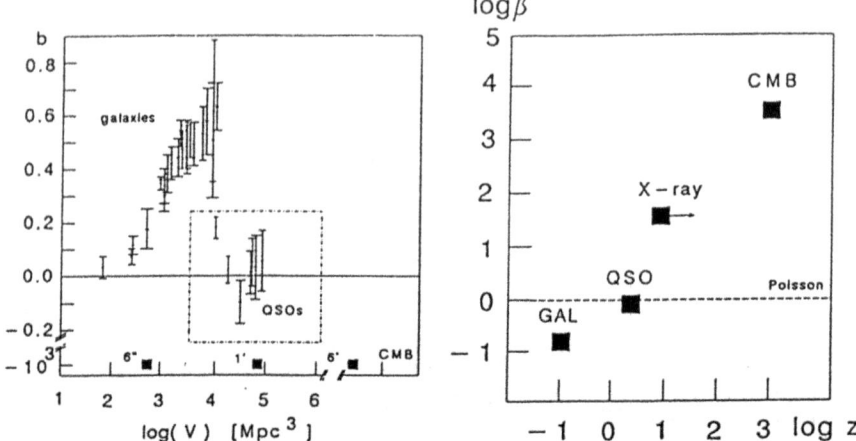

Fig. 2. Fluctuation parameter b versus volume V for a volume-limited MRSP galaxy and quasar (QSOs) subsample in field No. 411, supplemented by b-values of the CMB fluctuations.

Fig. 3. Evolution of clustering.

Fig. 3 shows the clustering parameter β as a function of z. With comparable cell sizes as apparent from *Fig. 2* the β - z - relation suggests an evolutionary trend from more homogeneous distribution via Poisson distribution towards large inhomogeneities on the scale of 10^4 Mpc^3. The scarcity of data on other background radiations permits the inclusion of only one further point, that of the X-ray background (Gursky and Schwartz 1977). The z-value is a lower limit, the reported volume size is larger than the others in the present sample. With all the caution appropriate when conclusions are drawn from insufficient data, the diagram implies a steady evolution of the distribution of galaxy size masses from homogeneity via Poisson distribution towards clustering. The fluctuations of the various background radiations play a vital role in tracing the general course of evolution while spatial statistics in the realm of quasars and galaxies may reveal fine structures in the β - z - relation, permitting the derivation of associated mechanisms.

4. REFERENCES

Gericke, V., 1988. Workshop: *The Large-Scale Structure of the Universe*, eds. Seitter, W.C., Duerbeck, H., Tacke, M., *Lecture Notes in Physics* **310**, Springer, Heidelberg, p. 123.

Gursky, H., Schwartz, D.A., 1977. *Ann. Rev. Astr. Astrophys.* **15**, 541.

Horstmann, H., 1988. Workshop: *The Large-Scale Structure of the Universe*, eds. Seitter, W.C., Duerbeck, H., Tacke, M., *Lecture Notes in Physics* **310**, Springer, Heidelberg, p. 111.

Saslaw, W.C. 1985. *Astrophys. J.* **297**, 49.

Saslaw, W.C., Hamilton, A.J.S., 1984. *Astrophys. J.* **276**, 13.

Schuecker, P., 1988. Workshop: *The Large-Scale Structure of the Universe*, eds. Seitter, W.C., Duerbeck, H., Tacke, M., *Lecture Notes in Physics* **310**, Springer, Heidelberg, p. 160.

Schuecker, P., Horstmann, H., Seitter, W.C., 1988. *IAU Symp.***130**, *The Structure of the Universe*, eds. Audouze, J., Szalay, A., Kluwer, Dordrecht, p. 526.

NEUTRINO EMISSION FROM SUPERNOVAE IN THE PRESENCE OF MAGNETIC FIELDS

C. Sivaram
Indian Institute of Astrophysics, Bangalore, India

The observation of the neutrino burst from SN1987A by the Kamiokande and IBM groups has provided not only an opportunity to test theories of stellar collapse but also gives information on the fundamental properties of the neutrino. In the SN1987A case, the density and temperature of the source and the neutrino propagation in the medium are vastly different from those of usual terrestrial and solar conditions.

Coherent scattering considerably reduces the mean free path while increasing opacity, so that the neutrinos emerge out on a diffusion time scale of $t_D \simeq 0.1 (\rho/10^{12} \text{ g cm}^{-3})$, i.e. ten seconds for $\rho \simeq 10^{14}$ g cm^{-3}, rather than on a hydrodynamical time scale of the collapse which is a few milliseconds. The fact that the burst lasted about ten seconds is evidence for such diffusion and multiple scatterings which reduce the neutrino energies to several MeV (as observed) rather than the Fermi energy of 200 MeV. Again, if the collapsing star traps a large amount of magnetic flux, the magnetic field during a collapse could have values well above $B > 10^8$ Gauss, relativistic neutrinos $(v \simeq c)$ with small magnetic moment can flip helicity in a magnetic field (i.e. left handed neutrinos precess into right handed components) at a rate:

$$\Gamma_{\text{LR}} \simeq \frac{2}{\pi} \frac{\mu_\nu B c}{\hbar},$$

where B is the magnetic field normal to the neutrino path.

For neutrino propagation in intergalactic (10^{-9} G) and galactic (10^{-6} G) magnetic fields, the requirement that they do not flip helicity on the way (i.e. $\Gamma_{\text{LR}} = 1/t_{LMC} \simeq \times 10^{-13} \text{s}^{-1}$) constrains $\mu_\nu < 10^{-20} \mu_B$ (μ_B being the Bohr magneton).

In the very unlikely situation that the magnetic moment of neutrino is $> 10^{-10}\mu_B$ (required by the VVO mechanism) then SN1987A neutrinos would have flipped helicities several times, so that they reach the earth as equal mixture of left and right handed helicities. This implies that one would detect only about half the neutrinos emitted, so that for N_f flavours the number flux is reduced by $1/2 \ N_f$.

F. Caffo et al. (eds.), Astronomy, Cosmology and Fundamental Physics, 471–472.
© 1989 by Kluwer Academic Publishers.

However, if at the neutrino source (i.e. the proto neutron star) the magnetic fi
rises to $> 10^8$ G, then for a neutrino diffusion time scale of 10 s (i.e. the timesc
of neutrino emission) the above equation implies that the neutrinos could flip helic
even for a neutrino magnetic moment as small as $10^{-24}\mu_B$. This in the WSG thee
would imply $M_\nu < 10^{-5}$ eV. So even for a magnetic moment $< 10^{-19}\mu_B$, only 1/2 N_s
detectable neutrinos would emerge out of the collapsing core.

REFERENCES

Shapiro, S., Wasserman, I. 1981, *Nature* **289**, 657
Voloshin, M.B., *et al.* 1982, *I.T.E.P. Report* No. 86-82

INFLATION-PRODUCED, LARGE-SCALE MAGNETIC FIELDS[1]

Lawrence M. Widrow

Abstract   We study the production of large-scale ($\sim Mpc$) magnetic fields in inflationary Universe models. In the usual electromagnetic gauge theory, the photon field is conformally invariant and the magnetic fields that are produced during an inflationary epoch are uninterestingly small. We have considered four models in which the conformal invariance of electromagnetism is broken. The primeval magnetic fields which result can have astrophysically interesting strengths, but are very model-dependent.

I. Introduction   Magnetic fields are observed throughout the Universe[2]. For example, there are large scale (≥ 10 kpc) coherent magnetic fields with a typical strength of $B \sim 3 \times 10^{-6}$ G in spiral galaxies such as our own. The energy density in these fields ρ_B ($= B^2/8\pi$) relative to the energy density in the cosmic microwave background radiation (CMBR) is $r = \rho_B/\rho_\gamma \sim 1$. Many theorist believe that these fields are generated and maintained by dynamo action[3]. A galactic dynamo can amplify a seed magnetic field by a factor of $exp(O(30))$. This implies that a seed magnetic field of $O(3 \times 10^{-19})$ [i.e., $r \simeq 10^{-34}$] is required. We also note that primeval magnetic fields are necessary in initiate electromagnetic currents in superconducting cosmic strings[4] (if such objects exist). The field strength required to initiate astrophysically interesting currents is $r \geq 10^{-10}$.

During most of its history, the Universe has been a good conductor. This implies that magnetic flux is conserved: $Ba^2 \sim cons't$ or that $\rho_B \propto a^{-4}$ where a is the cosmic scale factor. The dimensionless ratio $r \equiv (B^2/8\pi)/\rho_\gamma$ is therefore approximately constant and provides a convenient measure of magnetic field strength. Though a variety of theories have been proposed to explain the origin of these primeval fields [5] it is fair to say that at the present, no truly satisfactory mechanism has been suggested.

We believe that inflation[6] is a prime candidate for explaining the origin of primordial magnetic fields for three basic reasons: (1) Microphysical processes that operate on scales less than the Hubble radius during the de Sitter phase of inflation can lead to very long wavelength effects later on. (2) Such microphysical processes may in fact exist. De Sitter-produced quantum mechanical fluctuations excite modes with wavelengths of order the Hubble radius. The energy density in the mode with physical wavelength $\lambda_{phys} \simeq H^{-1}$ is $d\rho/dk \sim H^3$. (3) During inflation (and perhaps during much of the reheating phase which follows inflation) the Universe is devoid of charged plasma and is not a good conductor so that r can increase.

There is however one major obstacle. A pure $U(1)$ gauge theory with the standard Lagrangian, $\mathcal{L} = -\frac{1}{4}F_{\mu\nu}F^{\mu\nu}$, is conformally invariant. From this it follows that B always decreases as $1/a^2$ regardless of plasma effects. During inflation, the total energy density in the Universe ($\equiv \rho_{tot}$) is dominated by vacuum energy and is roughly constant. Therefore, the energy density in any inflation-

M. Caffo et al. (eds.), Astronomy, Cosmology and Fundamental Physics, 473–475.

produced magnetic field relative to ρ_{tot} will be greatly suppressed. Thus, the conformal invariance of electromagnetism must be broken if significant fields are to be produced. We have studied a number of ways of doing this. For example, we consider nonstandard gravitational couplings of the photon field A_μ to gravity such as RA^2, $R_{\mu\nu}A^\mu A^\nu$, $R_{\mu\nu\lambda\kappa}F^{\mu\nu}F^{\lambda\kappa}/m^2$, $R_{\mu\nu}F^{\mu\kappa}F^\nu_\lambda/m^2$, and $RF^{\mu\nu}F^{\mu\nu}/m^2$ (here m^2 is some mass scale squared). We also consider coupling of the photon to a charged field which is not conformally coupled and to the axion. The details of our calculations can be found in reference 1. Here, we review the calculations for the gravitational mass terms RA^2 and $R_{\mu\nu}A^\mu A^\nu$.

II. Review of New Inflation[7]

We assume a Friedmann-Robertson-Walker Universe where the stress energy is described by a perfect fluid with an equation of state $p = \gamma\rho$. The Hubble length $H^{-1} \equiv a/\dot{a}$ defines the length over which microphysics (e.g. quantum fluctuations) can operate. Astrophysical scales grow as a, whereas the Hubble length scales as $H^{-1} \propto a^{3(1+\gamma)/2}$. During either the radiation- or matter-dominated phases of the standard big-bang model, the Hubble radius grows faster than a while during inflation, the Hubble radius is essentially constant and grows more slowly than a. Astrophysical scales can therefore begin subhorizon sized (where physical processes can operate), exit the horizon during inflation and then later reenter the horizon during the radiation- or matter-dominated phases. The evolution of a given length scale in the inflationary Universe is illustrated in Fig. 1.

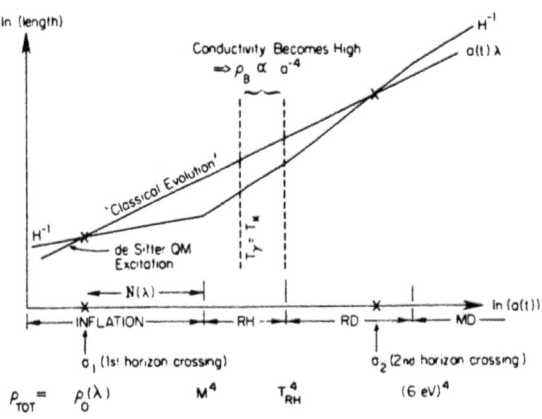

FIG. 1. Summary of the evolution of the mode with wavelength λ. The Universe is assumed to proceed through inflation, reheating (RH), radiation domination (RD), and matter domination (MD). The physical wavelength [$\equiv a(t)\lambda$] starts out subhorizon-sized and then crosses outside the Hubble radius $N(\lambda)$ e-folds before the end of inflaiton. The conductivity of the Universe becomes high some time during reheating. The fluctuation reenter the horizon when $a(t) = a_2$.

III. RA² Terms

Consider the Lagrangian

$$\mathcal{L} = -\frac{1}{4}F_{\mu\nu}F^{\mu\nu} - \frac{b}{2}RA^2 - \frac{c}{2}R_{\mu\nu}A^\mu A^\nu \tag{3.1}$$

The equations of motion for the electric and magnetic fields are

$$\nabla^\mu F_{\mu\nu} - bRA_\nu - cR^\mu_\nu A_\mu = 0 \qquad \partial_\mu F_{\lambda\kappa} + \partial_\kappa F_{\mu\lambda} + \partial_\lambda F_{\kappa\mu} = 0 \tag{3.2}$$

are where $F_{\mu\nu}$ is a^2 times the usual Maxwell tensor (i.e., $F_{12} = a^2 B_z$ etc.). The equations of motion can be combined to give

$$\ddot{\vec{F}}_k + k^2\vec{F}_k + \frac{n}{\eta^2}\vec{F}_k = 0. \tag{3.3}$$

where the dot denotes $d/d\eta$, and the conformal time η is related to the clock time t by through the equation $dt = a(\eta)d\eta$. $n \equiv \eta^2\left(6b\frac{\ddot{a}}{a} + c\left(\frac{\ddot{a}}{a} + \left(\frac{\dot{a}}{a}\right)^2\right)\right)$ and $F_k(\eta) \equiv a^2 \int d^3x e^{i\vec{k}\cdot\vec{x}}\vec{B}(\vec{x}, \eta)$ is a measure of the magnetic flux associated with the comoving scale $\lambda \sim k^{-1}$.

For modes well outside the horizon, $|\vec{F}_k| \propto \eta^{m\pm}$ where $m_\pm = 1/2(1 \pm \sqrt{1-4n})$. For $b = c = 0$ (standard electromagnetic theory) the energy density associated with a given mode always scales as a^{-4}. For other choices of b and c it is easy to compute behavior of F as a function of $a(\eta)$. It is then straightforward to compute the energy density in the magnetic field. With $p \equiv 1/2(1 - \sqrt{1 - 48b - 12c})$ and $q \equiv 1/2(1 + \sqrt{1 - 48b - 24c})$ we find that

$$r \simeq (7 \times 10^{25})^{-(p+2)} \left(\frac{M}{m_{pl}} \right)^{4(q-p)/3} \left(\frac{T_{RH}}{m_{pl}} \right)^{2(2q-p)/3} \left(\frac{T_*}{m_{pl}} \right)^{-8q/3} \left(\frac{\lambda}{1\ Mpc} \right)^{-2(p+2)} \quad (3.4)$$

In the above equation, T_* is the temperature at which plasma effects become dominant, so that for $T \leq T_*$, $r \simeq$ constant. T_* typically occurs during reheating. $\lambda = 1\ Mpc$ roughly corresponds to galaxy-sized fields. In Table I we give the results for r for various choices of the parameters and $\lambda = 1\ Mpc$. We see that for a wide range of choices for p and q and M and T, the strength of the large-scale magnetic fields generated can be astrophysically interesting. The physics underlying the above result is that during inflation, the photon acquires a negative mass. This in turn implies an instability and therefore growth in the magnetic field.

IV. Summary The generation of primeval magnetic fields through inflation appears to be a very attractive possibility. In this work we have investigated the production of these fields in the inflationary Universe. Specifically, we have studied theories in which the conformal invariance of electromagnetism is broken. Large scale magnetic fields are most easily produced by introducing into the usual Maxwell Lagrangian, RA^2 terms. However, these terms are unmotivated and are 'repulsive' to the theorist as they break the gauge invariance of electromagnetism. The other methods considered have been less successful but work is still in progress.

Table I Results for $r = (\rho_B/\rho_\gamma)|_{1\ Mpc}$, the magnetic field energy density on a comoving scale of 1 Mpc, relative to the CMBR for the 'RA^2' model. The dependence of r upon comoving scale λ is: $r \propto \lambda^{-n}$.

p	q	$T_{RH}(GeV)$	$M(GeV)$	$T_*(GeV)$	$\log(r)|_{1Mpc}$	n
-1	2	10^9	10^{17}	$10^{12.3}$	-57	2.0
-1	2	10^{17}	10^{17}	10^{17}	-56	2.0
-2	3	10^9	10^{17}	$10^{12.3}$	-13	0.0
-2	3	10^{17}	10^{17}	10^{17}	-8	0.0

References

1. This paper is based on M. S. Turner and L. M. Widrow, *Phys. Rev.* **D37** 2743 (1988).
2. The structure of magnetic fields in spiral galaxies is reviewed by Y. Sofue, M. Fujimoto, and R. Wielebinski, *Ann. Rev. Astron. Astrophys.* **24**, 459 (1986); For a discussion of magnetic fields in other astrophysical objects, see E.N. Parker, *Cosmical Magnetic Fields* (Clarendon Press, Oxford, 1979); Ya.B. Zel'dovich, A.A. Ruzmaikin, and D.D. Sokoloff, *Magnetic Fields in Astrophysics* (Gordon & Breach, New York, 1983); M. J. Rees, *Quar. J. R. Astron. Soc.*, in press (1987).
3. E. N. Parker, *Astrophys. J.* **163**, 255 (1971).
4. E. Witten, *Nucl. Phys.* **B249**, 557 (1985).
5. E. R. Harrison, *Mon. Not. R. Astron. Soc.* **147**, 279 (1970); **165**, 185 (1973); *Phys. Rev. Lett.* **30**, 188 (1973); A. Vilenkin and D. Leahy, *Astrophys. J.* **248**, 13 (1981); *Astrophys. J.* **254**, 77 (1982); C. Hogan, *Phys. Rev. Lett.* **51**, 1488 (1983).
6. A.H. Guth, *Phys. Rev.* **D23**, 347 (1981); A.D. Linde, *Phys. Lett.* **108B**, 389 (1982); A. Albrecht and P.J. Steinhardt, *Phys. Rev. Lett.* **48**, 1220 (1982).
7. For an up to date review of inflation, see, e.g., M.S. Turner, in *The Architecture of the Fundamental Interactions at Short Distances*, edited by P. Ramond and R. Stora (North-Holland, Amsterdam, 1987), pp. 577-620.

A NEW DETECTOR SCHEME FOR AXIONS

K. Zioutas[1] and Y. Semertzidis[2]
1) Nuclear and Elementary Particle Physics Section, University of Thessaloniki, Greece
2) Department of Physics and Astronomy, University of Rochester, NY, USA

ABSTRACT. The detection of axions via their coupling to electrons in resonant atomic or molecular transitions is considered. The possible involvement of this process in astrophysics is also discussed. The anomalous population observed in astrophysics masers may be explained through axion-induced reactions giving new constraints for the axion mass. A monoenergetic cosmic axion spectrum is predicted, being of potential interest for relevant experiments.

The few experiments [1] performed so far to study the coupling of axion to electron through the two-body decay of orthopositronium gave an upper limit for the coupling strength $\tilde{\alpha} < 10^{-10}$, for $m_a \gtrsim 100\,\text{keV}$.

Since the presently expected value for m_a is in the 1–100 GHz region, atomic or molecular transitions certainly come into question [2]. Threshold effects might even enhance their identification. An axion behaves like a magnetic dipole photon and can carry the following values for spin and parity [3]: 0^-, 1^+, … . The $0^+ \rightarrow 0^-$ transitions are no longer excluded if the axion participates in the decay. Atomic transitions are mostly E1 transitions, and the lifetime of magnetic transitions is increased by a factor of $\sim 10^4$. The associated problem is due to the dominant non-radiative decays, which can be suppressed by lowering the pressure. However, a better solution is offered by laser/maser devices, which have a forbidden line (magnetic instead of electric dipole transition). We mention Nd:glass, ruby, iodine, He/Ne, … lasers. Powerful klystrons are of potential interest as well. The number of oscillations taking place between the two laser states defines obviously the production rate of the axions, which are supposed to be produced in competition with the photons. No laser output is needed, although it has to be as powerful as possible in order to achieve the highest sensitivity. However, the thermal population of high spin atomic levels ($J \leqslant 30$) is of interest even at room temperature. Also, the gas excitation by a powerful maser, klystron, etc., tuned on a specific transition, can be another axion source. The detection of the axions produced in this way can be performed with a conventional microwave cavity [4], converting axions to photons, tuned exactly on the frequency of the atomic or molecular transition utilized for their production. Therefore, we can integrate by a much longer time and have greater sensitivity.

A gas tube for resonant axion absorption can also be used to convert axions to photons. The maximum cross-section is not negligible: $\sigma \approx (\lambda^2/2\pi) \cdot (\tilde{\alpha}/\alpha)$. The corresponding cross-section for axions can be of the order of $10^{-35}\,\text{cm}^2$. This value can be improved by many orders of magnitude by suppressing Doppler broadening effects etc. The most important advantage of the proposed detection system is that *the axion mass does not have to be known in order to have resonance conditions.* Such experiments can give at least first limits if we adopt in our calculations an axion to electron coupling strength $\tilde{\alpha} \gtrsim 10^{-21}$, and for $m_a > 10^{-5}$ eV being given only by the energy of the transition involved.

M. Caffo et al. (eds.), Astronomy, Cosmology and Fundamental Physics, 476–477.
© *1989 by Kluwer Academic Publishers.*

Table 1: Molecular microwave transitions allowing 0^- or 1^+ spin-parity change [5].
F = total angular momentum, ν = rest frequency.
(F = 3,4,5 $\to$ 3,4,5 is short for 3 $\to$ 3, 4 $\to$ 4, 5 $\to$ 5.)

Molecule	F $\to$ F transition	ν [GHz]	Molecule	F $\to$ F transition	ν [GHz]
OH	1,2 $\to$ 1,2	1.66	HNCO	1 $\to$ 1	21.98
	2,3 $\to$ 2,3	6.03			
	3,4 $\to$ 3,4	13.43	H_2CNH	1,2 $\to$ 1,2	5.29
	4,5 $\to$ 4,5	23.82			
	1 $\to$ 1	4.75	HCCCN	1 $\to$ 1	9.1
	1,2 $\to$ 1,2	7.80		1,2 $\to$ 1,2	18.2
	2,3 $\to$ 2,3	8.1			
			NH_2CHO	1,2 $\to$ 1,2	1.54
NS	3/2,5/2 $\to$ 3/2,5/2	11.52		1,2,3 $\to$ 1,2,3	4.62
				2,3,4 $\to$ 2,3,4	9.24
CN	1/2,3/2 $\to$ 1/2,3/2	113.5			
CH	1 $\to$ 1	3.33	CH_2CHCN	1,2,3 $\to$ 1,2,3	1.37
N_2H^+	1 $\to$ 1	93.17	CH_3NH_2	1,2 $\to$ 1,2	8.78
HCN	1 $\to$ 1	88.63		4,5,6 $\to$ 4,5,6	73.04
$H^{13}CN$	1,2 $\to$ 1,2	172.68		3,4,5 $\to$ 3,4,5	86.07
NH_3	1,2 $\to$ 1,2	23.69	HC_5N	1,2 $\to$ 1,2	5.32
	2,3 $\to$ 2,3	23.72			
	2,3,4 $\to$ 2,3,4	23.87			

We must stress that the proposed scheme might have consequences in astrophysics, where molecular clouds play a dominant role. Rotational excitations of some molecules or radicals (Table 1) which are encountered in interstellar space, give rise to $\Delta F = 0$ hyperfine structure transitions; these fit in any case either with 0^- or 1^+ quantum number changes. Therefore, the cosmic axion spectrum must have not only a thermalized component ($\beta \approx 10^{-3}$) but a rich spectrum of monoenergetic lines, very similar to the electromagnetic part of the cosmic radiation. This fact has to be taken into account in those experiments searching for axions of cosmic origin. The celebrated 21 cm M1 hydrogen line might be of particular interest within this work. Also in cosmic masers the anomalous population of the upper level or the pumping mechanism itself can be due to the axion participation, because of the quantum numbers involved. We must point out that all cosmic masers have $\nu > 1612$ MHz, whilst the hydrogen maser is missing in the Universe. Supposing that the postulated axion is the origin for these phenomena in the cosmos, its mass must be $1.4 < m_a < 1.6$ GHz.

REFERENCES

[1] G. Carboni and W. Dahme, Phys. Lett. 123B (1983) 349.
    U. Amaldi et al., Phys. Lett. 153B (1985) 444.
[2] K. Zioutas and Y. Semertzidis, Phys. Lett. 130A (1988) 94.
[3] S. Weinberg, Phys. Rev. Lett. 40 (1978) 223.
[4] S. De Panfilis et al., Phys. Rev. Lett. 59 (1987) 839.
[5] K.H. Hellwege, in Landolt-Börnstein VI/2C (1982) 72.

INDEX

Numbers in **boldface** indicate that the item is discussed throughout the entire article which starts on the mentioned pagenumber.

This index was compiled by Roeland van der Marel, Leiden Observatory, the Netherlands.